SECOND EDITION

SOILS AND ENVIRONMENTAL QUALITY

SECOND EDITION

SOILS AND ENVIRONMENTAL QUALITY

GARY M. PIERZYNSKI

DEPARTMENT OF AGRONOMY
KANSAS STATE UNIVERSITY
MANHATTAN, KANSAS

J. THOMAS SIMS

DEPARTMENT OF PLANT AND SOIL SCIENCE
UNIVERSITY OF DELAWARE
NEWARK, DELAWARE

GEORGE F. VANCE

DEPARTMENT OF RENEWABLE NATURAL RESOURCES
UNIVERSITY OF WYOMING
LARAMIE, WYOMING

CRC Press
Boca Raton   London   New York   Washington, D.C.

Library of Congress Cataloging-in-Publication Data

Pierzynski, Gary M.
    Soils and environmental quality / by Gary M. Pierzynski, J. Thomas Sims, George F.
Vance. -- 2nd ed.
        p.   cm.
    Includes bibliographical references and index.
    ISBN 0-8493-0022-3
    1. Soil pollution. 2. Soil remediation. 3. Soils. I. Sims, J. T. II. Vance, George F. III.
Title.

TD878 .P54 2000
628.5′5--dc21
                                                      99-057039
                                                         CIP

© 2000 by CRC Press LLC

No claim to original U.S. Government works
International Standard Book Number 0-8493-0022-3
Library of Congress Card Number 99-057039
Printed in the United States of America 1 2 3 4 5 6 7 8 9 0
Printed on acid-free paper

Preface

Studies of the environment include such areas as conducting experiments on the fundamental chemistry of environmentally relevant reactions, counting wildlife, evaluating
human health effects of contaminants, and formulating expert opinions on the political
impact of environmental policies. To people with intellectual curiosity, environmental
science is a topic without bounds. This book explores environmental quality from the
aspect of soil science. Soils play a vital role in the quality of our environment. For example,
soils impact the quality and quantity of our food, support the various biomes throughout
the world, and serve as foundations for our structures, as well as interact with the hydrosphere and atmosphere. An understanding of soil properties and processes is therefore
critical to the evaluation of how many contaminants, as well as essential nutrients, behave
in the soil environment. Soils can be a source, a sink, or an interacting medium for the
many contaminants that impact humans, plants, wildlife, and other organisms.

In this book we first provide an overview of basic soil science, hydrology, atmospheric
chemistry, the classification of pollutants, and the fundamentals of soil and water analysis.
Nitrogen, phosphorus, sulfur, trace elements, organic chemicals, global climate change,
acid deposition, and remediation of contaminated soils and groundwaters and surface
waters are discussed in depth. There are also comprehensive discussions on the role of
soils in the biogeochemical cycling of major elements and compounds of environmental
concern. Interactions of potential pollutants with soils and the aquatic and atmospheric
environments are emphasized. Methods of soil management or remediation to minimize
or correct pollution are presented. The concept of risk assessment is reviewed using
several contemporary examples such as pesticide concentrations in drinking water and
contamination of soils by trace elements in organic wastes.

This is the second edition of *Soils and Environmental Quality* and the authors feel that
substantial updating, improvements, and additions have been made to the first edition.
We have added case studies (presented as environmental quality issues/events boxes),
example problems, and problem sections throughout the book, expanded discussion of
many topics, and separated the presentation of remediation as a stand-alone chapter. We
have added a discussion of radionuclides, acid mine drainage, and methodologies. This
book was written for use as a text for an upper-level undergraduate course in soil and/or
environmental sciences, at a level that requires the reader to have a basic knowledge of
chemistry. With appropriate supplementation, we believe that the book could be used as
a basis for a graduate-level course as well. Individuals with an interest in soil science,
environmental engineering, forestry, environmental science, chemistry, biology, geology,
or geography will also find this book useful. Instructors are strongly encouraged to
incorporate discussions of environmentally related current events into their classroom
activities as many of these topics are covered in this book. Current events also illustrate
the political, economic, and regulatory aspects of environmental issues; demonstrate the
human side of environmental problems; provide examples of the use and misuse of the
scientific method; and present possible media bias. In addition, the use of environmental

quality issues/events boxes can stimulate students to think through the complexities of environmental issues when a decision has to be made. We would appreciate comments and suggestions that you, as readers of this book, feel would assist us in improving future editions. We incorporated many such suggestions into this second edition.

Gary M. Pierzynski
J. Thomas Sims
George F. Vance

Authors

Gary M. Pierzynski is a professor of soil and environmental chemistry in the Department of Agronomy at Kansas State University, Manhattan. He received his B.S. in crop and soil science (1982) and his M.S. in soil environmental chemistry (1985) from the Department of Crop and Soil Sciences at Michigan State University, East Lansing. He earned his Ph.D. in soil chemistry (1989) from the Department of Agronomy at The Ohio State University, Columbus.

Dr. Pierzynski's research interests include trace element chemistry, remediation of trace element–contaminated soils, phosphorus bioavailability, water quality, risk assessment, and land application of residual materials. His professional activities include serving as an associate editor for the *Journal of Environmental Quality*; soil and environmental quality division chair for the Soil Science Society of America; USDA-NRI panel manager for the Soils and Soil Biology Program; vice-chairperson for the Soil Remediation Subcommission of the International Union of Soil Science; co-chair of the USDA Chemistry and Bioavailability of Waste Constituents in Soils regional research committee; member and chair of the technical and organizing committees for the International Conference on the Biogeochemistry of Trace Elements Conference Series; peer-reviewer for EPA risk assessment efforts; and technical advisor for citizen groups in the Tri-State Mining Region. He has given over 25 invited presentations. Dr. Pierzynski teaches courses on environmental quality — the course outline of which served as the starting point for this book — plant nutrient sources, uptake and cycling, and advanced soil chemistry. He has received both the Faculty of the Semester and Outstanding Academic Advisor Awards from the College of Agriculture at Kansas State University.

J. Thomas (Tom) Sims is a professor of soil and environmental chemistry in the Department of Plant and Soil Sciences at the University of Delaware, Newark. Dr. Sims is also the director of the Delaware Water Resources Center. He received his B.S. in agronomy (1976) and his M.S. in soil fertility (1979) from the University of Georgia. He earned his Ph.D. in soil chemistry (1982) from Michigan State University. He is a fellow of the Soil Science Society of America and the American Society of Agronomy and recipient of the Outstanding Research Award from the Northeast Branch of the American Society of Agronomy and Soil Science Society of America.

Dr. Sims teaches undergraduate and graduate courses in environmental soil management and advanced soil fertility and conducts an active research program directed toward many of the environmental issues faced by agriculture in the rapidly urbanizing northeastern U.S. His research has focused on the development of nitrogen and phosphorus management programs that maximize crop yields while minimizing the environmental impact of fertilizers and animal manures on groundwaters and surface waters. Other research has evaluated the potential use of sludge composts, coal ash, and other industrial by-products as soil amendments. Again, the goal has been to develop environmentally sound management programs for these materials, based on their reactions in the soil and effects on plant growth and water quality. In his role as director of the University of Delaware Soil Testing Program, he has developed and evaluated soil tests for environmental purposes such as soil nitrate testing, environmental soil tests and field rating systems for phosphorus, and soil testing strategies for heavy metals in waste-amended soils.

 **George F. Vance** is a professor of soil and environmental science in the Department of Renewable Natural Resources at the University of Wyoming. He received his B.S. in crop and soil sciences (1981) and his M.S. in soil pedology/soil organic chemistry (1985) from the Department of Crop and Soil Sciences at Michigan State University. He earned his Ph.D. in environmental soil chemistry (1990) from the Department of Agronomy and was a postdoctoral research scientist in the Department of Forestry, both at the University of Illinois.

Dr. Vance regularly teaches courses in environmental quality and chemistry of the soil environment. He also has taught courses on analytical methods for ecosystems research, soil organic chemistry, bioremediation and soil genesis, morphology and classification. He assisted in developing the University of Wyoming environment and natural resources program and coordinated the formation of a soil science graduate degree and soil science/water resource option. He is currently the soil science degree coordinator and has served as the soil science section leader. As a result of his teaching and advising efforts, Dr. Vance has been recognized as one of the University Outstanding Teachers and has received the Excellence in Advising Special Recognition Award for the past 4 years.

Dr. Vance's research has focused on natural resource issues, land-use practices, and soil and environmental chemistry, including selenium problems, water quality assessment, waste utilization, disturbed land reclamation/revegetation, acid deposition, pesticide and nutrient mobility and fate, soil organic matter characterization, forest nutrient cycling, sorption of organic chemicals, and land-use planning with geographic information systems. He has authored or coauthored over 250 technical publications and presented or copresented more than 200 presentations.

Dr. Vance has served as president, Western Society of Soil Science; division chair, Soil Science Society of America (SSSA); president and western region chair, American Society for Surface Mining and Reclamation (ASSMR); chairperson, Regional Research Committees; Wyoming research coordinator, EPA Region 7&8 Hazardous Substance Research Center; co-chairman of the 1995 National ASSMR meeting; panel member, USDA grant program; and associate editor, *Journal of Environmental Quality* and *Clays and Clay Minerals Journal*. He has been an ARCPAC Certified Professional Soil Scientist since 1992. Dr. Vance was named the ASSMR 1998 Reclamation Researcher of the Year and is a fellow of the American Society of Agronomy and the Soil Science Society of America.

Acknowledgments

The authors would like to acknowledge Drs. R. L. Chaney, M. B. David, L. E. Erickson, J. M. Ham, D. J. Kazmar, S. D. Miller, L. C. Munn, A. L. Page, J. A. Ryan, J. Schmidt, A. N. Sharpley, J. G. Skousen, P. D. Stahl, B. Vasilas, and Kimberly Williams for their reviews of portions of this book. Their comments were extremely valuable and helped improve this book immeasurably. Tim Brewer and Barbara Broge assisted with many figures. Special thanks to Sarah Blair who worked tirelessly on 27 figures in the book. In particular, Figures 1.2, 7.7, and 11.11 are original works by Sarah Blair; she also made the global and soil element cycles uniform throughout the book.

Dedicated to the families of the authors

Joy, Garrison, and Jeanne
Connie, Amy, Ethan, and Katy
Maureen, Christy, Emily, and Malaki

whose love, patience, and support made this book a reality

Contents

Contents

chapter one

Introduction to environmental quality

Contents

1.1   Introduction

Environmental quality is a broad topic that can span many disciplines. Engineers, geographers, architects, lawyers, business persons, health professionals, biologists, and geologists, to name a few, could all be concerned with environmental quality in one way or another. This book takes the perspective of a soil scientist on environmental quality. Soil scientists have a unique vantage point for evaluating environmental quality because the global cycles of many substances that are potential pollutants involve soils. Many inputs used in agricultural production (fertilizers, pesticides) can be pollutants if not managed properly. Soils in nonagricultural settings can be polluted by atmospheric depositions, by improper disposal of wastes and by-products, or by accident, as in the case of chemical spills. Soil itself can be a contaminant. Soils also interact with the hydrologic cycle and the atmosphere by serving as a source or sink for various constituents in water or air. Ultimately, everyone is dependent on soils and their role in providing food and a clean environment in which to live.

There is a strong relationship between the production of food and environmental quality, and this relationship is quite different from the relationship between the production of manufactured goods and environmental quality. For example, if the production of television sets presented an unacceptable risk to the environment, then society could decide not to produce television sets anymore, while the same cannot be said for most food products. We must continue to produce food, and therefore society must take responsibility for the potentially hazardous materials used in production agriculture. Note the use of the word *society.* Ideally, all individuals should understand the basic food production processes since we are all food consumers. Included in this understanding would be the risk vs. benefit issues associated with inputs in the food production processes. The use of fertilizers and pesticides and the application of by-products to soils are some of the practices that can have very positive effects on food production and, at the same time, present serious threats to humans and the environment.

The work of soil scientists is not confined to production agriculture, however, because soil scientists are concerned with both rural and urban issues. Many activities in society can contaminate soils and lead to environmental degradation. Mining and smelting of nonferrous metals, coal mining, production and use of petroleum products, and nuclear waste are just a few activities that impact all citizens and the soil resource. Hazardous materials are used in many industrial processes and even in our own homes. We must realize that as long as hazardous materials are stored, transported, utilized, or produced as by-products, the chance exists that there will be accidental releases into the environment. In addition, most of the major atmospheric gases associated with global climate change undergo major interactions with soils.

Soil science is a broad topic and while we will not be able to cover all aspects in this book, sufficient material will be presented, along with the basics of hydrology and atmospheric science, to allow the reader to understand most environmental issues involving soils. A working knowledge of the environmental issues is critical if an individual wants to have meaningful input into decisions that impact the environment. The input can come in a variety of ways, including responding during the public comment period for new environmental legislation or for planned remedial actions for Superfund sites, participating at public hearings and meetings, and even supporting candidates for public office who advocate an environmental policy that is agreeable to you. One goal of this book is to present facts and issues as objectively as possible and to point out how biases may influence how information is presented.

1.2 Environmentalism

Environmental quality may seem to be a new topic when, in fact, the current emphasis on environmental quality is a renewed interest in an old topic. Modern-day environmentalism is nearly 40 years old if one considers the publication of *Silent Spring* by Rachel Carson in 1962 as the beginning of widespread environmental awareness by the general public. The U.S. Environmental Protection Agency (EPA) is almost 30 years old, as are a number of scientific journals that have the sole purpose of reporting on environmental research. The 30th anniversary of Earth Day is upon us. Several factors are likely responsible for the renewed interest in environmentalism. We are becoming increasingly aware that exposure to various substances may cause human health problems ranging from death to subclinical effects such as attention deficit syndrome. The natural tendency is to eliminate or reduce our exposure to substances that we suspect may adversely affect human health. We are also becoming increasingly aware of the effects of various substances on a variety of organisms in the environment, and we have become more sympathetic to these indirect effects. Overall, there is increased awareness of the impact of human activities on the world's natural resources. Finally, our ability to detect contaminants in soil, food, water, and air has greatly improved in recent years, both in terms of the type and concentration of a contaminant that can be detected. Concentrations in the micrograms per kilogram ($\mu g/kg$) or micrograms per liter ($\mu g/L$) (parts per billion, ppb) range are routinely measured with concentrations in the nanograms per kilogram (ng/kg) or nano-grams per liter (ng/L) (parts per trillion, ppt) range possible for some substances. This expanded capability allows a more accurate inventory of our environment and the influ-ences of our activities on the environment. These factors provide both the means and the motive for maintaining public interest in the environment.

An environmentalist is a person who works toward solving environmental problems, and environmentalism is an attitude by a group of individuals, perhaps all of society, that the environment takes a high priority in the decision-making process. A more philosoph-ical approach categorizes an individual's attitudes about the environment into one of three groups: egocentric, in which an individual's actions are guided solely by concern for him or herself; homocentric, meaning concern for the human species; and ecocentric, meaning an overall concern for the environment. Society as a whole and most individuals have clearly progressed slowly from egocentric attitudes toward more ecocentric attitudes.

Environmentalism has been a social movement. Eric Hoffer, in *The True Believer*, identified three stages of social movements. The first stage is called *people of words* (updated to be gender neutral). In this stage the social movement is discussed and debated within a small segment of society generally described as the intellectuals. Late in the first stage, the social movement may be described in various writings. Rachel Carson was a person of words. The second stage is called *fanaticism* and is characterized by an increase in awareness by society of the movement through various attention-grabbing, often illegal, actions performed by zealous believers. Environmental fanaticism may include various means of disrupting the production (e.g., vandalism) from an indus-try that is believed to be polluting the environment or even something as drastic as bombings or kidnapping. The final stage is called *practical people of action* or, by some, institutionalism. At this stage there is widespread awareness and support for the move-ment. Protests utilize legal means and the potential for violence decreases. Organized boycotts may be employed, for example. Committees or even agencies may be formed. Ecocentric attitudes would prevail in the context of environmentalism. Eric Hoffer (1951) summarizes by stating, "A movement is pioneered by men of words, materialized by fanatics, and consolidated by men of action."

Environmentalism as a whole has clearly progressed to the institutional stage with the federal EPA and comparable units of government in each state. Within environmen-talism, issues continue to arise that pass through some or all of the stages of a social movement. Recent attempts by members of environmental groups to physically block the passage of whaling ships would be an example of fanaticism. The issue of global climate change due to increases in atmospheric concentrations of greenhouse gases is late in the people of words stage with no indication of fanaticism as of yet.

The protection and study of soils has been an integral part of the environmental movement. Regulations exist at both the national and state levels to protect soils from contamination. The concept of soil quality assessment is being developed by scientists and offers a conceptual and quantitative framework by which the overall impact of soil management on the "health" of the soil can be determined.

1.3   Studying the environment: the scientific method

The *scientific method* provides a set of rules by which the scientific community conducts investigations such that experimental design, collection of data, and the interpretation of data are done in a systematic and objective fashion. There is no agreed-upon definition of the scientific method in the sense that the same steps are followed in every scientific investigation. There are some factors common to all investigations, however, that must be present before a particular study is acceptable to the scientific community.

   The scientific method can be viewed as a deductive process, that is, one that proceeds from the general to the specific. In this case, the first step is an *observation*. An observation could simply be poor plant growth or an elevated concentration of a heavy metal in a soil sample. Observations can be made with any of the five senses or with instrumentation; scientific instruments merely enhance the senses. Observations must be verifiable. Other investigators must be able to reproduce the conditions associated with the observation and obtain the same effect. If alternative techniques and tests are used in the verification, they add more validity to the observation. Observations that stand the test of verification become accepted as scientific facts.

   A considerable amount of science is involved with making and verifying observations and much useful information comes from this. Simple observations, however, do not explain mechanisms; that is, what caused the observed effect. For example, the application of coal fly ash to soil at high enough rates will inhibit plant growth. That is a verifiable observation. Coal fly ash can alter soil pH and it contains boron (B) and soluble salts, any of which can inhibit plant growth by itself. The second step in the scientific method is the *formulation of hypotheses* and the *testing of hypotheses with controlled experiments*. A *hypothesis* is a plausible explanation for the observation that is tentatively accepted and may serve as a basis for further investigation. *Controlled experiments* are replicated experiments in which all factors except the one in question are held constant. "The B in the fly ash inhibits the growth of plants when fly ash is applied to soil" is a reasonable hypothesis. A suitable experiment could be designed to determine if the hypothesis is true or false. Progress would be made regardless of the outcome of the experiment since we would know that B inhibited plant growth or if something else did. If the hypothesis was true, the next logical step would be to ask why B inhibited plant growth. The process continues.

   One observation that has been verified and is causing considerable debate is the fact that the concentration of gases that absorb heat (radiatively active or greenhouse gases) in the atmosphere have been increasing compared with preindustrial times. Some scientists believe that the increased concentrations of radiatively active gases will cause the average temperature of the atmosphere to increase, which will likely cause changes in the climate. Unfortunately, there is no easy experiment to conduct to test this hypothesis. Historical weather data provide some information, but we also know that the climate has varied considerably over time, the climate has been both warmer and colder than present conditions, and these changes occurred without influence from human activities. Computer models can simulate the effects of increasing concentrations of radiatively active gases on the climate, but our confidence in the models is limited. Hence, the global climate change issue is stretching the limits of science and the scientific method at the moment. Some people are comfortable extrapolating the scientific facts to conclude that global climate change is already occurring, and society needs to act quickly to slow the changes, while others remain complacent, believing that science has not yet proved that there is a problem.

   The replication of treatments in controlled experiments is an important point. Replications provide a means of determining if treatment effects are real (statistically significant) or due to chance (not statistically significant). The more replications that are used, the more certain the investigator can be that treatment effects did or did not occur.

   Other types of scientific investigations do not lend themselves to controlled experiments. Human health studies are a perfect example. Physicians may observe that patients with a certain type of cancer have been exposed to a particular chemical, but they cannot test the appropriate hypothesis with controlled experiments using human subjects. Epidemiological studies may be useful in these situations. *Epidemiology* is the study of the occurrence and nonoccurrence of disease in a population without the benefit of controlled experiments. If comparable groups (similar distributions of age, sex, etc.) that have been exposed to various levels of the chemical in question can be identified, then statistical comparisons of the frequency of diseases between the groups can be made. Epidemiological studies are not as good as controlled experiments and this can lead to some interesting statements. Tobacco companies have defended themselves with the argument that no one has ever conclusively proved that the use of tobacco products causes cancer. Technically they are correct because there have been no controlled experiments with humans. Health officials cite the epidemiological studies that show a strong association between tobacco use and the incidence of cancer. Individuals are left to decide which body of evidence they select to dictate their own behavior.

Most scientific investigations end at this point. Occasionally, a theory is proposed that attempts to explain or predict a large number of observations or facts. The theory of evolution is an example. Theories are also tested with hypotheses and appropriate experimentation. Then there are natural laws and basic principles. Exceptions to natural laws and basic principles have never been observed. The law of gravity and the conservation of energy are examples without exception.

Publication of the results of scientific investigations in refereed journals is an integral part of the scientific method. The process begins with the submission of a manuscript to a journal editor who then sends it out to several reviewers. The reviewers evaluate the work for uniqueness, adherence to the scientific method, statistical analysis, quality of writing, and overall scientific vigor before it is accepted for publication in the journal. Then the manuscript appears in the journal and all readers are free to accept, challenge, or attempt to verify the findings. When environmental issues are argued in the courtroom, technical experts often rely on research published in refereed journals to support their arguments. Recall the interest in the 1980s in the idea of cold fusion. In this instance, the results of an experiment were released prior to the work going through the publication process. A great deal of media and government attention was generated because of the far-reaching ramifications of cold fusion. When the details of the experiment were published, no one was able to verify the results, and the investigators and their institution suffered considerable embarrassment.

The importance of the scientific method cannot be overstated. Environmental issues can be emotionally charged because they can influence human or animal health and involve large sums of money. Decisions or actions based on poor information can result in wasted effort, wasted money, and needless regulations. Being objective requires that one evaluate information. Recognize the source of information and whether it is a product of the scientific method. Recognize information that may be biased because of vested or emotional interests. The scientific method provides a mechanism by which the attainment of knowledge can proceed in an unbiased fashion.

1.4   Environmental science and the general public

Responsibility for the environment ultimately belongs to society. Society elects the officials who promulgate the environmental regulations and who decide how tax dollars will be spent on environmental research. The responsibility is not straightforward, however, as many complicated interactions exist. Environmental regulations directly impact industries that provide jobs for society. The costs of compliance with environmental regulations by private businesses and governmental units themselves are passed on to consumers through increased prices and taxes. These interactions would tend to make for weaker environmental regulations. Countering this tendency are the fears of society about the effects of pollutants on humans and a growing respect for the environment in which they live.

Consider also the plight of countries that are not as economically well off as the U.S. and other economic powers. In these situations the cost of environmentally friendly practices is simply beyond reach and the economic gains and the immediate need to produce food are given a much higher priority than environmental matters. Since many of the human health effects from environmental contamination take years to develop, it is not difficult for people to delay consideration of environmental matters in light of more immediate concerns.

The media play a role by transferring knowledge from the scientific community to the general public. The media report information that may be of interest to their audience provided the information comes from what the media perceive as a credible source. Little consideration is given to whether the information is a product of a portion of or all of the scientific method. Often, the general public has difficulty placing a single media story in perspective because a synthesis of the broader picture is usually not presented. For example, numerous news stories have appeared about the possible link between exposure to pesticides and cancer. Some suggest a link exists; others cast doubt. Given the large numbers of pesticides and the many different types of cancer, it is not surprising that people are confused. In addition, some media outlets have a tendency to use sensationalism in their coverage of environmental matters.

Be aware of the difference between an *environmental event* and an *environmental issue*. An environmental event is an important occurrence related to the environment, whereas an environmental issue is a point or question related to the environment that is to be debated or resolved. A chemical spill is an environmental event, while the response to the spill, which might include actions that need to be taken to prevent additional spills, is an environmental issue. Reporting of issues and events should be objective, but objectivity is compromised most often when it comes to issues. Once again, the ability to evaluate information is the key to a complete understanding of environmental concerns.

One must also keep in mind that personal feelings about the environment involve value judgments. While individual values may vary widely, people should not expect to impose their own values onto others, and they will generally be better off respecting the values of others even when the values differ from their own. Value judgments must be taken into account even as we define pollution and contamination.

1.5   Defining pollution and contamination

Before a general discussion of soils and environmental quality can be presented, a definition of the terms *pollutant* and *pollution* must be given. Interestingly, definitions of these terms will vary from individual to individual. Consulting a dictionary yields synonyms such as impure, unclean, dirty, harmful, or contaminated. Although these words are appropriate, they do not provide a working definition that is useful for studying the environment.

Part of the problem in defining pollutant or pollution is that an agreement must be reached on what constitutes acceptable use of materials that may cause pollution. For example, some would consider the use of pesticides acceptable if they are reasonably certain that the effect of the pesticide is only that which was intended, whereas others would consider the use of any pesticides unacceptable. In the first case, the pesticide is a pollutant only if undesirable side effects occur, while in the second case, the pesticide is always considered a pollutant. These are the value judgments that were discussed earlier. What constitutes an acceptable level of pollution is another way to view this problem. Attitudes will range from none (ecocentric) to any level that will not harm me (egocentric).

A second part of the problem in defining pollutant or pollution is the distinction between *anthropogenic sources* and *natural sources*. A volcano may place a greater quantity of noxious gases and particulate matter into the atmosphere than the combined output from a large number of electric power plants, but some would not consider the output from the volcano as a pollutant because of its natural origin. Similarly, heavy metal mining activities may pollute soils with heavy metals, yet soils with high concentrations of heavy metals can occur naturally because of their proximity to metal ore deposits.

It should be apparent that there are value judgments associated with defining pollutant or pollution and it is impossible to change this situation. A reasonable working definition of a pollutant, taking into account the aforementioned problems, would be *a chemical or material out of place or present at higher than normal concentrations that has adverse effects on any organism*. The implications of this definition are that the pesticides applied to agricultural soils are not pollutants provided they do not move below the rooting zone of the crop or run off the site and impact another area. They would become pollutants if they occurred off site at concentrations high enough to cause harm to an organism, although the possibility still exists that the pesticide will have adverse effects on nontarget organisms that live in or frequent the soil that receives the pesticide application. The volcanic output and the naturally occurring, metal-contaminated soils represent pollutants because a chemical or material is present at higher than normal concentrations and can adversely affect living organisms. Miller (1991) states, "Any undesirable change in the characteristics of the air, water, soil and food that can adversely affect the health, survival, or activities of humans or other living organisms is called pollution," which also takes into account noise or thermal pollution in addition to pollution caused by chemicals or materials. Both definitions allow for value judgments by using the subjective phrases "out of place" and "undesirable change."

The term *contaminated* is often used synonymously with *polluted*, although subtle differences in the definitions would indicate that these terms are not interchangeable. *Contaminated* implies that the concentration of a substance is higher than would naturally occur but does not necessarily mean that the substance is causing any harm. *Polluted* also refers to a situation in which the concentration of a substance is higher than would naturally occur but also indicates that the substance is causing harm of some type, as noted in the two previously stated definitions. By this reasoning, a soil could be contaminated but not polluted.

The terms *toxic waste, hazardous waste,* or *hazardous substance* are often heard and, obviously, such materials can be pollutants. The added distinctions of toxic or hazardous are used for substances that can be acutely or chronically toxic to humans, as opposed to pollutants like sediment or phosphorus (P), which are not. There is no regulatory classification for toxic waste from a legal standpoint in most states. Hazardous waste, defined as part of the Resource Conservation and Recovery Act (RCRA), is solid waste that causes harm because of its quantity; concentration; or physical, chemical, or infectious characteristics. Harm is explained as significantly contributing to an increase in mortality or serious illness or presenting a significant hazard to humans or the environment if improperly treated, stored, transported, or disposed. When most of us think of hazardous waste, we have images of a dark, pasty substance handled by an individual in full protective gear. Most of us, however, have hazardous substances stored under our kitchen sinks in the form of drain cleaners, scouring powders, glass cleaners, and bleach.

Pollution is often broadly categorized according to its source. *Point-source pollution*, as the name implies, is pollution with a clearly identifiable point of discharge. The outflow from a wastewater treatment plant or a smokestack would be an example. *Nonpoint-source pollution* is pollution without an obvious single point of discharge. Surface runoff of a commonly used lawn herbicide would be an example. The implications for control are quite different. Controlling nonpoint-source pollution can be difficult because of the large areas involved and the need to deal with multiple landowners and sources. One of the successes of the early versions of the Clean Water Act (CWA) was a reduction in P discharges to surface waters from numerous point sources. These sources were primarily wastewater treatment plants. Federal matching money was made available and many communities upgraded their treatment plants from primary to secondary or advanced facilities. Sensitive water bodies, such as Lake Erie and Chesapeake Bay, benefited greatly from this legislation. However, P in surface waters remains a significant problem today. Phosphorus pollution is primarily of nonpoint-source origin such as agricultural uses of commercial fertilizers and animal manures. This contribution to the P problem will be much more difficult to solve. This topic is discussed in more detail in Chapter 5.

1.6   Classifying and characterizing pollutants

Table 1.1 presents a broad classification scheme for pollutants based on their general characteristics or uses. Each pollutant category can impact more than one medium. Each category can represent more than one process or member. The trace elements, for example, comprise more than 20 different elements. Acidification consists of several unrelated reaction pathways, some occurring in the atmosphere and others in the lithosphere, which produce acidic products that negatively impact their respective endpoints. Conversely, sediments are a relatively straightforward category with a single member.

The *nutrients* category primarily reflects the negative impacts of nitrogen (N) and P when present at high relative concentrations. Phosphorus in surface waters and nitrates in surface waters and groundwaters are the main indicators of environmental problems. Agricultural production practices are responsible for a fair share of the problems associated with excessive nutrients. Significant contributions are also made from private home and horticultural uses of fertilizers, naturally occurring sources, sewage treatment facilities, septic tanks, food-processing plants, and livestock. From an ecological viewpoint, high nutrient concentrations in surface waters lead to eutrophication, the enhancement of phytoplankton growth because of nutrient enrichment. Generally, phytoplankton are P limited in freshwater systems and N limited in marine systems. High nitrate concentrations in drinking water can also be a direct health threat to humans and animals.

The *pesticides* category represents a wide range of mostly organically based chemicals used to control pests such as weeds, insects, rodents, or plant pathogens. Once again, agricultural production practices account for the majority of pesticide use, with private home and horticultural uses also contributing. Pesticides are normally released intentionally into the environment at low levels, although accidental spills involving large quantities or high concentrations do occur.

Hazardous organic chemicals represent the large number of organic chemicals, other than pesticides, that are commonly used as fuels and as materials for a variety of industrial processes. These chemicals are acutely or chronically toxic to humans or other organisms when improperly administered, used, or disposed. Many of the hazardous organic chemicals have properties analogous to pesticides. However, this class of compounds is generally not intentionally released into the environment and soil contamination problems usually involve improper use, disposal, or accidental spills over small areas with high concentrations of the chemical in question. Contrast this to the environmental concerns for general-use pesticides, which generally involve applications at low concentrations to large areas as an accepted practice.

Hazardous materials include a large number of materials that generally fit the definition of hazardous waste, but are not included in the other categories. Hazardous waste was defined earlier but generally has properties that make the material harmful because it is ignitable, corrosive, reactive, or toxic.

The *acidification* category includes several unrelated processes. Acid precipitation is mainly the end result of the conversion of the oxides of N and sulfur (S) into their respective acids in the atmosphere. The concern is with precipitation having a lower than

Table 1.1 Classification of Potential Pollutants, the Impacted Media, and Primary Environmental Symptoms or Concerns

Pollutant or environmental concern	Examples	Medium impacted				Primary environmental symptom or concern
		Soil	Water Ground	Water Surface	Air	
Nutrients	N and P in commercial fertilizers, manures, biosolids, wastewater treatment effluent	*	*	*		Eutrophication, contaminated drinking water
Pesticides	Insecticides, herbicides, fungicides, etc.	*	*	*	*	Ecological risks, contaminated drinking water, human health
Hazardous organic chemicals	Fuels, solvents, volatile organic compounds	*	*	*	*	Ecological risks, contaminated drinking water, human health
Hazardous materials	Strong acids and bases	*	*	*	*	Acute exposure
Acidification	Acid precipitation, acid mine drainage	*	*	*	*	Degradation of structures, ecological risk
Salinity or sodicity	Saline irrigation water, saltwater intrusion, road salt	*	*	*		Loss of soil productivity
Trace elements	Heavy metals, elements normally present at low concentrations in soils or plants	*	*	*		Human health concerns, ecological risk
Sediments	Eroded soil in surface waters			*		Turbidity, eutrophication
Particulates	Soot, dust from wind erosion, volcanic dusts, ash				*	Visibility, respirable particles
Greenhouse gases	Carbon dioxide, methane, and other radiatively active gases				*	Global climate change
Smog-forming compounds	Ozone, secondary products of fuel combustion				*	Human health concerns

normal pH and its impact on the environment. Acid mine drainage represents another source of acidification where water has been acidified by the weathering and oxidation of sulfide minerals, primarily pyrite (FeS_2). Such water can have a pH as low as 2.0 and can have significant impacts on soil or surface water systems. Another example is that of soils being acidified after heavy use of ammoniacal N fertilizers (from the nitrification process) without neutralization of the acidity with liming materials.

The primary problems associated with *salinity* and *sodicity* are reductions in plant productivity due to water stress caused by the increased osmotic potential and changes in soil physical properties (dispersion with a resultant reduction in permeability) when salt or sodium concentrations in soils become too high. Irrigation with water containing high salt concentrations and salts with a high level of sodium (Na) relative to that of calcium (Ca) and magnesium (Mg) is the chief cause of these problems. Runoff from roads de-iced with sodium chloride (NaCl) onto roadside soils is another lesser source of salinity and sodicity.

Trace elements are elements that are normally present in relatively low concentrations in soils or plants. They may be essential for growth and development of humans or other organisms, although many are not. Trace elements of concern as pollutants are those that cause acute or chronic health problems in humans, animals, plants, or aquatic organisms when present above critical threshold concentrations. Examples include cadmium (Cd), copper (Cu), manganese (Mn), and zinc (Zn).

Sediments represent soil particles that have eroded from the landscape and have been carried to surface waters. Areas most susceptible to erosion include construction sites, recently tilled farmland, or overgrazed pastures. Sediments can physically block light transmission through water, which can alter the ecology of the body of water; can be enriched in P; can accelerate eutrophication; and can act as sources or sinks for a variety of water pollutants. Sediments can also accumulate in surface bodies of water inhibiting navigation and recreational activities and reducing the longevity of dam structures.

Particulates, *greenhouse gases*, and *smog-forming compounds* are air pollutants. Particulates are relatively inert particles generally consisting of carbon (C), soil, volcanic ash, etc. suspended in the atmosphere. Greenhouse gases are the gases responsible for the greenhouse effect. The *greenhouse effect* is an increase in the concentration of various gases in the atmosphere, primarily carbon dioxide (CO_2), methane (CH_4), nitrous oxide (N_2O), and chlorofluorocarbons (CFCs), that makes it more difficult for radiated heat to escape the atmosphere and that results in an increase in the mean global temperature above the temperature that would exist without the influence of humans. Some greenhouse gases are anthropogenic and others are naturally occurring, although their concentrations are increasing in the atmosphere due to human activities. Certain aspects of food production may also increase the concentrations of some greenhouse gases. Smog-forming compounds are the ingredients for the complex process of smog production.

Two important characteristics of pollutants in the environment are *persistence* and *residence time*. Persistence refers to the length of time a given pollutant remains unmodified while present in the soil, water, or air. Basically, this refers to the resistance of a substance to be broken down into less complex substances by abiotic or biotic processes and is sometimes quantified by a half-life term. Substances that are readily broken down have a low persistence, and vice versa. Residence time is the length of time it takes a pollutant to move from one compartment to another in the environment. Example compartments would be the atmosphere, soil, groundwater, surface water, or any other location that can be defined and studied.

Persistence and residence time play major roles in pollution management and control. For example, we know that organic chemicals that are halogenated (e.g., Cl, F, Br) are much less likely to be decomposed by soil microorganisms and therefore will have a greater persistence in the soil environment compared with nonhalogenated organic chemicals. This has led pesticide manufacturers to design nonhalogenated pesticides, as opposed to compounds like dichlorodiphenyltrichloroethane (DDT) that have very long persistence times. We know that soils have varying capacities for adsorption of both inorganic and organic substances. Those substances that are not readily adsorbed by soil will have a short residence time in the soil environment and will likely become groundwater contaminants, while substances that are readily adsorbed by soils will have a long residence time and will be lost from the soil system primarily by soil erosion. Elements, of course, cannot be broken down into simpler entities and will be persistent regardless of residence time. Residence time is not an issue for substances that are not persistent. Efforts to address acid precipitation by controlling the release of S and N oxides to the atmosphere have been successful because of the relatively short residence times for these substances in the atmosphere.

Nutrients, pesticides, hazardous organic chemicals, trace elements, acidification, and greenhouse gases will be discussed in more detail in individual chapters. Soils play a key role in the biogeochemical cycling of these materials or processes. The topics of sediments, particulates, salinity, and smog-forming compounds are beyond the scope of this book.

1.7 Human exposure to soil contaminants

In addition to the technical aspects involved in studying the relationship of soils to environmental quality, a topic of increasing importance is that of *risk assessment*. *Risk* is the chance of injury, loss, or damage. In the context of environmental science, risk assessment is the process used to estimate quantitatively the risks associated with exposure of any organism to various substances in the environment. Risk assessment can provide the basis for environmental regulations, although the information can be ignored if society chooses to do so.

Figure 1.1 presents a schematic of the pathways for human exposure to contaminants in soils. While the thrust of this book is not entirely on the human receptor, this figure focuses the discussion for the remainder of the book by identifying the key interactions between the soil and an organism. Comparable diagrams could be produced for any organism.

The major component of the figure is the human food chain, where crops are consumed directly by people or fed to livestock and the animal products are then used by people as food. However, many different exposure pathways are also illustrated including direct ingestion of soil, inhalation of dust, and ingestion of water. Contaminants can be applied

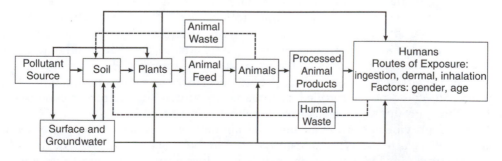

Figure 1.1 Pathways for human exposure to pollutants applied to soils. (Adapted from Brams et al., 1989.)

to soils, enter water supplies directly, or fall directly on or be directly absorbed by vegetation. Once in the soil, contaminants can leach and move to groundwater or be lost from the soil surface with runoff and enter surface waters. These water resources can be used to irrigate soils, to water livestock, or as a source of drinking water for humans. The soil itself can be ingested directly by humans or animals as well as be a source of respirable particles (dust). Children exhibit a significant amount of hand-to-mouth activity at certain ages and can consume considerable quantities of soil or indirectly consume soil through house dust. A small percentage of children exhibit pica behavior (having an abnormal craving to eat nonfood items, especially soil or paint chips) and environmental regulations written to protect these children will need to be much more stringent compared with regulations written to protect against typical soil ingestion rates.

It is necessary to differentiate the various steps a contaminant goes through as it progresses from the soil through the food chain to the human receptor. As a contaminant is transferred from one compartment to another, various processes can influence the amount that will eventually reach the human endpoint. As was discussed earlier, soils will adsorb some substances strongly while having little adsorption ability for other substances. Substances that do not adsorb strongly to soils will generally be more available for plant uptake, which will occur if the plant has a mechanism for moving that substance into the plant. For example, regulations pertaining to maximum permissible concentrations of lead (Pb) and Cd in soils allow much higher concentrations of Pb than Cd even though Pb is much more toxic to humans than Cd. This is because, at equal concentrations of soil Pb and Cd, Pb concentrations in plants will be much lower than Cd and, consequently, food chain transfer of Pb will be less. Similarly, contaminants can be partitioned into different body parts within an animal. Fat, organ meats, and muscle tissue will have much different concentrations of a particular contaminant. Polychlorinated biphenyls (PCBs), for example, are fat soluble and strongly partition into fatty tissue as compared with muscle tissue. Some investigators use transfer coefficients to describe the change in contaminant concentration as it is transferred from soil to plant, plant to animal feed, etc. Obviously each contaminant would require a unique set of transfer coefficients, which can have values ranging from 0 to 1000 or even higher. This approach is commonly used in risk assessments where the potential for food chain transfer of contaminants is a concern.

Figure 1.1 also depicts the recycling of human and animal wastes through soils, primarily as municipal biosolids (sewage sludge) and animal manures. This recycling usually occurs as a land application of the residual materials and is an opportunity to recycle plant nutrients and to improve the soil through the addition of organic carbon, but also must be managed properly to avoid overapplication of nutrients and inadvertent application of excessive amounts of contaminants. Biosolids are produced from the treatment of wastewater that may contain industrial inputs as well as human waste, thus creating the potential for contaminants to enter the soil system. Similarly, livestock are often fed or given supplements in addition to locally produced feed which creates the potential for external inputs to enter the soil system. Scientists generally believe that land application of biosolids and manures offers far more advantages than disadvantages and the risk from contamination is negligible when managed properly. However, recent action by the U.S. Department of Agriculture (USDA) declaring that crops grown on biosolid-amended soils could not be certified as "organically grown" represented a concession to special-interest groups that were concerned about this issue.

Table 1.2 provides estimates of the total amount of livestock waste, biosolids, and municipal solid waste (MSW) compost produced in the U.S. each year. The quantities are large, particularly for livestock. As the livestock industry produces more animals in confined facilities, the problems associated with efficiently recycling the nutrients in the waste have increased dramatically. It is estimated that the value of the nutrients in the manures

Table 1.2 Mass of Wastes Produced from Treatment of Wastewater, Composting of Municipal Solid Waste, and from Livestock in the U.S.

Source	Estimated amount produced (million Mg/year)
Biosolids	7.2
Municipal solid waste compost	0.13
Beef cattle	105
Dairy cattle	36
Swine	15
Chickens	121
Turkeys	5.4
Sheep	1.8

Source: Logan, T. T. et al., in *Proceedings of the 4th Joint WEF and AWWA Conference on Biosolids and Residuals Management*, Water Environment Federation, Alexandria, VA, 1995, 15-13 to 15-28.

approaches $3 billion per year and could supply 15% of the N and 43% of the P needed by crops, so the potential savings is not trivial. In some areas there is insufficient land base to apply the manures without overapplying nutrients, and environmental problems are created. Economics also plays a role as the nutrient content of the materials is quite low and the transportation costs quickly exceed the value of the nutrients.

Figure 1.1 also provides an opportunity to introduce the concept of *bioavailability*. A general definition is that the bioavailability of a substance is related to the possibility of it causing an effect, positive or negative, on an organism. A general example would be a therapeutic drug administered to patients in different forms, say, as a capsule compared with a tablet. If absorption of the drug by the patient (the organism and the effect) was greater for one form compared to the other at equivalent doses, we would say that the bioavailability of that form was greater. For soils we can define bioavailability as the fraction of the total amount of a substance in a soil that can cause an effect, positive or negative, on an organism. We know that at equal concentrations some substances are more likely to cause an effect than others, and for a given substance the likelihood of an effect will vary depending on the soil containing the substance. In our previous example of plant uptake of Pb and Cd, we would say that the bioavailability of Cd to plants is much greater than Pb, as shown in Figure 1.2. Further, we know that two soils with the same concentration of a given substance can produce different effects on an organism. The example illustrated in Figure 1.2 is comparable with the previous general example used for bioavailability; that is, the ingestion of equivalent amounts of a substance will produce differing effects. In this case the same amount of Pb present in two different soils produces differing blood Pb concentrations in the child after accidental ingestion by hand-to-mouth activity. The bioavailability of Pb in soils is an important parameter in communities with Pb-contaminated soils where exposure to children has significant health ramifications. The bioavailability concept can be applied in many of the intercompartmental transfers shown in Figure 1.1. We have already illustrated the soil-to-plant transfer and the direct ingestion of soil transfer.

Figure 1.1 and the concept of bioavailability can be used as a starting point for a discussion of remediation of contaminated soils. Knowing both the characteristics of the soil and the contaminant, one can determine where the contaminant is likely to move next (surface water, groundwater, air), whether movement through part or all of the food chain is likely, or whether direct contact with the soil is necessary for exposure to humans or other organisms. Soil removal, a commonly employed remediation technology, obviously eliminates all transfer of the contaminant from the soil to any other compartment. *In situ*

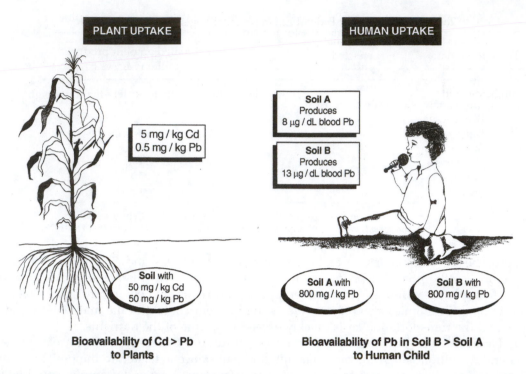

Figure 1.2   The bioavailability concept for soil contaminants. Two different trace elements can have different bioavailabilities in the same soil, and the same trace element can have different bioavailabilities in two different soils. (Drawing by Sarah Blair.)

stabilization does not remove the contaminant from the soil but changes the form and bioavailability of the substance such that transfer to another compartment is unlikely, and can even reduce uptake of the contaminant by an organism if that soil is ingested. Vegetation can have a variety of effects depending on the identity of the contaminant. For metal-contaminated sites that did not previously have vegetation, the establishment of vegetation (vegetative remediation or phytostabilization) will reduce runoff losses of soil and wind erosion, thus reducing exposure of aquatic ecosystems to the metals and reducing the risk for humans from inhalation of dust. For soils contaminated with organic chemicals, vegetation can promote the biological decomposition of the substance and even promote volatilization (phytoremediation). These topics will be developed in more detail in Chapter 11.

1.8   Federal environmental legislation

There are numerous significant pieces of federal environmental legislation that serve to protect our food, air, and water; ensure prudent use of pesticides; govern the handling of hazardous waste; and provide a mechanism for cleaning up contaminated sites. Some of this legislation is outlined in Table 1.3. The original year of enactment is given, although most legislation in this table has been updated or amended several times.

The *Federal Insecticide, Fungicide, and Rodenticide Act* (FIFRA) regulates the labeling of pesticides, where the label is a legal document that describes the appropriate use of a particular pesticide. The labeling process forces companies to conduct required environmental testing of pesticides to demonstrate that the compound has benefits to society that outweigh any potential risks.

Table 1.3 Significant Federal Environmental Legislation in the U.S.

Federal regulation	Year originally enacted	Description
Federal Insecticide, Fungicide, and Rodenticide Act (FIFRA)	1947	Regulates labeling of pesticides; requires that pesticides have benefits to society that outweigh potential harm to the environment
Clean Water Act (CWA)	1948	Controls discharges into navigable waters; provides NPDES permits and TMDLs; primary authority for water pollution programs
Clean Air Act (CAA)	1970	Established national ambient air quality standards (NAAQS) to protect human health; has forced automobile manufacturers to reduce auto emissions; addresses acid precipitation issues and power plant emissions
National Environmental Policy Act (NEPA)	1970	Provides national policy to try to prevent or reduce damage to the environment; forces federal agencies to assess potential enviornmental impacts of major programs, including federally funded local programs and research
Safe Drinking Water Act (SDWA)	1974	Regulates drinking water quality in public water systems; provides MCLs as water quality standards
Resource Conservation and Recovery Act (RCRA)	1976	Defines hazardous waste; forces companies to be accountable for hazardous waste from "cradle to grave"
Toxic Substances Control Act (TSCA)	1976	Allows for toxic substances control program by the EPA; controls labelling and disposal of PCBs and inspection and removal of asbestos
Comprehensive Environmental Response, Compensation, and Liability Act (CERCLA)	1980	Provides authority to respond to releases of hazardous waste; provides funds for cleanup of contaminated sites that pose an imminent threat to human health; makes property owners liable for cleaning up hazardous wastes; the original Superfund law
Superfund Amendments and Reauthorization (SARA)	1986	Continuation of CERCLA and Superfund process; includes emergency release and other notification requirements and community right-to-know provisions
Food Quality Protection Act (FQPA)	1996	Modifies FIFRA and Food, Drug, and Cosmetic Act (FDCA); eliminates Delaney clause for processed foods; facilitates labeling of minor use and reduced risk pesticides; promotes integrated pest management techniques

The Clean Water Act (CWA) works to ensure the integrity of surface waters throughout the U.S. Two enforcement tools available to the EPA through the CWA are the national pollution discharge elimination system (NPDES) permits and the total maximum daily loading (TMDL) limits that can be imposed on a body of water. The NPDES permits control discharges from point sources of contaminants and must be renewed periodically. The EPA can shut down operations that are not in compliance with the permit or change the conditions of the permit through the renewal process. The TMDL process establishes a maximum allowable loading for a particular contaminant in a body of water and then works with all point and nonpoint-sources of that contaminant to ensure that the TMDL is not exceeded.

One of the major accomplishments of the *Clean Air Act* (CAA) has been to reduce emissions of carbon monoxide (CO), hydrocarbons, and nitrogen oxide (NO) from vehicles, which has greatly improved air quality in urban areas. In addition, the act has been used to reduce emissions of sulfur dioxide (SO_2) and NO from coal-fired power plants, thereby reducing acid precipitation problems.

The *National Environmental Policy Act* (NEPA) was instrumental in creating a policy for the federal government dictating that it would act to prevent or reduce damage to the environment. Any actions by the federal government were subject to NEPA, including federally funded research grants.

The *Safe Drinking Water Act* (SDWA) regulates water treated by public water systems and uses the maximum contaminant levels (MCLs) to ensure that the concentrations of contaminants do not exceed critical values.

The *Resource Conservation and Recovery Act* (RCRA) defines hazardous waste and the harm that it may cause. It also forces companies to be responsible for hazardous wastes from their creation to their ultimate disposal, the cradle-to-grave concept.

The *Toxic Substances Control Act* (TSCA) created a toxic substances control program with the EPA that was essentially a reporting mechanism that creates a toxic substances inventory. Substances not regulated by FIFRA or the *Food, Drug, and Cosmetic Act* (FDCA) are covered under TSCA. The TSCA provides specific controls for PCBs and asbestos and dictates that any new substance is considered hazardous until proven otherwise.

The *Comprehensive Environmental Response, Compensation, and Liability Act* (CERCLA) and the *Superfund Amendments and Reauthorization Act* (SARA) represent what is commonly known as the Superfund Program. This program created the National Priorities List (NPL), covering sites in the U.S. that need remediation. If those sites present an imminent threat to human health, the program has funds available for remediation. This legislation also assigns liability for cleanup costs to the generator of the hazardous waste and to current landowners. One cannot escape the cost of remediation by selling the property and one can buy into liability by purchasing property. The latter characteristic has forced prospective buyers of property to conduct environmental audits to ensure they are not buying into a large cleanup expense. This has also inhibited the sale of older industrial properties and led to the *brownfields* issue, or abandoned industrial properties that prospective buyers are afraid to purchase because of potential liability. The EPA has recently begun to address this issue by waiving liability so that the brownfield sites are developed for new industries rather than remaining unused.

More recently, the *Food Quality Protection Act* (FQPA) has made significant modifications to FIFRA and FDCA. The FQPA now specifically promotes integrated pest management techniques, as an attempt to reduce pesticide use, and facilitates labeling of minor use and reduced risk pesticides. The FQPA also eliminated the *Delaney clause,* an absolute risk standard, for processed foods. The Delaney clause stated that pesticide residues were simply not allowed in processed foods if those pesticides were thought to be carcinogenic. While the logic of this seems sound, the requirement was difficult for the food processors to meet since pesticides are routinely used in food production and toxicology did not support the need for such a strict standard. The Delaney clause was replaced with a negligible risk standard that allows low levels of pesticide residues, similar to the way that raw foods are regulated.

One can trace the change in public attitudes about the environment through the evolution of the environmental legislation. Prior to the publication of *Silent Spring* in 1962 by Rachel Carson there was little widespread appreciation for the potential environmental impact of pesticides or other toxic substances. The resulting increase in public awareness led to the passage of some major legislation in the 1970s. This decade produced NEPA, the CAA, a significant amendment to the CWA in 1972, RCRA, and TSCA. All of these were instrumental and tremendously successful in reducing the amount of pollutants that were released into the environment. In most areas, water and air quality are better now than they were 20 years ago and the management of hazardous substances is strictly controlled because of these important pieces of environmental legislation. The next step was the passage of CERCLA and SARA in the 1980s, which addressed cleaning up contamination that had already occurred. The recent passage of the FQPA indicates we are willing to revisit environmental legislation and make needed changes. There will always be disagreements on how strict environmental legislation should be, but there is little doubt that the legislation that has been implemented has had very positive impacts on the environment.

1.9   Major environmental issues in soil science: a summary

An underlying theme in this book is *soil quality,* a concept that is still being developed by soil scientists and which will be described in more detail in Chapters 3 and 9. The Soil Science Society of America uses the following definition:

> The capacity of a specific kind of soil to function within natural or managed ecosystem boundaries, to sustain plant and animal productivity, maintain or enhance water quality, and support human health and habitation.

For the soil functions, we consider issues such as the ability of a soil to produce quality food, fiber, or feed; the construction properties and limitations; the ability to support habitation and recreation; and the ability to maintain an ecosystem or desired land use. The ability of a soil to produce quality food, fiber, or feed refers to the contribution the soil makes to the capability of a site to have profitable production of crops that are free of harmful substances. Construction properties and limitations, topics beyond the scope of this book, refer primarily to chemical and mineralogical characteristics that influence soil properties such as shrink–swell capacity and corrosiveness which, in turn, determine suitability for basements or buried materials. Supporting habitation and recreation implies that one can have the soils around homes, schools, and parks without concern for negative effects from contamination. The ability to maintain an ecosystem or desired land use recognizes the role of soils in the development and maintenance of ecosystems such as a tropical rain forest. When these soils are changed, say, by exposing the soil to erosion and oxidation of organic matter by clearing the rain forest for crop or lumber production, they may lose their ability to maintain the tropical rain forest and we would note that the soil quality had been reduced. Similarly, if poor quality irrigation water is used and soils become saline, they can no longer maintain our desired land use of crop production.

The quantification of soil quality is a difficult task at best. In the broadest sense, the physical and chemical properties of a soil determine its quality. *Physical properties* such as bulk density and texture influence aeration, permeability, infiltration capacity, water-holding capacity, or constructive properties and can be quantified and related to quality. *Chemical properties* are the concentrations of organic and inorganic constituents that determine characteristics such as soil fertility, biological activity, degree of contamination, salinity, corrosiveness, or shrink–swell potential and are also quantifiable and related to soil quality. Indeed, much of this book discusses the impact of excessive concentrations of nutrients, pesticides, hazardous substances, or trace elements on the function of soils. Much of the work in the soil quality area has focused on the crop production aspects of soil functions, but the assessment of soil quality relative to contamination has not been developed very much. Soil quality assessments are relative and it is easy to see how a change in soil properties, such as increasing contaminant concentrations, would be taken as a reduction in soil quality.

One could assemble a very long list of individual environmental problems related to soil science. The list would include such problems as leaking underground storage tanks, trace element–contaminated soils, saline soils, eroded soils, and acidified soils. It is probably more instructive to present categories of the major environmental issues in soil science. Most of the issues fall into one or more of the following categories:

1. Reductions in soil quality because of unacceptable concentrations of pollutants. This category includes soils that either directly or through food chain transfer expose humans or other organisms to pollutants that may cause direct detrimental effects. This category is quite extensive and includes problems associated with the pesticides, hazardous substances, and trace elements pollutant categories listed in Table 1.1.
2. Reductions in soil quality that limit soil function. Eroded, acidified, or salt-affected soils that can no longer support a desired land use or ecosystem fall into this category. Direct detrimental effects to humans or other organisms due to exposure to pollutants are generally not an issue in this case.
3. Soils as a source of contaminants. This category includes leaching and runoff losses of various chemicals or materials from the landscape. Here the presence of a substance in the soil is not the primary problem but rather the effects of the substance on the environment as it leaves the original point of application is of concern. Thus, a reduction in soil quality as indicated by chemical analysis is not necessarily the issue but rather a conflict between soil function or use and the surrounding environment. Horticultural and agricultural uses of pesticides and nutrients would be prime examples here. Solving these problems ultimately requires an understanding of all risks and a prioritization of land uses and the desired quality of the environment.

Societal responses to environmental soil science issues are varied. Regulations are written that prevent soil contamination or that control the use of substances that are considered pollutants. Research is conducted on methods for remediating contaminated soils and on understanding the fate and transport of contaminants in the environment. Government programs are used to prevent degradation of soil quality by erosion. Still, considerable work needs to be completed with regard to understanding the interaction of the soil environment with potential pollutants and on utilizing the soil resource for the benefit of society while maintaining, or improving, environmental quality.

As we study environmental science, part of the task is familiarizing ourselves with the technical aspects, such as the scientific method, nomenclature, and processes, and part of the task is appreciating the objectivity, philosophical approaches, and even moral questions required to have a complete understanding of environmental issues. Soils play a major role in the cycling of many environmental contaminants, and soil science serves as a useful discipline from which to study the environment. To do so we must have a basic understanding of soils, hydrology, and the atmosphere, which this book will attempt to provide. The major classes of soil pollutants will then be identified, followed by a detailed discussion of each. Risk assessment, as related to soils, is the topic of the final chapter.

Problems

1.1 List and describe the steps in the scientific method. How does epidemiology fit in with the traditional approach to experimentation?

1.2 Discuss how knowledge gained through the scientific method is vital to our understanding of the environment. How does the scientific method interact with the willingness of society to pass environmental legislation and to spend public resources to remediate contaminated sites?

1.3 Discuss environmentalism as a social movement. What are the stages of a social movement and what stage or stages are evident for environmentalism?

1.4 Describe the difference between an environmental issue and an environmental event.

1.5 Describe ecocentric, homocentric, and egocentric attitudes regarding the environment. Which type best describes your attitude toward the environment?

1.6 Discuss several definitions of pollutant or pollution. Which do you prefer and why? How do value judgments affect our definition of pollution? Write your own definition of pollution.

1.7 What is the difference between soil or water that is contaminated vs. that which is polluted?

1.8 Describe the various pollutant categories and give examples of each. Which involve the soil resource in some way?

1.9 Define bioavailability and discuss why the bioavailability of soil contaminants is important in our study of environmental quality.

1.10 Which pieces of federal legislation resulted in major reductions in the amount of pollutants released in the environment in the U.S. and started the trend of improving environmental conditions that continues today?

1.11 Describe the major environmental issues in soil science.

References

Brams, E., W. Anthony, and L. Witherspoon. 1989. Biological monitoring of an agricultural food chain: soil cadmium and lead in ruminant tissues, *J. Environ. Qual.*, 18, 317.

Hoffer, E. 1951. *The True Believer. Thoughts on the Nature of Mass Movements*, Harper & Row, New York.

Logan, T. J., G. M. Pierzynski, and R. B. Pepperman. 1995. National markets for organic products including biosolids: opportunities, competition and constraints, in *Proceedings of the 4th Joint WEF and AWWA Conference on Biosolids and Residuals Management*, Water Environment Federation, Alexandria, VA, 15-13 to 15-28.

Miller, G. T. 1991. *Environmental Science: Sustaining the Earth*, Wadsworth Publishing, Belmont, CA.

Supplementary reading

Carson, R. 1962. *Silent Spring*, Houghton Mifflin, Boston.

Hillel, D. J. 1991. *Out of the Earth: Civilization and the Life of the Soil*, Macmillan, New York.

Marco, G. J., R. M. Hollingworth, and W. Durham, Eds. 1987. *Silent Spring Revisited*, The American Chemical Society, Washington, D.C., 214 pp.

Rousseau, D. L. 1992. Case studies in pathological science, *Am. Sci.*, 80, 54–63.

chapter two

Our environment: atmosphere and hydrosphere

Contents

2.1   Introduction

Our environment comprises natural wonders that provide for the ingredients of life, including oxygen (O_2), water (H_2O), and nutrients, e.g., nitrogen (N), phosphorus (P), sulfur (S). These fundamental needs are met through the air we breathe, the fluids we drink, and the foods we eat. The atmosphere contains essential gases such as O_2, carbon dioxide (CO_2), and N_2 that are needed to sustain our existence. The water we drink comes from surface water and groundwater supplies, all of which have been cycling over millions of years. Production of crops for human and animal consumption relies on our ability to plow, seed, and cultivate our lands. Plants require CO_2, H_2O, and nutrients, suggesting there is a need to understand how the atmosphere and hydrosphere sustain plant growth and support animal and human life.

The part of the planet that supports living organisms is described as the *biosphere*. Thus, the biosphere comprises all lifeforms and their general surroundings, which includes most hydrosphere and soil ecosystems (see Chapter 3 for more details of the soil environment). Our atmosphere is primarily composed of nonliving substances and is therefore not considered part of the biosphere. However, the atmosphere, as well as most surficial environments, influences the ecology of most areas. Biosphere impacts, both positive and negative, are often the result of lifestyles that rely on natural resource utilization that, in turn, can affect the quality of our atmosphere and hydrosphere. This chapter reviews the basic characteristics of both the atmosphere and the hydrosphere, and examines how these two spheres are important for human subsistence as well as their relationship with environmental quality. To appreciate the complexity of evaluating water quality issues, we have also included a section that describes the rationale, procedures, and interpretation involved in environmental testing of surface waters and groundwaters.

2.2   Atmosphere

We live in a time when concern for air quality is growing due to the increasing amounts of pollutants that are added to the atmosphere daily. The atmosphere also plays an important role in nutrient and contaminant transport processes. Our current atmosphere provides us with protection against harmful solar and cosmic radiation, moderates surface temperatures, and is a major component of the hydrologic cycle. The study of the atmosphere and its phenomena is called *meteorology* and involves interactions between the atmosphere and the earth's land and ocean surfaces, and various influences on living systems.

Significant characteristics of the atmosphere include layers and boundaries, pressure and density, temperature and heat, wind circulation and patterns, chemical composition and reactions, moisture and precipitation, and atmospheric pollution. While each of these areas is important to the understanding of our atmosphere, we will only briefly discuss some of these topics. Additional information can be obtained from sources listed in the references and supplementary reading sections located at the end of this chapter.

2.2.1   Atmospheric layers and their properties

The atmosphere consists of several layers (i.e., troposphere, stratosphere, mesosphere, thermosphere, and exosphere) and transitional zones (i.e., tropopause, stratopause, and mesopause), each of which has distinct properties (Figure 2.1). Of the major layers, only the troposphere and stratosphere, which comprise the lower atmosphere (0 to 50 km), are generally considered of great significance to humans. This is because the troposphere is

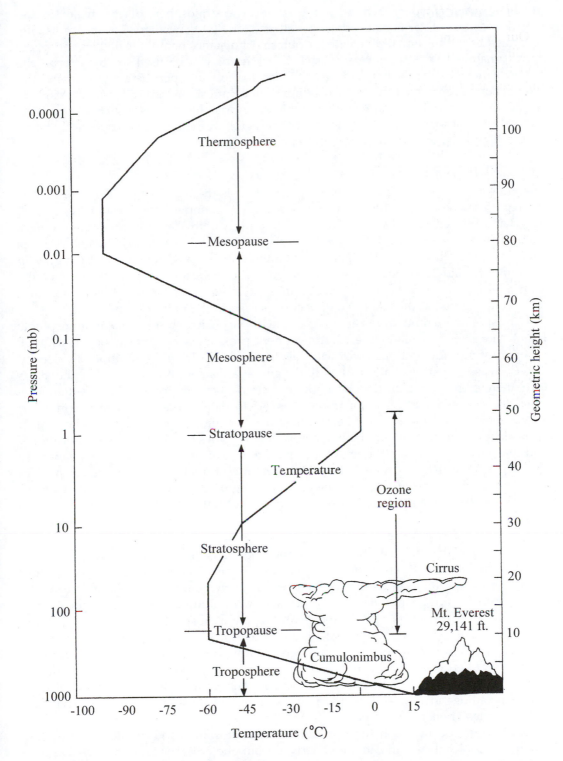

Figure 2.1 Variation in pressure and temperature in atmospheric zones to an altitude of 100 km. (Modified from Schlesinger, W. H., *Biogeochemistry: An Analysis of Global Change*, 2nd ed., Academic Press, San Diego, CA, 1997. With permission.)

involved in many biogeochemical processes and the stratosphere plays a major role in global transport of various materials as well as ozone (O_3) chemistry. Layers are usually separated by a transitional zone with the outermost distance or altitude of the atmosphere approximately 1000 km above sea level. Other terms can be used to characterize the atmosphere including the homosphere (i.e., combination of troposphere, stratosphere, and mesosphere layers which have a fairly uniform air composition throughout), heterosphere (i.e., both thermosphere and exosphere which comprise lighter gases such as hydrogen and helium), or ionosphere (i.e., the electrified region in the upper atmosphere that contains a high level of ions and free electrons).

2.2.1.1 Troposphere

The atmospheric layer closest to the earth's surface is the troposphere (Figure 2.1). It is the region that comprises the weather conditions (i.e., temperature, precipitation, wind, etc.) that influence our daily lives. The upper boundary of this layer is variable and ranges from 18 km above sea level at the equator to about 8 km above sea level at higher latitudes closer to the earth's poles. At a particular latitude, the troposphere is usually higher in the summer and lower in the winter seasons. The troposphere contains most of the atmospheric mass and includes 99% of atmospheric water vapor, which is highest and more concentrated at the equator and decreases toward the polar regions.

Pressure and temperature variations in the troposphere and other major atmospheric regions are also shown in Figure 2.1. Atmospheric pressure is an approximate logarithmic function of altitude and decreases with height above sea level. In addition, air density also decreases logarithmically with altitude, and at a distance of about 5.5 km above sea level there exists only half the total amount of atmospheric molecules. Average temperatures decrease with altitude in the troposphere (15 to –60°C) because of a decline in water vapor content and the level of heat radiation reflecting and emitted from the earth's surface; however, air temperature in the troposphere varies considerably with time of day and season, latitude, and altitude. A transitional zone known as the tropopause separates the troposphere from the stratosphere and is unique in that temperature is constant in this zone with increasing altitude. This layer is a region of cold temperature that acts as a protective barrier, reducing the loss of water to the stratosphere.

2.2.1.2 Stratosphere

The next major layer above the troposphere and tropopause is the stratosphere. It is located approximately 20 to 45 km above sea level (see Figure 2.1). Whereas the temperature decreases with altitude in the troposphere and is fairly constant in the tropopause, the stratospheric temperature (–60 to 0°C) increases with altitude due to the absorption of ultraviolet (UV) radiation by O_3. About 90% of the atmosphere's O_3 content is in the stratosphere; however, the water content of this layer is very low. Absorbed solar energy increases the temperature of the stratosphere. Although the maximum O_3 concentration (~12 ppm O_3) is generally just above the tropopause at approximately 25 km, the highest temperatures are found in the upper stratosphere. The cause of this is the greater absorption of incoming radiation and the lower atmospheric density at this altitude, which results in lower rates of energy transfer.

Ozone production is a natural process in the stratosphere and results from the combination of molecular O_2 and atomic O derived from photodissociation reactions. Destruction of O_3 also occurs through collisions with other atoms and molecules. Gases that are especially destructive to O_3 include NO and NO_2, known collectively as nitrogen oxides (NO_x), and chlorofluorocarbons (CFCs); NO_x is the result of natural and human activities, CFCs are synthetically produced. A simplified illustration of how CFCs react in the

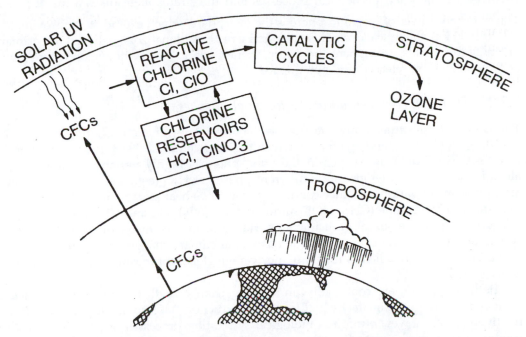

Figure 2.2   Role of CFCs and the degradation products derived from reactions with solar radiation in the reduction of ozone in the stratosphere. (Modified from Shen, T. L. et al., in *Composition, Chemistry, and Climate of the Atmosphere*, Van Nostrand Reinhold, New York, 1995. With permission.)

stratosphere is shown in Figure 2.2. With a loss of O_3 in this layer there would be lower absorption capacities for UV solar radiation, which, in turn, could influence the temperature in the troposphere, causing increased damage to plant life and greater incidence of skin cancer in humans (see Chapter 10 for more information on global climate change and acid deposition impacts involving NO_x and O_3). Another transitional zone exists between the stratosphere and the next highest layer, called the stratopause.

2.2.1.3   Mesosphere

Above the stratosphere and stratopause is the mesosphere, which extends from approximately 50 to 80 km above sea level (see Figure 2.1). Within this layer, concentrations of both O_3 and water vapors are minuscule, there is very little air, and atmospheric pressure is low. Temperature again decreases with altitude (0 to –95°C) because of a lack of molecules that absorb heat energy from solar radiation. About 99.9% of the atmospheric mass is below the mesosphere. Although the percentage of N_2 and O_2 in this layer is similar to that of air at sea level, the low density of this air would not provide sufficient O_2 for humans and suffocation would occur in minutes.

2.2.1.4   Thermosphere

The thermosphere is the *hot layer* above the mesosphere that is separated by the transitional mesopause zone (see Figure 2.1). This layer extends from about 90 to 500 km above sea level. Temperature of this layer (–95 to over 1200°C) cannot be determined directly due to the low density of the air. In fact, an air molecule in the lower thermosphere may travel a distance of 1 km before colliding with another molecule; in the upper thermosphere molecules may travel 10 km before a collision occurs.

2.2.1.5   Exosphere

The outermost layer of the atmosphere is the exosphere, which extends to approximately 960 to 1000 km above sea level. This layer acts as a transitional zone between the primary layers of the earth's atmosphere and interplanetary space. This is the region where atoms and molecules are released into outer space.

2.2.2   Chemical composition of the lower atmosphere

The chemical constituents making up the majority of lower atmospheric gases are N_2 (78.08%), O_2 (20.95%), and argon (Ar), which together comprise 99.9% (by volume) of the composition of air. Water vapor in the atmosphere is highly variable, changing with altitude, latitude, and temperature; different phases of H_2O (e.g., solid, liquid, and gas) are important to various weather conditions as well as heat storage and release. Carbon dioxide (CO_2), methane (CH_4), NO_x, sulfur dioxide (SO_2), O_3, and CFCs, although only minor in abundance, are of particular importance to the chemistry of the earth's atmosphere and to environmental issues such as global climate change and acidic deposition (see Chapter 10). A detailed summary of the average global composition of the atmosphere is given in Table 2.1.

Both N_2 and O_2 are continually undergoing change, with their rate of destruction balanced by an approximately equal level of production. Molecular nitrogen gas (N_2) is relatively inert; however, through biological N fixation processes organic N forms are produced that are sources of nutrients and structural components to living organisms. Conversely, decomposition of plant and animal remains, and biochemical reduction of soluble N species, can release N_2 back into the atmosphere. Oxygen is used during decomposition, respiration (O_2 is inhaled, CO_2 is exhaled), and combustion processes. During plant growth, sunlight provides energy for the production of sugars and O_2 from the combination of CO_2 and H_2O, a process known as photosynthesis. Approximately 1% of the sun's energy that reaches the earth's surface is used during photosynthesis; most of the incoming solar energy is utilized in the evaporation of water that reenters the atmosphere for future dissipation of the trapped energy in the form of heat.

The chemical composition of the lower atmosphere is relatively homogeneous, except in areas close to the earth's surface that are affected by air pollution. Mixing takes place in the troposphere due to winds and the rising of warm air that develops near the earth's surface, which results in a relatively constant blending (dilution and dispersion of the different chemical constituents) of the atmosphere. Inversions (i.e., temperature increases with altitude) will prevent mixing and cause the air to retain pollutants, a condition that is sometimes noticed during the night and earlier morning when wood smoke from chimneys does not dissipate readily. Topographic lows and urban areas are also known to prevent atmospheric mixing that homogenizes the local air mass. Within the stratosphere, only minor changes in chemical composition occur over time because very little vertical mixing takes place.

2.2.3   Atmospheric cycles

Several elements and compounds have atmospheric cycles that are part of their overall interchange among soil, hydrosphere, and biosphere ecosystems. Some of the more important elements include C, N, and S; cycles of N and S are discussed in Chapters 4 and 6, respectively. Chapter 10 provides information on the CO_2 cycle and its role in the greenhouse effect. In this section we will describe the basics of important atmospheric cycles, and provide general information to help understand their interaction with the soil, water, and biosphere.

Table 2.1   Global Average for Chemical Constituents in the Atmosphere

Chemical constituent	Common name	Percent	Approximate mass (kg)
	Dry air	100	5.12×10^{18}
Permanent gases			
N_2	Nitrogen	78.08	3.9×10^{18}
O_2	Oxygen	20.95	1.2×10^{18}
Ar	Argon	0.93	6.6×10^{16}
Ne	Neon	1.8×10^{-3}	6.5×10^{13}
He	Helium	5.2×10^{-4}	3.7×10^{12}
H_2	Hydrogen	5×10^{-5}	1.8×10^{11}
Variable gases (most of these values are estimates)			
H_2O	Water vapor	0.1–4	1.70×10^{16}
CO_2	Carbon dioxide	3.5×10^{-2}	2.45×10^{15}
CH_4	Methane	1.5×10^{-4}	4.3×10^{12}
N_2O	Nitrous oxide	3×10^{-5}	2×10^{12}
CO	Carbon monoxide	1×10^{-5}	6×10^{11}
NH_3	Ammonia	1×10^{-6}	3×10^{10}
NO_x	Nitrogen oxides	1×10^{-7}	8×10^{9}
SO_2	Sulfur dioxide	2×10^{-8}	2×10^{9}
H_2S	Hydrogen sulfide	2×10^{-8}	1×10^{9}
O_3	Ozone	0.01–25 ppm	3×10^{12}
CFCs	Chlorofluorocarbons	0.002 ppm	Variable

Note: Total atmospheric mass equal to 5.14×10^{18} kg.

Sources: Walker, 1977; Ahrens, 1994; Salstein, 1995.

Oxygen plays an important role in elemental cycles (i.e., C, N, P, S, and some trace elements), particularly for those elements that have atmospheric components (C, N, and S). Oxygen is a key element in atmospheric, geochemical, and life processes. Figure 2.3 indicates some of the various chemical reactions and fluxes that are involved in the O_2 cycle. Atmospheric O_2 (1.18×10^{21} g) represents the largest O_2 pool, which at present is in steady state due to consumption and production processes. It has been suggested that all of the O_2 in the atmosphere has been cycled through photosynthetic organisms such as plants and certain microorganisms. A large pool of oxygen exists in the lithosphere in reduced forms that are slowly released by weathering reactions. Oxygen is also consumed during the burning of fossil fuels, aerobic degradation of organic matter, as well as the oxidative weathering of soil, rocks, and minerals.

Both O_2 and CO_2 cycles are regulated to a large extent by living organisms. The turnover rate or lifetime of O_2 and CO_2 in the atmosphere is related primarily to photosynthesis and respiration processes. The time required to cycle O_2 and CO_2 through the atmosphere is different due to variation in the size of their pools. The *mean residence time* (MRT) is a measure of the time it takes a substance to cycle through a particular pool. For example, the MRT of atmospheric O_2 as it cycles through the biosphere is approximately 3000 years, whereas the CO_2 MRT is only about 5 years.

Biogeochemical cycles of nutrients and trace elements can be extremely complex and are often studied by examining individual ecosystems (i.e., atmosphere, hydrosphere, lithosphere, and biosphere). Of primary interest are the transfer rates, MRTs, and fates of nutrients, trace elements, and sometimes organic pollutants in one or more of the various ecosystems. Because atmospheric processes influence the cycling of these materials, it is important to understand the relationships among the various constituent pools, their

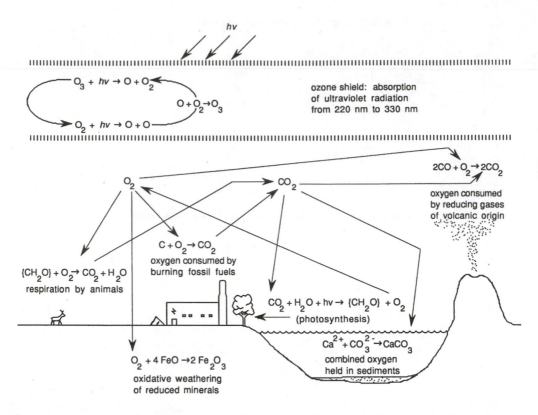

Figure 2.3   Examples of various oxygen and carbon dioxide reactions that occur in the environment. (Modified from Manahan, 1991.)

reactions, and fluxes, which will be examined in subsequent chapters for N, S, trace elements, and organic chemicals. The influence of human activity on biogeochemical cycles has stimulated several studies that seek to determine what the future may hold for humankind and environmental quality.

2.2.4   *Temperature and precipitation patterns*

Climate patterns can be quite varied on both regional and global scales. Average climatic parameters have also changed with time due to both natural (e.g., volcanic eruptions) and anthropogenic inputs. For example, there is sufficient evidence today to conclude that the global average temperature has been increasing over the past 100 years (see Chapter 10). While there is a certain amount of annual variation in the global temperature, Figure 2.4 shows that there is a definite increasing trend for warmer temperatures in recent years. Annual precipitation patterns are also variable, with reports of occurrences of 1-in-100-year rainfall events heard often. The global distribution of precipitation is generally related to atmospheric circulation patterns and the location of mountain ranges and high plateaus. We know that areas in the equatorial region are usually the wettest and the polar regions, leeward sides of mountains, and some subtropical high pressure areas are typically the driest.

Minimum, maximum, and ranges in temperatures and rainfall patterns across the U.S. are related to many climatic and geographic variables. For example, the topography along a transect from the Pacific Ocean into western Nevada can cause enormous discrepancies in the annual precipitation received within different regions of the

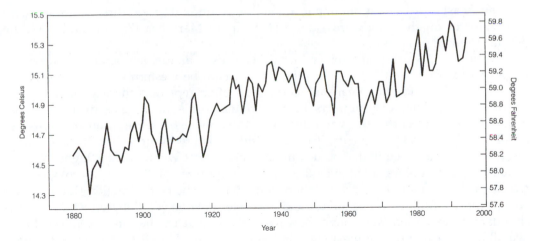

Figure 2.4   Change in the global average temperature from 1880 to 1994. Note the large variability between adjacent years (e.g., as much as 0.2°C) and the distinctive increasing trend over time. There has been an approximate 1.0°C increase in the global average temperature since 1880. (From McKinney, M.L. and R.M. Schoch, *Environmental Science: Systems and Solutions*, Jones and Bartlett Publishers, Sudbury, MA, 1998. With permission.)

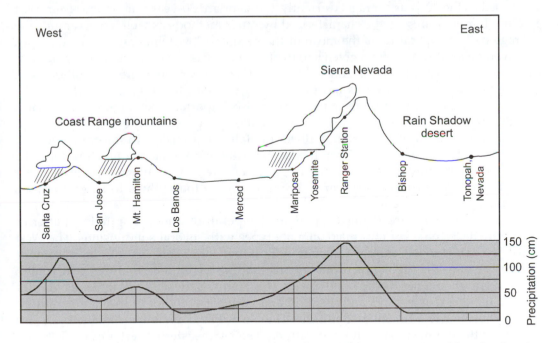

Figure 2.5   Precipitation patterns along a transect from California east into Nevada showing distinct topographical differences and rain shadow effects. (Modified from Ahrens, C. D., *Meteorology Today: An Introduction to Weather, Climate, and the Environment*, 5th ed., West Publishing, St. Paul, MN, 1994. With permission.)

transect (Figure 2.5). The figure illustrates how topography results in several "rain shadow" effects along the transect. These are regions on the lee side of mountains that receive conspicuously less precipitation than is obtained on the windward side of mountain ranges. There is a distinct difference in the amount of precipitation on the west and east sides of the mountain range, with amounts two, three, four, or more times greater on the windward side, e.g., compare Mariposa (70 cm) and Yosemite (90 cm) to Bishop (20 cm).

Within the U.S., there are tremendous differences in annual temperature and precipitation going from the West Coast to the East Coast. The western states receive winter precipitation from Pacific Ocean–fed storms that often replenish the loss of water that occurs during the hot, dry summers. Intense winter storms traveling eastward can cause major problems in central U.S. states due to heavy snowfalls; intense summer storms also result in serious flooding events. Because of the moisture brought in from the Gulf of Mexico, most U.S. precipitation occurs in the eastern states where there is higher average annual precipitation. As noted in Figure 2.6, the western region (San Francisco) generally receives greater amounts of precipitation in the winter, the central region (Kansas City) obtains mostly summer precipitation, and the eastern region (Washington, D.C.) receives abundant precipitation year round; however, there are years that are exceptions to the overall average. Temperature extremes in the U.S. are evidenced by the hot, dry desert conditions in some parts of the west to the cold, snowy mountainous regions of Alaska and the Rocky Mountains. Record high and low temperatures for the U.S. are 57°C (134°F) in Death Valley, CA and –62°C (–80°F) and –57°C (–70°F) in Prospect Creek, AK and Rogers Pass, MT, respectively.

Typically, temperature and precipitation characteristics for a given location are expressed as the 30-year average. Currently, the standard 30-year time span is from 1971 to 2000. Temperatures can be characterized by other methods as well. For example, the average daily temperature is the mean of the high and low temperatures for that day. The average temperature for a given length of time is the mean of the daily temperatures over that time period. The average high and low temperatures are also often used. Average annual precipitation is most often used to characterize precipitation, but the distribution of that moisture over the year is also important (see Figure 2.6). Finally, rainfall intensity plays a major role in the partitioning of water into surface runoff or water that infiltrates into the soil. A 10-cm rainfall event received uniformly over 8 h has an average intensity of 1.25 cm/h and will usually not generate much runoff. The same amount of rainfall received in 1 h, given an intensity of 10 cm/h, would likely cause considerable runoff. Some areas are more prone than others to short-duration, high-intensity storm events.

Climatic regions are defined based on temperature and precipitation patterns. Within the U.S. there are four major climate types with several subdivisions. The West Coast is characterized as moist climate with either long, cool or hot, dry summers. Further east are the highlands or mountains that have cold temperatures year round, and include the Sierra Nevada and Rocky Mountains. Between the highlands are dry climate regions that are arid and semiarid and vary between hot or cool temperatures, which is primarily a function of altitude. The Great Plains region is classified as a dry climate with cool mean annual temperatures. The northeastern quarter of the U.S. has a moist climate with severe winters and humid summers, whereas the southeastern quarter, the largest climatic region in the U.S., has a moist climate with mild winters and humid subtropical conditions.

2.2.5   *Atmospheric pollution*

Due to the magnitude of N_2 and O_2 in the atmosphere, these elements remain relatively constant. The percentages of the other atmospheric constituents, however, change with time as a result of natural and anthropogenic emissions. Atmospheric pollutants tend to

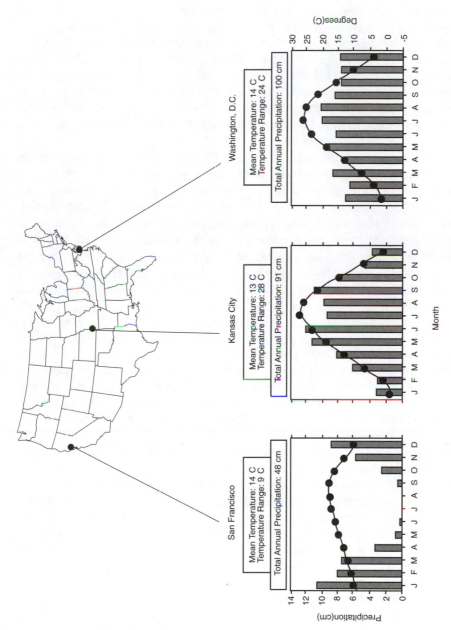

Figure 2.6   Differences in monthly temperature (solid circles) and precipitation (bars) in West Coast, central, and East Coast cities. Note the distinct difference in precipitation, with San Francisco receiving the majority of its precipitation in the winter, Kansas City in the summer, and Washington, D.C. receiving fairly constant precipitation year round.

be higher over source areas such as cities and fossil fuel-burning power plants. However, there are also natural sources that emit gases; swamps and anaerobic environments release CH_4 and hydrogen sulfide (H_2S), and wildfires can cause the emission of CO_2, carbon monoxide (CO), CH_4, and carbonyl sulfide (COS). Agricultural practices are partially responsible for the release of CO and N_2O from the cultivation and fertilization of soils. Even domestic ruminant animals contribute approximately 60 to 100 metric tons of CH_4 per year, which accounts for 15% of the CH_4 emitted globally.

Humans have had a profound effect on the composition of the atmosphere. Some of the trace gases that have increased due to human activities include CO_2, CH_4, N_2O, CO, and CFCs. These gases can either directly or indirectly increase the amount of absorption of infrared radiation, thus intensifying the greenhouse effect (see Chapter 10 for further discussion on global climate change). Although it may not be readily apparent at the present time, significant climate change may result from pollutant gases and particulates that are entering the atmosphere from burning fossil fuels, removal of natural vegetation, release of CFCs, and other human activities.

The atmospheric CO_2 concentration has increased by about 25% over concentrations determined from preindustrial times, around 1800. Between 1958 and 1998, the level of atmospheric CO_2 rose from 316 to 365 ppmv, a rate of increase of over 1 ppmv/year. Under the scenario of "business as usual," it has been predicted that CO_2 concentration will double that of the preindustrial level by the middle of this century. Business as usual refers to our continued reliance on burning fossil fuels that produce CO_2 and clearcutting of temperate and tropical forests, which decreases CO_2 consumption by photosynthesis. Greater levels of atmospheric CO_2 and other greenhouse gases are expected to increase the global mean temperature by 1.5 to 5°C sometime between 2025 and 2050 (see Chapter 10).

In addition to atmospheric gases, particulate matter is also present in the atmosphere and is made up of various organic or inorganic materials consisting of liquids or solids. Biological materials such as viruses, bacteria, spores, and pollen grains can also be classified as particulate matter if suspended in the atmosphere. Particulate matter is generally smaller than 0.5 mm and can be derived from several natural and anthropogenic sources. Particles that are less than 1 μm are capable of being retained in the atmosphere and transported long distances. Particles in the range of 0.001 to 10 μm are common in and around pollution source areas such as cities, highways, industrial and power plants. Winds can greatly reduce the concentration of these materials within short distances of their sources.

Airborne particles can originate from explosions, breakdown of materials by grinding action, volcanic activity, and wind erosion. Surface mining relies on explosives to loosen the underlying rock so that it can be moved or processed. Large amounts of particulate matter can result from wind erosion in arid and semiarid areas. As an example, windblown soil particles from arid regions have been estimated to contribute 10^{15} g of particulate matter to the atmosphere each year. Of this, 20% is less than 1 μm and can be transported over long distances.

Trace elements are added to the atmosphere by several processes. Fossil fuel and coal burning, smelting of Fe and nonferrous metals, volcanic ash, and wind erosion are responsible for increasing atmospheric concentrations of such elements as gold (Au), bromine (Br), cadmium (Cd), lead (Pb), selenium (Se), tin (Sn), and tellurium (Te) by as much as four orders of magnitude above normal levels. Concentrations of trace elements can be bioaccumulated by some plants and microorganisms, possibly increasing the potential harmful effects of trace elements. Trace element effects are discussed further in Chapter 7.

Environmental quality issues/events
Chlorinated gases and the ozone hole

Concerns associated with stratospheric O_3 deletion are most evident over Antarctica (South Pole); however, a smaller O_3 thinning is also apparent in the Arctic (North Pole). British scientists collecting data in the Antarctic since 1950 have discovered a seasonal O_3 fluctuation that probably started in the 1960s and began progressively to cause thinning of the O_3 layer, especially during the spring, with the lowest levels in mid-October (remember that spring in the Southern Hemisphere is fall in the Northern Hemisphere). Research conducted using high-altitude airplanes, satellites, and instrumented balloons released over the Antarctic discovered that there was nearly a 40% reduction of O_3 in the spring of 1984 that increased to about a 70% depletion by 1993. During an extreme period in 1993, the size of the area with at least 50% O_3 depletion was approximately three times as large as the continental U.S. On a global scale, 5% or more of the earth's O_3 has been depleted since the early 1980s.

Chlorinated gases are believed to be the primary substances that cause destruction of the O_3 layer. One chlorinated gas (hydrogen chloride, HCl), which has been released in large quantities in space shuttle exhaust, can destroy O_3. Researchers discovered in 1972 that the shuttle's emission of HCl as it traveled through the stratosphere was a potential cause of O_3 destruction and 2 years later they hypothesized that other chlorinated gases such as CFCs could also contribute to the problem. Chlorofluorocarbons (e.g., Freon™) were first developed in the 1930s by du Pont Industries as a supercoolant that, at the time, was deemed to be inexpensive, noncorrosive, nonflammable, and nonreactive (e.g., chemically stable). Freon has been used as a coolant for refrigerators and air conditioners, as well as an aerosol spray propellant for spray paints, hair spray, whipped topping stabilizer, cosmetics, and insecticides.

Within the Antarctic stratosphere, ice crystals form in the winter months that retain both O_3 and CFC molecules on their surfaces. During spring warming, the O_3 and CFCs that have accumulated are released as ice crystals melt. CFC molecules that are exposed to UV radiation in the presence of O_3 can lose chlorine (Cl) atoms, a process known as photodissociation. Chlorine is extreme effective at destroying O_3 molecules according to the following catalytic reaction series:

$$Cl + O_3 \rightarrow ClO + O_2 \tag{2.1}$$

$$ClO + O \rightarrow Cl + O_2 \tag{2.2}$$

with the net result being

$$O_3 + O \rightarrow 2O_2 \tag{2.3}$$

It is essential to note that all of the important O_3-destroying catalytic cycles include the chlorine monoxide (ClO) species. An additional reaction involving ClO that results in regeneration of chlorine atoms is

$$ClO + ClO \rightarrow Cl_2O_2 \tag{2.4}$$

$$Cl_2O_2 + h\nu \rightarrow Cl + ClO_2 \qquad\qquad (2.5)$$

$$ClO_2 \rightarrow Cl + O_2 \qquad\qquad (2.6)$$

One Cl atom photodissociated from a CFC molecule can catalyze thousands of O_3-destroying cycles, resulting in the degradation of about 100,000 O_3 molecules. With an estimated 20 million Mg of CFCs produced since its introduction, and with about 90% released into the atmosphere, it is no wonder the use of CFCs had to be reduced.

In recognizing the potential problems associated with CFCs, industry started to produce CFC-free propellants for spray cans and assisted in supporting an intense research program aimed at finding alternatives to the use of CFCs as refrigerants. An international conference in 1987 resulted in the "Montreal Protocol on Substances That Deplete the Ozone Layer" that was accepted by many nations as a way to curtail the problem. As shown in Figure 2.7, the original protocol, which mandated a reduction in the consumption of five CFCs to 50% of 1986 concentrations and three Br-containing halocarbons to 1986 concentrations by the year 2000, was deemed insufficient for reducing the O_3-damaging substances. Revisions to the Montreal protocol were negotiated in London in 1990 and again in Copenhagen in 1992. From Figure 2.7 it is apparent that declines in global O_3 levels will continue for some time to come, and that the "ozone hole" over Antarctica will continue to develop every year for most of the 21st century.

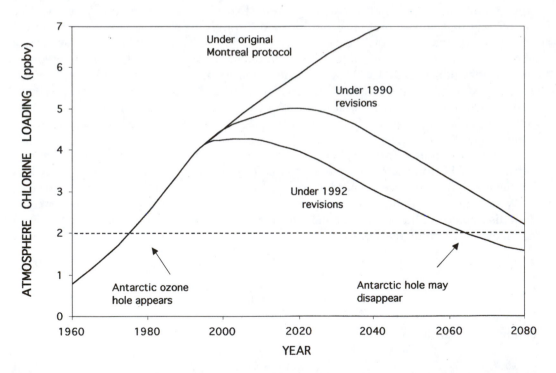

Figure 2.7   Atmospheric Cl concentrations (1960 to 1990) and the expected change due to the Montreal protocol for reducing Cl loadings. Note that 2 ppbv Cl is the theorized stratospheric limit that is required for the appearance of the Antarctic O_3 hole.

2.3   Hydrosphere

The hydrosphere includes water bodies such as rivers, streams, lakes, and oceans as well as soil water, groundwater, glaciers, and polar ice caps (Figure 2.8). Although approximately 70% of the earth's surface is covered with water, an enormous amount of water is found below ground. By far the largest source of water is the oceans and seas, followed by ice (polar ice caps and glaciers) and then groundwater. Freshwater and saline lakes and inland seas represent the largest pools of liquid water on land, with rivers and streams comprising only a small fraction of the world's water. The amount of water retained in soils is approximately 50 times that in rivers and streams. While the atmosphere contains only a small fraction of the total amount of water held in other pools, the quantity of water that passes through the atmosphere is immense and extremely important.

The study of the chemical, physical, and biological properties and reactions of water bodies is called *hydrology*; *limnology* is the study of freshwater systems; *oceanography* the study of the oceans; and *meteorology* the study of climate and weather, which is highly dependent on water in the atmosphere as noted in the previous section. *Geohydrology* is the study of water in geological systems such as aquifers and groundwater environments. Soil scientists also study the chemical, physical, and biological properties of soil ecosystems in which water plays a dominant role.

2.3.1   Properties of water

Water is essential to all forms of life on earth. It is also the central component to several soil processes. Ice can physically break rocks into small particles that can then be further weathered by chemical processes. Dissolution of soil minerals and the migration of the dissolved materials is a continual process that is driven by water leaching through soils. In areas of low rainfall, weathering and translocation of dissolved constituents is relatively slow compared with high rainfall areas. Transport of contaminants from soils to groundwaters or surface waters is generally accelerated as the amount of water that percolates through the soil increases.

Water, which is often referred to as the universal solvent, is essential for the transport of nutrients, gases, and organic compounds in the soil environment. Some of the unique

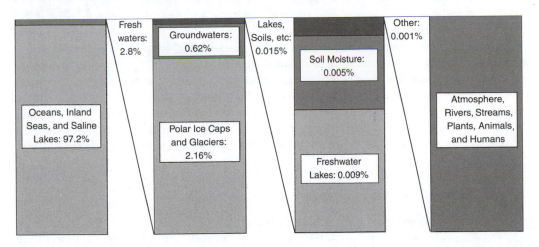

Figure 2.8   Categorization of hydrosphere pools ranging from the dominant sources on the left to those that comprise much smaller sources on the right.

Table 2.2   Properties of Water and Their Significance

Property	Significance	Comments
Solvent	Essential for many biochemical, chemical geological, and atmospheric processes	Ubiquitous substance
Density	Allows ice to float on water	Maximum density at 4°C
Dielectric constant	Reason most ionic substances at least partially dissolve in water	Highest of all pure liquids
Surface tension	Produces unequal attraction forces between two phases	Highest of all liquids
Heat of evaporation	Controls rate of heat and water transfer between water and atmosphere	Highest of all substances
Latent heat of fusion	Stabilizes temperature change at freezing point	Highest of all liquids except NH_3
Heat capacity	Balances temperature changes	Highest of all liquids except NH_3
Transparency	Allows transfer of sunlight to great depths in water bodies where it is required by photosynthetic organisms	Colorless substance

Source: Manahan, 1991.

properties of water are listed in Table 2.2. Many of these properties are the result of the molecular structure of water and its ability to form hydrogen (H) bonds. Hydrogen bonding allows water molecules to interact with one another and form clusters or liquid crystals. These interactions give water unique characteristics that set it apart from other molecules of similar size and weight. For example, CH_4 (molecular weight of 16) changes from a solid to a liquid at –182°C and from a liquid to a gas at –55°C, whereas phase changes for water (molecular weight of 18) occur at 0°C and 100°C for solids to liquid and liquid to gas transformations, respectively. Without the strong interactions between water molecules, water would be a gas at temperatures higher than –200°C and there would be no liquid forms of water on earth!

Water expands when it freezes; no other common liquid has this characteristic. Therefore, ice has a lower density than liquid water, which is the reason ice forms on the top of lakes and icebergs float in the cold oceans and seas. Imagine the consequences if the density of ice were higher than liquid water. What would happen to freshwater lakes if they froze solid? From a soil science point of view, freezing and thawing play an important role in soil formation, erosion, and structural problems (e.g., cracking and heaving). In addition, the high specific heat of water results in greater amounts of energy required to raise the temperature of water. This causes water to heat up and cool down at a much slower rate than all other liquids, except ammonia (NH_3). Because water is such a good solvent, and has the ability to dissolve most substances, there is a greater chance for contaminants to remain soluble for longer periods of time once they are dissolved in a water body.

2.3.2   Components of the hydrologic cycle

Water transfer or movement from one environment to another governs the hydrologic cycle (Figure 2.9). Water enters the atmosphere primarily through evapotranspiration processes (e.g., combination of water evaporation from soils and transpiration by plants) and is returned to ocean and land surfaces in the form of rain, fog, hail, and snow. The rate of water transferred from one pool to another, which is called *water flux*, is shown in Figure 2.10. Generally, the quantity of water leaving a water source is compensated by

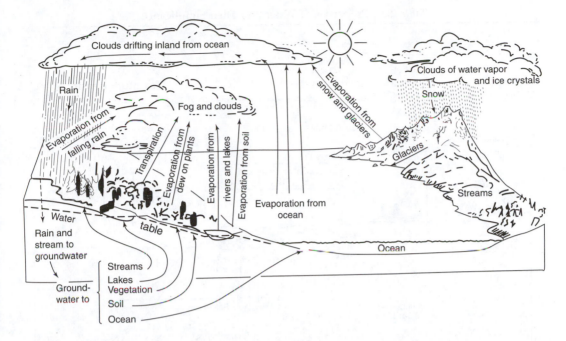

Figure 2.9   Illustration of the various pools and transfer processes that occur in the hydrologic cycle. (Modified from Gilluly, J. et al., *Principles of Geology*, 4th ed., W. H. Freeman, San Francisco, 1975. With permission.)

the amount of water entering the source. The rates shown are estimates for current water fluxes that have changed during the evolution of the earth.

Water-soluble pollutants entering pools with low MRTs may be rapidly transported to another ecosystem over a short period of time. The *pollution potential* of a substance is related to its water-soluble characteristics, harmful nature, and the MRT of a particular aquatic system. Thus, if an accident resulted in a contaminant spillage into a river (or other water body with a low MRT), and the contaminant is highly soluble, it could rapidly be transferred to another pool. If the MRT in the receiving pool is large, then the contaminant may remain in this pool for a longer period of time, causing long-term problems.

Example problem 2.1

The MRT for water in various pools can be calculated if it is assumed that input is equal to output, and if the mass of the water in the pool and the rates at which water is entering and exiting the pool are known (i.e., MRT = mass/flux). Using the information in Figure 2.10 we can calculate MRT for the various pools. For example, the MRT for water in the atmosphere would be

$$\text{MRT} = \frac{13,000 \text{ km}^3}{513,000 \text{ km}^3/\text{year}} = 0.025 \text{ years or } 9.2 \text{ days} \qquad (2.7)$$

The fast rate at which water moves from land to oceans in rivers results in a MRT for streams and rivers of

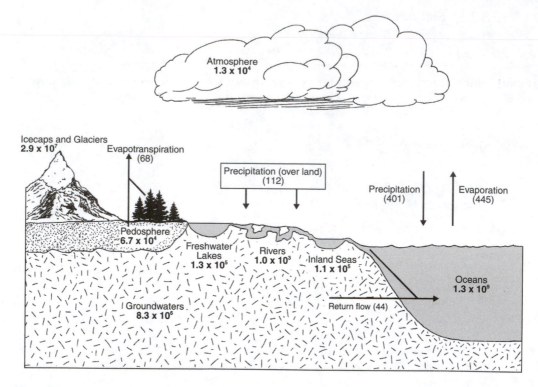

Figure 2.10   Content of water in various hydrosphere pools (bold numbers, km³) and global transfer rates (in parentheses, 10³ km³/year) for water movement in the hydrologic cycle. (Information obtained from the U.S. Geological Survey Web site — http://water.usgs.gov/data.html.)

$$MRT = \frac{1000 \text{ km}^3}{44{,}000 \text{ km}^3/\text{year}} = 0.023 \text{ years or } 8.3 \text{ days} \qquad (2.8)$$

As a comparison, the MRT of water in the ocean is

$$MRT = \frac{1{,}320{,}000{,}000 \text{ km}^3}{445{,}000 \text{ km}^3/\text{year}} \approx 3{,}000 \text{ years} \qquad (2.9)$$

whereas MRTs of water in lakes and groundwater systems have been estimated to be tens of years to hundreds or thousands of years, respectively.

2.3.2.1   Inland surface water

Inland surface waters include streams, rivers, and lakes. In general, water entering lakes, bays, and estuaries comes from the surrounding area, which is known as its watershed or drainage basin. Sources of water entering streams and rivers can include rainfall, surface runoff during periods of high rainfall, lateral water movement below the soil surface due to topography or stratified layers of different textures, water stored in adjacent wetlands areas, or groundwaters. Several natural (e.g., climate, vegetation, physiography, geology) and human (e.g., urbanization, agriculture, deforestation) factors influence the quality and quantity of water in inland surface water bodies.

2.3.2.2   Soil moisture

Infiltrating water can move below the soil surface into a region known as the vadose zone (unsaturated region) or move even deeper into the groundwater zone (saturated region) (Figure 2.11). The upper surface of the groundwater zone is called the water table, which fluctuates depending on the amount of water received by, or depleted from, the groundwater zone. The capillary fringe is the area above the water table where water in small pores is drawn upward by capillary action. Water movement in soils is due to a combination of (1) hydraulic gradients and (2) the ease with which water moves through soil or rock (hydraulic conductivity) (see Chapter 3 for more information on soil water).

2.3.2.3   Groundwater

Groundwater movement responds to gradients, which are a function of gravitational forces and the permeability of substrata materials. Substrata are characterized by their porosity and permeability that together represent the degree of void space and resistance to water movement. Thus, groundwater moves faster in coarse-textured substrata and as the slope

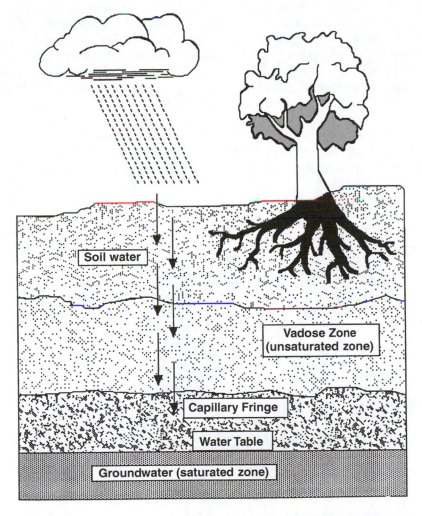

Figure 2.11   Saturated and unsaturated subsurface soil water zones. Unless drainage is restricted, most soils are unsaturated and hold water by adsorption and capillary forces.

of the water table increases. Aquifers are groundwater systems that have sufficient porosity and permeability to supply enough water for a specific purpose. In order for an aquifer to be useful, it must be able to store, transmit, and yield sufficient amounts of good quality water. Aquifers are classified as either confined (located under an impermeable substrata material) or unconfined (unrestricted above and having a water table). There may be enough pressure built up in a confined aquifer to create artesian conditions. Regions of substrata material that have low permeability and will not yield sufficient amounts of water to be practically useful are called *aquicludes* or *aquitards*.

2.3.3   Water use

In 1995, freshwater use from surface waters and groundwater sources such as rivers, lakes, reservoirs, and wells amounted to approximately 1300 billion L/day in the U.S. (Figure 2.12). The two primary users of this large amount of water were agriculture (39%) and the thermoelectric power industry (39%), followed by municipalities (12%) and other industries (6%). Livestock, domestic home use, mining, and commercial uses consumed approximately 1% each. Although it appears that thermoelectric power production requires a large amount of water, most of the water used is for cooling purposes; the heated water is then discharged for other uses. Because our main concern in this book is with soils and environmental quality we will focus primarily on agricultural and livestock water use (see Tapping into the Ogallala aquifer, p. 45, for further information).

Irrigation water use in the U.S. increased steadily from 1965 to 1980. Although irrigation water use is dependent on factors such as precipitation, water availability, energy costs, farm commodity prices, application technologies, and conservation practices, the total amount of water used for irrigation actually decreased from 1980 to 1995, even though the total irrigated area remained consistent at about 23.5 million ha. In the 19 western states, irrigated lands and water use declined during this period due to alterations in land use caused by urban growth, increased dry land farming, and ownership of water rights shifting from agriculture to municipalities. On the other hand, there has been an increase

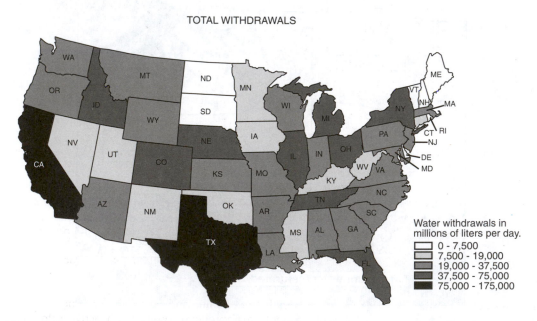

TOTAL WITHDRAWALS

Water withdrawals in millions of liters per day.
- 0 - 7,500
- 7,500 - 19,000
- 19,000 - 37,500
- 37,500 - 75,000
- 75,000 - 175,000

Figure 2.12   Amount of water used per day by individual states in 1995. (Modified from Solley et al., 1995.)

in irrigated lands in the eastern U.S. during this same period. Water use throughout the U.S. today averages about 0.64 hectare-meters (i.e., 2.1 acre-feet), down from more than 0.76 hectare-meters (2.5 acre-feet) in 1980. Of the 23.5 million ha irrigated in 1995, 55, 42, and 3% used flood, sprinkler, and drip irrigation, respectively, which utilized 63.0, 36.5, and 0.5% from surface water, groundwater, and treated wastewater sources. The western U.S. accounts for approximately 47% of the nation's total freshwater withdrawal, and with 90% of this water used for irrigation, the West accounts for about 70% of the nation's irrigation water use.

Because of its large population it is not surprising that California is the largest consumer of fresh water in the U.S. at approximately 142 billion L/day (11% of total U.S. freshwater use). However, most of California's water is used to irrigate 3.85 million ha, which amounts to 17% of the total U.S. irrigation water use; Nebraska (3.02 million ha) and Texas (2.56 million ha) are the next largest states with irrigated lands. It may be surprising to some that most livestock water use (20.7 billion L/day) is not for cattle, poultry, or hogs, but is primarily for fish farms. Idaho is the largest consumer of livestock water withdrawals at 26% of the total U.S. livestock water use. Most of Idaho's livestock water is used for fish farms (5.4 billion L/day) that produce about 80% of the world's farm-raised trout. Catfish farming also consumes much of our water used for livestock production. Louisiana uses 50 times more water for fish farming than is consumed in producing meat, poultry, and milk from other animals.

How much water do we use for domestic purposes? In the U.S., from 200 to 750 L (average of 300 L/day) are used per person on a daily basis for such in-home activities as drinking, cooking, bathing, washing clothes and dishes, toilet flushing, and lawn and garden watering. Since 1990, the amount of domestic water use has decreased by about 5% overall, even though during this same time there has been an increase in population. Decreased water use has been attributed to greater public education on the awareness of conservation practices such as use of low-flow toilets, showerheads, and faucets; xeriscaping (landscaping with plants that have low-water-use requirements); and proper lawn watering to name a few. Reductions in water use result in energy and cost savings because municipalities generally charge homeowners for sewage treatment based on their domestic water consumption.

2.3.4   Water pollution

There are several types of substances that can affect water quality. Water pollution can occur by a substance either directly or indirectly impacting a water system (Table 2.3).

Table 2.3   Different Classes of Water Pollutants and Their Causes

Water pollutant class	Contributions
Oxygen-consuming wastes	Decaying plant and animal remains
Plant nutrients	N and P from fertilizers
Inorganic chemicals	Toxic metals and acidic substances from mining operations and various industrial wastes
Organic chemicals	Petroleum products, pesticides, and materials from organic wastes industrial operations
Sediments	Fine soil and sediment particles that reduce solar radiation into water bodies, reducing food production
Infectious agents	Bacteria and viruses
Radioactive substances	Waste materials from mining and processing of radioactive substances or from improper disposal of radioactive isotopes
Industrial thermal energy	Cooling waters from thermoelectric power facilities

The substances include inorganic, organic, and biological materials, of which some have a direct impact on water quality, whereas others indirectly cause chemical, physical, or biological changes. Substances that can impact water quality, and which will be discussed later in this book, include N, P, and S, trace elements, pesticides, acid rain, and greenhouse gases. Although suspended organic matter and sediments are generally present in most streams and lakes, these materials can cause water degradation by increasing the biological oxygen demand or decreasing light penetration by increasing water turbidity. Even changes in water temperature due to thermal discharges near industrial and power plants can alter biotic diversity in rivers and lakes. Additional water pollutants such as radionuclides, carcinogens, pathogens, and petroleum wastes are also important in the context of environmental quality but will only receive limited coverage in this book.

When organic matter such as municipal biosolids and animal manures are added to surface waters, a rapid decline in available O_2 can occur. Oxygen is consumed in the biological decomposition of the added organic matter, and by oxidation of other reduced inorganic compounds (i.e., ammonium (NH_4^+), ferrous iron (Fe^{2+}), and sulfite (SO_3^{2-})) present in the added material. This results in lower O_2 availability for higher forms of aquatic life. Two measures that are used to estimate the quality of surface waters are *biochemical (or biological) oxygen demand* (BOD) and *chemical oxygen demand* (COD). BOD is a measure of the amount of O_2 consumed by microorganisms over a 5-day period, whereas COD indicates how much oxidizable material there is in a water sample by its chemical reaction with dichromate (Cr_2O_7). Values for COD are often higher than BOD, depending on the nature and quantity of oxidizable material in the water sample.

Biological communities (fish, plants, microorganisms, etc.) in surface waters are also impacted by conditions that are influenced by pH and salt concentrations. Mining activities can have a considerable effect on the quality of surface waters, as well as the land and air in the surrounding environment. Oxidation of reduced S substances can lead to acid mine drainage, which can be deleterious to plants, animal, and microorganisms. Acid mine drainage and some irrigation flow-through waters can also increase surface water salinity. Refer to Chapter 6 for more information on the effects of mining.

2.3.4.1 Eutrophication

Water quality is, and will continue to be, a major economic and environmental issue. The process of eutrophication is one example of why water quality is so important. For example, excessive nutrient additions to surface waters can lead to enhanced algal growth, decreased dissolved O_2, and reduced water transparency (Table 2.4). Eutrophication of lakes, streams, rivers, and other surface water bodies occurs when excessive amounts of

Table 2.4 Comparison of General Characteristics and Selected Properties of Oligotrophic and Eutrophic Water Bodies

Characteristic	Oligotrophic	Eutrophic
Nutrient status	Low	High
Algal blooms	Rare	Common
Biomass	Low	High
Aquatic diversity	High	Low
Dissolved oxygen (saturation %)	>80	<10
Total N (μg/L)	<200	>500
Total P (μg/L)	4–10	>30
Chlorophyll (mg/L)	1–3	>8
Turbidity (m) (Secchi disk transparency)	6–12	<2

Table 2.5   Water Quality Problems Associated with Eutrophication

Water quality problem	Contributing factors from eutrophication
Water safety, taste, odor	Nutrients, suspended sediments degrade water quality and increase cost and difficulty of drinking water purification; anoxic conditions and toxins produced in algal blooms can cause fish kills and make water unsafe for birds and livestock
Low species diversity	Stimulated growth of certain organisms decreases number and size of population of other species; with time lake becomes dominated by algae and coarse, rapid-growing fish; high quality edible fish, submerged macrophytes, and benthic organisms disappear
Impairment of recreational use and navigation	Increased sedimentation decreases lake depth; enhanced vegetative growth blocks navigable waterways; decaying algal biomass produces surface scums and odors, and increases populations of insect pests

nutrients, such as N and P, are added to the ecosystem. As nutrient inputs to surface waters gradually increase, the trophic state of the water body passes through four trophic stages: *oligotrophic, mesotrophic, eutrophic,* and *hypereutrophic.* At each stage, progressive changes in the ecology of water bodies occur, usually negatively affecting their economic and recreational uses. Some of the water quality problems associated with eutrophication are summarized in Table 2.5. The two nutrients that are the primary concern — N and P — and their role in eutrophication are discussed extensively in Chapters 4 and 5.

2.3.4.2   Erosion and sediments

Runoff occurs when surface water (e.g., precipitation or irrigation) exceeds the infiltration rate of a soil, causing water to run downhill due to gravity, often carrying loose soil material in the overland flow. Soil erosion due to water can result in several on-site and off-site problems (Table 2.6). Sediment loading to surface water bodies is a major water quality problem due to siltation and the potential nutrients, pesticides, and organic matter adsorbed to these materials. Other physical concerns related to increased particulate introduction are decreased light penetration from increased turbidity that reduces the growth of benthic plants and the buildup of materials due to sediment deposition that results in a loss of water-storage capacity in reservoirs, lakes, and wetlands.

Predicting soil erosion due to rainfall has been an active area of research for nearly 50 years. Since 1970, modeling efforts have utilized the Universal Soil Loss Equation (USLE), and more recently the Revised Universal Soil Loss Equation (RUSLE), to evaluate soil loss caused by rainfall and runoff. Both equations are defined by:

$$A = R \cdot K \cdot L \cdot S \cdot C \cdot P \qquad (2.10)$$

where A is the estimated annual soil loss per unit area that can be averaged over many years; R is the rainfall erosivity factor that depends on intensity, quantity, and duration; K is the soil erodibility factor, which is a function of soil texture, organic matter content, and structure that influence stability and permeability; L and S are topographic factors, length and steepness, respectively; C is the surface-cover factor based on whether a soil is vegetated, mulched, or bare; and P is the management practice factor related to an erosion control application. While the use of USLE or RUSLE is generally simple, its original development and calibration considered specific conditions that are not truly universal. Improvements in RUSLE have resulted in information related to each factor being updated, expanded, and enhanced.

Table 2.6 On-site and Off-site Damages Caused by Water Erosion

On-site	Off-site
Decrease in organic matter and plant nutrients	Sediment input to water bodies
Loss of topsoil and water-holding capacity	Source of soil and water pollution
Reduction in soil structure	Turbidity of surface waters
Impaired soil biotic communities	Destruction of off-site ecosystems
Reduced crop productivity	Damage to roads and other structures

In addition to using RUSLE for soil loss estimates, parameters such as the erodibility index (EI) can be determined from the following equation:

$$EI = \frac{R \cdot K \cdot L \cdot S}{T} \qquad (2.11)$$

which has been used in the USDA Conservation Compliance Plan (CCP) to determine if an area warrants federal assistance for erosion control. The value T is an erosion tolerance level that is usually set at 11.2 Mg/ha; however, levels of 2 to 4 Mg/ha have been recommended by some individuals for greater erosion control. A value of EI $\geq$ 8 indicates the land is highly erodible and in need of special management practices to minimize soil loss, which would also reduce the chance of surface water contamination by eroded sediments.

Surface runoff and eroded materials that enter water bodies usually result in a reduction in water quality. The concentration of soluble materials in surface runoff is generally a function of their solubility. Concentrations of contaminants in the transported soil-sorbed chemicals in overland flow, however, are usually greater than the bulk soil source concentrations. To determine the importance of sorbed chemicals in the eroding sediments, an "enrichment ratio," E_R, can be calculated that describes the ratio of the concentration of the chemical in the eroded material to that in the bulk soil:

$$E_R = \frac{\text{mg chemical/kg eroded soil}}{\text{mg chemical/kg bulk soil}} \qquad (2.12)$$

Values for E_R are generally greater than 1 and are dependent on soil type, erosion mechanism, and total mass of soil eroded. Organic matter and fine clay particles have greater surface areas and adsorption capacities. Erosion of these materials would usually result in higher E_R values. Rainfall detachment processes are also responsible for producing higher values of E_R than simple sediment detachment from surface runoff, primarily because rainfall detachment results in smaller particles. Over time, a constant erosion event will produce sediments that have larger values of E_R at the initial stages of precipitation that then taper off to E_R values closer to 1 in the later stages of the event.

The runoff coefficient (RC) is also used to characterize surface runoff:

$$RC = \frac{\text{volume of runoff/unit area}}{\text{volume of rainfall}} \qquad (2.13)$$

The RC is most often used for a single storm event and expresses the fraction (or percent) of the precipitation that leaves the area through overland flow. An area with an impermeable surface such as a parking lot obviously has little infiltration potential and would have an RC near 100%, whereas a freshly tilled field would have an RC less than 5%. The RC is influenced by many factors including the inherent infiltration rate of the soil, the slope, rainfall intensity, and antecedent moisture conditions. The total amount of sediment or contaminant leaving an area is a product of the volume of water leaving as overland flow and the concentration of sediment or contaminant in the water.

Environmental quality issues/events

Tapping into the Ogallala aquifer

Underneath much of the central and southern Great Plains states is the extensive Ogallala "High Plains" aquifer (Figure 2.13), which is one of the world's largest aquifers (450,000 km²). In 1990, it was estimated that the Ogallala contained 400 million hectare-meters of water, approximately equivalent to the amount of water in Lake Huron. The percentage of Ogallala water that resides under different states is 65, 12, 10, 4, 3.5, 2, 2, and 1.5% beneath Nebraska, Texas, Kansas, Colorado, Oklahoma, Wyoming, South Dakota, and New Mexico, respectively. The Ogallala has been the main provider of irrigation water in the agricultural region known as the "breadbasket of the world." Over 6 million ha are irrigated from the aquifer and yield a significant portion of U.S. agricultural products

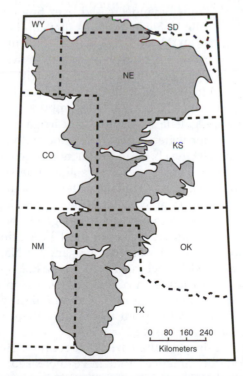

Figure 2.13   Location of the Ogallala High Plains aquifer that underlies eight states.

used for domestic human consumption (e.g., flour and cotton) and animals (e.g., feedlots), as well as for foreign exports. More Ogallala water is used annually for irrigation purposes than the entire flow of the Colorado River.

Prior to the 20th century western agricultural movement, minor amounts of Ogallala aquifer discharge occurred in playa basins and from streams, seeps, and springs located primarily along the eastern boundary of the High Plains. However, in the early 1900s, agriculture within the Ogallala region started utilizing water from the aquifer. After a slow increase in water use, the number of wells tapping into the Ogallala aquifer, and the amount of water used, increased rapidly from the mid-1940s into the 1970s. For example, in west Texas the number of irrigation wells in 1914, 1937, 1954, 1971, and 1978 was approximately 140, 1,200, 28,000, 66,000, and 75,000, respectively.

Currently, almost 97% of the water pumped from the Ogallala is for irrigation; about two thirds of U.S. irrigated land is located in the High Plains region. During the early years, there was very little consideration given to conserving water, and irrigation practices included unlined ditches and inefficient sprinkler systems. It was not unusual for these types of practices to result in less than 50% water-use efficiencies due to high evaporation rates and percolation below the rooting zone. As the number of wells tapping into the Ogallala water source increased each year, the aquifer water table increased in depth. Between 1950 and 1980, annual water use for irrigation increased from about 0.5 million to 3 million hectare-meters, although there has been a decrease in water use since the early 1980s. In some areas the water table dropped at rates of 15 to 100 cm/year; a large region in Texas and smaller areas in Kansas and Oklahoma have experienced overall declines in the water table by greater than 30 m. The consequence of a lower water table meant new wells had to be drilled deeper and some of the old wells that went dry either had to be abandoned or deepened. The increased costs associated with drilling deeper wells and the greater energy required to pump water from these lower depths caused reductions in agricultural profits with time.

Water yields from Ogallala wells have decreased over time because the water table has declined. A loss in yield results in fewer hectares that can be irrigated from a well. As an example, a well that delivers 3000 L/min can irrigate 65 ha (e.g., 160 acres, a quarter-section, or a 1/4 square mile) using a center-pivot irrigation system. In some parts of Texas, water yields have declined to less than 1000 L/min with the average number of hectares irrigated per well at 25 in 1980. In 1990, 30,000 million L of water per day were withdrawn from the Ogallala, of which Nebraska's consumption was 17,000 million L/day. As noted earlier, almost 97% of Ogallala water use is for irrigated agriculture.

The quality of Ogallala aquifer water is generally good with a total dissolved solids concentration usually less than 500 mg/L. Crop production is generally not limited when irrigation waters contain 500 to 1500 mg/L dissolved solids, unless sodium (Na) is a major constituent of the dissolved substances. Ogallala waters in southern Texas have been reported to contain dissolved solids at concentrations exceeding 3000 mg/L, which requires special management practices to be used for irrigation purposes. There are also some areas where Ogallala water quality does not meet EPA drinking water standards because of high concentrations of total dissolved solids and/or Na, Cl, SO_4, NO_3, Se, and fluorine (F). Some waters sampled in Nebraska and Kansas also contained measurable concentrations of the herbicides Atrazine and 2,4-D (2,4-dichlorophenoxyacetic acid).

Water in the Ogallala aquifer flows at a rate of about 50 m/year in a northwest to southeast direction. As water is deleted from the aquifer, there is not enough natural recharge from precipitation to compensate for the loss. At one time it was estimated that the Ogallala would be drained in the early 21st century. However, some regions have implemented conservation practices that have increased water-use efficiencies from less than 50% in the mid-1970s to 75% in 1990, and have further improved water use by installing state-of-the-art low-pressure, dropline center-pivot irrigation systems that are close to 95% efficient and underground drip line irrigation systems that essentially eliminate evaporative losses and are nearly 100% efficient. The USDA Conservation Reserve Program (CRP) has also decreased the number of irrigated hectares, particularly lands in areas that suffer from declining well yields, increased energy costs, lower farm profits, and poor water quality. This program resulted in a substantial decrease in the number of irrigated hectares; for example, there was a two thirds reduction in Texas High Plains irrigated lands between 1979 and 1989. However, with a loss of CRP funding, many of the lands enrolled in this program will again be used for irrigated agriculture purposes. In western Kansas, an Irrigation Research Project (IRP) was developed to evaluate solutions to the loss of the Ogallala water resource. The IRP will examine, evaluate, and develop (1) innovative irrigation systems; (2) profitable cultural practices; (3) economic feasibility of nontraditional crops; (4) management strategies to optimize water use; and (5) computer programs and models, in hopes of ensuring future economic stability to the region.

2.4   Environmental testing practices for waters

One of the major goals of environmentally sound soil management practices is the protection of water quality. Preventing nutrients, organics (synthetic and natural), pathogens, sediments, and other pollutants from entering our groundwaters and/or surface waters is a national priority. Because of this, legislative mandates involving national standards have been established for the concentrations of many pollutants in groundwaters and surface waters. Waters that exceed these concentrations are considered impaired and in need of some form of corrective action to make them safe as drinking waters or to ensure that they are fishable and swimmable. Consequently, accurate water analysis methods are essential in efforts to prevent or reverse water pollution. Information related to the remediation of contaminated waters is discussed further in Chapter 11.

Water samples collected to monitor water quality are typically analyzed for organic and inorganic chemical composition, physical properties, and biological organisms according to well-established protocols for sampling, storage, and analysis. It is beyond the scope of this chapter to describe all aspects of water analysis in detail. However, an overview of key principles is provided below and references that give detailed information on the standard approaches to collect and analyze water samples are found at the end of the chapter.

2.4.1   Legislative efforts related to water quality issues

Legislation in the U.S. mandating efforts to identify waters that are polluted, eliminate further discharge of pollutants, and require actions to restore water quality dates to the 1940s. The Clean Water Act (CWA), enacted in 1948 as the Water Pollution Control Act (WPCA), and later amended in 1977, is the main authority for all water pollution control actions at the federal level. This act established the National Pollution Discharge Elimination System (NPDES) permitting process to eliminate point-source pollution of the nation's navigable freshwaters. The 1977 amendment also authorized federal legislation to regulate the land application of sewage sludges (municipal biosolids) that resulted when wastewaters were treated to remove pollutants prior to effluent discharge into surface waters. Most recently, the CWA has been used as the legal basis for lawsuits filed by a consortium of environmental groups against the EPA to force state governments to restore surface water quality. Settlement of these lawsuits has resulted in many states being required to define total maximum daily loads (TMDLs) of pollutants that cannot be exceeded if surface waters are to remain fishable and swimmable. Note that a TMDL does not refer to the actual *concentration* of a pollutant in a water body (e.g., in mg pollutant/L), but to the *daily load* of the pollutant to the water body from a defined geographic area (e.g., kg pollutant entering the water body/day from a watershed or subwatershed). For example, TMDLs of 193 kg N/day and 13 kg P/day have been proposed for Delaware's Indian River Bay, a national estuary considered highly eutrophic due to point- and nonpoint-source pollution. Current nutrient loads to this bay have been estimated as 1285 kg N/day and 38 kg P/day, indicating a need for substantial reductions in nutrient loading from all sources to meet the criteria established in the CWA.

Another piece of legislation, the Safe Drinking Water Act (SDWA) of 1974, was passed to protect U.S. drinking water supplies by establishing maximum contaminant levels (MCLs) above which waters are considered unsafe for human consumption. The SDWA also defines enforcement standards that must be used by individual states as they determine the minimum water treatment needed to improve water quality. Examples of some MCLs are given in Table 2.7.

2.4.2   Collection, storage, handling, and analysis of water samples

Groundwaters, surface waters (fresh, estuarine, and saline), and wastewaters from municipalities and industries are the most common types of waters regularly analyzed in water quality monitoring programs. The parameters determined in a water analysis depend on the goal of the analysis. For example, if the goal is to determine if water is safe for human or animal consumption, the most common tests are for nitrate-nitrogen (NO_3-N), trace metals, pathogens, and organic chemicals. If the objective is to determine the potential impact of wastewater discharged from a point source on an aquatic ecosystem, analyses may include nutrients, BOD, metals, and alkalinity. If surface waters are being tested to determine if they are eutrophic, they may be analyzed chemically for nutrients (N and P), O_2, chlorophyll (a measure of algal productivity), and pH; physically for turbidity or depth of light penetration using a Secchi disk; and biologically for the rate of phytoplankton production. If the water or wastewater is being tested to determine its suitability for irrigation, analyses may include pH, alkalinity, salinity, and specific ions that are known to cause soil and plant problems, e.g., Na, Se, boron (B). A summary of the most common analyses conducted when testing groundwaters, surface waters, or wastewaters is provided in Table 2.8. Many other analyses can be conducted on waters, ranging from tests for hardness, Cl, and radioactivity to tests of the toxicity of the water to aquatic biota (e.g., fish, insects, mollusks, crustaceans). Readers are referred to *Standard Methods for the Examination of Water and Wastewater* (APHA, AWWA, WPCF, 1998) for a thorough description of water-testing methods.

As with any form of analysis, collection of a representative water sample is the first step in an effective water quality monitoring program. An important consideration for the collection of any type of water sample is the recognition that water composition is not constant within a water body. For example, vertical stratification can occur in lakes and reservoirs, flowing waters constantly alter water composition in rivers and streams, tidal influences can change the concentration of pollutants in estuaries, and seasonal variations induce temperature changes in the water column. Therefore, water-sampling protocols

Table 2.7 Examples of Maximum Contaminant Levels (MCLs) for Drinking Waters

Contaminant	MCL[a] (mg/L)	MCLG[b] (mg/L)
Inorganics		
Arsenic (As)	0.006	0.006
Cadmium (Cd)	0.005	0.005
Chlorine (Cl)	4	4
Chromium (total Cr)	0.1	0.1
Copper (at tap Cu)	TT[c]	1.3
Cyanide (CN)	0.2	0.2
Fluoride (F)	4	4
Lead (Pb)	0.015	0
Mercury (inorganic Hg)	0.002	0.002
Nickel (Ni)	0.1	0.1
Nitrate (as N)	10	10
Selenium (Se)	0.05	0.05
Sulfate (SO_4^{2-})	500	500
Thallium (Tl)	0.002	0.0005
Radionuclides		
Radon (Rn)	300 pCi/L	0
Uranium (U)	20 µg/L	0
Organics		
Alachlor	0.002	0
Atrazine	0.003	0.003
Benzene	0.005	0
Carbofuran	0.04	0.04
Carbon tetrachloride	0.005	0
Chlordane	0.002	0
2,4-D	0.07	0.07
Dinoseb	0.007	0.007
Dioxin	10^{-8}	0
Endrin	0.002	0.002
Ethylene dibromide	0.00005	0
Heptachlor	0.0004	0
Lindane	0.0002	0.0002
Methoxychlor	0.04	0.04
Pentachlorophenol	0.001	0
PCBs	0.0005	0
Tetrachloroethylene	0.005	0
Toluene	1	1
Trichloethylene	0.005	0
Vinyl chloride	0.002	0
Xylenes	10	10
Microbiological		
Giardia lamblia	TT	0
Legionella	TT	0
Total coliforms	TT	0
Viruses	TT	0

[a] MCL = Maximum contaminant level. Maximum permissible level of a contaminant in water which is delivered to any user of a public water system.

[b] MCLG = Maximum contaminant level goal. A nonenforceable concentration of a drinking water contaminant that is protective of adverse human health effects and allows an adequate margin of safety.

[c] TT = treatment technique. Lowest value that can be achieved using best available technology.

Source: U.S. Environmental Protection Agency Office of Water, Washington, D.C., 1996.

Table 2.8 Some of the More Common Parameters Measured in Groundwaters,
Surface Waters, and Wastewaters

Physical	Inorganic constituents	Organic constituents	Microbiological
Color	Metallic elements:	Methane	Coliforms
Conductivity	Al, Ag, As, Ca, Cd,	Oil and grease	Fecal streptococcus
Odor	Cr, Cu, Fe, Mg, Mn,	Organic and volatile acids	*Camplyobacter jejuni*
Oxygen transfer	Na, Ni, Pb, Se, Sr,	Organic C	*Salmonella*
Solids	Zn	BOD	*Shigella*
Salinity	Nonmetallic	COD	*Legionella*
Taste	constituents:	Pesticides	Enteric viruses
Temperature	Acidity, alkalinity,	Phenols	Iron and sulfur
Turbidity	B, CO_2, Cl, F, I,	Surfactants	bacteria
	NH_4, NO_2, NO_3,		Fungi, actinomycetes,
	dissolved O_2, pH, P,		nematodes
	Si, SO_4		

must carefully consider both spatial and temporal variability to ensure that the goals of the monitoring program are achieved.

When sampling surface waters it is important to consider sample location, sampling depth, and sampling frequency. If the impact of a point source on water quality is being monitored, samples should be taken at various points downstream from the point of discharge at measured time intervals after the influx of the pollutant. It is also important to sample different depths in the water body because variations in density and solubility can cause some pollutants to mix more thoroughly with waters than others. If nonpoint-source pollution is the major concern, a fixed sampling station near the outlet of a watershed is often used to monitor water quality continuously. In both situations there is a need to sample across the width and throughout the depth of the water body. In lakes and reservoirs where the residence time of the water is longer than in flowing waters, and where temperature differences with depth can be significant, sampling by depth and season is particularly critical to understand when and where pollutants are most likely to cause water quality problems. In flowing waters there is a need to sample at several points across the width of the river or stream because flow rates and water depths generally decrease near shore. Measuring flow velocity, as well as chemical composition, is a necessary step if the goal is to determine the mass (load) of pollutant emitted from a watershed to a water body.

Groundwater sampling programs should consider most of the points mentioned above for surface waters. The main difference is that groundwater samples are collected from monitoring wells drilled to different depths in underground aquifers. Wells must be installed according to established protocols for accurate results, and groundwater samples must be handled very carefully to prevent changes that can occur when subsurface water is exposed to the atmosphere. It is also necessary to purge sampling wells prior to collection of the actual sample to be measured by pumping and discarding known volumes of water. This is because water that collects in the wells may differ in composition compared with water in the aquifer because in the well it has been exposed to different temperatures and atmospheric conditions. As with surface water samples, depth and spatial distribution of groundwater sampling points is important to determine accurately the extent of pollution of an aquifer and the direction of pollutant movement. In many cases groundwater sampling wells are "nested" at different depths and located at multiple points in transects along the path of groundwater movement.

The collection of surface runoff demands special attention to determine the origin of the runoff accurately. A watershed is an area of land that drains its surface runoff to a specific point, often a river or other body of water. Obviously, the water in a river partially comprises surface runoff, but the river also contains contributions from groundwater sources and is influenced by in-stream processes such as settling of coarse particles and streambed erosion. A sample of water from the river, therefore, tells us little about the characteristics of surface runoff. To assess the characteristics of surface runoff accurately it is important to collect samples of the water before it reaches the receiving body of water. It is also imperative that the area of land that is contributing water to the collection point be known. Surface runoff from natural watersheds can be collected provided there is a single point of discharge. Alternatively, small watersheds (i.e., 1 to 10,000 m²) can be created by constructing barriers to prevent surface runoff from outside the area of interest and to direct surface runoff of interest to a common collection point.

Surface runoff is a precipitation event–driven process. Runoff occurs during and shortly after the precipitation event occurs. A water sample can be collected that represents the entire volume of runoff or samples can be collected periodically during the runoff event. Specialized equipment is required for this latter collection process, which is particularly challenging because of the large volumes of water that are generated during rainfall events. For example, 1 cm of runoff over a hectare of land will result in a volume of 100,000 L. When runoff is collected from large areas, small flow-weighted samples are gathered that represent the entire volume of water.

Water samples are also subject to a number of changes after sampling that can alter their physical, chemical, and biological properties. Standard techniques for the proper storage and handling of water samples have been developed to ensure that analyses are not biased by changes in factors such as pH, oxidation–reduction status, or temperature after collection and prior to analysis. For example, samples to be analyzed for cations (e.g., Ca^{2+}, Cu^{2+}, K^+, Mg^{2+}, Na^+, Ni^{2+}, Pb^{2+}, Zn^{2+}) should be placed in a high density polythylene (HDPE) plastic bottle and acidified to a pH of <2.0 using nitric or sulfuric acid. This is done to prevent adsorption of the cations to the storage container and to minimize the precipitation of cations as insoluble carbonates or hydroxides during storage. Water samples that will be analyzed for anions (e.g., NO_3, Cl, F, SO_4, PO_4) should be stored in HDPE bottles and refrigerated at temperatures <4°C. Organic chemicals (e.g., pesticides, nonvolatile hydrocarbons) are normally stored in amber glass bottles at <4°C to prevent photooxidation and microbial decomposition. Reducing agents are often added to these samples to prevent oxidation of the organic compounds during storage. Even if preserved properly, water samples cannot be stored indefinitely. Volatile or highly degradable substances in waters, such as O_2, PO_4, NO_3, cyanide (CN), should be analyzed quickly (<48 h). Water samples to be tested for pesticides and other organic chemicals should not be stored for more than 7 days; however, metals can be analyzed as much as 6 months after sampling if properly acidified and appropriate storage practices are followed.

A substance in a water sample may be present both in a dissolved state as well as adsorbed to sediment particles in the water. Phosphorus, for example, may be present in the water as dissolved P or as sediment-bound P (sometimes called particulate P). In a standard water analysis, the sample is filtered through a 0.45-μm filter. Anything that passes through the filter is considered to be in the dissolved state while that retained on the filter is described as sediment bound. Sediment-bound P is determined by the difference in total P concentration of the unfiltered sample and the P concentration in the filtered sample.

Chemical, physical, and biological analyses of water samples must follow strict, well-defined protocols. A myriad of methods are available to determine concentrations of inorganic, organic, and biological constituents in waters accurately and reproducibly. Detailed information on the proper means to handle and analyze water samples is available in several publications given in the references section at the end of this chapter.

2.4.3   Interpretation of water analysis results

Interpretation is the process by which data obtained in a water analysis are used to develop recommendations on the suitability of the water for the desired end use. In some cases the interpretation process is simple and straightforward. For example, if a water analysis of a private drinking water well shows that it exceeds the established MCLs for a pollutant (e.g., NO_3-N), the recommendation would be to stop using the well as a drinking water source or to install purification technologies that can remove the pollutant from the water. Similarly, if wastewater discharged from a point source exceeds the regulatory criteria for BOD and nutrients, the recommendation would be to cease discharging the wastewater until steps have been taken to reduce these pollutants to acceptable levels.

In most cases, however, interpretations of water analyses can be quite complicated and require additional information for proper interpretation, which is particularly true for nonpoint-source pollution. If a lake, river, estuary, or aquifer exceeds the desired concentrations of a pollutant and no obvious point source exists, the interpretation process is very complex. Efforts must be taken to characterize all potential nonpoint-sources that contribute the pollutant to the water body, the pathways by which the pollutant moves from land to the water, any channel processes that can affect the concentration of the pollutant, and the potential for bottom sediments in the water body to act as an ongoing or intermittent source (or sink) of the pollutant. It is not enough to measure the concentration of the pollutant in the water and determine if it exceeds standards established to protect public health or ecosystem stability. Society expects those involved in water pollution control to go farther and identify the causes of the pollution and the measures that can be taken to reduce the impacts of the pollutants on water quality. Further complicating the interpretation process is the fact that it is rare to find that only one pollutant is impacting water quality. In most cases, multiple pollutants are entering groundwaters and surface waters in different quantities, at variable rates, and from several locations in the watershed. Interactions between pollutants are also common. For example, while N and P both contribute to eutrophication, the ratio of N:P is more important than the actual concentration of either nutrient (see Chapter 5). Similarly, one cannot accurately assess the extent of water quality degradation by measuring only one pollutant or by evaluating pollutant concentrations separately. This is clearly illustrated in Table 2.4, which shows that several criteria (dissolved O_2, nutrients, light penetration, biomass, etc.) must be considered together to determine the extent of surface water eutrophication. Research has also shown that mixtures of metals and surfactants can be more toxic to fish than metals alone because the surfactants reduce the surface tension on gill membranes, increasing the permeability of the gill to the metals.

A good example of current efforts to interpret water quality analyses is the National Water Quality Assessment Program (NAWQA) conducted by the U.S. Geological Survey (USGS). The NAWQA program began in 1991 when the U.S. Congress charged USGS "to work with other federal, state, and local agencies to understand the spatial extent of water quality, how water quality changes with time, and how human activities and natural factors affect water quality across the Nation" (U.S. Geological Survey, 1999). The NAWQA program is assessing water quality in more than 50 major river basins and aquifer systems in the U.S. that affect >60% of the population in watersheds that cover about 50% of the total U.S. land area. Some preliminary results of this assessment are as follows:

- Streams and groundwater in basins with significant agricultural or urban development, or mixed land uses, almost always contain complex mixtures of nutrients and pesticides.
- Concentrations of N and P commonly exceed levels that can contribute to excessive plant growth in streams.
- Nitrate generally does not pose a health risk for residents whose drinking water comes from streams or aquifers buried relatively deeply beneath the land. Health risks increase in those aquifers located in geologic settings that enable rapid movement of water. This finding raises potential concerns for human health, particularly in rural areas where shallow groundwater is used for domestic water supply.
- At least one pesticide was found in almost every water and fish sample collected from streams and in more than one half of shallow wells sampled in agricultural and urban areas. Concentrations of individual pesticides in samples from wells and as annual averages in streams were almost always lower than EPA drinking water standards. Effects of pesticides on aquatic life are a concern based on U.S. and Canadian guidelines.

As illustrated by the USGS NAWQA study, interpretation of water analyses results is a complex, somewhat daunting process, but one that is essential to the safe and efficient use of groundwaters, surface waters, and wastewaters. The most effective recommendations will be those that are based on multidisciplinary evaluations of the chemical, physical, and biological properties of waters. Integrating water analyses into predictive models that can assess the effects of land management on water quality is needed in the long term to determine the most effective means to preserve and restore water quality.

Problems

2.1 Explain how the biosphere is related to the atmosphere and hydrosphere. Provide examples of how changes in environmental quality, i.e., global climate change, water pollution, etc., may impact different biological species such as trees, fish, and deer populations and habitat migration.

2.2 Although the troposphere is primarily responsible for weather conditions that we experience on a daily, seasonal, and yearly basis, what other properties and influences are important in the role of climate?

2.3 Which of the troposphere gases is the most variable and why? Explain why atmospheric oxygen (O_2) is more important to humans than atmospheric nitrogen (N_2).

2.4 How important is the atmosphere in the cycling of pollutants? What kind of materials are more susceptible to long-range transport and deposition? Why do you think gases trapped in polar ice are important to the study of environmental quality?

2.5 Explain why topography can influence vegetative communities. Be sure to include a discussion of how slope aspect (e.g., south- vs. north-facing slopes) modifies the mesoclimatic (e.g., local temperature, winds, rainfall) and soil (e.g., moisture, temperature, organic matter content) conditions that affect vegetation dynamics.

2.6 Select three cities, one each from the West Coast, the midcontinent, and the East Coast, and track their weather conditions (e.g., high and low temperatures, precipitation, and any other weather-related conditions) for a week. Describe the changes that occurred during the weekly evaluation and provide reasons the climate varied. You can find temperature and precipitation data on Web sites on the Internet.

2.7 What would happen if ice (solid phase) had a density lower than water (liquid phase)? Think of aquatic ecosystems and the impacts that would occur if these environments froze from the bottom up.

2.8 Calculate the MRT and pollution potential for a contaminant in a pond that has a volume of 0.1 km^3, an outlet flow rate of 2 m^3/s, and an inorganic contaminant (1 kg) that is highly soluble and toxic to fish. Compare your answer to a lake with a volume of 0.5 km^3 and a flow rate of 50 m^3/s, with 10 kg of the same contaminant.

2.9 Provide some examples of conservation methods that could decrease water use by (a) the home owner; (b) irrigated agricultural sector; (c) city parks and recreational departments; and (d) golf course operations.

2.10 What impacts could urbanization have on the water quality of a river that flows through the development? How do cities prevent the contamination of their drinking water sources? Are water pollution problems always associated with human activities?

2.11 The enrichment ratio for N [$E_R(N)$] in a 10-ha agricultural area is 4 during a storm event. Assuming the total soil N concentration is 1200 mg/kg, how much N would be added to the stream if the storm resulted in erosion of 1 Mg soil/ha from the field? If the farmer implements a conservation practice that reduces the amount of erosion by 50%, determine the reduction in the amount of N entering the stream considering a similar type of storm event.

2.12 Runoff is collected from small plots 2 × 2 m in size. A thunderstorm delivers 5 cm of rain and 30 L of runoff is collected. Calculate the runoff coefficient (RC).

2.13 Runoff is collected from an area 1 ha in size. The soil in this small watershed contains 600 mg/kg total P. A storm delivers 2.5 cm of rain and the runoff collector device indicates 50,000 L of runoff was generated. A representative sample of the runoff has 6000 mg/L of sediment and the sediment contains 1000 mg P/kg. Calculate the runoff coefficient (RC), enrichment ratio for P, the amount of sediment lost (in kg sediment/ha), and the amount of sediment-bound P lost (in kg P/ha).

2.14 Explain the difference between a TMDL and an MCL. What is the value of establishing TMDLs for use in watershed-scale water quality protection efforts?

2.15 Water quality monitoring programs must follow strict quality assurance and quality control (QA/QC) protocols for sample collection, storage, and handling. Explain the justification for the time, effort, and expense needed to follow rigorous QA/QC protocols.

2.16 You are asked to determine if the water in a newly drilled well is safe for use as drinking water because of concerns in the region about NO$_3$-N contamination of the aquifer. You collect a sample, bring it back to the laboratory, dilute it tenfold with deionized water, analyze it colorimetrically for NO$_3$-N, and obtain the following results, along with data from the standard curve you prepared using solutions of known NO$_3$-N concentrations:

Concentration of NO$_3$-N in standard solution, mg NO$_3$-N/L	Absorbance
0	0
0.25	0.042
0.5	0.080
1.0	0.159
2.0	0.308
5.0	0.803

Absorbance of the diluted well water sample equals 0.094. Show how you would determine the concentration of NO$_3$-N in the water sample graphically and also by use of a linear regression equation. Is the well water safe for use as a drinking water supply (compare your results with Table 2.7)?

2.17 Locate the USGS National Water Quality Assessment Program report "The Quality of Our Nation's Waters: Nutrients and Pesticides" on the USGS home page on the Internet (http://www.water.usgs.gov/pubs/circ/circ1225). Review the section on computer modeling of nitrate contamination of groundwaters on page 51. Explain the use and value of this modeling effort to the national effort to improve groundwater quality.

References

Ahrens, C. D. 1994. *Meteorology Today: An Introduction to Weather, Climate, and the Environment*, 5th ed., West Publishing, St. Paul, MN.

American Public Health Association, American Water Works Association, and Water Pollution Control Federation. 1998. *Standard Methods for the Examination of Water and Wastewater*, 20th ed., American Public Health Association, Washington, D.C.

Berner, E. K. and R. A. Berner. 1987. *The Global Water Cycle*, Prentice-Hall, Englewood Cliffs, NJ.

Gilluly, J., A. C. Waters, and A. O. Woodford. 1975. *Principles of Geology*, 4th ed., W.H. Freeman, San Francisco, CA.

Hites, R. A. and S. J. Eisenreich. 1987. *Sources and Fates of Aquatic Pollutants*, American Chemical Society, Washington, D.C.

Knopman, D. S. and R. A. Smith. 1993. Twenty years of the Clean Water Act. *Environ.* 35:17–34.

Manahan, S. E. 1991. *Environmental Chemistry*, 5th ed., Lewis Publishers, Chelsea, MI.

McKinney, M.L. and R.M. Schock. 1998. *Environmental Science: Systems and Solutions*, Jones and Bartlett Publishers, Sudbury, MA.

Salstein, D. A. 1995. Mean properties of the atmosphere, in *Composition, Chemistry, and Climate of the Atmosphere*, H. B. Singh, Ed., Van Nostrand Reinhold, New York, 19–49.

Schlesinger, W. H. 1997. *Biogeochemistry: An Analysis of Global Change*, 2nd ed., Academic Press, San Diego, CA.

Shen, T. L., P. J. Wooldridge, and M. J. Molina. 1995. Stratosphere pollution and ozone depletion, in *Composition, Chemistry, and Climate of the Atmosphere*, H. B. Singh, Ed., Van Nostrand Reinhold, New York, 394–442.

Sloane, C. S. and T. W. Tesche, Eds. 1991. *Atmospheric Chemistry: Models and Predictions for Climate and Air Quality*, Lewis Publishers, Chelsea, MI.

Solley, W. B., R. R. Pierce, and H. A. Perlman. 1995. Estimated use of water in the United States in 1995, U.S. Geological Survey Website. Available at http://water.usgs.gov/watuse/pdf1995/html/.

Tegart, W. J. G., G. W. Sheldon, and D. C. Griffiths. 1990. *Climate Change: The IPCC Impacts Assessment*, Australian Government Publishing Service, Canberra.

U.S. Environmental Protection Agency. 1983. Methods for Chemical Analysis of Water and Wastes, rev. ed., March, EPA-600/4-79-020, U.S. EPA, Cincinnati, OH.

U.S. Environmental Protection Agency. 1986. Quality Criteria for Water — 1986, EPA 440/5-86-001, U.S. EPA, Washington, D.C.

U.S. Environmental Protection Agency. 1989. *Drinking Water Health Advisory: Pesticides*, Lewis Publishers, Chelsea, MI.

U.S. Environmental Protection Agency. 1996. Drinking Water Regulations and Health Advisories, U.S. EPA Report 822-B-96, U.S. EPA, Washington, D.C.

U.S. Geological Survey. 1999. The Quality of Our Nation's Waters: Nutrients and Pesticides, USGS Circ. No. 1225, U.S. Geological Survey, Reston, VA.

Walker, J. C. G. 1977. *Evolution of the Atmosphere*, Macmillan, New York.

Supplementary reading

Goldman, C. R. and A. J. Horne. 1983. *Limnology*, McGraw-Hill, New York.

Nelson, D. W. and R. H. Dowdy. 1988. Methods for groundwater quality studies, in *Proc. National Workshop on Groundwater Quality*, Arlington, VA, USDA-ARS, Lincoln, NE.

Nielsen, D. M. 1991. *A Practical Handbook of Ground Water Monitoring*, Lewis Publishers, Boca Raton, FL.

Reisner, M. 1993. *Cadillac Desert: The American West and Its Disappearing Water*, revised and updated, Penguin Books, New York.

Singh, H. B., Ed. 1995. *Composition, Chemistry, and Climate of the Atmosphere*, Van Nostrand Reinhold, New York.

Singh, V. P., Ed. 1995. *Environmental Hydrology*, Kluwer Academic Publishers, Norwell, MA.

Van der Leeden, F., F. L. Troise, and D. K. Todd. 1990. *The Water Encyclopedia*, 2nd ed., Lewis Publishers, Chelsea, MI.

Ward, A. D. and W. J. Elliot, Eds. 1995. *Environmental Hydrology*, CRC Press, Boca Raton, FL.

chapter three

Our environment: soil ecosystems

Contents

3.1   Introduction

The term *soil* often has separate meanings to individuals from different scientific disciplines: to the agronomist or botanist, soil is best defined as a medium for plant growth;
to the engineer, soil refers to the loose material that lies between the ground surface and
solid rock; and, to the soil scientist, soil is described as the unconsolidated mineral and
organic matter at the earth's surface that has been altered by pedogenetic (soil-forming)
processes. Although there is no uniform, comprehensive definition for soil, it is apparent
that the functions of soil are numerous. Soils are dynamic ecosystems that support plant
life by providing the essential requirements for growth, including nutrients, water (H_2O),
oxygen (O_2), and support. Soil is also essential to humans because it provides natural
resources (e.g., food, fiber, construction materials), support for dwellings and roads, and
a means of recycling or detoxifying waste materials that are produced daily. Therefore, it
is imperative that we appreciate soil properties, functions, and how soils should be managed in order to maintain a sustainable ecosystem.

Various physical, chemical, and biological processes are constantly changing soils
over geologic time. In the following sections on the soil environment, we will delineate
the important features and define the terms commonly used to describe soils. To
categorize soils for land-use purposes, one must understand the general properties of
soils, as well as how to characterize select physical, chemical, and biological factors
important to soil health and land management. Therefore, one should know the specific
components of a successful soil testing program that apply to land use programs where
plant growth or crop yields are of greatest interest. From an environmental perspective,
we must also be able to evaluate some specific considerations that relate to *environmental soil testing* where protection of human health and environmental quality are the
primary concerns.

3.2   The soil environment

Soils are defined according to Singer and Munns (1999) as:

> Complex biogeochemical materials on which plants may grow; hav-
> ing structural and biological properties that distinguish them from
> the rock and sediments from which they normally originate; con-
> sisting of dynamic ecological systems that provide plants with sup-
> port, water, nutrients, and air; supporting a large population of
> microorganisms that recycle the materials of life; sustaining all eco-
> systems on land, including the entire human population, by provid-
> ing food, fiber, water, building materials and sites for construction
> and waste disposal; and protecting groundwaters by filtering toxic
> chemicals and disease organisms from wastewater.

Thus, soils and soil materials are used for agricultural, engineering, environmental, and
other important purposes. Some of the more common agricultural uses of soil include
croplands, grazing lands, pastures, forests, and for various horticultural purposes. Nonagricultural uses of soils involve recreation, foundations, building materials, and for waste
disposal. Within the continental U.S., use of nonfederal lands in the early 1990s (in millions
of hectares) amounted to 27% range (162), 27% forest (160), 26% crops (155), 8% pastures
(53), 6% urban (37), and 6% other (36). With the increase in urbanization, some of our
most productive lands, those classified as prime farmlands, have been converted to nonagricultural uses. In a 10-year period from 1982 to 1992, 6.1 million ha of rural lands were

used for urban development, of which 1.6 million ha were prime farmlands. With an increasing loss of our most productive soils, we are faced with using less productive lands to produce the foods required by animals and humans. Land-use-planning efforts that determine where we can best locate subdivisions, produce crops, and dispose of our wastes are becoming important to sustaining a balanced environment.

Soil pollution is often thought of as resulting from chemical contamination such as contaminant spills, or the use of excessive amounts of pesticides and fertilizers that can result in surface water and/or groundwater contamination. There are, however, other forms of soil pollution or degradation, including erosion, compaction, and salinity. Soils have often been damaged when used for on-site land disposal of waste chemicals and unwanted materials (see Chapter 11 for more information on reclamation). Most soils are capable, to some degree, of adsorbing and neutralizing many pollutants to harmless levels through chemical and biochemical processes. There are, however, limits to the ability of a soil to accept wastes without some negative effect on the environment. To appreciate the use of our soil resources, we must understand the basic properties of soils and how they interact to regulate the health and quality of our lands.

3.2.1   Soil physical properties

Soils contain solids, liquids, and gases. The soil physical properties that are of primary interest include the composition and arrangement of solids and how movement of liquids and gases is affected by the solids. It should be noted that soil color and temperature are also considered physical properties of soils. The arrangement of soil solids determines the amount of open volume, or *pore space*, that a soil possesses. Organic matter can significantly influence a soil's infiltration rate, water-holding capacity, permeability, aggregate stability, and consistence, even though soils contain much lower contents of organic matter than mineral materials. In this section we discuss the nature of soil solids and their importance to the movement of soil solutions and soil gases. Migration of soluble and gaseous pollutants are thus controlled to a large extent by the physical properties of a soil.

3.2.1.1   Particle size

Soils are composed of solid materials ranging in size from stones to fine clays (Table 3.1). The larger materials, called coarse or mineral fragments, including stones, cobbles, and gravels, are chemically and physically weathered over long periods of time to form the smaller soil particles of sand, silt, and clay. Soil particles are defined on the basis of their diameter, although rarely do these particles exist as spherical objects. Clay minerals, for instance, are three-dimensional, layered structures that commonly have a platelike appearance. Soil particle sizes often differ with the classification schemes used by different groups; the U.S. Department of Agriculture system (Soil Survey Staff, 1975; 1992) will be used throughout this book.

There are 12 soil textural classes that are defined by the relative proportion of sand, silt, and clay that makes up a soil sample (Figure 3.1). There are two generally used methods for determining soil texture: (1) the field method done by hand and (2) mechanical analysis. The field method is taught in introductory soil science courses and will not be discussed here. Using the mechanical analysis method to determine the sand, silt, and clay requires the removal of course fragments by sieving the soil through a 2-mm sieve. Chemical treatments are also necessary to remove cementing agents such as organic matter and carbonates. The percent sand, silt, and clay should always total 100%, and once known, the soil texture can be found by using the textural triangle (Figure 3.1).

Table 3.1 Size Classification of Soil Particles According to the
U.S. Department of Agriculture System

Soil particles	Diameter (mm)	Comparison
Coarse fragments		
Stones	>254	Greater than 10 in.
Cobbles	75–254	3 to 10 in.
Gravel	2–75	0.08 to 3 in.
Soil particles		
Sand	2.0–0.05	
Very coarse	2.0–1.0	Thickness of a nickel
Coarse	1.0–0.5	Size of pencil lead
Medium	0.5–0.25	Salt crystal
Fine	0.25–0.10	Flat side of a book page
Very Fine	0.10–0.05	Nearly invisible to the eye
Silt	0.05–0.002	
Coarse	0.05–0.02	Root hair
Medium	0.02–0.01	Nematode
Fine	0.01–0.002	Fungi
Clay	<0.002	
Coarse	0.002–0.0002	Bacteria
Fine	<0.0002	Viruses

Example problem 3.1

What is the textural class of a soil that contains:

40% sand + 40% silt + 20% clay?

From the textural triangle, this soil would be classified as a loam. Only two of the soil particle percentages are actually needed to determine the soil textural class since the point at which the two meet does not change when the third particle percentage is used.

Soil texture is often considered a basic property of the soil because the textural class of a particular soil generally remains unchanged over time equivalent to the human lifespan. This is one reason soil descriptions used in soil surveys record the soil textural class of each horizon. However, a disturbance of an area, such as water or wind erosion, could alter the textural class of the soil surface of both the soil being eroded and the soil where the erosional deposition occurs. Over long time periods (geologic time), weathering and translocation of soil materials may change soil texture.

3.2.1.2 *Aggregates and soil structure*

Soil particles that are held together by chemical and physical forces form stable *aggregates*. Natural aggregates are called *peds*. Collectively, the type of soil aggregates or peds defines *soil structure*. Soil structure influences the amount of water that enters a soil (*infiltration*) and gas diffusion at the soil surface. Soil structure also plays an important role in the movement of liquid and gaseous substances through soils. The *porosity* of a soil is a function of its structure.

Soil structure is classified based on type, size, and grade. Common structure types include granular, platy, subangular and angular blocky, prismatic, and columnar shapes

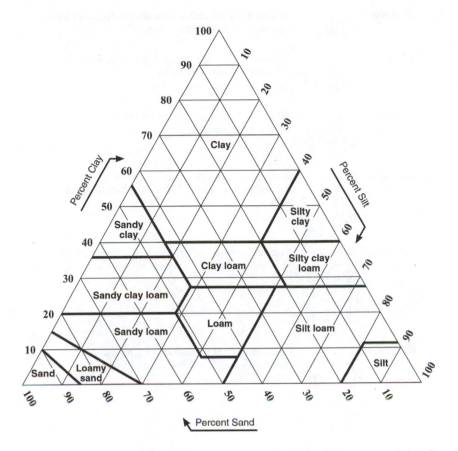

Figure 3.1 Textural triangle indicating the range in sand, silt, and clay composition for each soil textural class. The three corners of the triangle represent 100% of the primary soil particle-size classes. Note that a soil sample containing equal parts of sand, silt, and clay would be classified as a clay loam.

(Figure 3.2). Granular structure is representative of A horizons. Platy structure, although not very common, is found in E horizons of forest soils or in some clay-compacted B horizons. The blocky, prismatic, and columnar structures are often found in B horizons; angular and subangular blocky structures are typical of soils in humid regions and prismatic and columnar structures are common to soils in arid and semiarid regions. Structural size classes vary with the type of structure considered, and range from very fine to very coarse. Structural grade is determined by observing the soil structure in place in a soil pit and is related to the overall structural development of a soil. Structural grades include weak, moderate, and strong classifications.

3.2.1.3 Particle and bulk density

Particle and bulk density measurements are useful for estimating the type of soil minerals present and the degree of soil compaction, respectively. *Particle density* is the mass of a particle per volume (Mg/m^3 or g/cm^3), and the volume of pore space and weight of water are not included in particle density measurements. Common soil minerals (quartz, feldspars, micas, and clay minerals) have particle densities between 2.60 and 2.75 Mg/m^3, but a value often used to represent the average soil particle density is 2.65 Mg/m^3. *Bulk density* is a measure of the mass per volume (Mg/m^3) of a soil. Undisturbed soils are usually used for bulk density measurements so that a true representation of the amount of solid present

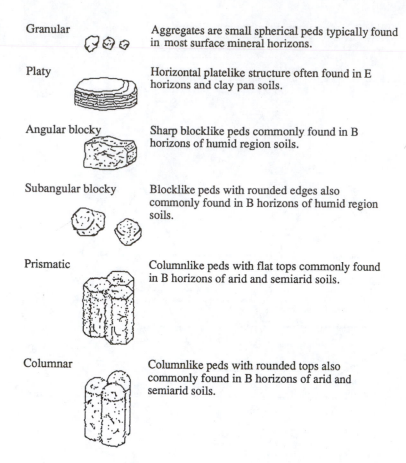

Granular — Aggregates are small spherical peds typically found in most surface mineral horizons.

Platy — Horizontal platelike structure often found in E horizons and clay pan soils.

Angular blocky — Sharp blocklike peds commonly found in B horizons of humid region soils.

Subangular blocky — Blocklike peds with rounded edges also commonly found in B horizons of humid region soils.

Prismatic — Columnlike peds with flat tops commonly found in B horizons of arid and semiarid soils.

Columnar — Columnlike peds with rounded tops also commonly found in B horizons of arid and semiarid soils.

Figure 3.2   Soil structural types, descriptions, and their location in the soil profile.

in a particular soil volume can be calculated; with disturbed soils, natural soil pore spaces are destroyed. Bulk density is calculated on an oven-dry weight basis and does not take into consideration the amount of water present in the soil at the time of sampling. Mineral soils, unless developed in volcanic ash, generally have bulk densities greater than 1.0 Mg/m^3 with a common bulk density of a soil with a loam texture being 1.3 Mg/m^3. Soils with high densities (e.g., 2.0 Mg/m^3) will likely have slow water infiltration and permeability, which can result in ponding or surface runoff. Organic soils typically have bulk densities less than 1.0 Mg/m^3, whereas mineral soils derived from volcanic ash often have bulk densities in the range of 0.3 to 0.85 Mg/m^3.

3.2.1.4   Soil minerals

Soil minerals are classified as primary and secondary minerals according to their origin. *Primary minerals* are those that formed during the cooling of molten rock and are predominately silicate minerals (Table 3.2). Igneous rocks are composed entirely of primary minerals; metamorphic and sedimentary rocks can contain various amounts of primary and secondary minerals. *Secondary minerals* are formed in soils from soluble products derived from the weathering of rocks. *Clay minerals* are one of the most important soil secondary minerals due to their large surface area and reactivity with ionic and dissolved organic compounds, and will be discussed in more detail in the next section. Carbonates and sulfates accumulate in B horizons and are often the dominant secondary minerals present in soils in arid and semiarid regions.

Table 3.2   Common Primary and Secondary Minerals Found in Soils

Mineral	Chemical formula	Weatherability
Primary minerals		
Quartz	SiO_2	Most resistant
Muscovite	$KAl_3Si_3O_{10}(OH)_2$	
Microcline	$KAlSi_3O_8$	
Orthoclase	$KAlSi_3O_8$	
Biotite	$KAl(Mg,Fe)_3Si_3O_{10}(OH)_2$	
Albite	$NaAlSi_3O_8$	
Hornblende	$Ca_2Al_2Mg_2Fe_3Si_6O_{22}(OH)_2$	
Augite	$Ca_2(Al,Fe)_4(Mg, Fe)_4Si_6O_{24}$	
Anorthite	$CaAl_2Si_2O_8$	
Olivine	$(Mg,Fe)_2SiO_4$	Least resistant
Secondary minerals		
Geothite	$FeOOH$	Most resistant
Hematite	Fe_2O_3	
Gibbsite	$Al(OH)_3$	
Clay minerals	Aluminosilicates	
Dolomite	$CaMg(CO_3)_2$	
Calcite	$CaCO_3$	
Gypsum	$CaSO_4 \cdot H_2O$	Least resistant

Source: Brady, N. C., *The Nature and Properties of Soils*, 10th ed., Macmillan Press, New York, 1990.

3.2.1.5   Organic carbon

Carbon (C) is present in all soils in the form of soil organic matter; however, soils formed in arid regions can contain significant amounts of inorganic C as carbonate minerals. Soil organic matter contents vary from less than 1% in coarse-textured soils and soils of arid regions to nearly 100% in some poorly drained organic soils; typical farmland topsoils may contain 2 to 10% organic matter. Organic matter influences soil physical, chemical, and biological properties. Soil structure is often improved with the addition of organic materials such as animal manures, municipal biosolids and composts, and from crop residues that are returned to the soil. Further discussion on the role of soil C is presented in Section 3.2.2.

3.2.1.6   Soil water

Soils hold water in pore spaces by the cohesive and adhesive nature of water and soil particle surfaces. *Cohesion forces* are the result of water molecule polarity and hydrogen bonding, which attracts water molecules to one another. *Adhesion forces* are responsible for attracting water molecules to soil mineral and organic matter surfaces. These forces allow water to move upward in soils by capillary action, or along surfaces of soil particles as water-films.

Several terms are used to describe soil water, including water content and water potential. *Water content* is a measure of the amount of water in a soil, and is usually expressed on a percent basis. The water content of a soil is determined by knowing the weight of both the water and solid materials contained in a soil sample. Each of these weights can easily be determined in the laboratory using an oven and balance.

Example problem 3.2

The percent soil water content (SWC) is calculated as follows:

$$\%SWC = [(\text{wet soil wt} - \text{oven-dried soil wt})/(\text{oven-dried soil wt})] \times 100 \qquad (3.1)$$

For a soil with a wet weight of 160 g and a dry weight of 120 g, the %SWC would be:

$$\%SWC = [(160\ g - 120\ g) / 120] \times 100 = 33.3\% \qquad (3.2)$$

Soil *water potential* is a measure of the strength, or energy, with which water is held by the soil. Water moves in soils from areas of high water potential to low water potential; water potential is in turn related to soil moisture content, textural class, structure, salt content, and organic matter content. Clays have lower water potentials than sands when both have the same water content. *Total water potential* is the overall effect due to a combination of several potentials, of which matric, pressure, gravity, and solute potentials are the most important. *Matric potential* (also known as tension or suction potential) represents the interaction of water with soil surfaces and the tendency of small pores to retain water more strongly than large pores. In well-drained soils with low soluble salts, soil water potential is nearly equal to the matric potential. *Pressure* and *gravity water potentials* are related to external forces that are exerted on soil water. Pressure potential is due to atmospheric or gas pressure effects, and gravity potential is a result of gravity's pull on soil water. *Solute*, or *osmotic*, *potential* is due to the tendency of water to move from dilute to concentrated solutions.

Water moves in soils as a vapor or a liquid. *Vapor flow* through a soil is generally a slow process. Water vapor is present in all unsaturated soils and moves by diffusion within the soil due to vapor pressure and temperature gradients. Soil water movement is classified as either saturated or unsaturated flow depending on the soil moisture content (see Figure 2.11). Saturated flow occurs in soils where the void space is filled with water. Subsurface horizons can become saturated if water movement is restricted, for example, in soils with a high water table, a clay pan, or in stratified soils. Unsaturated flow occurs whenever void spaces are partially filled with air. In both saturated and unsaturated conditions, water flow is a function of the driving forces acting on the water (hydraulic gradient) and the ability of the soil to allow water movement (hydraulic conductivity).

Water that infiltrates into soils can be stored, transferred to streams, rivers, lakes, oceans, or become part of the groundwater pool. Surface runoff occurs when rainfall cannot be absorbed by the soil because the rate of infiltration is slow or the soil becomes saturated. Water that falls on land surfaces can be returned to the atmosphere by evapotranspiration, which is a combination of evaporation from soil or plant surfaces and transpiration from plants. Runoff can result in soil erosion and pollution through the transport of soluble and particulate-bound nutrients and pesticides (see Chapter 2).

3.2.2   Soil chemical properties

Mineral solubility, soil reactions (e.g., pH), cation and anion exchange, buffering effects, and nutrient availability are major chemical properties of soils, which are determined primarily by the nature and quantity of the clay minerals and organic matter present. A knowledge of the chemistry of soil solutions is important in understanding how best to handle waste materials such as animal manures, municipal biosolids, and food by-products.

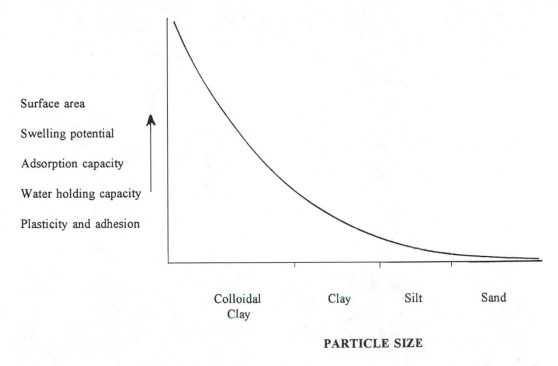

Surface area

Swelling potential

Adsorption capacity

Water holding capacity

Plasticity and adhesion

Colloidal Clay   Clay   Silt   Sand

PARTICLE SIZE

Figure 3.3 Relationship between particle size (using an equivalent weight or volume of material) and several soil chemical and physical properties.

In this section we will examine clay mineral and organic matter properties, and describe the soil reactions that are controlled by these materials.

3.2.2.1 Clay minerals

Layered aluminosilicate minerals, better known as clay minerals, have a profound influence on many soil chemical reactions because of their high "active" surface area (Figure 3.3). The term *active* refers to charges that develop on clay mineral surfaces and the ability of some types of clay minerals to expand. Clay minerals should not be confused with clay-sized particles because these latter materials can also include particles of quartz, calcite, and gypsum that are 2 μm or less.

The clay minerals of most interest are crystalline and have regular layers of tetrahedral and octahedral sheets (Figure 3.4). Tetrahedral sheets are comprised of silicon and oxygen atoms with three out of every four oxygen atoms shared between adjacent tetrahedra. These shared oxygens are referred to as basal oxygens of the tetrahedral sheet; the unshared oxygen is called the apical oxygen. Octahedral sheets are of two types — dioctahedral and trioctahedral. Dioctahedral sheets have two out of every three octahedral sites occupied, most often by the trivalent aluminum (Al^{3+}) cation. Trioctahedral sheets have all octahedral sites occupied with divalent cations, most commonly magnesium (Mg^{2+}).

The layered silicate clay minerals have structures that are either 1:1, 2:1, or 2:1:1 layers of tetrahedral and octahedral sheets. The 1:1 clay minerals have one tetrahedral and one octahedral sheet held together by the sharing of the apical tetrahedral oxygen (Figure 3.5). The 2:1 clay minerals have an octahedral sheet sandwiched between two tetrahedral sheets (Figure 3.5). The 2:1:1 layered clay minerals are similar to 2:1 clays with an additional dioctahedral or trioctahedral sheet between 2:1 layers.

The principal clay minerals found in soils are listed in Table 3.3. Kaolinite and halloysite are 1:1 clay minerals that have silicon (Si) in their tetrahedral sites and Al in

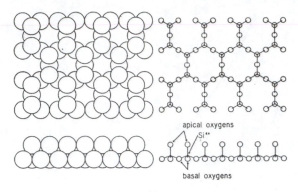

Tetrahedral sheet

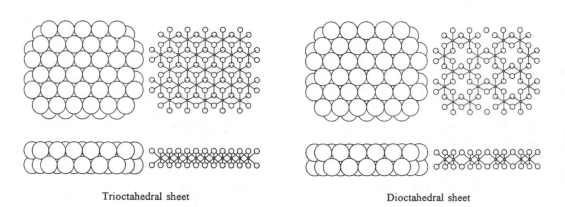

Trioctahedral sheet   Dioctahedral sheet

Figure 3.4   Two representations (space-filled and bond arrangement) of tetrahedral and octahedral sheets. Octahedral sheets can be trioctahedral or dioctahedral depending on the number of sites filled. (Adapted from Schultze, D. G., in *Minerals in the Soil Environment*, J. B. Dixon and S. B. Weed, Eds., Soil Science Society of America, Madison, WI, 1989. With permission.)

octahedral sites. The layers of the 1:1 clay minerals are held rather tightly together through electrostatic bonds between the basal oxygen of the tetrahedral sheets and the hydroxyls of the adjacent octahedral sheet. Kaolinite is a common mineral in soils whereas halloysite is less stable and is often transformed to kaolinite over time.

Isomorphic substitution occurs when an element substitutes for another in the mineral structure, such as Al^{3+} substituting for Si^{4+}. If an element of a lower charge substitutes for an element of a higher charge, a permanent negative charge develops in the clay mineral. Both kaolinite (Figure 3.6) and halloysite have very little surface charge due to low isomorphic substitution in their structures. Illite, or hydrous mica, is a 2:1 nonexpanding, dioctahedral clay mineral which has potassium (K) in the *interlayer* spaces (i.e., area between the 2:1 layers). The K atom is held tightly in the interlayers and is commonly referred to as being *nonexchangeable*. Montmorillonite is one of several types of clay minerals in the *smectite* family (Figure 3.6). Smectites are 2:1 clay minerals that have low to moderate isomorphic substitution. Because of the low permanent charge, smectites are capable of expanding, thus increasing the amount of exposed surface area. The high surface area together with the amount of charged sites give smectites the ability to adsorb cations. *Cation exchange capacity* is the amount of exchange sites that can adsorb and release cations. Montmorillonite is a common smectite in most soils. Vermiculite (Figure 3.6) is a

PLANE OF IONS   SHEETS, LAYERS

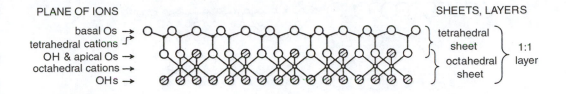

1:1 Layer - one tetrahedral sheet and one octahedral sheet

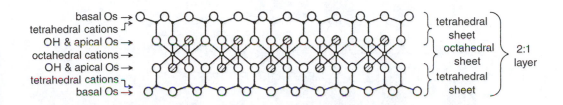

2:1 Layer - two tetrahedral sheets with an octahedral sheet in the middle

Figure 3.5 Structural representation of 1:1 and 2:1 layers showing the chemical bonds between elements. (Adapted from Schultze, D. G., in *Minerals in the Soil Environment*, J. B. Dixon and S. B. Weed, Eds., Soil Science Society of America, Madison, WI, 1989. With permission.)

Table 3.3 Common Clay Minerals Found in Soils Throughout the World

Mineral	General formula[a]
1:1	
Kaolinite	$Al_4Si_4O_{10}(OH)_8$
Halloysite	$Al_4Si_4O_{10}(OH)_8\ 4H_2O$
2:1	
Illite	$Al_4(Si_6Al_2)O_{20}(OH)_4 \cdot nH_2O$
Smectites	
Montmorillonite	$M_x(Al_{4-x}Mg_x)Si_8O_{20}(OH)_4$
Beidellite	$M_xAl_4(Si_{8-x}Al_x)O_{20}(OH)_4$
Nontronite	$M_xFe_4(Si_{8-x}Al_x)O_{20}(OH)_4$
Vermiculite	$Mg(Al,Fe,Mg)_4(Si_6Al_2)O_{20}(OH)_4 \cdot nH_2O$
2:1:1	
Chlorite	$Mg_6(OH)_{12}\ (Mg_5Al)(Si_6Al_2)O_{20}(OH_4)$

[a] M represents a monovalent cation.

2:1 clay mineral which has high isomorphic substitution. The interlayers of vermiculite contain exchangeable Mg^{2+} ions. However, since vermiculite has a higher layer charge, interlayer Mg^{2+} in vermiculite is held more tightly than is interlayer Mg^{2+} in smectite. Although vermiculite has a high cation exchange capacity, when saturated extensively with K^+ or NH_4^+ ions, vermiculite becomes nonexpanding. Chlorite (Figure 3.6) is also

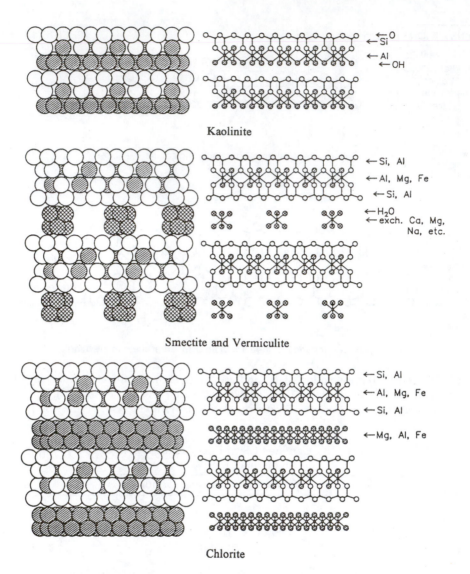

Figure 3.6 Structural representation of kaolinite (1:1), smectite and vermiculite (2:1), and chlorite (2:1:1) clay minerals. (Adapted from Schultze, D. G., in *Minerals in the Soil Environment*, J. B. Dixon and S. B. Weed, Eds., Soil Science Society of America, Madison, WI, 1989. With permission.)

prevalent in many soils. The interlayer space between the 2:1 layers in chlorite is occupied by Mg octahedral sheets, thus making this a 2:1:1 layered silicate. Chlorites have low surface area due to the attraction of the brucite layer (Mg octahedral sheet) by the 2:1 layers, which reduces their potential cation exchange capacity.

Figure 3.7 describes the general conditions that influence the formation of clay minerals and oxides. Some primary aluminosilicate minerals are physically and chemically weathered to produce clay minerals. Muscovite altered to illite is a good example of this. Both muscovite and illite have 2:1 structures; however, physical breakdown, loss of K (–K), and a slight alteration in the muscovite structure produces illite which has a lower cation exchange capacity and greater swelling potential. Some clay minerals are recrystallized products of ions released during the breakdown of other minerals. Formation of kaolinite, a 1:1 clay mineral, is the result of recrystallization because there are no primary or

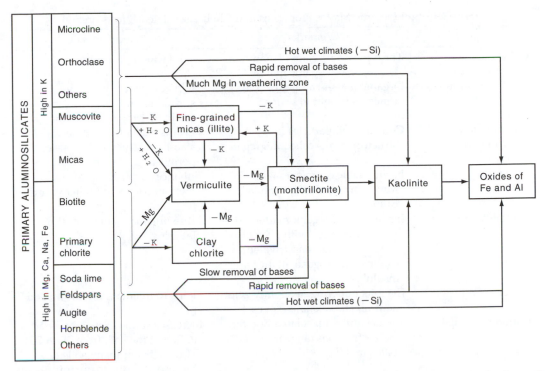

Figure 3.7   Sequence for the formation of clay minerals and Fe and Al oxides. Note the loss of soluble cations in the genesis of clay minerals. (From Brady, N. C. and Weil, R. R., *Elements of the Nature and Properties of Soils,* Prentice-Hall, Upper Saddle River, NJ, 2000. With permission.)

secondary minerals with analogous 1:1 structures. As noted in Figure 3.7, kaolinite forms directly from the ions released from the weathering of primary aluminosilicates or smectite if accompanied by a rapid loss of base cations. Intense weathering is responsible for the loss of Si and the production of Fe and Al oxides.

3.2.2.2   Soil organic matter
Organic matter plays an important role in the chemistry of soils. Soil properties associated with soil organic matter include soil structure, macro- and micronutrient supply, cation exchange capacity, and pH buffering. Organic matter is also a source of C and energy for microorganisms.

Soil organic matter is comprised of decomposed plant and animal residues. It is a highly complex mixture of C compounds that also contain nitrogen (N), sulfur (S), and phosphorus (P). Organic matter is made up of humic substances and biochemical compounds. Humic substances are operationally defined based on their solubility characteristics: humic acids are soluble in alkaline but not acid solutions, fulvic acids are soluble in acid and alkaline solutions, and humin is the insoluble material that remains after humic and fulvic acid extraction. Biochemical compounds include identifiable organic compounds such as organic acids, proteins, polysaccharides, sugars, and lipids. General properties of soil organic matter and their effect on soils are listed in Table 3.4.

From an environmental quality standpoint, organic matter can be both beneficial and detrimental. Soil organic matter can adsorb trace element pollutants, e.g., lead (Pb), cadmium (Cd), copper (Cu), which will reduce the chance of contamination of surface waters and groundwaters. Another advantage is the adsorption of pesticides and other organic chemicals. This reduces the possibility of pesticide carryover effects, prevents

Table 3.4 Soil Organic Matter Properties and Their Associated Effects on Soil

Property	Remarks	Effects on soil
Color	The typical dark color of many soils is caused by organic matter	May facilitate warming
Water retention	Organic matter can hold up to 20 times its weight in water	Helps prevent drying and shrinking; improves moisture-retaining properties of sandy soils
Combination with clay minerals	Cements soil particles into structural units called aggregates	Permits exchange of gases; stabilizes structure; increases permeability
Chelation	Forms stable complexes with Cu^{2+}, Mn^{2+}, Zn^{2+}, and other polyvalent cations	Enhances availability of micronutrients to higher plants
Solubility in water	Insolubility of organic matter is due to its association with clay; also, salts of divalent and trivalent cations with organic matter are insoluble	Little organic matter is lost by leaching
Buffer action	Exhibits buffering in slightly acid, neutral, and alkaline ranges	Helps to maintain a uniform soil pH
Cation exchange	Total acidities of isolated organic matter fractions range from 300 to 1400 cmol/kg	Increases cation exchange capacity (CEC) of the soil; from 20 to 70% of the CEC of many soils (e.g., Mollisols) is due to organic matter
Mineralization	Decomposition of organic matter yields CO_2, NH_4^+, NO_3^-, PO_4^{3-}, and SO_4^{2-}	Source of nutrients for plant growth
Combination with xenobiotics	Affects bioactivity, persistence, and biodegradability of pesticides	Modifies application rates of pesticides for effective control

Source: Stevenson, F. J., *Humus Chemistry: Genesis, Composition, Reactions*, 2nd ed., John Wiley & Sons, New York, 1994. With permission.

contamination of the environment, and enhances both biological and nonbiological degradation of certain pesticides and organic chemicals. In addition, organic matter is known for its capacity to sorb gases such as NO, NO_2, and CO.

Although there are many benefits of soil organic matter, there are also detrimental effects that occur under certain situations. Soils with high organic matter contents may require higher pesticide application rates for effective control. Water contamination is then a concern if these pesticides are leached or transported by wind or water erosion. Soils also have a finite capacity to adsorb trace elements and should not receive extensive applications of organic wastes such as municipal biosolids and animal manures that contain high levels of trace elements. Applications of wastes containing trace elements beyond recommended rates (see Chapters 7 and 9) can lead to levels in soils that may be toxic to plants, and possibly to animals and humans upon consumption of foods grown in the contaminated soils.

3.2.2.3 Ion-exchange

Ion-exchange is one of the most significant functions of soils. When isomorphic substitution occurs with cations of lower charge substituting for cations of high charge, a permanent charge develops. Charged sites can also develop as a function of pH with different clay minerals, metal oxides, and organic matter. These charged sites are the result of ionization (H^+ dissociation) or protonation of uncharged sites; ionization results in a negative-charged site and protonation a positive-charged site. Both of these reactions are dependent on pH and are called pH-dependent charge. As the pH increases, the cation exchange capacity of the soil is generally greater due to an increase in the amount of pH-dependent charged sites. Under acid soil conditions, some clay minerals, metal oxides, and organic matter will have positively charged, anion exchange sites. Inorganic and organic ions having charges that are opposite of the exchange site are attracted or adsorbed to the soil surface by electrostatic processes.

Both clay minerals and organic matter have the ability to adsorb soluble chemicals from the soil solution. This is an important means by which plant nutrients are retained in crop rooting zones. On the other hand, adsorption of dissolved chemical species by colloidal-size clay minerals and dissolved organic matter can enhance the leaching and transport of heavy metals, organic chemicals (e.g., pesticides), and other potential pollutants.

3.2.3   Soil biological properties

Soil biological communities vary from soil to soil and range from plant roots, rodents, worms, and insects, which are usually visible to the eye, to microorganisms (bacteria, actinomycetes, fungi, algae, and protozoa) that are so small they require a microscope to see them. Factors such as rainfall, temperature, vegetation, and physical and chemical properties of soils influence the type and number of living organisms in a soil community. The relative number and total biomass (weight of organism per unit volume) are given in Table 3.5 for a surface soil (note the extremely large number of microorganisms).

Organisms inhabiting soil environments can be grouped into two broad categories. These categories include the *autotrophs*, which assimilate C from CO_2 and obtain energy from sunlight or through the oxidation of inorganic compounds, and the *heterotrophs* which use organic C as a source of energy and C. The autotrophs are considered producers because of their ability to convert CO_2 and energy from the sun into organic C products, a process called *photosynthesis*. Only vascular plants and some bacteria and algae are considered producers. Heterotrophs, on the other hand, are regarded as *consumers* and *decomposers*. Soil animals and most microorganisms fall into this category.

3.2.3.1   Plants

The part of the planet that supports life is described as the biosphere. Thus, the biosphere comprises all living organisms and their general surroundings, and includes most hydro-sphere and soil ecosystems. Living organisms that are considered essential to biosphere communities include aquatic and terrestrial plants, animals, and microorganisms. Ecosystems that comprise different parts of the biosphere are geographic entities defined by natural features. Although we typically confine ourselves to regions that have political boundaries, such as cities, states, and countries, environmental quality and protection should be a concern to all of us. Biosphere impacts, both positive and negative, are often the result of lifestyles that rely on natural resources.

Aboveground plant parts are a source of food for consumers such as grazing animals and humans. Belowground plant parts (i.e., roots, tubers, and other organs) are a source of food for humans and animals as well as soil consumers and decomposers, and influence the type and activity of microorganisms living in and around the plant roots. The *rhizosphere* is the area around roots that is influenced by the presence of the roots; it often has from 10 to 100 times more microorganisms than the bulk soil. Root exudates include organic and inorganic substances that provide nutrients for the rhizosphere microorganisms.

Symbiotic associations between plant roots and microorganisms result in a variety of beneficial effects to both the plant and the microbes. Nodules that form on leguminous plant roots are caused by bacteria that are capable of converting atmospheric N_2 into N compounds utilizable by the plant (see Section 4.2.4). Mycorrhizal fungi form symbiotic

Table 3.5   Estimated Number and Biomass of Soil Animals and Microorganisms in Surface Horizons

Organisms	Abundance per meter^3	Abundance per gram	Biomass (kg/HFS)
Soil animals			
Earthworms	200–2,000	< 1	110–1,100
Nematodes	10^7–10^8	10^4–10^5	11–110
Others	10^4–10^6	Variable	17–170
Microorganisms			
Bacteria	10^{14}–10^{15}	10^8–10^9	450–4,500
Actinomycetes	10^{13}–10^{14}	10^7–10^8	450–4,500
Fungi	10^{11}–10^{12}	10^5–10^6	1,120–11,200
Algae	10^{10}–10^{11}	10^4–10^5	56–560
Protozoa	10^{10}–10^{11}	10^4–10^5	17–170

Note: Biomass values based on live weight per hectare furrow slice (HFS).

Source: Brady, N. C., *The Nature and Properties of Soils*, 10th ed., Macmillan Press, New York, 1990.

relationships with plant roots. Plants benefit from this association by increased absorption of nutrients (e.g., N, P, S, and micronutrients) and water.

Plants and their natural functions can remediate sites by processes that involve the stabilization, degradation, removal, and/or containment of environmental contaminants. Trees, grasses, legumes, and aquatic plants have potential to be used in the phytoremediation of areas contaminated with toxic metals, hazardous organic compounds, and even radionuclides, often at much lower costs than conventional remediation methods such as soil washing, incineration, and various physical and chemical processes (see Chapter 11 for more information on remediation technologies). Because phytoremediation is a natural process, many of the commonly used practices in agricultural, horticultural, and landscape ecology can be extended to the cultivation, harvesting, and processing of plants used for phytoremediation. According to Flathman et al. (1999) the major phytoremediation technologies include: "enhanced rhizosphere degradation; phytoextraction (including rhizofiltration and/or phytovolatilization); phytodegradation alone or with phytovolatilization; phytostabilization; and alternative cover (i.e., vegetative cap)." These different remediation technologies are becoming more common primarily because research is now showing that costs associated with phytoremediation can be as little as half that of conventional methods. For example, costs to remediate a ton of petroleum hydrocarbon–contaminated material would be $10 to $35 for phytoremediation, $50 to $150 for *in situ* bioremediation, $20 to $200 for soil venting, $80 to $200 for soil washing, $120 to $300 for indirect thermal treatment, and more than $200 for incineration and other technologies. The cleanup of NO_3-contaminated groundwaters with poplar trees was estimated, for a 5-year period, to be $225,000, in comparison to pump-and-treat groundwater using reverse osmosis at $660,000. The use of phytoremediation in a creosote-contaminated site is discussed in the environmental quality issues/events box, Restoration of a contaminated wood-preservative site in Laramie, WY, in Chapter 8 (see p. 301).

3.2.3.2   Soil animals

Soil animals include all animals that in one way or another influence soil properties. Large animals such as cattle and deer can influence soils primarily through overgrazing, which influences the vegetative community structure, function, and health. Effects of overgrazing include reduced ground cover, altered plant species composition, soil compaction, and possibly soil erosion. Small burrowing animals can have a great impact on soil properties. Gophers, shrews, prairie dogs, badgers, moles, ground squirrels, and mice are some of the burrowing animals that can profoundly influence the development of soils by mixing topsoil and subsoil materials. Tunnels and holes produced by burrowing animals will increase the infiltration of water and air. Subterranean soil animals include arthropods (e.g., ants, beetles, centipedes, millipedes, mites, spiders, springtails), worms (e.g., nematodes, earthworms), and protozoa. Protozoa and nematodes are generally the most abundant of all soil animals (Table 3.5 and Figure 3.8); protozoa are, in a strict sense, classified as microanimals and will therefore be discussed in the next section.

Several arthropods are involved in the decomposition of plant materials and also assist in mixing the fragmented vegetative parts into the soil. Millipedes are primarily saprophytic and consume dead organic materials, whereas ants and termites, as well as prairie dogs, can mix large volumes of soil through their tunneling efforts.

Earthworms play an important role in soil formation and breakdown of organic residues. They are usually found in the organic-rich surface horizon, but some species burrow as deeply as 6 m. Earthworms ingest organic matter and soil, which are then decomposed by digestive enzymes and grinding action as they pass through the worm's gut. Each day, the weight of soil material ingested and excreted by earthworms can be equivalent to their body weight. The amount of earthworm casts (excreted material)

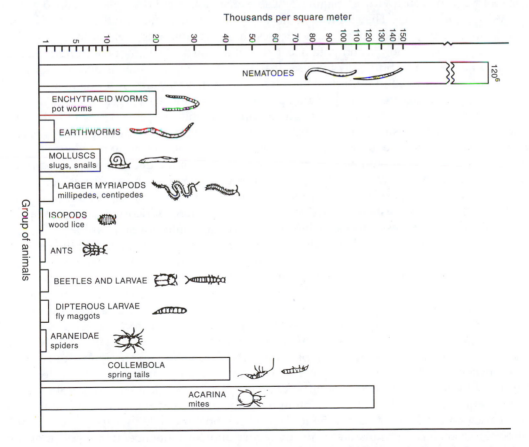

Figure 3.8   Illustration of the types of soil animals and their numbers determined in a square meter of a grassland soil. (From Paul, E. A. and Clark, F. E., *Soil Microbiology and Biochemistry*, Academic Press, San Diego, CA, 1989. With permission.)

produced yearly in a hectare of land has been estimated to range from 70 to 250 Mg/ha. These casts have been shown to increase the availability of the nutrient elements N, P, K, calcium (Ca), and Mg compared with original soil material. *Lumbricus terrestris*, the common earthworm, is one of over 1800 worm species known worldwide.

Nematodes are microscopic, unsegmented roundworms that are classified based on their feeding habits. The most common are those that feed on decaying organic matter; however, others prey on bacteria, fungi, algae, or other nematodes, or they infect plant roots. One method of controlling the latter type of nematodes is by soil fumigation, although this method of control can be very costly and potentially harmful to the environment if not managed carefully. Use of crop rotation has been shown to be a very effective practice for controlling not only pests such as the nematodes, but also insect and disease problems and is also beneficial to soil fertility. Innovative efforts have been used to control the sugar beet nematode (*Heterodera schachtii*) by including trap crops, usually radish species, that allow colonization by the nematode but also impair normal reproduction.

3.2.3.3 Soil microorganisms

Soil microorganisms include bacteria, actinomycetes, fungi, algae, and protozoa. *Viruses*, which are restricted to growing only in other living cells, are molecules of RNA or DNA within protein coats and are discussed here only because they play an important part in the microbiology of soils. Viruses require a viable metabolizing host organism in order to reproduce or duplicate. They induce the host organism to manufacture the necessary components for virus reproduction. Many viruses can cause diseases such as those found in plants (e.g., tobacco mosaic and potato leaf roll) and animals (e.g., foot and mouth disease and bovine leukemia). Survival of viruses that are pathogenic to humans and animals, and which are known to be in municipal biosolids and animal manures, is a potential concern when these materials are used in a land application program (see Chapters 9 and 11).

Bacteria are the most numerous of all soil microorganisms (Table 3.5), and individual bacterial cells are the smallest and most difficult to see under the microscope. A handful of soil can contain several billion bacterial cells. The ability of bacteria to reproduce rapidly and adapt to new environmental situations is important to the decomposition and transformation of both natural and anthropogenic products. Some of the functions performed either entirely or in part by bacteria include nutrient cycling, decomposition of organic materials, N fixation, pesticide detoxification, and oxidation–reduction reactions. Without bacteria to mediate these processes, life as we know it would not be possible.

Bacteria are classified as either autotrophs and heterotrophs, depending on their sources of energy and C. They can be further grouped as aerobes (require O_2), anaerobes (do not require O_2), and facultative anaerobes (grow in the presence or absence of O_2). The type and abundance of any one group of bacteria depends on such factors as available nutrients and soil environmental conditions. Some of the aerobic and anaerobic bacteria form spores that allow them to survive under adverse environmental conditions during dry and high temperature periods. Soil conditions that affect the growth of bacteria are many; however, the most important factors are the O_2, temperature, moisture, acidity, and inorganic and organic nutrient status of the soil environment. Bacteria and fungi are usually active in aerated soils, but in ecosystems containing little or no O_2, most of the biological and chemical transformations are governed by bacteria. Optimum temperature and moisture requirements for most bacteria are 20 to 40°C and 50 to 75% of the moisture-holding capacity of the soil, although specialized groups are active at higher and lower temperatures and moisture contents. Extremely high and low pH (i.e., alkaline and acid conditions) are generally not suitable for most bacteria, but there are specialized bacteria that are active in soils with pH levels less than 3 and others above pH 10.

Actinomycetes physiologically resemble bacteria; however, their slender branched filaments also resemble filamentous fungi. In most soils, bacteria and fungi are more numerous than actinomycetes (Table 3.5), but in some warm climate soils, actinomycete biomass may exceed that of others. While actinomycetes may not be as important as bacteria and fungi for carrying out biochemical processes, they are known to be involved in the decomposition of some resistant components of plant and animal tissues (e.g., cellulose, chitin, and phospholipids), synthesis of humic-like substances from the conversion of plant remains and leaf litter (i.e., green manures, compost piles, and animal manures), infection of plants, animals, and humans, and excretion of antibiotics or production of enzymes that can influence soil community composition. The "earthy" aroma of soils is largely caused by a volatile compound produced by actinomycetes.

Fungi obtain their energy and C from the decomposition of plant and animal remains, or soil organic matter. Two groups of fungi common to soils are the unicellular organisms called yeasts and the multicellular filamentous organisms, such as molds, mildews, smuts, and rusts, which are the most prevalent. Filamentous fungi are abundant in well-aerated fertile soils, whereas yeast inhabit anaerobic environments. The filamentous bodies, which individually are called *hyphae* and collectively *mycelia*, are found interwoven among soil particles, organic matter, and plant roots. Mushrooms are the true reproductive structures of a filamentous fungi that may have extensive hyphae below ground. Fungi also perform several functions in soil, including decomposing plant and animal organic substances, binding of soil particles into aggregates, forming symbiotic (mycorrhizal) associations with plants, and acting as a predator in controlling certain microorganisms and soil animals. In acid surface layers and forest soils, fungi make up the majority of the soil biomass and are most active in organic matter decomposition.

Algae, like vascular plants and some bacteria, are capable of performing photosynthesis because they contain chlorophyll. In soil with moist and fertile conditions, algae can potentially produce several hundred kilograms of organic substances per hectare on an annual basis. Blue-green algae are capable of fixing atmospheric N_2 into organic N compounds. Lichens, which are a symbiotic association of an algae and a fungus, are commonly the first colonizers of bare rock and soil parent material. Organic acids synthesized and released by lichens are known to weather rock surfaces, which plays an important role in the early stages of soil development in some areas.

Protozoa are unicellular animals that are primarily microscopic in size. Some protozoa, however, are known to reach macroscopic dimensions. They feed on decomposing organic matter and organisms such as bacteria and sometimes other protozoa. Protozoa populations are often related to bacteria numbers; as bacteria increase following the addition of organic residues to soils, protozoa increase as well.

3.2.4   Soil development

Soil is a natural, three-dimensional array of vertically differentiated material at the surface of the earth's crust. The variation in soils throughout the world is a function of five soil-forming factors (Jenny, 1941), which can be expressed as:

$$Soil = f(pm, r, cl, o, t) \tag{3.3}$$

with *pm, r, cl, o*, and *t* representing parent material, relief (topography), climate, organisms, and time, respectively. Equation 3.3 indicates that the formation of a particular soil is determined by the amount of time a parent material located on a specific landscape has been affected by climate and organisms. *Soil genesis* is the process in which soil develops from parent materials and includes the physical and chemical weathering of parent material particles, physical movement of the particles, mineral alterations and transformations, addition of organic matter, and formation of horizons. Soils vary in horizon type and thickness, texture, structure, color, and other characteristics.

Parent materials are mineral and organic materials that are chemically, physically, and biologically weathered to form soils. They are classified as either residual or transported materials. Residual parent materials are formed from the weathering of confined rocks and consist of igneous, sedimentary, and metamorphic rocks that vary in hardness, color, mineralogy, particle size, and crystallinity. Transported parent materials have been carried by wind or water; examples of these materials include eolian deposits of windblown sand, silt, and clay. Rapidly moving water or slowly moving glaciers are both capable of transporting large amounts of materials. Water-transported materials are generally sorted by particle size, whereas glacial deposits, known as till, are often a mixture of particle sizes.

Relief, also called topography, refers to the angle and length of the slope and proximity of the surface to the water table. The amount of water that infiltrates or runs off a soil is often determined by the position of the soil in the landscape. The greater the angle and the longer the slope, the higher the probability for erosion. In addition, the direction of the slope (slope aspect) determines the amount of solar radiation received, which influences soil and air temperatures. In midlatitudes where the effects of slope aspect are greatest, the slope aspect often determines the type of vegetation growing on south- and north-facing sides of mountains. In the Rocky Mountains, forests can often be observed on north-facing mountain slopes due to cooler temperatures and higher soil moisture levels, whereas rangeland containing sagebrush typically dominates south-facing slopes.

In discussing *climate* effects on soil formation, we are primarily concerned with the influence of precipitation and temperature on weathering and degradation rates, and translocation processes. The form (rain and snow) and amount of precipitation determines how much water may enter a soil; however, the topography of an area can cause water to concentrate on portions of the landscape. Also, water may erode the soil surface in which case soil degradation occurs. Weathering and degradation rates are generally faster at higher temperatures, so a combination of increased precipitation and warmer temperatures will enhance soil formation and organic matter decay. Snowmelt can also increase the leaching of dissolved inorganic and organic chemicals that have built up in the soil profile over the course of the winter.

Plants, animals, and microorganisms are considered part of the *organism* soil-forming factor. Plants cycle nutrients taken from the atmosphere and soil, converting them to plant tissue, which, after dying, is decomposed by soil microorganisms and recycled back to the soil. Organic matter is derived from both plant decomposition and synthesis reactions by soil microorganisms and plays an important role in weathering, degradation, and translocation processes. Some soil organic compounds can influence pH and movement of dissolved ions to lower depths. Respiration of carbon dioxide (CO_2) by plant roots and microorganisms can also increase acidity (e.g., carbonic acid (H_2CO_3) formation) resulting in lower soil pH, which occurs predominantly in areas where intense respiration occurs. Physical mixing of soil particles and organic matter by soil animals can either enhance or decrease soil formation. Ants are capable of bringing coarse-textured material up from below the soil surface to build their mounds. Worms are excellent soil-forming animals known for their ability to enhance soil fertility and physical properties.

Soils are often described based on their morphological characteristics. Soils contain layers, approximately parallel to the land surface, called *horizons* that are distinguished from one another due to differences in texture, color, structure, or other chemical and/or physical properties that set it apart from the other horizons. Horizons can be formed from the loss (*eluviation*) of material such as organic matter, clay, or iron (Fe), and Al, or from the accumulation (*illuviation*) of these constituents. Horizons are described, first, by a *master horizon*

Table 3.6 Master Soil Horizons and Subordinate Designations

Properties or characteristics

Master horizons

O	Surface horizon dominated by organic matter
A	Mineral horizon with organic matter accumulation at the surface or just below the O
E	Horizon leached of organic mater, clay, and sesquioxides
B	Horizon with an accumulation of organic matter, clay, sesquioxides, carbonates, gypsum, and/or silica formed below the O, A or E
C	Unconsolidated material underlying A or B relatively unaffected by pedogenesis
R	Consolidated rock having little or no evidence of weathering

Subordinate designations

a	Highly decomposed organic matter
b	Buried soil horizon
c	Concretions or nodules
d	High-density unconsolidated material
e	Partially decomposed organic matter
f	Frozen soil (permafrost)
g	Gleying (high water table)
h	Illuvial accumulation of organic matter
i	Relatively undecomposed organic matter
k	Accumulation of carbonates
m	Cementation
n	Accumulation of sodium
o	Accumulation of Fe and Al oxides
p	Plowed or disturbed
q	Accumulation of silica
r	Weathered or soft bedrock
s	Accumulation of sesquioxides
ss	Presence of slickensides
t	Accumulation of silicate clays
v	Plinthite (red-colored Fe material)
w	Weakly developed (color or structure)
x	Fragipan (high density)
y	Accumulation of gypsum
z	Accumulation of salts

Source: Munn and Vance (1998). (Unpublished source.)

designation using the capital letters O, A, E, B, C, or R and, second, according to the soil horizon characteristics, by a *subordinate designation* (Table 3.6). Surface horizons are classified as either O or A horizons depending on their organic matter content. In highly eroded areas where soil surface materials have been removed, E, B, C, or R horizons may be exposed. Soils with E horizons are very common in forest soils. The E horizon is usually located below the O or A horizon, and forms as a result of the loss of considerable amounts of organic matter, clay, or Fe and Al compounds (sesquioxides). An E horizon is generally lighter in color than the horizons above and below. Horizons that have accumulated constituents translocated from the horizon above are called B horizons. Parent material, the geologic material from which the soil has formed in, is designated as a C horizon. Unconsolidated rock is classified as an R horizon. In addition to the five master horizons described above, examination of some soils indicate there are some horizons that appear to have characteristics similar to the horizons above and below. These are called *transitional horizons* and are designated by two capital letters such as AB or E/B.

Table 3.7 Diagnostic Horizons, Designations, and Their Properties;
Diagnostic Horizons Are Used to Classify Soils at High Levels of Soil Taxonomy

Diagnostic horizon	Designations	Properties
		Epipedons
Ochric	A	Light-colored, low organic matter that does not meet criteria of other epipedons
Mollic	A	Thick, dark-colored, well-structured with high base saturation >50%
Umbric	A	Similar to mollic but with low base saturation <50%
Melanic	A	Like a mollic except developed on volcanic tephra
Anthropic	A	Anthropogenic mollic-like horizon high in P
Plaggen	A	Anthropogenic thick horizon (>50 cm) developed from long-term manure applications
Histic	O	Organic horizon formed in poorly drained areas
		Subsurface horizons
Argillic	Bt	Accumulation of silicate clays as evidenced by clay films
Agric	A or B	Accumulation of organic matter, clay, or silt below the plow layer
Albic	E	Light-colored exuvial horizon which has lost organic matter and clays
Calcic	Bk	Accumulation of calcite ($CaCO_3$) or dolomite ($CaCO_3 \cdot MgCO_3$)
Cambic	Bw	Weakly developed horizon
Duripan	Bm	Hardpan due to cementing with silica
Fragipan	Bx	Weakly cemented, brittle layer
Gypsic	By	Accumulation of gypsum
Natric	Btn	Argillic horizon high in sodium having columnar or prismatic structure
Oxic	Bo	Highly weathered horizon with accumulation of Fe and Al oxides and nonexpanding silicate clays
Petrocalcic	Bk	Cemented calcic horizon
Petrogypsic	By	Cemented gypsic horizon
Placic	B	Thin cemented plan that is black to red and held together by Fe, Mn, and/or organic matter
Salic	Bz	Accumulation of salts
Sombic	B	Accumulation of organic matter low in base saturation
Spodic	Bh, Bhs, Bs	Accumulation of organic matter and sesquioxides
Sulfuric	B	Highly acid soil with sulfur-containing mottles

Source: Munn and Vance (1998). (Unpublished source.)

Horizons are further subdivided to differentiate the major properties or characteristics that set them apart from other horizons. The use of the subordinated designation allows us to distinguish, for example, a B horizon that has accumulated clay material (Bt) from a B horizon that is only weakly developed (Bw) (Table 3.7). Not all subordinate designations are used with each of the master horizons. When describing the O horizon, one should always determine the decomposition state of the organic matter to classify the horizon correctly as an Oa, Oe, or Oi, which represent sapric, hemic, and fibric materials, respectively. These three subordinate designations are reserved solely for organic horizons. The subordinate designation p (plow layer) is only used with the master horizon letter A. Most of the subordinate designations are used to classify B horizons further. It is relatively easy to determine which of the lowercase letters are used for B and sometimes C horizons since they represent processes of accumulation or changes from the geologic parent material of the soil.

3.2.5 Soil classification and land use

Several soil classification systems are used throughout the world. The one used in the U.S., and several other countries, is *Soil Taxonomy*, which was developed by the U.S. Department of Agriculture (Soil Survey Staff, 1975; 1992). Soil Taxonomy was designed to group soils according to morphological characteristics and environmental properties such as diagnostic horizons, soil texture, soil structure, soil color, soil mineralogy, soil moisture regime, and soil temperature regime. It is a hierarchical system that contains six categories of classification ranging from a general grouping of all soils to a more specific, highly detailed grouping. The six categories, listed according to increasing detail, are *order, suborder, great group, subgroup, family,* and *series*.

The initial step in classifying soils is to determine the diagnostic horizons. Diagnostic horizons are of two groups: *epipedons* — those that form at the soil surface — and *subsurface diagnostic horizons* — those that form below the soil surface (see Table 3.7 for horizon designations and properties). Of the seven epipedon horizons listed in Table 3.7, only one, the histic epipedon, which is designated as an O horizon, is composed primarily of organic matter. Of the 17 subsurface diagnostic horizons, the argillic, albic, cambic, and spodic horizons are common to soils in humid regions of the U.S. and Canada. Soils in arid parts of the U.S. often have one of the following diagnostic subsurface horizons: calcic, gypsic, natric, or petrocalcic. Knowledge of diagnostic horizons is essential for classifying soils.

Soils with similar types of diagnostic horizons are classified into 12 soil orders (Table 3.8). Knowing the soil order provides a general picture of what morphological and possible chemical and physical properties a soil possesses. The more detailed categories in Soil Taxonomy provide more specific information about the soil. The family name of a soil is the most useful for direct interpretation of the properties of the soil and the environment where it is located. Soil series are subdivisions of families, and are distinguished based on specific profile characteristics.

Table 3.8 Twelve Soil Orders and Their Diagnostic Features According to Soil Taxonomy

Soil order	Common diagnostic horizon	Diagnostic features
Alfisol	Argillic/Natric	Base saturation >50%; no mollic, oxic, or spodic
Andisol	Melanic/Ochric with andic properties	Volcanic ash–derived soil; noncrystalline or poorly crystalline minerals
Aridisol	Ochric plus Calcic	Dry soil; argillic, cambic, or natric common
Entisol	Ochric no diagnostic subsurface horizons	Nominal profile development
Gelisols	Histic/Ochric with permafrost	Must have permafrost within upper 100 cm
Histosol	Histic	Organic soil; peat or bog
Inceptisol	Cambic	Few diagnostic features; ochric or umbric horizons common
Mollisol	Mollic	Dark soil with high base saturation; no oxic or spodic
Oxisol	Oxic	Highly weathered; no argillic or spodic
Spodosol	Spodic	Accumulation of organic matter and sesquioxides; albic horizon common
Ultisol	Argillic	Base saturation <50%; no oxic or spodic
Vertisol	Ochric/mollic	Contains high amounts of swelling clay causing deep cracks when dry

Source: Munn and Vance (1998). (Unpublished source.)

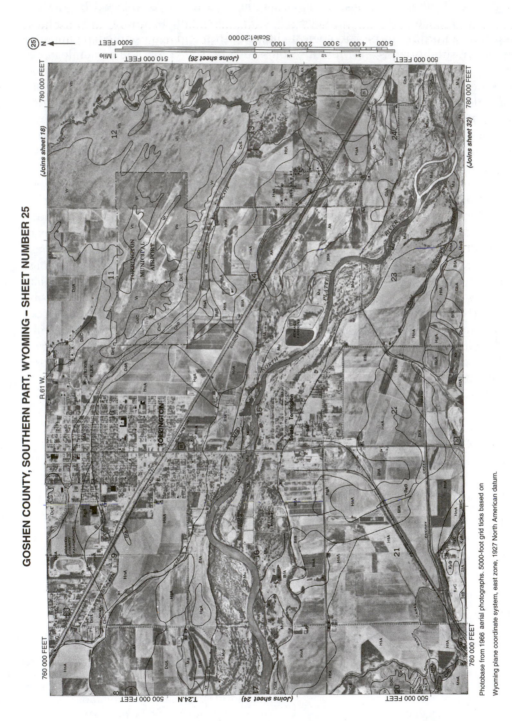

Figure 3.9 Example of a map sheet taken from the Soil Survey of Goshen County, Wyoming. The mapping units listed on the map represent the type of soil association, which is a landscape that has a distinctive pattern of soils, and slope of the area. For example, the city of Torrington is located primarily on map unit HnA, which represents Haverson and McCook loams on 0 to 3% slopes. (From Soil Survey of Goshen County, Wyoming, Southern Part by F. Stephens, E.F. Brunkow, C.J. Fox, H.B. Ravenholt,

Soil survey reports (Figure 3.9) are valuable sources of information for land-use planning. They contain information that can help make land-use decisions for agricultural and nonagricultural purposes. In addition to delineating soil by map units, soil surveys contain detailed information for land-use decision making regarding the management of soils, croplands, woodlands, recreation areas, and wetlands. Soil surveys are also useful for evaluating soils for their suitability for irrigation and their drainage potential. Limitations for various land uses are listed in interpretive tables. These tables, along with the soil description, are a necessity for initial planning of waste disposal sites such as for septic tanks, sewage lagoons, and sanitary landfills.

Geographical Information Systems (GIS), which have recently been broadened to include geographic information science, have been applied to land management decisions by using the data layers and database information to query for specific characteristics and properties that may be beneficial or detrimental to a particular land use. A GIS can be used to compile, store, and retrieve spatially derived information that can be used to identify or explore specific environments or relationships within a geographic region, analyze data for decision making, provide data for application-specific models, and produce visual and numerical outputs that can be utilized for planning efforts. Data layers that can be developed and retrieved for future use include soils, vegetation, climatic conditions (e.g., precipitation and temperature parameters), water bodies, land-use type, topography (e.g., digital elevation models, DEMs), and others that have spatial coordinates. Land-use planning and precision farming are two areas that utilize GIS. Precision farming also relies on global positioning systems (GPS) to determine the exact location within a field for quantitative applications of such materials as fertilizers, liming products, and possibly organic wastes.

Environmental quality issues/events
Using a GIS to identify areas for alternative crop production

The Bighorn Basin in Wyoming is an important agricultural region that relies heavily on irrigation for the cultivation of sugar beet, barley, alfalfa, and dry bean crops. To diversify crop production in the basin, a GIS was used to develop a soils layer for the region, create continuous data layers of several climatic variables, catalog crop growth parameters for different agricultural crops, and combine growth parameters with environmental data to display areas for possible alternative crop production. Environmental traits of the basin (e.g., soils, temperature minimums and maximums, frost-free period, growing degree-days, etc.) will dictate where various crops can be produced. Both ARC/INFO (version 7.1.2) and ArcView (version 3.1) (ESRI, Redlands, CA) were used for GIS evaluation of 28 potential alternative crops. Maps developed in this project provide basin producers with potential alternatives to current crop production practices.

Because published soil survey data were not available for a majority of the basin, a GIS-derived soils map was generated using regional bedrock geology, surficial geology, and elevation properties. The available soil series were related to bedrock and surficial geology combinations for predicting soils for the rest of the basin. As these soil series did not account for the entire region, other geologic combinations were applied to the soil model utilizing GIS decision rules. Based on this method, every bedrock/surficial geology combination was assigned a soil family classification and a representative soil series was assigned to each family from series currently mapped in Wyoming.

The GIS-derived soil map included 19 different soil associations, with a large proportion of the land classified as Torrifluvents. Most of the irrigated agriculture in the Bighorn Basin is located in Torrifluvents. Soils in the area that are not suitable for crop production include those that are frigid, have high shrink–swell tendencies, possess too low a water-holding capacity, or are shallow to bedrock. The remaining soils are suitable for crop production (soils that are deep or have a texture conducive to plant growth) and represent 62% of the study area. The lowland areas comprise soils typically under agricultural production, while higher elevation soils are generally used for livestock grazing and timber harvesting. Creation of the soil data layer allowed for the use of textural classes in an alternative crop analysis.

The 18 weather stations existing in the basin have collected weather data for 30 years or longer. The climatic parameters recorded at these sites were extrapolated to predict conditions throughout the area. Using geostatistical techniques, the weather station locations and attributes were used to create continuous weather patterns for the basin. Sample variograms were created for each of the climatic variables to determine the model that could optimize the data. Model parameters such as nugget, sill, and range were estimated, and cross validation was performed to check the accuracy of the model selected. Once the correct model and parameters were chosen, kriging was performed to estimate the unsampled points. The climatic data were interpolated using the spline method to create 31 continuous GIS data layers that represented the Bighorn Basin climatic conditions. Examples included growing degree-days, June maximum temperature, and 90% chance of frost-free period. Climatic variable contour maps were related to basin topography; i.e., with an increase in elevation there were decreases in frost-free period, growing degree-days, and temperature, but there was an increase in precipitation. Basin elevation gradually increases from east to west.

General growth requirements for the different alternative crops were compiled and cataloged in a GIS format. Because specific crop variety information was unavailable, the growth parameters for each crop were a broad categorization of what each species would require for growth and development. Once the data were compiled into the GIS, analyses were performed using the GIS map algebra protocol. The parameters queried included such items as soil texture, length of frost-free season, maximum July temperature, minimum September temperature, and growing degree-days. Areas with the desired crop growth parameter combinations for each crop were identified and displayed as maps illustrating potential areas for alternative crop production.

Examples of areas suitable for the production of amaranth, buckwheat, canola, and faba bean are shown in Figure 3.10. Eight of the alternative crops modeled (amaranth, chickpea, onion, quinoa, safflower, sesame, sorghum, and sunflower) may potentially be cultivated in the eastern part of the basin. This is the warmest area and has the longest frost-free period. These crops require more heat to develop than the cool-season crops. Broccoli, buckwheat, canola, carrot, cauliflower, crambe, Kentucky bluegrass turf seed, leek, lentil, lettuce, mint, and radish are cool-season crops that prefer less heat during the growing period. Because of this preference for cooler temperatures, production of these crops may be possible in the western portion of the basin. Finally, eight of the crops investigated (asparagus, beet, cabbage, cowpea, field pea, faba bean, medic seed, and tall fescue turf seed) can potentially be cultivated throughout the area on suitable soils. This is due to the broad temperature range to which these crops are adapted.

GIS provides an excellent means for exploring the possibilities of cultivating new or alternative crops. The process of locating areas suitable for growth is rapid, once the

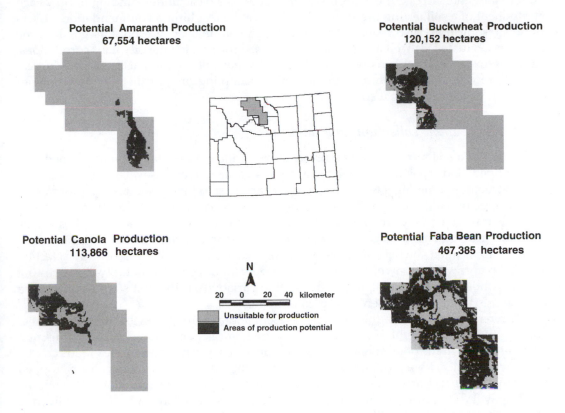

Figure 3.10   Potential production areas for the introduction of alternative crops, amaranth, buckwheat, canola, and faba bean, in the Bighorn Basin, Wyoming.

necessary information is entered into the GIS database. Crop diversification is important for environmental and economic reasons. However, before attempting the cultivation of alternative crops, additional parameters relating to the exact variety to be grown should be thoroughly investigated.

— *Source:* Young et al. (in press)

3.3   *Environmental testing practices for soils*

Soil testing has been defined as "rapid, chemical analyses to assess the plant-available nutrient status, salinity and elemental toxicity of a soil" (Peck and Soltanpour, 1990). Soil testing includes not only the actual analyses of soils but also the interpretation of the results and the development of recommendations for an increasingly diverse variety of land management programs including, but not limited to, production agriculture. In practice, soil testing is probably one of the most successful applications of the basic principles of soil science to all forms of land management.

A comprehensive soil testing program has four basic components: (1) soil sample collection and handling; (2) soil chemical, physical, or biological analysis; (3) interpretation of the analytical results of the soil test; and (4) development of economically and environmentally sound recommendations for lime, fertilizers, and other soil amendments.

Acceptance and successful use of soil testing requires a basic understanding of how each component of a soil testing program can influence the resulting recommendations. Users of soil testing must also realize that results of a soil analysis are only as good as the samples evaluated, as well as the information used for predicting soil–plant correlations. In many instances, problems with soil testing are not due to analytical errors or unreasonable recommendations but to inappropriate sampling or unrealistic expectations of what can be learned from a soil test.

3.3.1   Soil sample collection and handling

Collection of a soil sample that is representative of the entire area of interest, whether it is a cropped field, a pasture, rangeland, a lawn or garden, or even severely disturbed soils at construction or mining sites, is the most important step in any soil testing program. A high degree of natural variability in soil chemical, physical, and biological properties can exist even within a very small area and create differences that may have marked effects on plant growth. Soil management practices, such as the method of fertilizer or animal manure application, the type of tillage operations used, and even the year-to-year selection of what plants will be grown, can further contribute to the spatial variability in soil testing results. Understanding the nature of this spatial variability is the first step in collecting a soil sample that truly represents the area of interest.

A careful assessment of the soil types present and a thorough review of past management activities that might affect the soil properties measured in a soil test are thus prerequisite steps in the determination of the number of separate soil samples required at a site. Areas that are very different in soil type, which have been managed or cropped differently, or that clearly have some type of problem with plant growth should be sampled separately. While collection of multiple soil samples is time-consuming and costly, it is the best way to estimate nutrient needs accurately for a large area containing different soils or for areas that will be subdivided and used for different purposes.

The proper time of year to collect a soil sample and the frequency of soil sampling should also be considered carefully. Soil samples can be collected at any time of the year that the ground is not frozen. Routine soil tests for lime and fertilizer recommendations should be performed from 3 to 6 months prior to planting. This usually provides sufficient time for management decisions based on the results of the soil test to be made and implemented in a timely manner. For example, if a soil test recommends that limestone is needed to correct a problem with soil acidity, it is important to know this several months in advance of planting because of the time required for the limestone to react in the soil and raise the soil pH to the desired value for the plants to be grown. Soil testing well in advance of planting also allows time to change the type of plants to be grown if soil test results indicate that growing conditions are inappropriate for the plant(s) initially selected for the site. The frequency of soil testing varies somewhat with the plants to be grown and the nutrient management practices used. Ideally, soils should be sampled at the same time of year (e.g., spring or fall) and at 2- to 3-year intervals. Sampling at the same time of year minimizes the effect of natural seasonal variations on soil pH and plant nutrients.

Once the areas to be sampled have been identified, the next step in soil sampling is to determine the appropriate depth to sample. This decision depends on the type of soil test to be performed. Soil samples for routine soil tests of plant nutrient availability and lime requirement are usually obtained from the "topsoil," typically the top 15 to 20 cm of the soil. Some important exceptions to this general rule are soil nitrate tests (see Chapter 4), soil samples to estimate the effect of pH on herbicide activity in no-till cropping systems (0 to 2 cm), samples from pastures (0 to 10 cm), and, in certain situations, samples from subsoil horizons. When collecting subsoil samples, care should be taken to minimize contamination of the subsoil sample by topsoil. Such contamination can seriously influence the results obtained. In all cases it is important to ensure that the sample is collected at the appropriate depth. Sampling too deep or too shallow may lead to inappropriate conclusions about the distribution of a nutrient or nonessential element in the soil profile, and lead to inaccurate soil management recommendations.

3.3.2   *Methods of soil analysis*

Soil analysis in the laboratory is the most accurate and reproducible step in soil testing. Standard techniques for sample handling, preparation, and processing, as well as a regular program of quality control, ensure that instruments are running properly and that analyses are performed reliably each time. Numerous publications describe the analytical methods used by soil testing laboratories in detail (see supplementary reading section of this chapter). When soil samples are received by the soil testing laboratory, they are usually dried, ground, thoroughly mixed, and sieved to remove any particles larger than 2 mm in diameter. A typical "routine" soil test is then conducted, which usually includes soil water pH, lime requirement, organic matter content, and "available" soil nutrients, e.g., P, K, Ca, Mg, manganese (Mn), and zinc (Zn), as determined by a chemical extraction process.

Soil test "extractants" are typically dilute solutions or mixtures of acids, bases, salts, and chelates. Biological techniques (e.g., bioassays based on microbial growth) have been used but are generally too expensive and time-consuming for routine use. The most effective concentrations and relative proportions of the reagents in a chemical extracting solution were usually determined empirically by comparison of the amount of an element extracted from the soil with some type of biological response (e.g., yield) by a target organism (e.g., plants). The nature of the chemical solutions used in soil testing is illustrated by comparing the Mehlich 1 soil test, used in the southeastern and mid-Atlantic U.S., a dilute mixture of two strong acids ($0.05\ M$ HCl + $0.0125\ M$ H_2SO_4), the AB-DTPA soil test, used in the western U.S., a dilute combination of a base ($1\ M$ NH_4HCO_3) and a chelating agent ($0.005\ M$ DTPA), and the Bray P_1 soil test, used in the midwestern U.S., a dilute mixture of a strong acid ($0.025\ M$ HCl) and a complexing ion ($0.03\ M$ NH_4F). All tests are well accepted but their use should be confined to the region where research has shown them to be accurate.

Soil testing extractants developed for plant nutrients have also been used, with some success, to assess the risk of plant uptake of nonessential elements. Other soil testing methods are also used for these elements, although not to measure biological availability. Examples include measuring total sorbed metals to monitor the accumulation of elements in the soil up to some defined regulatory limit and use of the toxicity characteristic leaching procedure (TCLP), sometimes used to determine if a soil is sufficiently polluted with an element or organic compound to be considered a hazardous waste. Recent advances in environmental soil testing for some metals (i.e., Cd, Pb, Zn) have used extracting solutions that simulate the biological activity within the human digestive system, referred to as the physiologically based extraction test (PBET).

Many other chemical, physical, and biological soil tests are available to characterize the suitability of soils for various land uses. Readers are referred to *Methods of Soil Analysis* (Sparks, 1996) and *Soil Sampling and Methods of Analysis* (Carter, 1993) for a thorough discussion of available soil testing methods and analytical methods that are widely used.

3.3.3   *Soil testing interpretations and recommendations*

Interpretation of soil testing results is a process by which the analytical results of the test(s) are related to soil management decisions. For example, with plant nutrients the goal is to identify, based on the concentration of the nutrient in the soil, the probability of a profitable plant response to additions of that nutrient in fertilizers or other soil amendments (e.g., lime, animal manures, municipal biosolids, composts). Soil test categories (*Low, Medium, Optimum, High, Very High, Excessive*) relate the quantitative results of a soil test to the likelihood of a crop yield response to fertilization or manuring. Examples of the definitions of these soil test categories follow:

- *Low:* The nutrient concentration in the soil is inadequate for the growth of most plants and will very likely limit plant growth and yield. There is a high probability of a favorable economic response to additions of the nutrient.
- *Medium:* The nutrient concentration in the soil may be adequate for plant growth, but should be increased into the optimum range to ensure that plant growth and yield are not limited. There is a low to moderate probability of a favorable economic response to additions of the nutrient.
- *Optimum:* The nutrient concentration in the soil is in the range recommended for the growth of all plants. Since there is a very low probability of a favorable economic response, nutrient additions are rarely recommended.
- *Excessive:* The nutrient concentration in the soil is above the range recommended for the growth of all plants. Additions of the nutrient will be unprofitable, may have undesirable effects on plant growth, and are not recommended. Erosion, runoff, and leaching from soils that are excessive in P can have negative effects on surface water quality (see Chapter 5).

The final step in a soil testing program is the development of a site-specific recommendation. Plant nutrient recommendations are based not only on the soil test value for a particular nutrient but on other pertinent factors as well. Among these are the type of plant to be grown and the realistic yield goal for that plant in a given soil, prior soil management and cropping practices (e.g., use of animal manures, municipal biosolids, composts, legumes, liming materials), the nutrient source and application technique to be used, and the potential for nutrient accumulations or losses that may create environmental problems. Most soil testing recommendations for plant nutrients follow a *sufficiency level* philosophy. This approach is based on the well-documented fact that a measurable "critical" soil test level exists below which responses to nutrient additions are probable and above which they are not. Soil test calibration research is used to develop the quantitative relationship between soil test values for a nutrient and the probability of a profitable crop response to additions of that nutrient. Decades of practical experience and applied research have shown that soil test–based recommendations are the most economically and environmentally efficient means to identify the need for supplemental plant nutrients. However, the individual submitting a soil sample needs to be familiar with the soil management process. Slight adjustments in a soil test–based recommendation in accord with the experience of the farmer, home owner, Cooperative Extension agent, or crop consultant may be necessary and useful.

3.3.4   Environmental soil testing

As discussed above, soil testing has traditionally been used to evaluate the soil limitations to agronomic crop performance imposed by nutrient deficiencies, pH, soluble salts, etc., and to guide the recommendation process so that these limitations could be eliminated economically and without impacting the quality of our environment. In recent years, interest has developed in the area of *environmental soil testing*, defined here as "quantitative analysis of soils to determine if environmentally unacceptable levels of nutrients, nonessential elements or organic compounds are present." Environmental soil testing is a much more ambiguous process than agricultural soil testing because it is usually quite difficult to quantify the meaning of the term *environmentally unacceptable*. Absent a clear, quantitative measure of success, such as crop yield for agricultural soil testing, the entire process of soil testing, from sample collection to recommendation, becomes more diffuse and complex. Nevertheless, the rising interest in environmental protection in many areas of the world has prompted an increased effort to use soil testing to assess the risks posed by soils to other sectors of the environment, particularly groundwaters and surface waters.

In the broadest sense, the goals of environmental soil testing are the same as those of routine, agricultural soil testing — rapid, accurate, and reproducible soil analysis by the most appropriate methods — and a reasonable interpretation of results related to environmental risk. Some factors to consider when developing an environmental soil testing program are discussed next, using two reasonably common examples, potentially toxic trace elements and plant nutrients that are known to degrade water quality. In both cases it is critical to consider the conceptual differences in interpretation of an environmental soil test, compared to an agricultural interpretation (Figure 3.11).

3.3.4.1 Soil testing for potentially toxic trace elements

Soil testing for potentially toxic trace elements is an environmental issue because some plant nutrients (Cu, Mo, and Zn) and nonessential elements (arsenic (As), Cd, chromium

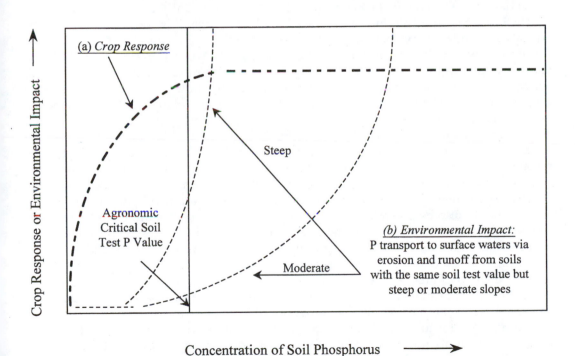

Figure 3.11   Illustration of the different nature of the relationships between the concentration of an element in the soil and (a) crop response to the element vs. (b) the potential for the element to impact the environment. In this example, the influence of soil P on water quality is affected by both the concentration in the soil and topography (slope) which influences soil, water, and P loss by erosion and runoff. (Adapted from Sims, J. T., Soil fertility evaluation, in *Handbook of Soil Science*, M. E. Sumner, Ed., CRC Press, Boca Raton, FL, 1999. With permission.)

(Cr), mercury (Hg), nickel (Ni), Pb, selenium (Se)) have been shown to be toxic to either plants, animals, or humans. Soils may naturally contain high concentrations of one or more of these elements (very unusual), or concentrations may increase as a result of some anthropogenic activity such as the intentional addition of wastes as beneficial soil amendments (e.g., animal manures, municipal biosolids, industrial by-products). Some soils may be highly polluted with toxic elements due to mismanagement of potentially beneficial wastes, by an accidental spill or discharge, or as a result of an industrial activity such as mining or smelting. Note that contamination and pollution are not the same. Contamination occurs when a substance is present at concentrations higher than would occur naturally but no adverse effect on an organism is apparent while pollution implies not only an elevated concentration in the soil, but clearly documented adverse effects on some organism (see Chapter 1).

Soil testing for potentially toxic trace elements begins with an understanding of the nature of the risk involved — what organisms may be affected, by what pathway, and by which elements. Primary areas of general concern are direct soil ingestion, phytotoxicity, plant uptake and food chain contamination, and water pollution from erosion, runoff, or leaching (see Chapter 12 for a discussion of risk assessment). Direct soil ingestion is normally only a concern for Pb and with young children who are most likely to ingest soil or inhale dust from high-Pb soils. Phytotoxicity is rarely an issue with Pb except in highly polluted soils at industrial sites, but it is a greater concern with Cu, Cr, Ni, and Zn. Food chain contamination and human health effects are most often associated with Cd and Hg and water quality concerns with As, Cu, Hg, and Se. Knowledge of the nature of the risk and the element of concern helps to determine the most effective soil sampling protocol. In the case of Pb in urban soils where human health is the concern it may be advisable to collect very shallow soil samples (<2 cm) since this is the soil depth most likely to be ingested. A similar depth would be useful if runoff or erosion is the issue because rainfall primarily interacts with only the uppermost few centimeters of the soil surface. However, if concerns exist about groundwater pollution, sampling into subsoil horizons, perhaps to a depth of 1 to 2 m, would be recommended to determine if elemental leaching has occured. For elements where the main concern is phytotoxicity and food chain contamination, the normal sampling depths associated with the crops to be grown are usually acceptable (e.g., 0 to 20 cm). If remediation of the site is the goal, either by soil removal, soil washing, or soil amendment, then systematic deep sampling (e.g., 0 to 5, 5 to 10, 10 to 20, 20 to 40, 40 to 60, 60 to 100 cm) is recommended to determine the depth of soil contamination and thus the extent of remediation required. Careful consideration should also be given to the soil sample handling and preparation steps to avoid contamination from any sampling and mixing tools or grinding and sieving devices (e.g., stainless steel, used in many electric grinding device contains Cr and Ni) and to protect the health and safety of the individual taking the sample.

The method of analysis to be used varies with the intent of the test. If the goal is to assess biological availability (e.g., plant uptake), then many dilute acid or chelate-based soil test extractants may be suitable, providing due consideration is given to the most appropriate test for the intended land use. However, if the goal is to quantify the extent of accumulation of an element, relative to normal soils or natural background levels, or to monitor this accumulation over time, methods that determine or approximate total elemental content are recommended. One example is EPA Method 3050, which successively digests a soil sample with concentrated nitric (HNO_3), and hydrochloric (HCL) acids and hydrogen peroxide (H_2O_2) to measure "total sorbed metals" by acidic dissolution of clays, oxides, and carbonates and oxidation of organic matter; elements associated with silicates are not dissolved. Therefore, for a true measure of total elemental content of a soil sample, complete digestion of the soil with strong acids (e.g., HNO_3-$HClO_4$ for Cd,

Hg, and Pb), by carbonate fusion (Cr, Ni), or by alkaline oxidation techniques (As, Se) is required. Given the costs and difficulty of measuring total elemental content of trace elements in soils, it is often advisable to use rapid soil testing methods as surrogate monitoring techniques.

Interpretations and recommendations for soil tests for potentially toxic trace elements are considerably more difficult than for agricultural systems, and are often very site and element specific. As mentioned above, the main difficulty lies with soil test calibration — establishing a quantitative relationship between an agreed-upon measure of environmental risk and the amount of an element measured by the soil test. In most cases this has been done by the use of complex risk assessment models that first identify the "target organism" of concern (e.g., human vs. plant) and then evaluate all possible pathways by which the target organism may be exposed to the risk factor (the toxic trace element). If possible, a "most sensitive pathway" is identified, defined as the lowest soil concentration at which an adverse effect on the organism would be likely to occur (e.g., soil ingestion vs. consumption of contaminated drinking water). Regulatory upper limits may then be established for that pathway, which can be monitored by the soil testing methods described above. This approach was used by EPA in the formulation of the national rule for the disposal and utilization of municipal biosolids, which established regulatory limits for the total amount of several trace elements that could be applied to agricultural soils (see Chapters 9 and 12).

3.3.4.2   Soil testing for plant nutrients with water quality impacts

Soil testing for plant nutrients that can degrade water quality is focused primarily on N and P because of their well-documented effects on groundwater (N) and surface (N, P) water quality. The principles, practices, and problems of soil N management and N testing are discussed in Chapter 4. Other than the methods described in that section (e.g., the pre-side-dress soil nitrate test (PSNT), which focus on identifying sites with an adequate N supply and thus avoiding unnecessary applications of fertilizers or organic sources of N (animal manures, municipal biosolids), there are no other approaches currently available for use in an environmental soil testing program for N. Environmental soil testing for P, however, is a considerably different matter. Growing international concerns about the role of P in the eutrophication of surface waters (see Chapter 2) have stimulated a large research effort on environmental soil testing for P (see Chapter 4). The focus of this effort has been the use of soil testing alone, or as a component of indexes and models, to identify soils that are most likely to be significant nonpoint-sources of P pollution of streams, rivers, ponds, lakes, and bays.

One approach proposed for environmental soil testing for P is simply to establish an upper critical limit for soil P using currently available agronomic soil testing methods (e.g., Bray P_1, Mehlich 1, Olsen P). Soils that exceeded the upper critical limit for soil test P would no longer receive P inputs from any source (e.g., fertilizers, animal manures, municipal biosolids) and would be targeted as priority areas for soil and water conservation practices to prevent P loss in erosion, runoff, and leaching. Two reasons are usually given to justify the need for this upper critical limit. First is that a growing body of research shows that soils that are overfertilized with P relative to crop requirements will create an increased risk of nonpoint-source pollution of surface waters. Second is the concern that continuing to apply P to soils well beyond values that are needed for crop production contradicts the principles of ecological health and sustainable agriculture. The rationale underlying this second concern is the fact that P is obtained from a finite natural resource base, at a cost to society, and that agricultural practices that waste this resource are inconsistent with sustainability. Despite these concerns, there has been a reluctance to establish upper critical limits for soil test P presumably because

(1) agronomic soil tests were not designed or calibrated for environmental purposes and (2) there may be an unjustified reliance upon soil test P alone by environmental regulatory agencies attempting to control nonpoint-source pollution of surface waters, ignoring the complex interaction between soil P and the transport processes that move P from soil to water.

Other approaches to environmental soil testing for P are direct measurement of soluble or biologically available P, estimating potentially desorbable P by Fe oxide strips, and characterizing the degree of P saturation of soils. While research has been promising, to date none of these approaches has received widespread acceptance as a means to quantify the environmental risk associated with high P soils. Readers are referred to the supplemental reading section of this chapter for more detailed information on this topic.

3.3.5   Limitations of soil testing

Soil testing is a valuable tool in any land-use program. Soil testing is not, however, a crystal ball, nor is it capable of providing absolute answers to all soil management questions. Some factors that affect the successful use of recommendations based on soil tests include:

- A soil test is only as good as the soil sample provided to the laboratory. A poorly collected sample can result in a recommendation that is completely inaccurate or inappropriate for the desired use of that soil. Carefully collecting a representative soil sample is essential.
- Since soil testing methods can definitely influence the analytical results obtained and the resultant recommendations, it is important to use the methods developed for your state or region. Submitting samples to a laboratory that uses methods unsuitable for local soils can lead to inaccurate recommendations.
- Unexpected occurrences related to climate, management, or other uncontrollable factors can require adjustments in soil test recommendations. Most soil test recommendations are based on multiyear, multisite research with soils, plants, and climatic conditions typical to the region. Occasionally, however, conditions during a particular growing season may significantly impact the way that plants respond to soils and soil amendments. Plants may exhibit nutrient deficiency symptoms early in the year, but not later as soils warm and roots penetrate into subsoils. Insect or disease pressures or drought conditions can reduce yields and thus nutrient requirements. The use of in-season soil tests or plant analyses can help users of soil testing make needed adjustments in a soil test recommendation.

3.4   Plant and organic waste analysis

In addition to soil testing, several other diagnostic procedures can improve the efficiency of any land-use program. The two most common practices are *plant analysis* and *organic waste analysis*. Some of the key factors to consider when using plant and organic wastes tests are summarized below.

3.4.1   Principles of plant analysis

Plant analysis is defined as "…the determination of the elemental composition of plants or a portion of the plant for elements essential for growth. It can also include determining elements that are detrimental to growth of animals or humans through our food chain"

(Munson and Nelson, 1990). Accurate interpretation of the results of a plant analysis requires that calibration data exist that relates nutrient or nonessential concentrations in the plant to uptake of the element from the soil and/or growth response to additions of the element. The most common elements determined in a routine plant analysis are the essential plant nutrients: N, P, K, Ca, Mg, S, boron (B), chlorine (Cl), Cu, Fe, Mn, molybdenum (Mo), and Zn. In most cases, plant analysis results are compared to optimum nutrient ranges, with analytical values expressed in percentages (for major nutrients) and parts per million (mg/kg; for micronutrients and nonessential trace elements), on a dry weight basis. Optimum nutrient ranges for many plants, especially agronomic plants, have been established by years of research and are found in many publications (see supplemental reading section).

As with soil testing, the collection, handling, and analysis of a plant sample must be done properly for useful results to be obtained. The main factors to consider when initiating a plant analysis program are summarized in Table 3.9. Specific guidelines for collecting samples for plant analysis including the plant part and number of plants to sample, the appropriate plant growth stage to collect, and any specific handling requirements for the sample can be obtained from the laboratory that will perform the analysis.

Plant analysis is most frequently used to confirm suspected deficiency situations and determine appropriate corrective applications (e.g., fertilization). However, research has also shown that plant analysis techniques have the potential as monitoring tools to track in-season plant N needs (much like the PSNT test) or as post-harvest tools for evaluating the success of N management for the current growing season (see Chapter 4). The leaf chlorophyll meter, an instrument that measures the "greenness" of the leaf and hence plant N status, has shown promise for use in management systems for corn where fertigation is possible. Likewise, the stalk NO_3^- test in which a 15-cm section of the lower portion of a corn stalk is analyzed at physiologic maturity for NO_3-N, has been shown to be a good indicator of optimal or overapplication of N to corn.

3.4.2   Principles of organic waste analysis

The most common types of organic wastes analyzed are animal manures and municipal biosolids, composts, and wastewaters from processing plants. Animal manures and municipal biosolids are extremely variable in composition, and the handling and analytical procedures used during analysis can greatly affect the accuracy and interpretation of the results. It is, therefore, important to consider carefully the factors described in Table 3.10 to ensure that a representative sample is obtained and analyzed properly.

Interpretation of the results of an organic waste analysis is a complex process that must reflect the effects of the waste treatment process, storage, handling, application, and soil/crop management on the availability of nutrients and nonessential elements. It is beyond the scope of this book to cover all the details involved in the interpretation and use of organic waste analyses. In many cases comprehensive, specific analyses of municipal and industrial organic wastes are required of the waste generators (e.g., municipal water treatment plant) and/or waste users by state or federal regulatory agencies prior to land application. In most cases, land application of biosolids is based on applying the agronomically optimum rate of N for the crop to be grown. However, federal regulations also establish the maximum concentration of trace elements and polychlorinated biphenyls (PCBs) permitted in biosolids used in land application programs, as well as the maximum annual and cumulative loading rate of these elements (see Chapter 9). State regulations may be similar to the federal rules or more restrictive. Key aspects of interpretation of organic waste analyses are provided in Table 3.10. Readers are referred to the supplemental reading section of this chapter for more information.

Table 3.9 Typical Components of a Modern Plant Analysis Program

Component	Comments
Sample collection	Accurate interpretation of a plant sample can only be accomplished by analyzing the appropriate part of the plant, collected at the correct stage of growth. Soil testing and plant analysis laboratories provide detailed sampling information for most plants. If sampling information is not available, the rule of thumb when collecting a plant sample is to select upper, mature leaves. Avoid leaves damaged by insects or disease or contaminated with fertilizers, dusts, or sprays. If possible, provide samples from healthy and deficient plants for comparative purposes. Be sure that these samples were collected at the same time, from the same plant parts, or the comparison will not be valid. After collection samples should be decontaminated by gently washing with deionized water or, if severely contaminated, a dilute (~0.1%) detergent solution. If a detergent solution is used, it must be followed by a thorough rinsing with pure water. The element most affected by contamination from soil or dust is Fe, hence if this is the element believed to be deficient, extreme care must be taken to eliminate surface contamination prior to analysis.
Sample preparation	Once contaminants have been removed, plant samples should be dried quickly at about 65 to 80°C (150 to 175°F). If samples cannot be dried immediately, store them under refrigeration (2 to 5°C) to avoid molding or organic matter decomposition. Samples should be dried in paper, cotton, or plastic mesh bags that allow adequate air movement during drying. After the samples have been dried, it is necessary to grind them to reduce the particle size and homogenize the sample prior to analysis. This is usually accomplished with a stainless steel grinding mill equipped with a 20- or 40-mesh screen. Dried, ground plant samples can be stored for long periods in glass or plastic vials, preferably at cool temperatures.
Sample analysis	Elemental analysis of a plant sample requires that the organic fraction of the sample be destroyed, leaving the mineral nutrients (P, K, S, Ca, Mg, Mn, Cu, Zn, Fe, B, etc.) either in solution or in a form that is readily dissolved. Two techniques have been developed for organic matter destruction, *wet digestion* and *dry ashing*. In wet digestion, plant samples are dissolved in concentrated mixtures of acids (HNO_3, $HClO_4$, H_2SO_4) at high temperatures. Wet digestion is generally preferred for all elements, particularly with samples where Al, Fe, Zn, and S are of interest. Wet digestion is not recommended for B analysis because of problems with loss of B during the digestion and contamination from borosilicate glassware, some chemical reagents and other sources. Dry ashing refers to high-temperature (500°C) oxidation of a plant sample in a muffle furnace or high-temperature oven. This destroys the organic matter by combustion and leaves the mineral elements in the residual "ash." After combustion the ash is dissolved in an acid solution. The solution is filtered to remove undissolved solids and is then ready for analysis. Dry ashing is suitable for most elements as long as oven temperatures are carefully monitored. Excessive temperatures can cause volatilization losses of K and other elements. Once in solution, a wide variety of instruments are capable of determining nutrient concentrations. The most commonly used are the *atomic absorption spectrophotometer* (AAS, individual element analysis) and the *inductively coupled plasma spectrometer* (ICP, capable of multielement analysis on one sample). For certain elements (N, P, K, S, Mg, B, Ca) automated colorimetric analysis is acceptable.

Source: Adapted from Sims and Gartley (1996).

Table 3.10　Factors to Consider in the Collection and Analysis of Organic Wastes

Component of organic waste testing program	Comments
Sampling and handling	The heterogeneity in organic waste composition, combined with changes in waste properties that occur during storage, means that great care is needed to obtain a representative sample. In general, the waste should be sampled as close to the time of application as possible to minimize any changes that might occur between the time of sampling and application. Multiple subsamples of the waste should be obtained and mixed thoroughly in a clean, plastic bucket. A representative subsample is then removed, placed in a plastic container, and immediately delivered to the testing laboratory. Avoid storing the sample for prolonged periods prior to analysis. Samples should be collected annually or whenever a change in any factor that might alter waste composition occurs.
Analysis and interpretation	Organic waste analyses should include, at minimum, total N, ammonium-N, and total P, K, Ca, and Mg. Additional analyses that can be conducted less frequently include total S, B, Mn, Cu, and Zn. Special tests that may occasionally be required are pH, soluble salts and some trace elements such as As, Mo, and Se. Total elemental concentrations are not used directly to determine the amount of plant-available nutrients in organic wastes because only a portion of the total amount of a nutrient in an animal waste will be available during a normal growing season. The application rates of organic wastes are usually based on the amount of plant-available N anticipated to be provided by the waste. However, the rates of other nutrients added when a waste is applied, particularly P, should be considered to avoid excessive accumulations or imbalances of nutrients in soils that could have negative effects on plant growth or environmental quality. For municipal biosolids land application is also regulated based on the concentrations and loading rates of several trace metals (see Chapters 9 and 12).

Source: Adapted from Sims and Gartley (1996).

Problems

3.1 You are invited to give two presentations, one to a group of elementary school children and another to your state legislators. The topic of your presentation is "Soil Environmental Quality." You are asked to describe first the meaning of soil so that everyone will understand your definition of soil. Write a brief description of what you would present to both groups. Would you define soil differently to both groups, and, if so, why?

3.2 Why is it important to understand soil properties and functions to evaluate environmental quality issues and events that pertain to different ecosystems?

3.3 Determine the textural class of soils that contain:
   a. 25% sand　　25% silt　　50% clay
   b. 50% sand　　25% silt　　25% clay
   c. 75% sand　　15% silt　　10% clay
   d. 25% sand　　50% silt　　25% clay
   e. 33% sand　　33% silt　　34% clay

What would be the textural class of these soils if each contained 5% organic matter?

3.4 Calculate the mass of soil (in kg) in 1 ha (10,000 m³) to a depth of 1 cm and a bulk density of 1.35 Mg/m³. Repeat this calculation for a 15-cm depth.

3.5 Describe the relationship between particle density and bulk density. How do these soil physical properties influence (a) the movement of water; (b) plant growth; (c) wind and water erosion; and (d) environmental quality?

3.6 How does soil organic matter affect the physical, chemical, and biological properties of soils? Explain the relationship between soil organic matter and (a) soil structure; (b) ion-exchange capacity; and (c) microbial activity.

3.7 What soil properties can influence the mobility of contaminants that could degrade surface water and groundwater systems?

3.8 A soil contains 50 mg/kg Zn. Calculate the mass of Zn in 1 ha of soil to a depth of 15 cm assuming a bulk density of 1.35 Mg/m³. Develop an equation that converts the concentration of a substance in soil (in mg/kg) to mass per hectare (in kg/ha) as the soil depth varies (in cm).

3.9 List some of the benefits to the use of plants for remediating a contaminated site. What are the limitations to using the phytoremediation technology in toxic metal surface–contaminated sites or in groundwaters polluted with hazardous organic solvents?

3.10 Describe three physical, chemical, and biological processes involved in soil development. Indicate the importance of each of these soil processes.

3.11 Why is a soil survey a valuable resource when considering the purchase of a building site, location of a septic system, production of a new or alternative crop, and implementation of a wildlife reestablishment program?

3.12 Describe how land-use planning and GIS can be used to assist a city, county, or state in governing the proper use of certain areas for agricultural production, landfills, wastewater treatment facilities, road construction, recreational areas, and long-term sustainable approaches to population growth and economics.

3.13 Soil sample collection is one of the most critical aspects of the soil testing process. Explain how you would decide where to collect soil samples for (a) a corn–soybean rotation on a 100-ha dairy farm; (b) a mine spoil area that is to be revegetated and used as pastureland for beef cattle; (c) a Pb-contaminated soil that must be remediated prior to use as parkland.

3.14 Chemical soil test extractants remove "plant available" nutrients from soils by dissolution, desorption, and chelation. How would you make 10 L of the following soil test extractants:

a. Mehlich 1: 0.05 M HCl + 0.0125 M H_2SO_4

b. Mehlich 3: 0.2 M CH_3COOH + 0.25 M NH_4NO_3 + 0.015 M NH_4F + 0.013 M HNO_3 + 0.001 M EDTA).

c. Ammonium bicarbonate-DTPA: 1 M NH_4HCO_3 + 0.005 M DTPA

3.15 Using Figure 3.11 discuss the concept of a soil test "critical value" — what it means and how its interpretation differs when crop yield is the main goal compared with protection of human health and environmental quality.

3.16 You digested a 0.5 g corn ear leaf sample in concentrated nitric-perchloric acid. The digested solution is filtered into a 50 mL volumetric flask and made to volume with deionized water. The concentration of P in the diluted solution is 30 mg P/L. What is the concentration of P in the corn ear leaf in mg/kg and percent?

3.17 Explain the difficulties likely to be encountered when interpreting an organic waste analysis and making a recommendation for the proper waste application rate.

References

Brady, N. C. *The Nature and Properties of Soils*, 10th ed., Macmillan Press, New York, 1990.

Brady, N. C. and R. R. Weil. 2000. *Elements of the Nature and Properties of Soil*, Prentice-Hall, Upper Saddle River, NJ.

Carter, M. R., Ed. 1993. *Soil Sampling and Methods of Analysis*. Lewis Publishers, Boca Raton, FL.

Coyne, M. S. 1999. *Soil Microbiology: An Exploratory Approach*, Delmar Publishers, New York.

Flathman, P. E., G. R. Lanza, and D. J. Glass. 1999. Introductory feature on phytoremediation, *Soil and Groundwater Cleanup*. February/March.

Jenny, Hans. 1941. *Factors of Soil Formation: A System of Quantitative Pedology*, McGraw-Hill; Dover, Mineola, NY.

Munn, L. C. and G. F. Vance. 1998. Genesis, Morphology and Classification of Soils, University of Wyoming Course Packet, Department of Renewable Natural Resources, University of Wyoming, Laramie, 111 pp.

Munson, R. D. and W. L. Nelson. 1990. Principles and practices in plant analysis, in *Soil Testing and Plant Analysis*, 3rd ed., R. L. Westerman, Ed., Soil Science Society of America, Madison, WI.

Paul, E. A. and F. E. Clark. 1989. *Soil Microbiology and Biochemistry*, Academic Press, San Diego, CA.

Peck, T. R. and P. N. Soltanpour. 1990. The principles of soil testing, in *Soil Testing and Plant Analysis*, 3rd ed., R. L. Westerman, Ed., Soil Science Society of America, Madison, WI, 1–8.

Schlesinger, W. H. 1997. *Biogeochemistry: An Analysis of Global Change*, 2nd ed., Academic Press, San Diego, CA.

Schulze, D. G. 1989. An introduction to soil mineralogy, in *Minerals in the Soil Environment*, 2nd ed., J. B. Dixon and S. B. Weed, Eds., Soil Science Society of America, Madison, WI, 1–34.

Sims, J. T. 1999. Soil fertility evaluation, in *Handbook of Soil Science*, M. E. Sumner, Ed., CRC Press, Boca Raton, FL.

Sims, J. T. and K. L. Gartley. 1996. Nutrient Management Handbook for Delaware, Coop. Bull. No. 59, University of Delaware, Newark.

Singer, M. J. and D. N. Munns. 1999. *Soils: An Introduction*, 4th ed., Prentice-Hall, Upper Saddle River, NJ.

Soil Survey Staff. 1975. Soil Taxonomy: A Basic System of Soil Classification for Making and Interpreting Soil Surveys, Agricultural Handbook No. 436, U.S. Government Printing Office, Washington, D.C.

Soil Survey Staff. 1992. *Keys to Soil Taxonomy*, 5th ed., Technical Monograph No. 19, Soil Management Support Services, Pocahontas Press, Blacksburg, VA.

Sparks, D. L., Ed. 1996. *Methods of Soil Analysis*, Soil Science Society of America, 1996. Madison, WI.

Spiedel, D. H. and A. F. Agnew. 1982. *The Natural Geochemistry of Our Environment*, Westview Press, Boulder, CO.

Stevenson, F. J. 1994. *Humus Chemistry: Genesis, Composition, Reactions*, 2nd ed., John Wiley & Sons, New York.

U.S. Department of Agriculture, Soil Conservation Service (in cooperation with the Wyoming Agricultural Experiment Station). 1971. Soil Survey of Goshen County, Wyoming: Southern Part.

Young, J. A., B. M. Christensen, M. S. Schaad, G. F. Vance, and L. C. Munn. GIS identification of potential alternative crops utilizing soil and climatic variables in the Bighorn Basin, Wyoming, *Am. J. Alternative Agric.* (in press).

Supplementary reading

Baker, D. E., D. R. Bouldin, H. A. Elliott, and J. R. Miller. 1985. Criteria and Recommendations for Land Application of Sewage Sludges in the Northeast, Northeastern Regional Bull. No. 851, Pennsylvania State University, University Park.

Black, C. A. 1993. *Soil Fertility Evaluation and Control*, Lewis Publishers, Boca Raton, FL.

Dahnke, W. C. and R. A. Olson. 1990. Soil test correlation, calibration, and recommendation, in *Soil Testing and Plant Analysis*, 3rd ed., R. L. Westerman, Ed., Soil Science Society of America, Madison, WI, 45–71.

Dixon, J. B. and S. B. Weed, Eds. 1989. *Minerals in the Soil Environment*, 2nd ed., Soil Science Society of America, Madison, WI.

James, D. W. and K. L. Wells. 1990. Soil sample collection and handling: technique based on source and degree of field variability, in *Soil Testing and Plant Analysis*, 3rd ed., R. L. Westerman, Ed., Soil Science Society of America, Madison, WI, 25–44.

Keller, E. A. 1988. *Environmental Geology*, Merrill, Columbus, OH.

Mills, H. A. and J. B. Jones, Jr. 1996. *Plant Analysis Handbook II*, MicroMacro Publishing, Athens, GA.

Olson, R. A., F. N. Anderson, K. D. Frank, P. H. Grabouski, G. W. Rehm, and C. A. Shapiro. 1987. Soil test interpretations: sufficiency vs. build-up and maintenance, in Soil Testing: Sampling, Correlation, Calibration, and Interpretation, Spec. Pub. No. 21, Soil Science Society of America, Madison, WI, 41–52.

Page, A. L., T. J. Logan, and J. A. Ryan, Eds. 1987. *Land Application of Sludge: Food Chain Implications*, Lewis Publishers, Chelsea, MI.

Plank, C. O., Ed. 1992. Plant Analysis Reference Procedures for the Southern Region of the United States. So. Coop. Ser. Bull. No. 368, Georgia Agricultural Experiment Station, Athens, GA.

Risser, J. A. and D. E. Baker. 1990. Testing soils for toxic metals, in *Soil Testing and Plant Analysis*, 3rd ed., R. L. Westerman, Ed., Soil Science Society of America, Madison, WI, 275–298.

Sims, J. T., S. D. Cunningham, and M. E. Sumner. 1997. Assessing soil quality for environmental purposes: roles and challenges for soil scientists, *J. Environ. Qual.*, 26:20–25.

Sims, J. T. and G. V. Johnson. 1991. Micronutrient soil tests, in *Micronutrients in Agriculture*, 2nd ed., Soil Science Society of America, Madison, WI.

Sims, J. T. and A. Wolf, Eds. 1995. Soil testing in the northeastern United States, *Soil Testing Procedures for the Northeastern United States*, 2nd ed., Northeastern Regional Bull. No. 493, University of Delaware, Newark.

Singer, M. J. and D. N. Munns. 1996. *Soils: An Introduction*, 3rd ed., Prentice-Hall, Englewood Cliffs, NJ.

Smith, M. A., Ed. 1985. *Contaminated Land: Reclamation and Treatment*, Plenum Press, New York.

U.S. Environmental Protection Agency. 1992. Sewage Sludge Use and Disposal Rule (40 CFR Part 503) — Fact Sheet, U.S. EPA, EPA-822-F-92-002, U.S Government Printing Office, Washington, D.C.

Westerman, R. L., Ed. 1990. *Soil Testing and Plant Analysis*, 3rd ed., Soil Science Society of America, Madison, WI.

chapter four

Soil nitrogen and environmental quality

Contents

4.1 Introduction: nitrogen and the environment

Nitrogen (N) is arguably the most important and yet most difficult to manage of all the plant nutrients. While absolutely vital to modern agriculture, it also has a number of serious environmental impacts, briefly summarized in Table 4.1. This chapter focuses on N in agricultural soils, emphasizing the issues of greatest importance to environmentally sound soil N management. Some situations that differ considerably from

Table 4.1 Summary of Environmental Problems Associated with Nitrogen

Environmental issue	Causative mechanisms and impacts
Human and animal health	
Methemoglobinemia	Consumption of high-nitrate drinking waters and food; particularly important for infants because it disrupts O_2 transport system in blood
Cancer	Exposure to nitrosoamines formed from reaction of amines with nitrosating agents; skin cancer increased by greater exposure to ultraviolet radiation resulting from destruction of the O_3 layer
Nitrate poisoning	Livestock ingestion of high-nitrate feed or water
Ecosystem damage	
Groundwater contamination	Nitrate leaching from fertilizers, manures, sludges, wastewaters, septic systems; can impact both human and animal health, and trophic state of surface waters
Eutrophication of surface waters	Soluble or sediment-bound N from erosion, surface runoff, or groundwater discharge enters surface waters; direct discharge of N from municipal and industrial wastewater treatment plants into surface water; atmospheric deposition on water quality and biological diversity of fresh waters
Acid rain and ammonia evolution and redeposition	Nitric acid originating from reaction of N oxides with moisture in atmosphere is returned to terrestrial ecosystem as acidic rainfall, snow, mists, fogs (wet deposition) or as particulates (dry deposition); damages sensitive vegetation, acidifies surface waters, and, as with eutrophication, can unfavorably alter biodiversity in lakes, streams, bays; ammonia evolved from concentrated animal feeding operations can acidify soils and alter species diversity in nearby woodlands
Stratospheric ozone depletion and global climate change	Nitrous oxides from burning of fossil fuels by industry, automobiles, and from denitrification of nitrate in soils are transported to the stratosphere where O_3 destruction occurs; ultraviolet radiation incident on earth's surface increases, as does global warming

Source: Adapted from Keeney (1982).

production agriculture (e.g., land reclamation, urban horticulture, forestry) will be reviewed to illustrate the range of approaches needed to manage soil N effectively in an ecosystem shared by cities, farms, and industries. We will address the following key questions:

- What is the basis for public concerns about the effects of N on human and animal health, its role in the pollution of groundwaters and surface waters, the formation of acid rain, and the destruction of the stratospheric ozone (O_3) layer?
- How can we use our knowledge of the many complex chemical and biological processes of the *soil N cycle* to improve our management of all N sources, from fertilizers to animal manures, municipal biosolids (sewage sludges), and industrial organics?

4.1.1 Origin and distribution of nitrogen in the environment

To understand fully the environmental problems caused by N, and to develop sensible, cost-effective approaches to N management, it is essential to have a basic understanding of the origin and cycling of N in the earth's four major "spheres": the lithosphere, hydrosphere, atmosphere, and biosphere (see Chapters 2 and 3 for discussion of these spheres). Most (>98%) of the earth's N is found in the lithosphere, either in the earth's core, in igneous and sedimentary rocks, oceanic sediments, or in soils. The remaining 2% is distributed between the atmosphere, hydrosphere, and biosphere. In the atmosphere N exists mainly as the inert gas N_2, which comprises 78% of atmospheric gases. In the hydrosphere, N occurs as dissolved organic or inorganic N. Nitrogen is also a vital component of the biosphere, which consists of living plants and animals. Nitrogen can be found in many different forms in these spheres, including molecular N, organic molecules, geologic materials, gases, and soluble ions.

Nitrogen is a very dynamic element, capable of being transformed biochemically or chemically through a series of processes conceptually summarized as the *nitrogen cycle* (Figure 4.1). Most N transformations involve the *oxidation* (loss of electrons) or *reduction* (gain in electrons) by the N atom by both biological and chemical means. The oxidation states of N in nature range from +5 in the nitrate anion (NO_3^-) to –3 for ammonia (NH_3) or ammonium (NH_4^+). The soil N cycle (Figure 4.2 and Section 4.3), is a subset of the

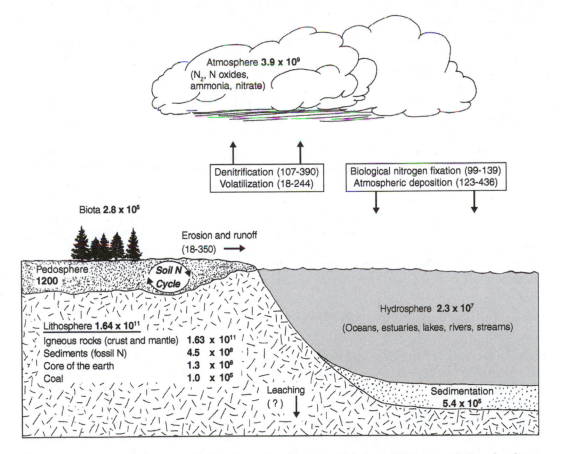

Figure 4.1 The global nitrogen cycle (units = Terrograms = Tg). Inset indicates soil N cycle, shown in more detail in Figure 4.2. (Data from Hauck and Tanji, 1982; Stevenson, 1982.)

overall N cycle. We seek to understand how management of the soil N cycle affects other segments of the global N cycle, such as groundwater aquifers in the hydrosphere.

In the broadest sense, there are three major natural inputs to the soil N cycle: *atmospheric deposition, biological N fixation,* and *weathering and decomposition*. Atmospheric deposition occurs as inorganic and organic N in precipitation or dry particulate matter. Biological N fixation is the conversion of gaseous atmospheric N_2 to NH_3 and then organic N by symbiotic and nonsymbiotic organisms. Weathering and decomposition reactions are those in which previously fixed or deposited organic or inorganic N is transformed from stable inorganic or organic N to more chemically and biologically active forms of N. The major natural processes by which N is lost from soils are by evolution as a gas (NH_3 *volatilization* and *bacterial* or *chemical denitrification* of NO_3^-) and by transport processes, as soluble N in waters percolating downward through soils (*leaching*) or in water and sediments moving across the soil surface (*erosion* and *runoff*). Soil N management primarily involves manipulating or supplementing (e.g., fertilization, manuring) natural processes to produce plants for food, fiber, or aesthetic purposes. Environmental concerns about N arise when one of these transformations results in the conversion and concentration of N in a form that can adversely affect the health or quality of an organism or an ecosystem. It is important to remember that, although only a few forms of N are now regarded as harmful to our environment, the processes of the N cycle regulate, on a global and local scale, the amount of N in each form. Controlling pollution caused by N, therefore, starts with an understanding of how we can control the N cycle in soils and other ecosystems.

Global estimates of long-term changes in the distribution of N among the four spheres are filled with uncertainty. However, human activities have clearly impacted this distribution, enriching some sectors of the earth's environment with N while simultaneously depleting others. Many fundamental aspects of modern civilization, such as agriculture, urbanization, industry, transportation, and water or waste treatment systems have the potential to significantly affect the distribution of N on a localized scale, and cumulatively, on a global scale. Unfortunately, the human-induced movement of an element, such as N, to a part of the environment where it can have a negative effect is often a synonym for pollution. Although, as seen in Table 4.1, there can be a wide variety of environmental impacts from the redistribution of N, from the perspective of soil N management, the forms of N of greatest importance are NO_3^-, NH_3, and the gaseous N oxides (N_2O, NO, and NO_2). Our understanding of the mechanisms by which NO_3^- leaching, NH_3 volatilization and redeposition to soils and waters, and nitrous oxide (N_2O) emissions from soils to the atmosphere result in pollution is far from complete. However, we do have a basic understanding of these processes and have developed many practical means to minimize the effects of N on soil, air, and water quality.

4.1.2 Nitrogen effects on human and animal health

There is little doubt that groundwater, and more specifically drinking water, contamination by NO_3^- is now the environmental issue of greatest concern for N management. The major human and animal health issues associated with the consumption of excessive NO_3^- in drinking waters, or even some foods, are *methemoglobinemia* ("blue-baby syndrome") and possible carcinogenic effects from another class of nitrogenous compounds, the nitrosamines. Methemoglobinemia is not caused directly by NO_3^- but occurs when NO_3^- is reduced to nitrite (NO_2^-) by bacteria found in the digestive tract of humans and animals. Nitrite can then oxidize the iron (Fe) in the hemoglobin molecule from Fe^{+2} to Fe^{+3}, forming methemoglobin, which cannot perform the essential oxygen transport functions of hemoglobin. This can result in a bluish coloration of the skin in infants, hence the origin of the term "blue baby syndrome." Methemoglobinemia is a much more serious problem for very young infants than adults because after the age of 3 to 6 months, the acidity in the human stomach suppresses the activity of the bacteria that transform NO_3^- to NO_2^-. Although documented cases of methemoglobinemia are extremely rare, the U.S. Environmental Protection Agency (EPA) has established a maximum contaminant level of 10 mg NO_3-N/L (45 mg NO_3^-/L) to protect the safety of U.S. drinking water supplies. Animals can also be susceptible to methemoglobinemia, although the health advisory level for livestock is much higher, ~40 mg NO_3-N/L (180 mg NO_3^-/L). The other major human health concern with N is the potential carcinogenic effect of nitrosamines, compounds with the general chemical structure $R_2N–N=O$, where R represents any carbon group. Nitrosamines are formed by the reaction, under highly acidic conditions (pH<4), of secondary and tertiary amines (R_2NH, R_3N) with nitrous acid anhydride (N_2O_3). Nitrosamines have produced tumors in laboratory animals, but conclusive evidence for a causative role of nitrosamines in human cancer does not exist.

4.1.3   Nitrogen and eutrophication

Eutrophication is defined as an increase in the nutrient status of natural waters that causes accelerated growth of algae or water plants, depletion of dissolved oxygen (O_2), increased turbidity, and a general degradation of water quality. Causes and management of eutrophication are discussed in more detail in Chapters 2 and 5, but the enrichment of lakes, ponds, bays, and estuaries by N and phosphorus (P) from surface runoff or groundwater discharge is known to be a contributing factor. The levels of N required to induce eutrophication in fresh and estuarine waters are much lower than the values associated with drinking water contamination. Although estimates vary, and depend considerably on the N:P ratio in the water, concentrations of 0.5 to 1.0 mg N/L are commonly used as threshold values for eutrophication. Marine environments, where salinity levels are greater, are more sensitive to eutrophication and thus have lower threshold levels of N (<0.6 mg N/L). It is important to remember that N concentrations in surface waters reflect not only agricultural inputs (primarily nonpoint in nature), but the inputs of N from direct discharge of wastewaters from municipalities, industry, and recreational developments. Other sources of the total pool of N in surface waters are (1) atmospheric deposition, both in precipitation (e.g., "acid rain" as HNO_3 and redeposition of volatilized NH_3) and as particulate matter (e.g., dusts from wind erosion and solid particles from industrial emissions), and (2) biological N fixation of atmospheric N_2 by aquatic organisms.

4.1.4   Atmospheric effects of nitrogen

Nitrogen has been shown to have two serious effects on the earth's atmosphere and, because of these atmospheric changes, on the quality of terrestrial and aquatic environments. Nitric acid (HNO_3), primarily caused by the release of nitrogen gases ($NO_x = NO_2 + NO$) to the atmosphere during the burning of fossil fuels, is a major component of *acid rain*. Studies have shown that acid rain (pH < 5.6), acid mist, or dry deposition of acidic particulates can negatively and seriously affect forest ecosystems and surface waters, but the impact on agricultural soils and crops to date has been minimal (see Chapter 10). Soil acidification can also be caused by the redeposition of NH_3 that volatilized from soils or, more significantly, from areas of highly concentrated animal feeding operations. When redeposited to soils NH_3 is rapidly converted to NO_3-N, a process that results in soil acidification (see Section 4.2.1). Acidification of forest soils in Germany by NH_3 originating from the highly concentrated swine, dairy, and poultry operations in the Netherlands has negatively affected some forest ecosystems, caused regional tensions, and increased efforts to reduce NH_3 losses from barns, feedlots, and lagoons. Additional information related to acidic deposition can be found in Chapter 10.

Nitrous oxides (N_2O) and NO_x have also been shown to cause the photo-oxidation of O_3 in the stratosphere, reducing the capacity of the O_3 layer to protect the earth from the intense ultraviolet radiation emitted by the sun and contributing to global warming. A simplified version of the reactions of N oxides with O_3 is shown below:

$$\underset{\textit{Photodissociation}}{N_2O + O \xrightarrow{\hspace{3cm}} 2\,NO} \tag{4.1}$$

$$\underset{\textit{Photo-oxidation}}{NO + O_3 \xrightarrow{\hspace{3cm}} N_2O + O_2} \tag{4.2}$$

In addition to the burning of coal, oil, and gasoline, a major nonpoint-source of NO is the process of biological *denitrification*, in which soil microorganisms reduce NO_3^- to NO and N_2O under oxygen-limited conditions. *Chemodenitrification* can also occur in soils, resulting in the evolution of N oxides under well-aerated conditions. These processes are discussed in detail in Section 4.2.2.

4.1.5   *Risk assessment for nitrogen pollution*

The environmental effects of N described above have been documented on local, regional, and global scales. Unlike some forms of pollution, the adverse impacts of N are not directly obvious or dramatic and the level of risk is not as clear. Given the limited nature of the resources available to mitigate all forms of pollution and the critical importance of N to food production, proper assessment of this risk is essential to prioritize efforts to remediate N pollution. The process used for risk assessment will be described in detail in Chapter 12, but certain points are clear at this time with regard to N pollution and soil management.

First, there is a clear public perception that agriculture contributes to N pollution of groundwaters and surface waters by improper fertilization and organic waste management. While scientific research has substantiated this concern, it has also indicated that groundwater contamination by NO_3^- is often localized in nature and associated with specific regional problems such as well-drained soils, shallow water tables, highly concentrated animal production, intensive irrigation, and waste or wastewater disposal by municipalities or industry. The combined weight of public perception and scientific documentation ensures that the risk will be addressed by scientific, advisory, and regulatory agencies. In essence, there is sufficient agreement among all parties that risk exists and that a significant commitment of resources to reduce its impacts is needed. In contrast, atmospheric effects of N oxides originating from agricultural soils are not as well understood. Also, the greater role of industry and urban areas in generating N oxides clearly dictates that, while emission of N oxides from soils should be minimized, our control efforts should first be directed toward nonagricultural sources.

Second, we must acknowledge that there are significant obstacles to reducing N pollution originating from both agricultural and nonagricultural soils. The pressures to produce increasing amounts of food with fixed amounts of arable land often result in the use of higher and often less efficient N rates as we substitute fertilizer N for soil N. The need to maintain farm profitability and the low costs of fertilizer N favor the use of insurance fertilization to overcome unexpected N losses that may be caused by uncontrollable climatic events. The nature of modern animal-based agriculture concentrates nutrients derived from soils, fertilizers, and organic N in other regions into areas without an adequate land base for proper N use. A similar scenario exists for organic wastes produced in cities (e.g., municipal biosolids). Conversion of grasslands and forests to agricultural land for the production of annual crops can release organic N stored for centuries over a relatively short period of time, enhancing the potential for enrichment of groundwaters and surface waters with NO_3^-. Recreational developments near sensitive water bodies can discharge N from septic systems into groundwaters and overfertilization of turf in urban areas can produce runoff high in NO_3^-. It is apparent that the technology, education, effort, and cost required to design and implement improvements in N use for each of these scenarios is formidable. Prioritization of risk thus becomes of critical importance as we consider allocation of resources.

Finally, it is also important to acknowledge the time required to correct the problem. Groundwaters that have been contaminated with NO_3^- by land-use practices over a 30-year period are likely to require several decades to "dilute" to acceptable concentrations with the low NO_3^- leachate coming from soils receiving improved management practices. This means that the cost of reducing N pollution of our natural waters must be borne by society for many years, and that a long-term, integrated effort by all responsible parties will be needed.

4.2   The soil nitrogen cycle

The *soil N cycle* (Figure 4.2) is a subset of the global N cycle and can be viewed as a conceptual summary of the interactions among the chemical, physical, and biological transformations of N in soils. This chapter focuses on environmental concerns related to soil N management, hence the interactions of the soil N cycle with the segments of the environment most sensitive to pollution by N (groundwaters, surface waters, the atmosphere) are emphasized. From

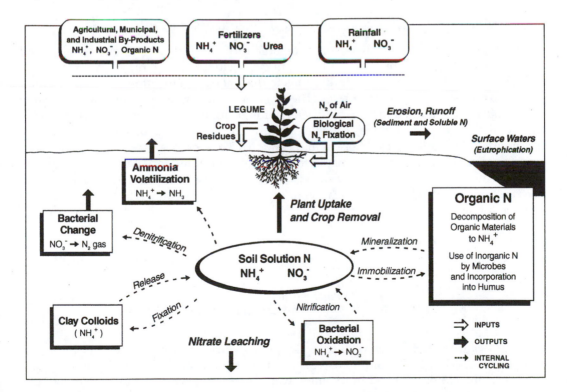

Figure 4.2   The soil nitrogen cycle.

this perspective, the key N transformations are the cycling of N between organic and inorganic forms (*mineralization* and *immobilization*), gaseous losses of N to the atmosphere (*ammonia volatilization* and *denitrification*), losses associated with water movement (*leaching* and *erosion*), and *biological N fixation*. Many of these reactions are controlled by soil micro-organisms that alter the form and oxidation states, and thus the fate, of N between N_2, N_2O, NO, NH_3/NH_4^+, NO_2^-, and NO_3^-. It is critical to recognize, from a management standpoint, how the relative importance of each reaction varies with soil and environmental conditions and when it is possible to exert a significant degree of control over a given reaction. Our ability to control an N transformation, however, will not only depend on the biology, chemistry, or physics of the process, but also on the intensity and economics of management. Farmers using irrigation, for example, have more control over the timing of N delivery to crops than those in dryland agriculture because of their ability to inject soluble N fertilizers into irrigation waters. This can improve the crop N use efficiency, reducing NO_3^- leaching. Industries involved in land reclamation or municipalities charged with biosolids disposal may increase the extent of waste processing (e.g., composting or lime-stabilization) due to greater regulation on organic waste use related to pathogens, metals, or organic pollutants. These changes in waste properties may then affect the N transformations likely to occur upon land application of the waste material.

Sound environmental management of N begins with an understanding of the major components of the soil N cycle. Enormous scientific effort has been expended to study N transformations in soils. The challenge now is to translate this knowledge into practical management programs that achieve both production and environmental goals. In the following sections the basic principles underlying each of the key N transformations are described, and in later sections of this chapter techniques that can be used to manage these N transformations will be related to agricultural, urban, and land reclamation issues.

4.2.1 *Mineralization, nitrification, and immobilization of nitrogen*

Mineralization refers to the conversion of organic forms of N (e.g., proteins, chitins, and amino sugars from microbial cell walls, and nucleic acids) to inorganic N, as ammonium-N (NH_4^+). The organic N may be indigenous to the soil or freshly added as crop residues, animal manures, or municipal wastes. The process is mediated by a diverse population of heterotrophic soil microorganisms (bacteria, fungi, actinomycetes) that produce a wide variety of extracellular enzymes capable of degrading proteins (proteinases, peptidases) and nonproteins (chitanases, kinases) into NH_4^+. These microbes use the energy derived from the oxidation of soil organic matter for metabolic activities and the N released during decomposition to produce the amino acids and proteins essential for population growth. The reactions involved in mineralization of organic N to inorganic NH_4^+ can be summarized as follows:

$$\text{Organic N} \xrightarrow{\textit{Proteolysis, Aminization}} \text{Amino-N (R–NH}_2\text{)} + CO_2 + \text{Energy, by-products} \qquad (4.3)$$

$$\text{Amino-N (R–NH}_2\text{)} \xrightarrow{\textit{Ammonification}} NH_3 + H_2O \rightarrow NH_4^+ + OH^- \qquad (4.4)$$

Once mineralized, NH_4^+ can be taken up by plants, nitrified, immobilized by soil micro-organisms, lost as a gas by ammonia volatilization, held as an exchangeable ion by clays or other soil colloids, or fixed in the interlayers of certain clay minerals. These potential fates of NH_4^+ are described below.

Mineralization of N from soil organic matter has been shown to provide a significant portion of the N requirement of many crops. Plants can absorb NH_4^+ directly from the soil solution and, in fact, studies have shown that NH_4^+ is taken up preferentially by some plants over other sources of N (e.g., NO_3^-). In general, however, NO_3^- uptake by plants is greater than NH_4^+ because NO_3^- is usually present at higher concentrations in the soil solution and moves more freely to plant roots by mass flow and diffusion than NH_4^+. Total N values for topsoil horizons of most mineral soils range from 0.05 to 0.15%. Under well-aerated conditions about 1 to 3% of this organic N will mineralize annually, producing ~15 to 70 kg N/ha/year. In comparison, fertilizer N recommendations for many annual, nonleguminous crops range from ~50 to 200 kg N/ha. Long-term use of animal manures or leguminous rotational crops such as alfalfa can greatly increase the amount of potentially mineralizable organic N in soils. As will be discussed in Section 4.4, this can result in marked reductions in fertilizer N requirements, an important consideration from both economic and environmental perspectives. The timing of N mineralization, relative to the timing of crop N uptake, can be as important a consideration as the amount of N mineralized. Most studies have shown that under optimum conditions N mineralization follows a curvilinear pattern, as illustrated in Figure 4.3a. The amount of *potentially mineralizable organic N* in soils (N_o) and the rate of N mineralization (k) have been successfully described by relatively simple first-order kinetic models that relate the change in mineralized N (N_m) in the soil with time (dN_m/dt) to the amount of mineralizable substrate (N_o) as follows:

$$dN_m/dt = k(N_o) \qquad (4.5)$$

Data from laboratory studies that measure the amount of NH_4^+ and NO_3^- leached or extracted from soils at differing time intervals (N_m, t) can be used with the integrated form of this equation to estimate N_o and k for different soils, as a function of soil horizonation, changing soil chemical and environmental conditions (pH, temperature, moisture), or as affected by long-term changes in soil N caused by differing tillage practices, fertilizer applications, or organic waste amendments. Ideally, if N_o and k are known, the amount of mineral N in the soil after a specified time interval could be predicted as

$$N_m = N_o\,(1 - e^{-kt}) \qquad (4.6)$$

Some research has shown that a model based on two pools of N_o (N_f = "fast" mineralization of readily decomposable organic N and N_s = slow mineralization of recalcitrant organic N) with separate rate constants for mineralization of each pool (k_f for N_f, k_s for N_s) better describes the results of N mineralization studies, particularly for waste-amended soils (Figure 4.3b):

$$N_m = N_f(1 - e^{-k_f t}) + N_s(1 - e^{-k_s t}) \qquad (4.7)$$

Nitrification is the conversion of NH_4^+ into NO_2^- and then NO_3^- by the actions of obligately aerobic, chemoautotrophic bacteria, i.e., obtain carbon (C) from carbon dioxide (CO_2) or carbonates and energy from the oxidation of NH_4^+ or NO_2^-. Initially, bacteria of the genera *Nitrosomonas*, *Nitrosospira*, or *Nitrosococcus* oxidize NH_4^+ to hydroxylamine (NH_2OH) and then, through several intermediate compounds that are not as well known, to NO_2^-. Key features of this step are the change in oxidation state of N from –3 to +3, and acidification of the soil by hydrogen ions produced when NH_4^+ is oxidized:

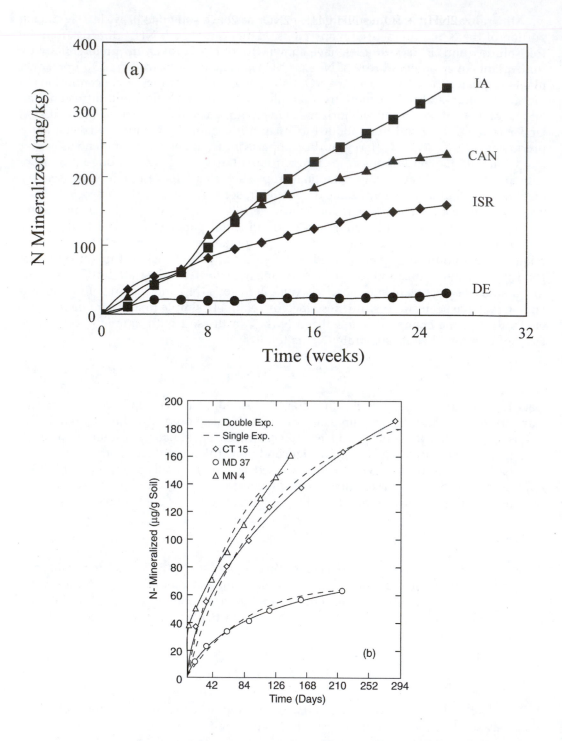

Figure 4.3 (a) Nitrogen mineralization patterns for different soils. [Adapted from laboratory studies conducted by Chae and Tabatabai, 1986 (Iowa, IA); Ellert and Bettany, 1988 (Canada, CAN); Hadas et al., 1983 (Israel, ISR); Sallade and Sims, 1992 (Delaware, DE).] (b) Comparison of single and multiple substrate models to simulate N mineralization in a sludge-amended soil. (From Deans, J. R. et al., *Soil Sci. Soc. Am. J.*, 50, 323, 1986. With permission.)

$$2NH_4^+ + 3O_2 \rightarrow NH_2OH \rightarrow 2NO_2^- + 2H_2O + 4H^+ + Energy \qquad (4.8)$$

In the next reaction, bacteria of the genera *Nitrobacter, Nitrospira,* or *Nitrococcus* continue the oxidative process, convert NO_2^- to NO_3^-, and change the oxidation state of N from +3 to +5:

$$2NO_2^- + O_2 \rightarrow 2NO_3^- + Energy \qquad (4.9)$$

Nitrate can then be used directly by plants or soil microorganisms or lost from the crop rooting zone by *denitrification, leaching,* or *erosion/runoff* (Sections 4.2.2 and 4.2.3). In most soils nitrification is a rapid process, somewhat unfortunate given the much greater mobility in soils of NO_3^- than NH_4^+. Chemical inhibitors of nitrification can delay this conversion process and have been shown in laboratory studies to be quite effective. To date, however, field research on the effectiveness and economic value of nitrification inhibitors for fertilizers and organic wastes has been inconclusive. The properties and use of some nitrification inhibitors are described in more detail in Section 4.4.2.

Immobilization is essentially the reverse of mineralization, and involves the assimilation of inorganic N (NH_3, NH_4^+, NO_2^-, NO_3^-) by soil microorganisms and the transformation of these mineral forms of N into organic compounds during microbial metabolism and growth. Plant uptake can also be viewed as a form of immobilization, and understanding or controlling the competition between plants and soil microorganisms for inorganic soil N is an important aspect of soil N management. As immobilization represents the formation of organic nitrogenous compounds, it will be controlled to a large extent by the availability of the C needed to produce amino acids and proteins. If a large supply of available C is present in the soil, relative to inorganic N, microbial growth and consumption of soluble N will be stimulated, thus enhancing the conversion of soluble N into biomass N. Lower ratios of available C to N will result in an excess of NH_4^+ or NO_3^- in the soil, relative to microbial requirements. The C:N ratio of native or added organic matter, along with environmental conditions that regulate microbial population growth, will thus control the amount of inorganic N available for plant uptake or other less desirable fates (leaching, denitrification). Stable soil organic matter has C:N ratios from 10:1 to 12:1, while soil organisms have C:N ratios from 5:1 to 8:1. Mineralization of soil organic matter provides adequate C and N for soil microorganisms and, as mentioned above, a small to moderate quantity of available N for plant uptake. Adding organic amendments with differing C:N ratios to soils, however, can cause significant but reasonably predictable changes in the amount of plant-available inorganic soil N, as illustrated by eight studies on N mineralization from crop residues with C:N ratios ranging from 8:1 to 80:1 (Figure 4.4a). A C:N ratio of ~25:1 (organic matter that is 40% C and 1.6% N) is commonly used as the ratio where mineralization and immobilization are in balance. Adding materials with wide (>30:1) C:N ratios (straw, sawdust, papermill sludges) can cause a rapid increase in microbial biomass and a depletion of available soil N to the point where N deficiency can occur in many plants. Conversely, some organic amendments (e.g., municipal biosolids, poultry manure) with very low C:N ratios can produce large excesses of soluble N and must be managed carefully to avoid N losses to sensitive parts of the environment. Composting is an effective way of stabilizing N in rapidly mineralizable organic wastes, as shown in Figure 4.4b. Careful attention to plant N nutrition is required when composts are used, however, as soils amended with composts may exhibit an initial period of N immobilization and then release N more slowly than the rate required by plants. The C:N ratio of added organic matter does not remain constant during the decomposition process as C from microbial respiration is evolved from the soil as CO_2. With time, therefore, the

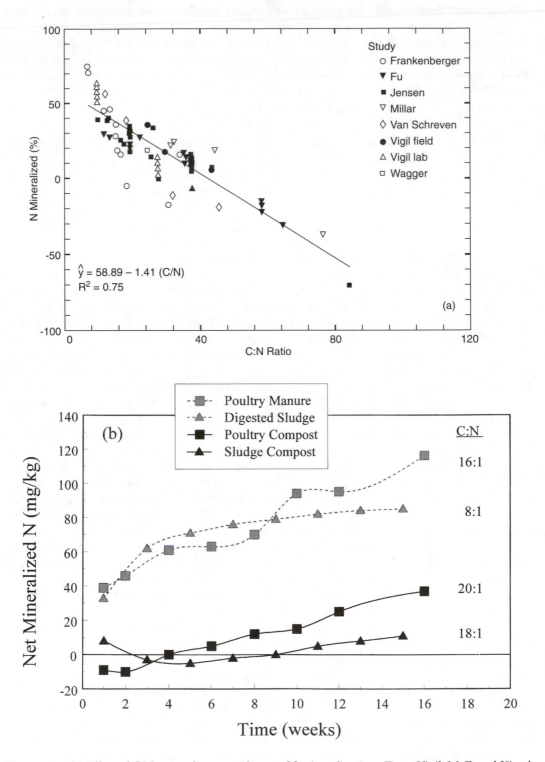

Figure 4.4   (a) Effect of C:N ratio of crop residues on N mineralization. (From Vigil, M. F. and Kissel, D. E., *Soil Sci. Soc. Am. J.*, 55, 757, 1991. With permission.) (b) Effect of composting digested sewage sludge or poultry wastes on N mineralization. (Adapted from Epstein et al., 1978; Sims et al., 1992.)

C:N ratio will decrease into the range where mineralization, not immobilization, predominates and the soil once again provides some available N for plant uptake.

Given the importance of soil microorganisms in mineralization, nitrification, and immobilization reactions, it is apparent that proper management of soil N, particularly from organic sources (crop residues, animal manures, municipal biosolids) requires an understanding of the soil and environmental factors that can affect the activity of microorganisms controlling these reactions. All parameters that affect biological activity (temperature, moisture, aeration, and soil pH) have been shown to influence the rate and extent of these three N transformations. "Optimum" conditions for each transformation have been broadly defined and vary slightly between mineralization–immobilization reactions and nitrification. Because a much wider variety of organisms participate in mineralization and immobilization, these processes are somewhat less sensitive to changing environmental conditions than nitrification. As an example, unlike nitrification, mineralization can proceed under anaerobic conditions and at much wider temperature ranges. This can result in an accumulation of NH_4^+ in flooded soils where nitrification is inhibited by a lack of O_2, or under extreme soil temperature regimes (<5°C or >40°C). For mineralization and immobilization, the optimum conditions are a temperature range of 40 to 60°C, with a Q_{10} (change in reaction rate when temperature increases 10°C) of about 2.0, and a soil moisture content of 50 to 75% of soil water-holding capacity, although the actual optimum moisture percentage varies with soil texture (Table 4.2). For nitrification, optimum conditions include temperatures of 30 to 35°C, a moisture content of 50 to 67% of soil water-holding capacity, and a pH between 6.6 and 8.0. Nitrifying organisms are more sensitive to excessive soil acidity and their activity decreases markedly when the soil pH is less than 5.0. Nitrate production has been observed in highly acidic mine soils and forest soils, suggesting that some nitrifiers have adapted to these unfavorable pH conditions.

4.2.2   Gaseous losses of nitrogen: ammonia volatilization and denitrification

Ammonia volatilization refers to the loss of NH_3 from the soil as a gas and is normally associated with high free NH_3 concentrations in the soil solution and high soil pH. Surface applications of ammoniacal fertilizers or readily decomposable organic wastes (animal manures, municipal biosolids) to soils can result in considerable N loss by NH_3 volatilization, particularly if the soil (or organic waste) is alkaline in nature. This reaction and

Table 4.2   Influence of Soil Temperature and Moisture on Nitrogen Mineralization

Soil temperature (°C)	N Mineralization rate constant (k)(week^{-1})	
	Range for 11 soils	Mean
5	0.007–0.015	0.009
15	0.010–0.022	0.014
25	0.019–0.047	0.029
35	0.044–0.071	0.055
	Total N mineralized at 35°C (mg/kg)	
Soil moisture tension (MPa)	Fine sandy loams	Loams to clay loams
0.01	39	71
0.03	36	67
0.20	29	50
1.50	26	43
Estimated optimum soil moisture content (%):	13	28

Sources: Adapted from Stanford et al., 1973 (temperature); Stanford and Epstein, 1974 (moisture).

the major factors influencing the magnitude and rate ($d[NH_3]/dt$) of N loss by volatilization can be summarized as follows:

$$NH_4^+ \leftrightarrow (NH_3)_{solution} + H^+ \qquad (4.10)$$

$$(NH_3)_{solution} \leftrightarrow (NH_3)_{gas} \qquad (4.11)$$

and

$$\frac{d[NH_3]}{dt} = [A] \times [K] \times [P_l - P_g] \qquad (4.12)$$

where

$\dfrac{d[NH_3]}{dt}$ = NH_3 loss with time (t)
A = area of soil–solution interface
K = mass transfer coefficient, a function of air velocity above the soil and the air and soil temperatures
P_l = partial pressure of NH_3 in soil solution
P_g = partial pressure of NH_3 in air above soil solution

Volatilization represents both the loss of a plant nutrient and a potential environmental/ecological impact as surface waters can be enriched by NH_3 volatilized from areas where organic wastes are concentrated (e.g., feedlots, manure lagoons) and soils can be acidified when the NH_3 deposited is nitrified. For example, a study in England (Pitcairn et al., 1999) found that NH_3 emissions from livestock operations adversely affected flora in nearby woodlands (<300 m from NH_3 source). The number of natural species in the woodland decreased and "weedy" species increased. A "critical depositional load" of >20 kg N/ha was identified, above which undesirable changes in species diversity occurred.

As described above, the key factors in the volatilization of NH_3 from soils are those that affect (1) the transfer of a gas between the soil solution and the atmosphere (area of solution–atmosphere interface, velocity of air across the soil surface) and (2) the general rate of a chemical reaction (temperature, partial pressure of NH_3 in both phases). Soil and management factors that control these two aspects of NH_3 volatilization include soil temperature, moisture, and texture, nature of N source, and methods of application of fertilizers or organic wastes (e.g., surface broadcast, injected, incorporated). Conditions associated with maximum volatilization losses of NH_3 will include surface applications of fertilizers or manures, high pH or calcareous soils, soils with low cation exchange capacities and therefore little ability to retain the NH_4^+ cation, and a warm, slightly moist environment. In general, the most effective method to reduce NH_3 volatilization losses is to incorporate the N source into the soil by tillage, injection, irrigation, or, through timing of application, by natural rainfall, as shown in Figure 4.5 and Table 4.3. Once incorporated, the cationic nature of NH_4^+ results in its electrostatic attraction to cation exchange sites on clays and organic matter, thus reducing NH_3 losses. Volatilization is a particular problem for pastures, turf, and no-tillage agriculture where surface applications of fertilizers and animal manures are required.

Urea [$CO(NH_2)_2$] fertilizers and uric acid found in animal manures and other organic wastes represent a special case of NH_3 volatilization. When urea is added to a soil it is

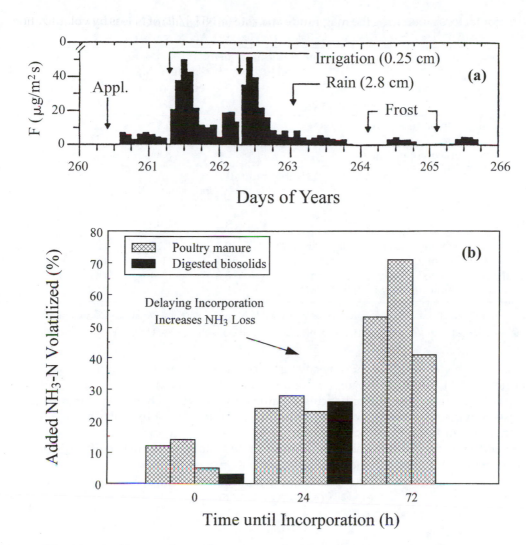

Figure 4.5 (a) Volatilization of NH_3 from urea applied to a moist, bare soil surface. Initial rate of volatilization (F) is slow. Application of low rate of irrigation (0.25 cm) dissolves urea, enhances NH_3 loss until soil surface dries (days 261 to 262). Second irrigation results in similar pattern of NH_3 loss as soil wets and dries. Rainfall (2.8 cm) on day 263 leaches NH_3 into soil and volatilization rate decreases markedly. (From McInnes, K. S. et al., *Agron. J.*, 78, 192, 1986. With permission.) (b) Effect of time between application of poultry manure and digested sewage sludge on NH_3 losses from soils. (Adapted from Donovan and Logan, 1983; Gartley and Sims, 1993.)

decomposed in a reaction that is catalyzed by the enzyme urease (urea amidohydrolase), as shown below:

$$\text{CO(NH}_2)_2 + 2H_2O \xrightarrow{\text{Urease}} (NH_4)_2CO_3 \text{ (unstable)} \rightarrow CO_2 + 2NH_3 + H_2O$$

(4.13)

Urease is produced by soil microorganisms involved in the decomposition of soil organic matter, in crop residues, and other organic materials. Urease acts by rapidly hydrolyzing

Table 4.3 Effect of Application Method and Rainfall on Corn Yields
Where Volatile Ammonia-Based Fertilizers (UAN, Urea) Were Used

Effect of application method

| | | Corn yield (Mg/ha) | | | |
| | | N rate (kg/ha) | | | |
N source	Application method	90	180	270	Mean
UAN	Broadcast spray	5.6	6.8	7.2	6.5
	Surface band	7.4	8.3	8.7	8.1
	Incorporated band	7.9	8.8	8.7	8.5

Effect of rainfall

| | | Corn yield (Mg/ha) | | | |
| | | N rate (kg/ha) | | | |
N source	Rainfall	50	101	202	Mean
NH_4NO_3	25 mm within 36 h of application	10.1	11.3	11.6	11.0
Urea	25 mm within 36 h of application	10.4	11.7	11.2	11.1
NH_4NO_3	None for 3 days then 3 mm	8.8	10.7	11.1	10.2
Urea	None for 3 days then 3 mm	8.5	9.2	10.4	9.4

Note: In rainfall effects study, NH_4NO_3 represents a stable N source with low volatilization potential.

Sources: Adapted from Touchton and Hargove, 1982 (application method); Fox and Hoffman, 1981 (rainfall effects).

the C–NH_2 bonds in the urea molecule. The rapid formation of a highly concentrated solution of NH_4^+ in the microenvironment surrounding adjacent urea granules increases the pH to about 8.5, greatly enhancing volatilization of NH_3. Chemical inhibitors of the urease enzyme have been developed to reduce the rate of urea hydrolysis and thus the potential for N losses by NH_3 volatilization.

Denitrification is defined as the reduction of NO_3^- to gaseous forms of N (NO, N_2O, N_2) by chemoautotrophic bacteria, as follows:

$$NO_3^- \rightarrow NO_2^- \rightarrow NO \rightarrow N_2O \rightarrow N_2$$

nitrate   nitrite   nitric   nitrous   dinitrogen
                      oxide   oxide

(4.14)

The bacteria responsible for denitrification are normally aerobic, but under anaerobic conditions certain bacteria can use NO_3^- as an alternative to O_2 as an acceptor of electrons produced during organic matter decomposition. During the denitrification process, N is reduced from an oxidation state of +5 in NO_3^- to 0 in N_2. The critical factors regulating the rate and duration of denitrification in soils are the availability of NO_3^- (the substrate) and C (source of energy and electrons) and the absence of O_2. For denitrification to occur, soils must first produce or be amended with NO_3^- and then enter an anaerobic period. Conversely, denitrification can be inhibited by any process that restricts NO_3^- production (e.g., slows nitrification) or enhances NO_3^- removal (leaching, plant uptake), or promotes aerobic conditions (artificial drainage, plant depletion of soil moisture).

As NO_3^- originates primarily by nitrification, an aerobic process, the conditions most conducive to denitrification are alternating aerobic and anaerobic cycles or adjacent aerobic and anaerobic zones. Soils that are periodically flooded or experience transitory anaerobic conditions due to heavy rains can accumulate pools of NO_3^- during an aerobic period that is lost by denitrification during the subsequent anaerobic cycle. Rapid O_2 consumption by soil microorganisms during the decomposition of organic matter can produce anaerobic zones adjacent to aerobic areas where NO_3^- is being produced. When NO_3^- moves into

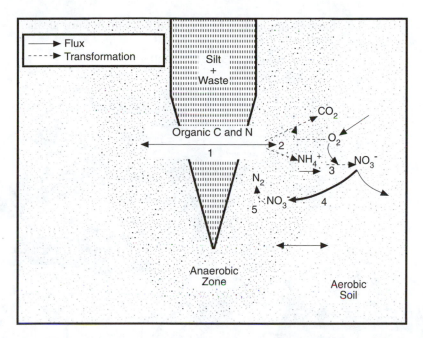

Figure 4.6 Denitrification induced by slit injection of a highly decomposable pharmaceutical waste. Concentrating available carbon stimulates high O_2 consumption, producing an anaerobic zone adjacent to an aerobic zone where nitrification occurs as waste mineralizes. (From Rice, C. W. et al., *Soil Sci. Soc. Am. J.*, 52, 102, 1988. With permission.)

the anaerobic zone by diffusion or mass flow, it can then be denitrified, as illustrated in Figure 4.6 for a soil receiving slit injections of a pharmaceutical waste. Common examples of anaerobic zones are microsites in the rhizosphere that have been enriched with root exudates, localized deposits of highly decomposable crop residues, and organic wastes injected into relatively small soil volumes, as opposed to a more uniform spatial distribution. An ecological example would be a wetland adjacent to an agricultural field or perhaps an artificial wetland used to treat wastewaters from a livestock operation (Figure 4.7). In these situations NO_3^- produced in the normally aerobic field soil moves by leaching and lateral flow into the wetland, dominated by flooded anaerobic conditions and high quantities of available C, and is then removed from drainage waters by denitrification.

Soil temperature and pH can also influence denitrification. Although denitrification can occur at temperatures between 2 and 75°C, the optimum temperature is ~30°C. Soil temperature affects not only the rate of microbial metabolism, but also influences chemical processes such as the rate of diffusion of O_2, NO_3^-, N_2O, and N_2 in the soil water and atmosphere. The optimum range in pH for denitrifying organisms is from 6.0 to 8.0, but as with nitrification, denitrification has also been measured in highly acidic soils.

4.2.3 Leaching and erosional losses of nitrogen

Nitrogen can be transported from soils into groundwaters and surface waters by *leaching, erosion,* or *runoff.* Losses of N by leaching occur mainly as NO_3^- because of the low capacity of most soils to retain this anion. In general, any downward movement of water through the soil profile will cause the leaching of NO_3^-, with the magnitude of the N loss proportional to the concentration of NO_3^- in the soil solution and the volume of leaching water. Leaching of NO_3^- is economically and environmentally undesirable. Nitrate that leaches below the crop rooting zone represents the loss of a valuable plant

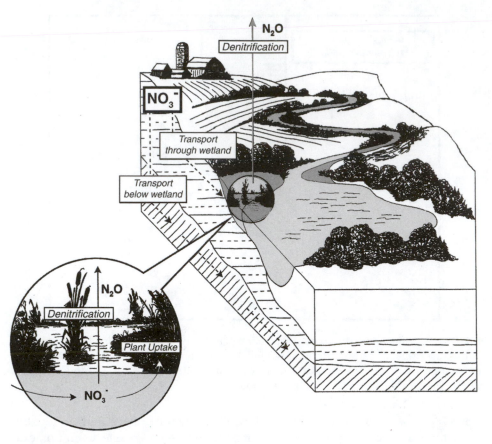

Figure 4.7 Wetlands as denitrifying zones capable of removing NO_3^- from runoff or subsurface drainage from agricultural fields.

nutrient and, as mentioned earlier, can contribute to pollution of groundwater aquifers and eutrophication of surface waters, as shown in Figure 4.8. Much of the research conducted with fertilizers, animal manures, and other by-products (e.g., municipal biosolids) has been directed toward reducing NO_3^- leaching, especially in humid regions. Nitrate pollution of groundwaters, however, is not a universal problem and is often regional or local in nature, as shown in Figure 4.9.

Situations most conducive to NO_3^- leaching and groundwater pollution include sandy, well-drained soils with shallow water tables in areas that receive high rainfall or intensive irrigation and frequent use of fertilizers, manures, or other N sources. Nitrate leaching concerns are not restricted to these conditions, however. Any situation involving over-application of N, organic waste storage areas (e.g., feedlots, lagoons), or intensive irrigation has the potential to cause significant NO_3^- leaching, regardless of soil type and climate. Chemical retention of NO_3^- in soil profiles, unlike other anions (phosphate, sulfate), is less important in reducing leaching, although some highly acidic subsoils have been shown to have significant anion exchange capacity. Denitrification in groundwaters or ground-water discharge areas (e.g., wetlands) may reduce leaching losses of N. Subsoil denitrification, however, is likely to be of little value in mitigating NO_3^- leaching, primarily because of the low available C levels in most subsoils. In any case, given the atmospheric impacts of N oxides, management techniques designed to control leaching by enhancing denitrification are certain to be examined carefully. Approaches to minimize NO_3^- leaching, therefore, are operational in nature and focus on controlling the timing of NO_3^- formation

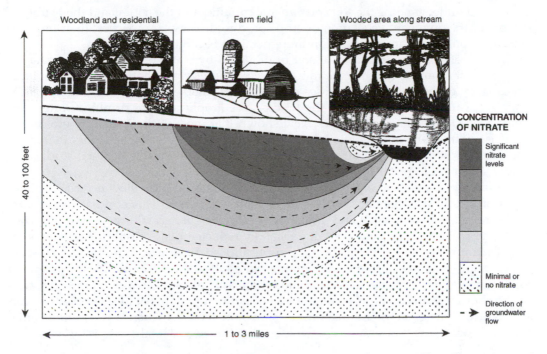

Figure 4.8 Generalized representation of NO$_3^-$ transport in groundwater aquifers, illustrating variable nature of NO$_3^-$ concentration with land use, aquifer depth, and distance to discharge area. (Adapted from Hamilton and Shedlock, 1992.)

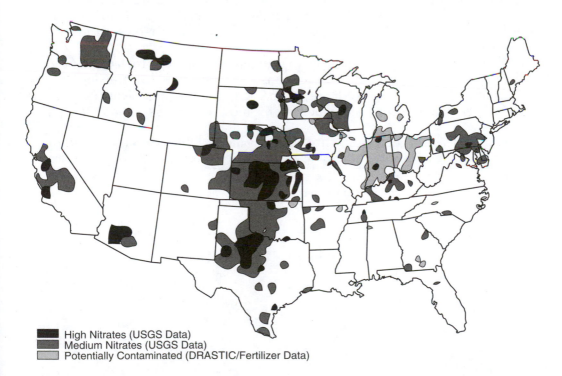

Figure 4.9 Potential for groundwater contamination by NO$_3^-$ in the U.S., illustrating regional nature of the problem. (Adapted from Nielsen and Lee, 1987.)

in soils, and understanding the soil and climatic conditions of the region and the N uptake patterns of the dominant vegetation or crops produced. These approaches, their implications, and constraints are discussed in more detail in Section 4.4.

Erosion refers to the transport of soil from a field by water or wind; *surface runoff* is the water lost from a field when the rate of precipitation or irrigation exceeds the infiltration capacity of the soil. Both processes can transport soluble inorganic N and organic N to surface waters and contribute to the process of eutrophication or drinking water contamination. Many watershed studies have shown that most of the N lost by erosion or runoff is sediment-bound organic N. Although the solubility of NO_3^- favors its loss in runoff as opposed to sediment transport, total N losses from most watershed studies are usually severalfold greater than soluble N.

Surface applications of organic wastes are undesirable because they increase the likelihood of soluble and organic N losses by erosion and runoff. This approach is usually not permitted with municipal and industrial wastes, but in agricultural operations, conservation practices designed to control erosion by reducing tillage result in application of animal manures to soil surfaces. Surface applications of manures can also occur when farmers apply manures during winter months, when the soil is frozen and less susceptible to equipment damage and erosion, and more time is available. The use of grassed waterways or border strips that trap sediment and accumulate soluble N in plant biomass can help reduce N losses in these situations, as shown in Table 4.4 where a cornstalk residue strip 2.7 m wide with 50% ground cover reduced sediment and total N losses by 70 to 80%.

4.2.4   Biological nitrogen fixation

Biological N fixation is the conversion of atmospheric N_2 into an organic form of N, either through symbiotic associations between plants and microorganisms, or independently by free-living organisms such as cyanobacteria ("blue-green algae") and certain heterotrophic bacteria. In the context of the global N cycle, biological N fixation represents the major N input to soils.

Symbiotic N fixation can occur between leguminous plants and bacteria, nonlegumes and actinomycetes, and plant–algal associations. In this symbiosis an N-fixing organism can enzymatically reduce N_2 to NH_3 using an energy supply (photosynthate) provided by the host. From an agricultural perspective, the most important symbiotic N fixation is

Table 4.4   Influence of Antecedent Soil Moisture Conditions on the Effectiveness of Residue Strips at Reducing Erosion and Runoff Losses of Nitrogen from a Bare Soil with 5% Slope

Strip width, residue cover, and antecedent moisture	Entering residue strip			Leaving residue strip		
	Sediment[a]	Runoff[a]	Total N[b]	Sediment[a]	Runoff[a]	Total N[b]
1.8 m and 27% cover						
Dry	9.5	171	22.5	6.0	244	16.5
Very wet	16.6	320	31.9	14.3	373	28.2
2.7 m and 50% cover						
Dry	22.7	284	45.3	4.9	280	14.1
Very wet	24.0	386	39.4	4.9	462	13.0

[a] kg/h/m of width.

[b] g/h/m of width.

Source: Adapted from Alberts et al., 1981.

that occurring between leguminous plants and bacteria of the genera *Rhizobium* and *Bradyrhizobium*. Although we focus on N fixation by legumes, it should be recognized that the ability of nonlegumes to fix atmospheric N_2 can be equally important to nonagricultural ecosystems and to certain crops. An example of nonleguminous, N fixation is the symbiotic relationship between actinomycetes of the genus *Frankia* and a wide variety of trees and woody shrubs that are important in soil formation, revegetation of disturbed, highly erodible forest soils, and as sources of fuel-wood. A nonleguminous, agricultural example of symbiotic N fixation is the association between the cyanobacterium *Anabaena* and the freshwater, free-floating fern *Azolla* found widely in flooded, tropical areas used for rice production. The *Anabaena* are located on the stem and fronds of *Azolla*, but unlike *Rhizobium* or *Frankia*, they do not directly convert N_2 into organic N within the plant. Rather, *Anabaena* produce soluble NH_4^+ that is taken up by *Azolla* roots growing in the water or upper portion of the sediment of a rice paddy. This fixed N can then be used directly by rice when the *Azolla* die and the organic N is mineralized, or in some countries, the *Azolla* is harvested and used as animal feed or an organic fertilizer.

Leguminous plants have been important components of agricultural crop production systems for centuries and represent a major pathway for the conversion of atmospheric N to soil N and then to a form that can be used by humans and animals. Over 14,000 species of legumes are known, of which more than 100 are commonly used in agriculture. In addition to their value as food crops, some legumes are grown as cover crops to provide N for a subsequent nonleguminous crop (e.g., clovers or vetches followed by corn). Environmental issues related to legumes will be discussed in more detail in Section 4.4.2, but are similar to those associated with other organic N sources, and primarily center on the management of legume N to avoid NO_3^- leaching, runoff and denitrification.

The process of symbiotic N fixation by legumes begins when bacteria living in the soil near the root system of a host plant attach themselves to cells of root hairs and in doing so induce a curling of the root hair around the bacteria. The bacteria then produce an infection thread that invades the root cell, allowing bacterial penetration and proliferation within the root cortex. In response to this bacterial invasion, the plant synthesizes a nodule, a protective structure that encapsulates the bacteria that have now changed both physical shape and metabolic function and which are now referred to as bacteroids (Figure 4.10). The bacteria–host plant relationship is often quite specific and a given species of

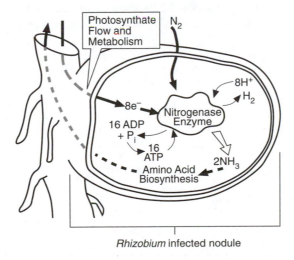

Figure 4.10   Schematic representation of the form and function of bacteroids in the nodules of legume roots.

Rhizobium or *Bradyrhizobium* will not infect or nodulate incompatible host plants. The nodule is comparable with other energy-converting organs in the plant, and possesses a membrane that regulates the entry and exit of metabolites to and from the bacteroids. In the N fixation reaction bacteroids use photosynthate-derived energy (ATP and electrons), O_2, and a specific enzyme, *nitrogenase*, to reduce N_2 to NH_4^+, which is then released into the host cell and used in amino acid synthesis. The overall equation for N fixation is

$$\text{Nitrogenase}$$
$$N_2 + 16\,ATP + 8e^- + 10H^+ \rightarrow 2\,NH_4^+ + H_2\,(g) + 16\,ADP + 16P_i \tag{4.15}$$

Each host plant–bacteria association has its own genetic potential for N fixation. For a given symbiosis, the actual amount of N fixed depends primarily on available soil N and the energy status of the plant, which in turn is influenced by such factors as carbohydrate supply, light intensity, soil nutrient status, soil temperature, and moisture. Examples of agriculturally important legumes and estimated amounts of N fixed under optimum conditions are given in Table 4.5.

A generalization of environmental factors important in N fixation would include a soil pH (near neutral), an adequate supply of key nutrients, P, potassium (K), calcium (Ca), copper (Cu), Fe, molybdenum (Mo), cobalt (Co), "optimum" temperature and moisture (varies with host plant), adequate aeration, and a low level of available soil N (NH_4^+ and NO_3^-). It is important to remember, however, that even under optimum conditions legumes do not obtain all of their N from biological N fixation. Soil N can provide as much as 50% of the total N in legumes with low N fixation rates. Also, if the concentration of NO_3^- in the soil solution exceeds ~1 mM, the process of N fixation is restricted. This can occur when legumes are planted in soils with high levels of residual N, perhaps resulting from previous fertilizer or organic waste applications. If supplied with sufficient N fertilizer, legumes will not nodulate and N fixation will not occur. This has led to debates over the economic value of "starter" N fertilizers for legumes (e.g., soybeans) as well as the environmental impact of applying organic N sources, such as animal manure and municipal biosolids, to legumes. The presence of a small amount of available N prior to effective nodulation may benefit the plant, but excessive fertilizer N will often reduce or prevent biological N fixation.

In addition to symbiotic processes, N fixation can occur with some species of "free-living" organisms. The major examples are the cyanobacteria (e.g., *Anabaena*, see above)

Table 4.5 Major Legume Crops Used in Agriculture
and Estimates of Annual Nitrogen Fixation

Crop and N-fixing bacteria		Annual N_2 fixation	
		Range (kg/ha/year)	Typical value (kg/ha/year)
Host plant	***Rhizobium***		
Alfalfa	*R. meliloti*	60–500	225
Clovers	*R. trifolii*	60–350	115
Peas, vetch	*R. leguminosarum*	90–180	100
Beans	*R. phaseoli*	20–100	45
Host plant	***Bradyrhizobium***		
Lupins	*B. lupinii*	150–170	160
Soybeans	*B. japonicum*	65–200	100
Cowpea	*B. parasponiae*	65–130	100

Note: *Rhizobium* are fast-growing symbiotic bacteria; *Bradyrhizobium* are slow-growing. Average values from various sources.

that are autotrophic (requiring only light, water, N_2, CO_2, and salts) and certain heterotrophic bacteria. Important nonsymbiotic, N-fixing bacteria include *Azotobacter* and *Beijerinckia*, aerobic saprophytes that obtain their energy from the oxidation of organic matter; *Azospirillum*, facultatively anaerobic bacteria; and *Clostridium*, anaerobic saprophytes. In general, because the amount of N fixed by these bacteria is small (5 to 50 kg N/ha) they are of little importance to most soil N management programs.

4.3   Sources of nitrogen

The production of food and fiber, the growth of plants for aesthetic purposes, and the reclamation and stabilization of lands disturbed by construction, mining, and other industrial or urban activities often require the addition of supplemental N to obtain optimum plant growth. Two broad categories of N sources exist, inorganic and organic. Inorganic N sources are predominantly commercial fertilizers, but also include limited quantities of mineral deposits and industrial by-products. Organic N sources commonly include animal manures, crop residues, municipal biosolids and wastewaters, and a wide variety of industrial organic wastes.

For most of recorded history, the major source of N added to soils was organic in nature, primarily as animal manures or crop residues. The first N fertilizer used commercially in the U.S., Peruvian guano, was also organic in nature, formed from centuries of deposition of excreta by seafowl along the South American coast. In the late 1800s and early 1900s, industrial processes were devised that could fix atmospheric N_2, converting it to either NH_3 gas, Ca cyanamide, or nitric acid. Later improvements in the efficiency of one such process, the Haber–Bosch method for the synthesis of NH_3 gas from atmospheric N_2, resulted in the availability of inexpensive, high analysis N fertilizer materials which began to change markedly the nature of agriculture on a global scale. Food production was no longer limited by the availability of soil, manure, or legume N because fertilizer N could now be used to greatly increase yields of most agricultural crops. Beyond this, the ease of handling commercial N fertilizers reduced the labor requirements of crop production and contributed to the development of larger, more specialized farms devoted in many cases to the production of fewer crops. Increased production capacity did not always equal increased efficiency of N use, however. As farmers, and those advising farmers, learned to use new fertilizer materials alone or in combination with organic N sources, new application equipment, and new cropping systems, they often proceeded without a real understanding of the potential environmental impacts involved. The uncertainty associated with these new production practices and the low costs of fertilizer N undoubtedly resulted in overfertilization with N and contributed to groundwater contamination by NO_3^-, especially in areas of intense fertilizer use.

Regardless of the form of N used or the nature of the soil–plant system, maximizing the efficiency of N recovery and minimizing the potential of the N source to pollute the environment are now fundamental goals of modern agriculture. To manage either type of N source properly, it is essential to understand how the physical, chemical, or biological properties of the material affect its handling, application, and fate among the many transformations of the soil N cycle. It is also important to understand that all environmental problems associated with improper N management can be caused by both inorganic and organic N sources. That is, despite the interest in sustainable agriculture and organic farming, research has found no intrinsic superiority associated with organic N sources. In fact, in many situations, the physical properties and heterogeneity in composition of organic wastes make them more difficult to manage successfully than inorganic N fertilizers.

The purpose of this section is to provide an overview of the production, composition, and characteristics of the major inorganic and organic sources of N used as soil amendments and to illustrate the key aspects related to sound environmental use of all N sources.

4.3.1   Inorganic sources of nitrogen

Most commonly used N fertilizers, summarized in Table 4.6, are produced from NH_3 gas synthesized by the Haber–Bosch process. This process uses natural gas (CH_4), atmospheric N_2, and steam (H_2O) to produce NH_3 gas as follows:

$$7CH_4 + 10H_2O + 8N_2 + 2O_2 \rightarrow 16NH_3 + 7CO_2 \qquad (4.16)$$

This clearly illustrates another environmental aspect of N use, its impact on natural resources, as the production of N fertilizers consumes natural gas, a finite and critically important natural resource. Efficient use of N fertilizers will thus enhance the longevity of natural gas supplies.

Once NH_3 has been synthesized it can be (1) used directly as anhydrous ammonia, a pressurized gas [NH_3]; (2) reacted with CO_2 to form urea [$CO(NH_2)_2$]; (3) oxidized to NO_3^- and reacted with more NH_3 to form ammonium nitrate [NH_4NO_3]; and (4) combined with sulfuric acid to produce ammonium sulfate [$(NH_4)_2SO_4$] or with various types of phosphoric acid to form ammonium phosphates such as diammonium phosphate [DAP, $(NH_4)_2HPO_4$] or monoammonium phosphate [MAP, $NH_4H_2PO_4$]. Further industrial processes can be used to produce N solutions such as urea-ammonium-nitrate [UAN] or aqua ammonia [NH_3], or controlled, slow-release, solid N fertilizers that are coated with resins (Osmocote, used in greenhouses and nurseries) or sulfur (S-coated urea) to delay their rate of dissolution in the soil. A wide variety of mixed fertilizers

Table 4.6   Properties of Major Commercial Nitrogen Fertilizers

N Source	Chemical composition	N (%)
Ammoniacal N sources		
Anhydrous ammonia	NH_3	82
Aqua ammonia	$NH_3 \cdot H_2O$	20–25
Ammonium chloride	NH_4Cl	25
Ammonium nitrate	NH_4NO_3	33
Ammonium sulfate	$(NH_4)_2SO_4$	21
Nitrate N sources		
Calcium nitrate	$Ca(NO_3)_2$	15
Potassium nitrate	KNO_3	13
Sodium nitrate	$NaNO_3$	16
Urea materials		
Urea	$CO(NH_2)_2$	45
Urea-ammonium-nitrate solutions	30–35% urea: 40–43% NH_4NO_3	28–32
Ureaform	Urea-formaldehyde	38
IBDU	Isobutylidene diurea	32
SCU	S-coated urea	36–38
N–P materials		
Monoammonium phosphate (MAP)	$NH_4H_2PO_4$	11
Diammonium phosphate (DAP)	$(NH_4)_2HPO_4$	18–21
Ammonium polyphosphates (liquid)	$(NH_4)_3HP_2O_7$	10–11

containing N, P, K, and other nutrients are also produced. Many urban situations (e.g., turf, home gardens) and land reclamation projects make extensive use of mixed fertilizers to provide N and enhance the overall nutritional status of low-fertility soils. Nitrate-N fertilizers such as Ca or K nitrate [$Ca(NO_3)_2$, KNO_3] are also available, but due to the production costs and lower efficiency of N recovery from NO_3-based materials are primarily used on specialty crops (vegetables, fruits, tobacco) or on crops that are sensitive to NH_3.

Economics, more than any other factor, often controls the N source selected. In general, fertilizers with higher N contents have lower costs of storage, transportation, handling, and application, and hence are more economic. However, the properties of some high-analysis N fertilizers can increase these costs. For example, anhydrous ammonia (NH_3), which has the highest N content of any fertilizer material (82% N), requires complex application equipment that must pressurize the gas, convert it to a liquid, and inject the liquefied NH_3 into the soil in a manner that reduces volatilization losses. Ammonia volatilization losses are also a major concern with urea, the highest analysis solid N fertilizer (45% N), and some type of incorporation is often required to maximize the efficiency of N recovery from this N source. The physical and chemical properties of N sources are not the only factors that influence source selection. In the U.S. there are marked regional preferences for N fertilizers, often related more to the fertilizer manufacturing and transportation infrastructures that have evolved with time rather than to specific requirements of regional cropping systems (Figure 4.11).

Other factors that influence the selection of an N fertilizer include crop management practices (crop type and rotation, tillage, irrigation), soil type and sensitivity to N losses, the effect of the N source on soil pH (nitrification is an acidifying process and N fertilizers vary in their potential to decrease soil pH), the need to supply other nutrients simultaneously, and the suitability of the material for existing application equipment. The implications of these factors for N management are discussed in more detail in Section 4.4.

4.3.2   Organic sources of nitrogen

A wide variety of organic materials are used as soil amendments, many of which contain appreciable quantities of N. Animal manures, municipal biosolids and wastewaters, composts of municipal solid waste or yard waste, food processing wastes, industrial organics, and crop and forest product residues are the dominant organic wastes produced worldwide. Estimates of the quantity of organic wastes produced annually in the U.S. exceed 400 million Mg per year. The magnitude of this problem is even more apparent when expressed on a "per-person" basis. For animal manures, municipal solid wastes and biosolids, industrial solid wastes and water treatment sludges, and silviculture residues, the amount generated is roughly equivalent to 900, 550, 450, and 400 kg per person per year (dry weight basis), respectively. Changes occurring in the U.S. and other countries are likely to alter the nature and distribution of organic wastes significantly. In many states landfilling of yard wastes (leaves, grass clippings), which accounted for ~20 to 25% of the landfill volume, is rapidly becoming an unacceptable practice, and is being replaced with the commercial production of yard waste composts. Similarly, a number of municipalities have developed co-composting facilities for biosolids and the organic fraction of municipal refuse. Production of these composts is certain to result in large increases in the amount of organic N applied to soils. Other changes include the production of composted or pelletized animal manures in areas where large excesses of manure, relative to arable land, exist. Movement of animal manures off the farm for use in urban areas, or in construction and reclamation projects,

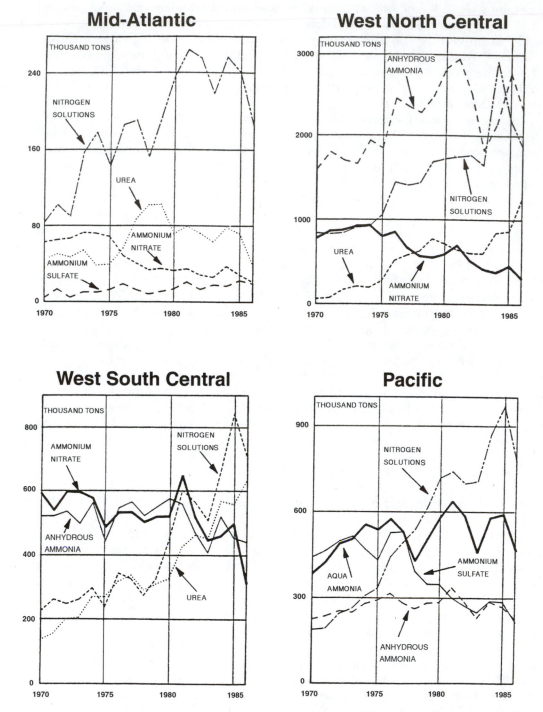

Figure 4.11 Regional preferences and trends in N fertilizer use for 1970 to 1985. (Adapted from Berry and Hargett, 1986.)

will likely result in an increased use of these materials in land application programs involving horticultural crops, turf, and revegetation of sites for use by domestic animals and wildlife.

Table 4.7   Representative Values for Nitrogen
and Availability for Selected Organic Wastes

Organic N source	Total N (%)	Organic N mineralized[a] (%)
Animal manures		
Beef	1.3–1.8	25–35
Dairy	2.5–3.0	25–40
Poultry	4.0–6.0	50–70
Swine	3.5–4.5	30–50
Biosolids		
Aerobic digestion	3.5–5.0	25–40
Anaerobic digestion	1.8–2.5	10–20
Composted	0.5–1.5	(−10)–10
Other wastes		
Fermentation wastes	3.0–8.0	20–50
Poultry processing wastes	4.0–8.0	40–60
Papermill sludges	0.2–1.0	(−20)–5

Note: Average values from various sources.

[a] Organic N mineralized estimated from laboratory incubation studies. Negative values
for composts and papermill sludges indicate that immobilization of N occurred.

Although all organic wastes contain N, the amount, forms, and availability of N can vary widely. Mean N contents and ranges are available for most organic wastes, but the considerable variability in total and inorganic (e.g., NH_3, NO_3) N contents of most wastes makes interpretation of analytical results for N or other elements an ongoing problem. Broad generalizations of the N content in organic wastes are probably justified as it has been well documented that certain organic wastes will consistently have higher total N contents than others, as shown in Table 4.7 for animal manures (poultry > swine > dairy > beef) and municipal biosolids (aerobic > anaerobic > composted). The wide range in total N content among similar types of wastes, however, can have significant implications for N loading to soils and crops, as shown in Figure 4.12. In this study, the amount of manure needed to provide a desired amount of N for corn was estimated for poultry manure based on analysis of manure samples from 17 different on-farm storage areas. When the predicted amount of N to be added was compared with the actual N applied, based on analysis of manure samples collected during application, overapplication of 10 to 20 kg N/Mg manure commonly occurred, as did underapplication of 5 to 10 kg N/Mg. Therefore, the accurate application of a recommended manure rate for corn (~5 Mg/ha), based on analysis of the manure, commonly resulted in the application of *excess* manure N approaching the total N requirement of the crop (~100 kg N/ha). Clearly, a more comprehensive approach than simple N analysis and equipment calibration is needed to avoid over- or underapplication of N from organic wastes.

In general, the key to effective use of organic N sources is an understanding of the factors that influence the extent and rate of conversion of organic N to forms that are available for plant uptake or loss to the environment. The availability of N in organic wastes will be influenced by their composition, largely controlled by waste production and storage practices, and by the chemical and biological changes they undergo following application to the soil. Much of the research in this area has been directed toward identifying the differences in N availability between various types of wastes and then using waste properties to predict N availability. As with waste composition, these studies have

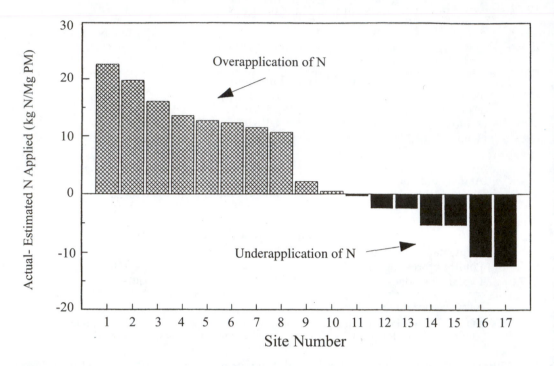

Figure 4.12   The difference between total N applied based on poultry manure (PM) samples collected during field application and the amount estimated to be applied based on laboratory analyses of stockpiled manure samples. Results of 17 field studies. (Adapted from Igo et al., 1991.)

shown that wastes can be broadly grouped in terms of organic N availability (Table 4.7). Certain wastes (e.g., poultry manure, aerobically digested biosolids) are not only higher in total N but will provide more N upon decomposition in the soil than other more stable wastes (e.g., composts, papermill sludges). Simple and complex approaches have been taken to predict N availability. Biosolids have been ranked according to N mineralization potential as waste-activated (40%) > raw and primary (25%) > anaerobically digested (15%) > composted (8%). For some wastes both total and NH_4-N must be included to estimate accurately the amount of potentially available N (PAN), as shown in a simple model developed for poultry manures:

$$PAN = [k_m(N_0) + e_f (NH_4\text{-}N + NO_3\text{-}N)] \tag{4.17}$$

where k_m is the percentage of added organic N (N_0) mineralized (k_m = 40 to 60%, depending on season of year manure was applied, e.g., winter vs. spring), and e_f is a factor reflecting the efficiency of recovery of NH_4-N (20 to 80%, depending on time until incorporation of manure) and NO_3-N added in the manure.

The use of organic wastes as N sources cannot be based on N availability alone. Other nutrients, or nonessential elements in these materials can determine not only the application rate but also their suitability for various end uses. At present, there are three main aspects of waste composition that affect short- and long-term use of organic wastes in land application programs: (1) P buildup to excessive levels in waste-amended soils; (2) the potential for heavy metal contamination of soils, crops, and waters; and (3) the possible adverse environmental impacts of organic pollutants found in organic wastes. Readers are referred to Chapters 5 (Phosphorus), 7 (Trace Elements), 8 (Organic Chemicals), and 12 (Risk Assessment) for more detailed discussions of these topics.

4.4 Principles of efficient nitrogen management

Efficiency can be defined as the ability to accomplish a task without the waste of time, energy, or resources. Nitrogen efficiency for a soil–plant system, in its simplest form, can be viewed as the ability to manage the time, energy, and resources needed to obtain an acceptable level of plant growth (the task) with minimal loss of N (the waste). Few, if any, natural systems are 100% efficient, and, given the complexity of the soil N cycle and the constraints imposed by time, labor, soil type, cropping practices, environmental conditions, and available resources, it is perhaps not surprising to find that efficiency values for the recovery of applied N in most cropping systems rarely exceed 60% and commonly range between 30 and 50% under normal management. Unfortunately, where N is concerned, inefficiency of recovery by one ecosystem (soil–plant), often results in the redistribution of NO_3^-, NH_3, or N_2O to another. This represents a second impact of N efficiency, the potential for adverse effects on another resource, the environment. Based on the above definition of efficiency, the key questions that must be addressed for N management are as follows:

- How do we define an acceptable level of plant growth, particularly in nonagricultural systems such as urban areas, forests, and land reclamation sites?
- How do we decide what degree of N "waste" is unacceptable, given the complex interactions between the politics and economics of plant production, the unpredictable cycling of N between soil, air, and water, and the varying degrees of ecosystem sensitivity to N pollution?
- Perhaps most important, how do we design more efficient systems for N use, and monitor them well enough to know if they are improvements over past practices?

Questions such as these rarely have straightforward answers; rather, we respond to them by the development of processes and management systems that continually evolve as our knowledge base grows. Acceptable levels of plant production must consider both global food shortages and the profit margins of individual farmers. In many urban areas and areas dominated by animal-based agriculture, they must also reflect urgent waste disposal needs caused by the concentration of nutrients in areas that do not have the land base to use them. If a soil has been amended with wastes for years and requires little if any N to produce the maximum yield attainable given other inflexible constraints (e.g., sunlight, rainfall), how do we reconcile the concept of acceptable plant growth with the continued generation of animal manure or municipal biosolids N? Similarly, environmental regulatory agencies have established unacceptable levels of N in drinking waters and surface waters, but in some areas research has shown that, due to the nature of the soils, climate, and limitations imposed by current technology, we cannot profitably produce grain crops without having drainage waters that exceed these standards. Can we improve efficiency enough by management or legislation to overcome the fundamental limitations on crop N recovery in these areas? Even more important, how will we determine if our new, and perhaps more expensive, practices are reducing the N pollution of groundwaters, surface waters, or the atmosphere?

Complex problems such as these require multidisciplinary management approaches. The central challenge faced at present by research, advisory, and regulatory agencies worldwide is the need to integrate the expertise of many disciplines to develop management plans that maximize the economic value of fertilizers and organic waste N, while minimizing the adverse environmental impacts. Given the complexity of the soil N cycle and its interactions with other ecosystems, coordinated planning by agronomists, soil scientists, horticulturists, silviculturists, atmospheric chemists, engineers, hydrogeologists, microbiologists, resource economists, and others will be needed. In this section we focus on the *process* of efficient N management, primarily for agricultural crops, although some references to forest soils, horticultural operations, turf, and land reclamation are given to illustrate key differences in N management required for these situations.

4.4.1 General principles of efficient nitrogen use

Approaches to assess the efficiency of N use by agricultural crops normally include both agronomic and environmental components. We seek to maximize agronomic efficiency by producing greater yields with less N and environmental efficiency by minimizing the escape of added N from the soil to an ecosystem sensitive to N pollution. Quantitative assessment mechanisms are therefore essential if we are to evaluate the success of existing and proposed management programs. Crop *N use efficiency* (NUE) from an agronomic perspective (production and economics) has traditionally been expressed in terms of *yield efficiency* (YE) or *N recovery efficiency* (NRE), defined as

$$YE = \frac{[\text{Crop yield}]_{+N \text{ Source}} - [\text{Crop yield}]_{\text{Soil alone}}}{\text{Total N added by N source}} \qquad (4.18)$$

$$NRE = \frac{[\text{Crop N uptake}]_{+N \text{ Source}} - [\text{Crop N uptake}]_{\text{Soil alone}}}{\text{Total N added by N source}} \qquad (4.19)$$

where: $[\text{Crop yield}]_{+N \text{ Source}}$ and $[\text{Crop N uptake}]_{+N \text{ Source}}$ refer to the yield and total N uptake, respectively, by a crop that has received additions of either fertilizer or organic N. $[\text{Crop yield}]_{\text{Soil alone}}$ and $[\text{Crop N uptake}]_{\text{Soil alone}}$ refer to the yield and total N uptake, respectively, by a crop grown in an unamended soil; these values represent the native N-supplying capacity of the soil.

Yield efficiency considers only the relationship between N fertilization and crop production and is primarily an economic assessment. Nitrogen recovery efficiency provides an estimate of applied N that, because it was not taken up by the crop, has been redistributed to some other component of the N cycle. Nitrogen recovery efficiency thus addresses both economic and environmental concerns, and, if combined with other data, such as changes in soil NO_3^- concentrations in the crop rooting zone, can help identify the fate of the unrecovered N. Several key points should be kept in mind when evaluating NRE values for differing crops and management practices. First, it is important to remember that unrecovered N is not necessarily lost from the soil–plant system; it may be temporarily immobilized as organic N, remain in the rooting zone as NH_4^+ or NO_3^-, or even be recycled through irrigation waters. In these situations the unrecovered N from one crop may be recovered through uptake by a subsequent crop. Second, as shown in Figure 4.13, the curvilinear nature of crop response (yield and N uptake) to applied N means that estimates of YE and NRE will vary with N rate. While, generally speaking, YE and NRE decrease and the potential for N loss to other ecosystems increases as N rate increases, the percentage of applied N that is not recovered by the crop definitely depends on the amount of N used. Finally, as most studies base NRE on the difference in N uptake between a fertilized crop and a nonfertilized control, it should be noted that applying N to a crop has been shown to affect the uptake of native soil

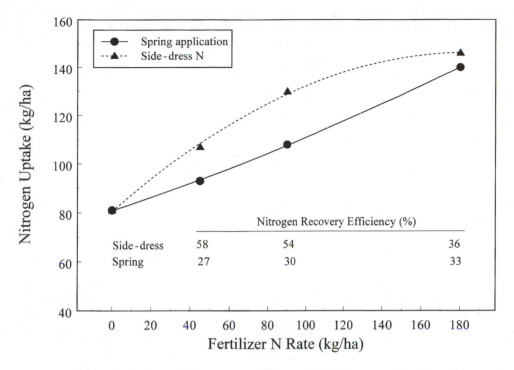

Figure 4.13 Nitrogen uptake and N recovery efficiency (NRE) for corn. Fertilizer N was either applied in a single spring application or in a split application using starter and side-dress fertilizers. (Adapted from Bock, 1984.)

N, both by altering soil N transformations and enhancing the ability of the crop rooting system to recover N.

The influences and interactions of soil type, crop management (tillage, irrigation, rotation, manure use), and method and timing of N application on NRE have been assessed for many N sources and agronomic crops (Table 4.8). The purpose of most studies has been to identify the magnitude of N loss in different environmental settings and under changing management practices. More sophisticated research has attempted to determine the specific fate of applied N (Figure 4.14), often relying on the use of ^{15}N-labeled fertilizers or crop residues to trace the movement of N.

The environmental efficiency of N use, while clearly important, is much more difficult to quantify, primarily because of the difficulty in establishing direct linkages between agricultural management practices and polluted resources. Most studies have used the NRE approach and the general assumption that if a management system has low NRE values (high amounts of unrecovered N) and the potential to impact the environment adversely, the system requires improvement. If some direct measurement of N pollution in the area dominated by the system is also available (e.g., high NO_3^- concentrations in wells), then the pressure to modify the system is particularly great. An example of such a scenario would be the use of poultry manure in crop production on the Delmarva Peninsula (Delaware–Maryland–Virginia), where data from well surveys indicated widespread contamination of shallow groundwater wells with NO_3^-. The soils in the region are coarse textured and well drained, rainfall is plentiful (~100 cm/year), and the use of overhead irrigation is increasing. The area contains one of the most highly concentrated poultry industries in the world, and an agriculture dominated by crops with high N requirements (corn, wheat, barley). Laboratory studies showed that poultry manure N was rapidly converted to NO_3^- in most soils, field studies with

Table 4.8 Representative Values for Nitrogen Use Efficiency of Corn

N Rate (kg/ha)	Treatments studied and N use efficiency (%)	
Delaware	Poultry manure	NH_4NO_3
84[a] (107[b])	50	62
168 (214)	33	57
252 (321)	34	50
Maryland	Minimum tillage	Plow tillage
90[a]	53	67
80	53	52
270	53	38
Vermont	Dairy manure + NH_4NO_3	NH_4NO_3 alone
0[a] (243[b])	41	—
56 (243)	35	97
112 (243)	18	93
168 (243)	16	68
Wisconsin	Biosolids: Site 1	Biosolids: Site 2
340[b]	17	21
680	19	17
1360	12	10
2720	9	6

Note: Nitrogen use (recovery) calculated using Equation 4.18.

[a] Rates of NH_4NO_3 added.

[b] Total amount of N added in manure or biosolids.

Source: Adapted from Sims, 1987 (Delaware); Meisinger et al., 1985 (Maryland); Jokela, 1992 (Vermont); Kelling et al., 1977 (Wisconsin).

irrigated corn found NRE values of 20 to 50%, and monitoring studies of wells in manured fields found NO_3^- concentrations in excess of current EPA drinking water standards (10 mg NO_3-N/L). Although it was recognized that other sources, primarily N fertilizers and rural septic systems, were contributing to groundwater NO_3^-, the potential for the poultry-based agriculture of the area to contaminate groundwaters and surface waters resulted in intensive efforts by advisory and regulatory agencies to improve crop N management in general and manure management in particular on this peninsula (see environmental quality issues/events box: Improving agricultural nitrogen management on the Delmarva Peninsula, p. 142).

In summary, improving the efficiency of N management will require an integrated approach, often regional in nature. Traditional approaches have combined the development and implementation of research-based *best management practices* (BMPs) for soil conservation and nutrient management with long-term basic research to alter fundamental aspects of the N cycle (Table 4.9). Recently, intense environmental pressures to control N pollution have raised the issue of mandatory, legislated controls on fertilizer N use. Many states with localized animal production facilities are reexamining the scale of management required to improve NUE, recognizing that practices which optimize N recovery in individual farm fields may be of little use if the nutrients generated on the farm exceed the capacity of the farm to use them. The following sections provide examples of techniques used to improve NUE in agriculture and nonagricultural systems, primarily at the field level. As N pollution reflects the combined effect of many farm fields as well as urban areas, municipal and industrial waste disposal sites, land reclamation projects, and rural septic systems, it is important to view these practices from a broader perspective and to integrate them into larger scale N management programs.

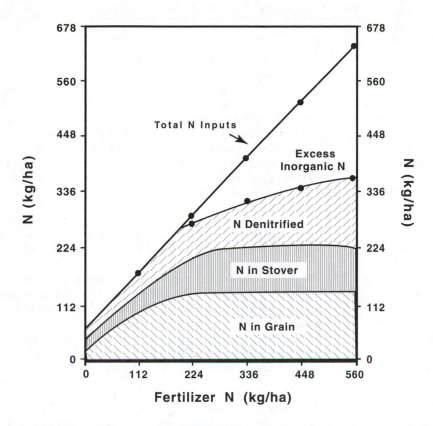

Figure 4.14 Influence of N rate on the fate of fertilizer N. (From Boswell, F. C. et al., in *Fertilizer Technology and Use*, O. P. Englestead, Ed., American Society of Agronomy, Madison, WI, 1985. With permission.)

4.4.2 Nitrogen management in agriculture

Current approaches to N management in agricultural cropping systems ordinarily begin with an assessment of the crop N requirement at a realistic yield goal. Fertilizer or organic waste management practices, based on local soil and climatic conditions, are then relied upon to minimize the excess amount of N required, as a result of anticipated N losses (system inefficiency), to attain optimum yields. The amount of supplemental N needed by a crop is related to the potential for excess N in the soil by the following equation, commonly referred to as a *soil nitrogen budget*:

$$N_f = [N_{up} + N_{ex}] - [N_{som} + N_{na}] \qquad (4.20)$$

where

N_f = amount of N needed from fertilizer, animal manure, etc.

N_{up} = crop N requirement at desired yield

N_{ex} = excess N lost by denitrification, erosion, leaching, or volatilization; varies as a function of the efficiency of each soil–crop system

N_{som} = N added from mineralization of soil organic matter, crop residues, previous applications of organic wastes, etc.

N_{na} = natural additions of N (rainfall, irrigation, atmospheric deposition)

Table 4.9   Summary of BMPs for Efficient Nitrogen Use

Management approach	Examples and comments
Soil, crop, and water management	
Soil and water conservation	Contour plowing, terracing, reduced tillage, improved irrigation management; all act to reduce erosion, runoff, and leaching of N
Cropping sequence and cover crops	Rotating legumes and nonlegumes to reduce need for N fertilizers; legumes and winter annual cover crops can "scavenge" residual soil N; benefits from crop rotations include economic stability, erosion control, reduced pest and disease pressure
Watershed management	Soil and water conservation and nutrient management supported by widespread educational programs, cost-sharing, and guidelines or regulations on irrigation, fertilizer use
Nutrient management	
Soil, plant, and waste testing	Recent advances in soil and plant N testing (e.g., the PSNT, LCM) provide opportunities for more efficient use of fertilizers and manures
Application timing and method	Split applications of N, fertigation, slow-release fertilizers, injections and deep placement of volatile N fertilizers; all directed toward improving synchrony between N availability and crop N uptake pattern
Fertilizer and waste technology	Nitrification and urease inhibitors improve efficiency of fertilizer N recovery; composting and pelletizing stabilize N in organic wastes and provide materials that can be handled and applied more efficiently
Fundamental changes in agriculture	
Legislation	Regulations, not guidelines, mandate amount and timing of N application from fertilizers or organic wastes; farm-scale nutrient budgets require farmers to find alternative uses for excess manure; most approaches are voluntary at present but legislation affecting nutrient management has been introduced and/or passed in several states
Genetic advances	Genetic alteration to introduce biological N fixation ability into nonlegumes (corn, wheat); increase N fixation capacity of legumes; advances in genetic manipulation of plants make these long-term goals of basic research in N fixation more feasible
Cropping patterns	Increased use of legumes, decreased production of cereal grains; more legumes would reduce fertilizer N use, but likely affect economics of production, dietary habits of consumers; pressures to convert to low-input, "sustainable" agriculture with less reliance on fertilizers and pesticides; low-input agriculture often labor-intensive, while the availability of farm labor is already inadequate in many urban societies; loss of cropland due to urbanization, desertification; as food production is highly dependent on amount of arable land, major losses of cultivated acreage would increase pressure to obtain higher yields on remaining land, requiring higher inputs, and increasing potential for nonpoint-source pollution; many urban areas and areas of intense animal production already have inadequate land available to use organic wastes they currently produce

Source: Adapted from Keeney (1982).

Minimizing N_{ex}, therefore, requires that we do not overestimate the crop yield potential for a particular soil, and thus both N_{up} and N_f, the crop N requirement and amount of fertilizer needed to attain the desired, hopefully realistic, yield. Nor should we underestimate the potential of the soil or other natural sources of N (N_{som}, N_{na}) to provide a significant percentage of N_{up}. Unfortunately, there are many examples where both types of errors have occurred, frequently resulting in groundwater contamination by NO_3^-. The issue of overapplication of N is complicated not only by overly optimistic estimates of potential yield, but by the relatively inexpensive nature of fertilizer N, and, when animal manures and municipal biosolids are involved, the continuing pressure to dispose of organic wastes, regardless of the true N requirement of the crop. The relationship between unrealistic yield goals and potential for groundwater contamination is clearly shown in Figure 4.15 for irrigated corn production in Nebraska, and the potential for a serious imbalance in a regional N budget is illustrated in Table 4.10 for the poultry and commercial fertilizer industries in Delaware.

In addition to overestimates of N_f, another serious problem for N use efficiency has been the failure to develop reliable tests for available soil N and N added in organic wastes so that farmers can accurately adjust fertilizer N rates to compensate for what the soil and other soil amendments provide. The lack of N-testing procedures has not been due to a lack of research effort, however, and recently major advances have been made in the area of soil and plant testing for N (see environmental quality issues/events box: Improving agricultural nitrogen management on the Delmarva Peninsula, p. 142).

An accurate soil test for N has been a long but elusive goal for soil scientists. The complex and dynamic nature of N cycling and its extreme sensitivity to often unpredictable climatic factors such as temperature and rainfall have made it difficult to use chemical extractants to estimate N availability in advance of planting as is commonly done for other plant nutrients, e.g., P, K, Ca, magnesium (Mg), manganese (Mn), Cu, and zinc (Zn). Similar problems have prevented the adoption of rapid chemical tests for available N in organic wastes. Residual tests for NO_3-N have had a history of success in arid-zone soils, but not in humid regions. In 1984 a significant breakthrough in soil N testing occurred that has shown the potential for markedly improving the efficiency of organic N sources for certain agronomic crops. The "pre-sidedress soil nitrate test" (PSNT) was conceived and first evaluated by F. R. Magdoff of the University of Vermont to address the problem of overfertilization of corn with N in the northeastern U.S., particularly in fields with histories of manure and legume use. The PSNT has four basic tenets, briefly summarized as follows:

1. All fertilizer N for corn, except a small amount banded at planting, should be sidedressed when the crop is beginning its period of maximum N uptake.
2. Soil and climatic conditions prior to sampling integrate the factors influencing the availability of N from the soil, crop residues, and from previous applications of organic wastes.
3. A rapid sample turnaround (<14 days) by a soil testing laboratory is possible.
4. Farmers will normally not sample to a depth > 30 cm.

The PSNT has since been evaluated in over 300 field studies in the northeastern and midwestern U.S. and has been repeatedly shown to be successful in identifying N-sufficient soils (Figure 4.16a). Some of the logistical difficulties associated with the need for a rapid sample analysis have been overcome by the development of "quicktest" kits and NO_3^- sensitive electrodes that can be used in the field. Even more encouraging are the results of a recent study with the leaf chlorophyll meter (LCM) that showed this extremely rapid, in-field measurement of leaf "greenness" was as accurate as the PSNT in identifying

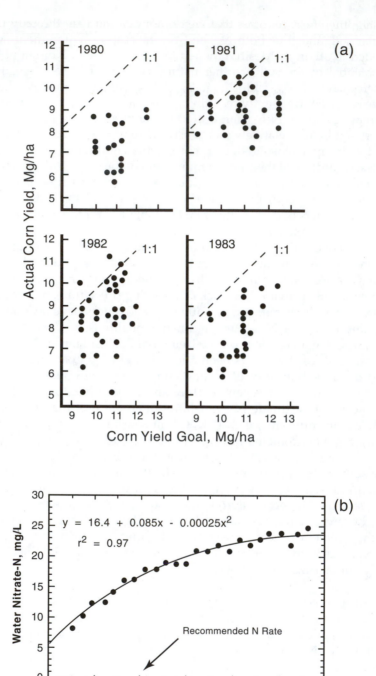

Figure 4.15   (a) Comparison of farmer yield goals for irrigated corn with actual yields obtained. Dashed line indicates 1:1 fit between yield goal and actual yield. (From Hergert, G. W., Status of residual nitrate-nitrogen soil tests in the United States of America, in *Soil Testing: Sampling, Correlation, Calibration, and Interpretation,* J. R. Brown, Ed., SSSA Spec. Pub. 21, Soil Science Society of America, Madison, WI, 73–88, 1987. With permission.) (b) Influence of deviation from recommended N rate on groundwater nitrate concentrations in Nebraska. (From Schepers, J. S. et al., *J. Environ. Qual.,* 20, 12, 1991. With permission.)

Table 4.10 Nitrogen Mass Balance for Delaware and Sussex County, DE[a]

Nitrogen produced or sold vs. crop nutrient requirements	Nutrient available or required (Mg/year)	
	Statewide	Sussex County
Nitrogen produced or sold: sources		
Commercial fertilizers	20,680	10,520
Poultry litter/manure	14,270	12,625
Other manures	1,180	390
Municipal biosolids	825	275
Total N produced or sold	36,955	23,810
Nitrogen inputs needed for optimum crop yields		
Corn	8,500	5,170
Soybeans	0	0
Wheat	3,175	1,725
Barley	1,375	680
Sorghum	580	220
Processing vegetables	2,790	1,270
Fresh-market vegetables	880	400
Hay	680	230
Total required by all crops	17,980	9,695
Annual nutrient balance		
Mg per state or county	18,975	14,115
kg N/ha	84	140

[a] Sussex County, DE is the site of Delaware's poultry industry. The county has approximately 100,000 ha of cropland and produces about 230 million chickens each year.

N-sufficient sites (Figure 4.16b). Another new approach to assess N sufficiency for corn is the late-season stalk nitrate test (Figure 4.16c). This "post-mortem" test uses the concentration of NO_3^- in the lower portion of the stalk at corn maturity to identify fields that received excessive N from fertilizers or manures. Recent research suggests that late-season leaf chlorophyll meter readings may be as accurate as the stalk nitrate test. If so, this would greatly decrease the logistical problems of late-season N testing and increase the use of this approach to monitor the success of an N management program.

The implications of these N tests are straightforward, but not simple. For most farmers, a PSNT soil sample would be taken and, if necessary, additional fertilizer applied via side-dressing. However, studies from soils commonly amended with animal wastes have shown that often little or no side-dress N is required, even when manure was not applied in the current year. This is illustrated in Figure 4.17 where the *economically optimum N* (EON) rates are compared for 11 fields with and without long-term histories of manure use. Economically optimum N rates are defined as the N rate where economic return on fertilizer N investment is maximized, based on assumed fertilizer costs and crop prices. The EON rates for manured and nonmanured fields were 34 and 128 kg N/ha, respectively, reflecting the greater N-supplying capacity of soils with long histories of manure applications. The greatest difficulty with the PSNT approach, apart from logistical problems associated with the rapid analytical turnaround, has been the presence of high percentages of soils that have been shown to need less manures or biosolids than are generated by the farm or municipality, or none at all. Simply put, these tests have shown that, particularly for animal-based agriculture, more N is produced than is needed by the farm, given the land available and the economics of waste handling and application. This once again illustrates the need for organic waste

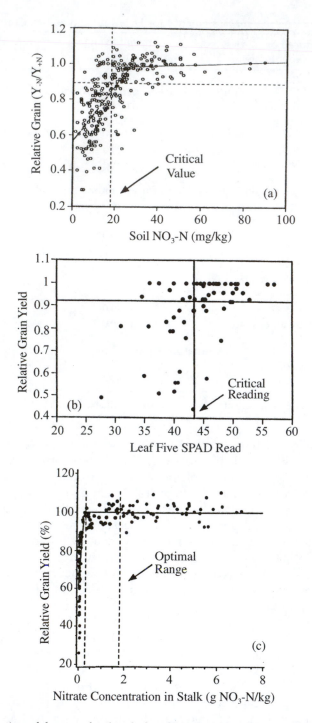

Figure 4.16 Illustration of the use of soil and plant N testing procedures to identify N-sufficient soils. (a) Relative yield of corn (Y_{-N}, yield in control (0 N) treatment, divided by Y_{+N}, yield from adequately fertilized treatment) vs. soil nitrate (0 to 30 cm) in the late spring (From Magdoff, F. R. et al., *Commun. Soil Sci. Plant Anal.*, 21, 1103, 1990. With permission.) (b) Relative yield of corn vs. leaf chlorophyll meter reading taken in late spring. SPAD refers to type of meter used to make reading. (From Piekielek, W. P. and Fox, R. H., *Agron. J.*, 84, 59, 1992. With permission.) (c) Relative yield of corn vs. nitrate concentration in the lower portion of the stalk at maturity. (From Binford, G. D. et al., *Agron. J.*, 82, 124, 1990. With permission.)

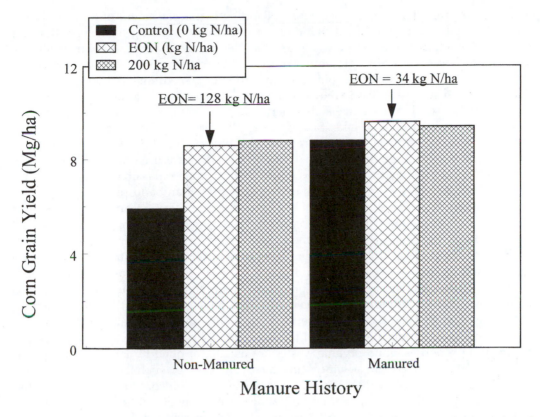

Figure 4.17 Yield response of corn to economically optimum N (EON) rates in fields with and without a history of manure applications. Numbers in bars are average yields, in Mg/ha, of all nonmanured or manured sites at EON rate. (Adapted from Roth and Fox, 1990.)

management at a larger scale, state or regional in scope, oriented toward redistribution of waste N to nutrient-deficient areas.

Once the need for and rate of supplemental N has been determined, either through soil and plant testing or a general knowledge of the cropping system, the next consider-ations are the source of N to be used and the method and timing of N application. Maximization of N efficiency requires that the best N source be applied in as timely a manner as possible, often through the use of multiple applications. As mentioned in Section 4.3, the primary consideration in selection of an N source is often economics, beginning with cost per unit of N. Other factors that influence the total cost of the N source include availability of the material (transportation costs), storage and handling (equipment and application costs), crop-specific fertilization requirements (e.g., the need for a mixed fertilizer), and any properties of the fertilizer material that may significantly affect its efficiency in the cropping system used. In animal-based agriculture where manure is constantly produced, other costs arise, such as the need for manure analysis, manure storage and treatment (e.g., lagoons, storage barns, composting facilities), and specialized equipment for manure application.

Selection of an N source can rarely be separated from selection of an N application method. Typical methods used to apply N fertilizers include *broadcasting, banding, injection,* and *fertigation*. Broadcasting refers to the uniform applications of fertilizer materials to the soil surface, either as solid granules or pellets or in a liquid spray. The broadcast fertilizer may then remain on the surface (no-tillage crops, turf, forages) or be incorporated by a tillage operation such as disking or plowing. Broadcast applications made to a growing crop are referred to as *top-dressing*. Banding is the placement of fertilizer N in a narrow band and can be done at planting; for row crops, banding fertilizers after crop emergence is referred to as *side-dressing*. Band applications of N at planting normally use N–P "starter" fertilizers with placement of the fertilizer approximately 5 cm below and 5 cm to the side of the seed to avoid salt injury to young seedlings. Banding at planting places an initial supply of N and P in a zone highly accessible to young root systems, and is particularly useful in no-tillage soils where cooler soil temperatures often delay initial root growth and nutrient uptake. Side-dress applications of N fertilizers can be done through injections, surface bands, sprays, or "dribbles." Injection is the placement of fertilizer in the soil through specialized subsurface application equipment and is most commonly used for anhydrous NH_3 and liquid fertilizer materials such as UAN solutions. Fertilizer injections are normally made at deeper depths than band placements, particularly for anhydrous NH_3 which reverts to a gas following injection and can be lost from the soil via volatilization if not placed correctly. Fertigation is the application of soluble N in irrigation waters and is not the same as *foliar fertilization* where fertilizers are sprayed directly on plant foliage. Only a limited number of N fertilizers are suitable for fertigation, due to the high degree of solubility and purity required to ensure complete dissolution and avoid clogging of application equipment. Urea-ammonium-nitrate solutions are normally preferred over anhydrous or aqua NH_3, which have higher volatilization potentials. Fertigation is most often used with overhead sprinklers or drip irrigation systems.

The method of application selected for an organic N source depends primarily upon its physical properties. Solid organic wastes are usually applied by large flail or spinner spreaders, although recent advances in waste processing, such as pelletizing, have increased the flexibility available to waste applicators. Liquid organic wastes are either injected or applied through wastewater irrigation systems. Incorporation of organic wastes is often mandated to avoid potential runoff of nutrient-rich solids or soluble organic materials into streams and lakes.

One of the most critical aspects of NUE is the timing of N application relative to crop N uptake. Most crops, particularly annual crops, have well-defined patterns of N accumulation, as illustrated for corn in Figure 4.18. Fertilizer application techniques that deliver supplemental N in close synchrony with N uptake will, in general, be the most efficient, particularly on soils that are highly sensitive to N losses. A generalized ranking of the relative efficiency of the most common N application techniques would be fertigation > banding ~ side-dress > surface broadcast. The greatest efficiency of N application normally results when several application techniques are combined. For irrigated corn this might involve a small amount of fertilizer N applied at planting, a side-dress application providing ~30 to 40% of the N requirement immediately prior to the period of most rapid N uptake, and several fertigations to supply the remainder of the crop requirement (Figure 4.18a). In a similar situation using organic wastes this could involve a low (suboptimum N rate) application rate of animal manure shortly before planting, followed by use of the PSNT to identify the side-dress N requirement (Figure 4.18b). There are many situations, however, where serious limitations to improved timing of application of an N source exist. One of the greatest of these limitations is the amount of time required for delayed or multiple applications of N during the part of the growing season when many other operations must be performed in a timely manner, including tillage, planting, herbicide application, harvest of other crops, and installation of irrigation systems. As an example, for many farmers applications of fertilizer or manure N during fall and winter months would be preferred, primarily because of the greater amount of free time available during this part of the year. However, the low efficiency of N recovery during these periods, because of leaching and denitrification losses that can occur prior to crop growth in the spring and summer, make fall applications undesirable both economically and environ-

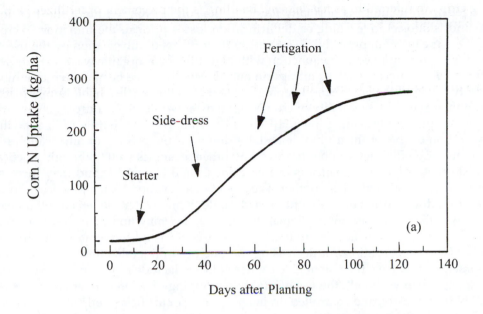

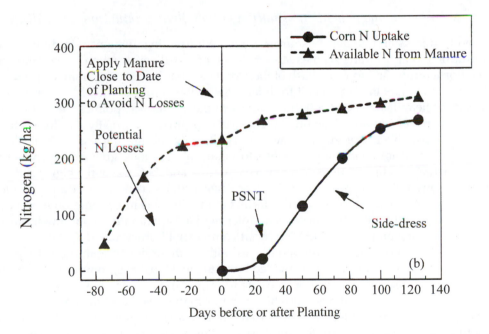

Figure 4.18 (a) Generalized representation of seasonal N uptake pattern for corn (kg/year), illustrating an efficient combination of N fertilizer techniques. (b) N uptake pattern for corn combined with typical rate of N release from poultry manure. Illustrates the potential for N losses from poorly timed manure applications and the timing of the PSNT, relative to side-dress N application.

mentally. Manure applications are often preferred during the fall or winter because the dry or frozen nature of the soil at this time can reduce the amount of soil compaction and erosion caused by heavy equipment in fields relative to wet, spring conditions.

The basic reason that applications far in advance of crop N uptake are undesirable is the rapid rate of nitrification in most soils. The NH_4^+ applied in most fertilizers or mineralized from manures is normally found in the soil as NO_3^- within a short period of time and is subject to leaching or denitrification losses prior to the initiation of crop uptake. One means to improve NUE is to delay the process of nitrification by the use of chemical inhibitors applied in conjunction with N fertilizers or organic wastes, as shown in Figure 4.19. Chemicals such as nitrapyrin and thiosulfate have been shown to inhibit the activity of nitrifying organisms, keeping applied N in the less leachable, ammoniacal form and increasing the possibility of crop N uptake. Slow-release N sources achieve the same goal by physically sealing the NH_4^+ within a resin, wax, or coating of S, delaying the dissolution of the fertilizer granule and, by doing so, the process of nitrification.

Nitrogen use efficiency can be improved by other means as well, although practical manipulations of N loss mechanisms can be difficult and expensive, and may increase one form of loss while reducing another. The use of conservation tillage practices can be expected to reduce erosion and runoff losses of N. Reducing water movement off a field, however, will likely increase infiltration and thus NO_3^- leaching and denitrification. Surface applications of wastes may also reduce soil–waste contact and accelerate waste drying, enhancing NH_3 volatilization but decreasing the rate of N mineralization. Other conservation practices that have the potential to reduce N losses include more efficient irrigation practices (e.g., drip vs. flood), the use of multiple cropping or winter cover crops to trap residual N from wastes, and controlled drainage systems or artificial wetlands to enhance denitrification in field border areas.

4.4.3   Nitrogen management in situations other than production agriculture

Nitrogen fertilizers and organic N sources are widely used in situations that differ greatly from production agriculture, such as land reclamation, road construction, urban and commercial horticulture, and forestry. Each of these end uses has its own constraints with regard to N use efficiency. Some are related to differences in the nature of the N cycle in natural, disturbed, or heavily amended soils, and others to the economics, politics, and logistics of soil–plant management. Reclamation and construction projects normally aim to produce a perennial, low-maintenance ground cover that will only require the use of N fertilizers initially, or in small, infrequent applications. Horticultural situations (greenhouses, nurseries, turf, ornamental plantings) are more intensive in nature and, because of the high cash value of the plants involved, rely heavily on fertilizer, often with less concern for the economic or environmental efficiency of N use. Forest fertilization with N is most common in the commercial forestry industry and represents an intermediate situation between low-input reclamation/construction projects and high-input commercial horticulture.

Organic N sources are commonly used in both of these nonagricultural settings, for several reasons. The organic matter provided by municipal biosolids, composts, and animal manures is often used in reclamation projects to improve the poor physical properties of highly disturbed soils. Greenhouses use high organic matter, "soilless" media almost exclusively to provide a light, well-aerated growth medium with high moisture-holding capacity. Both situations are good end uses for municipal organic wastes, such as biosolids composts, because they mainly use non-food-chain plants, thus concerns about dietary accumulations of potentially toxic waste constituents (e.g., heavy metals) are reduced.

Management of soil, fertilizer, and organic N in land reclamation and forestry situations begins with an understanding of the differences that can occur in N cycling and plant growth, relative to cultivated soils. These differences can be fundamental or practical in nature. At the most basic level, the microorganisms responsible for key N transformations such as mineralization, nitrification, and denitrification may be less efficient, inactive, or nonexistent in soils with extreme physical or chemical properties, such as highly acidic mine spoils or pine forests. From a practical standpoint, amending these soils with fertil-

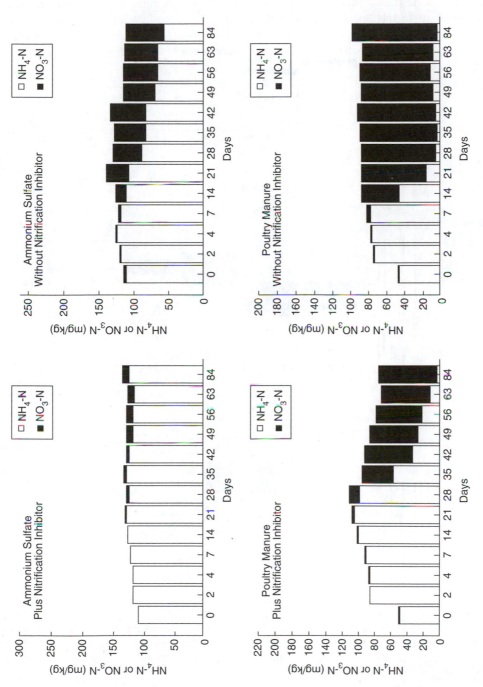

Figure 4.19 Illustration of the influence of nitrification inhibitors (thiosulfate) on the rate of nitrification of an inorganic N fertilizer (ammonium sulfate) and an organic N fertilizer (poultry manure). (Adapted from Sallade and Sims, 1992.)

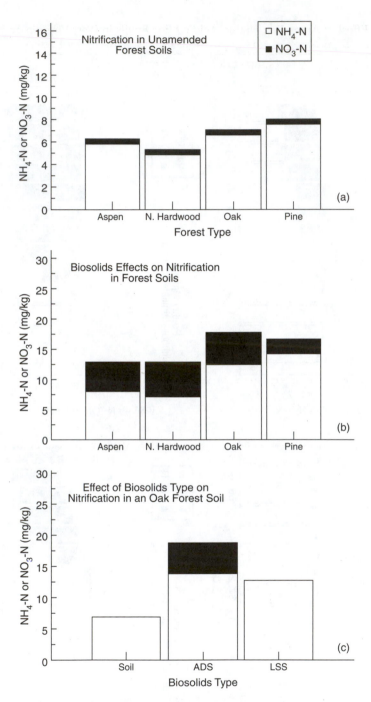

Figure 4.20   Nitrification in forest soils. (a) Effect of forest type on nitrification in unamended forest soils. (b) Amending forest soils with sewage sludge (biosolid) increases nitrification. (c) Sludge type affects extent of nitrification in an oak forest (ADS = anaerobically digested sludge, LSS = lime-stabilized sludge.) (Adapted from Burton et al., 1990.)

izer, organic N, or even small, intermittent doses of N as acid rain, can alter the activities of soil microbes and change the nature of N cycling, as seen in Figure 4.20 for nitrification

Table 4.11 Effect of Amending Highly Acidic Mine Spoils in West Virginia on Nitrification, Denitrification, and Populations of Nitrate-Reducing Bacteria

Mine spoil	pH	N mineralization potential[a] (μg N/kg/h)		Denitrification parameters[b]	
		NH_4-N	NO_3-N	DEA (μg N/kg/h)	MPN/g
Bald Knob					
Unamended	3.8	21	3	5	3
Amended	6.8	–12	38	11	180
Osage					
Unamended	2.7	33	8	11	6
Amended	5.2	–8	21	68	980

Note: The Bald Knob site was amended with lime and fertilizer, Osage with coal fly ash.

[a] Rate of change in NH_4-N and NO_3-N in mine spoil materials in a 30-day laboratory incubation study.

[b] Rate of denitrification of added NO_3 from mine spoils as determined in laboratory study of denitrifying enzyme activity (DEA). MPN is estimate of most probable number of nitrate reducers in mine spoil material based on five-tube assay.

Source: Adapted from Shirey and Sexstone (1989).

in biosolids-amended forest soils and in Table 4.11 for nitrification and denitrification in acid mine spoils.

Plant growth characteristics and N uptake patterns must also be carefully considered in land reclamation and forestry situations. In most cases the plants to be grown are perennials, not annuals, and hence have differing seasonal and long-term N uptake patterns. For example, a study of the use of inorganic fertilizers and poultry litters as N sources for managed pine forests in the southern U.S. showed that NO_3^- leaching losses could be considerable in early years because of the very low initial N uptake by young pine seedlings. Nitrate-N concentrations as high as 50 to 100 mg NO_3-N/L were measured below the rooting zone of pine seedlings. Delaying N applications until the trees were several years old and had better developed and deeper root systems was suggested as a management practice that could reduce NO_3^- leaching.

From a horticultural perspective, there has been increasing interest in the environmental implications of N management for greenhouses, nurseries, and turf. The soilless media used in many greenhouses and nurseries are not relied upon to provide much of the plant N requirement, serving mainly as a physical growth medium. Growers instead commonly use fertigation with high-N-concentration solutions combined with intensive watering to remove salts from containers. This has raised questions about the impact of drainage waters from these facilities on groundwater and surface water quality. More efficient fertilization programs are available and are being adopted in response to these environmental concerns. Examples include the use of slow-release fertilizers, "ebb-and-flow" fertigation systems where the nutrient solutions are recycled and reused, and on-site wastewater treatment systems. Similarly, improved N management programs for turf have been developed and promoted more intensively to home owners and those involved in industrial or recreational turf management (e.g., golf courses, parks, athletic fields). Avoiding N losses by leaching or runoff is only part of the concern with turf. Overfertilization with N increases the frequency of cutting and the volume of grass clippings that must be recycled or disposed of in landfills. As many municipalities are no longer accepting yard wastes, the pressure to avoid the use of excess N, and to recycle N "on-site" through proper cutting schedules, mulching, or composting has increased.

Environmental quality issues/events
*Improving agricultural nitrogen management
on the Delmarva Peninsula*

Background

The Delmarva (Delaware–Maryland–Virginia) Peninsula is the site of one of the most concentrated poultry industries in the U.S. Approximately 600 million broiler chickens are grown each year, primarily in five adjacent counties located on the southern part of the peninsula. The manures, litters, and composts produced by this industry have historically only been used as fertilizers for the grain crops grown as feed for the industry (corn, soybeans, wheat). Additionally, ~45 million bushels of corn and soybeans are imported to the area each year, an input of ~65 kg N/ha/year. Commercial fertilizer N is also widely used on the peninsula and, along with the by-products of the poultry industry, contributes to an annual areawide surplus of ~85 kg N/ha/year (140 kg N/ha/year in Sussex County, site of the poultry industry in Delaware; see Table 4.10). Nitrate contamination of the shallow groundwater aquifers used as drinking water supplies on Delmarva has been documented in several studies (Figure 4.21) and attributed in part to leaching of N from agricultural cropland receiving manures and fertilizers. Groundwaters also discharge via base flow into many important surface waters that are now eutrophic, including Delaware's Inland Bays, a national estuary.

State and federal agencies have responded to public concerns about nitrate contamination of groundwaters and surface waters by supporting years of research to develop BMPs that can reduce NO_3^- leaching. Despite this, substantial water quality improvements have not occurred, at least partially because the BMPs have not been widely implemented. In 1997, as a result of a lawsuit by environmental groups, a total maximum daily load (TMDL) agreement was signed between Delaware and the EPA. This agreement requires Delaware to establish maximum limits for pollutants (nutrients, sediments, pathogens, organics) that can be discharged into waters by point and nonpoint-sources and calls for the development of pollution control strategies to reduce pollutant levels below the TMDL. For example, reductions in N loading of 85% (point and non-point-sources) are needed to meet the TMDL for Delaware's Inland Bays. Achieving reductions of this magnitude will require wider implementation of current BMPs and the development and testing of new BMPs.

Current best management practices for nitrogen

The current BMPs for agricultural N management on Delmarva are based on research and subsequent "on-farm" evaluations conducted over a period of about 20 years. Some examples include:

- *Comprehensive nutrient management planning* — To determine, at the farm (or watershed) scale, whether there is a surplus or deficit of N on the farm and how to manage all N sources most efficiently. If an N surplus exists (common on many small to midsized poultry farms) alternatives to land application of animal manures are encouraged along with a review of the need for commercial fertilizer N.

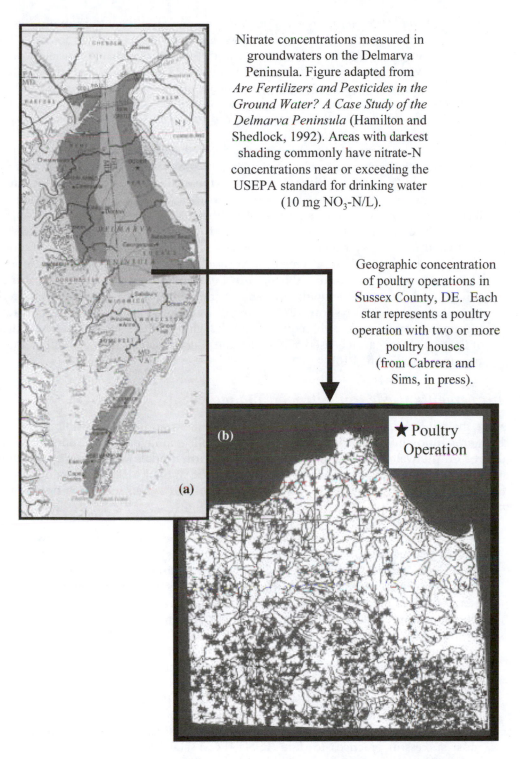

Nitrate concentrations measured in groundwaters on the Delmarva Peninsula. Figure adapted from *Are Fertilizers and Pesticides in the Ground Water? A Case Study of the Delmarva Peninsula* (Hamilton and Shedlock, 1992). Areas with darkest shading commonly have nitrate-N concentrations near or exceeding the USEPA standard for drinking water (10 mg NO_3-N/L).

Geographic concentration of poultry operations in Sussex County, DE. Each star represents a poultry operation with two or more poultry houses (from Cabrera and Sims, in press).

★ Poultry Operation

Figure 4.21   (a) Areas on the Delmarva Peninsula with elevated NO_3-N concentrations in groundwater (Adapted from Hamilton and Shedlock, 1992). (b) Geographic intensification of poultry production in Sussex County, DE. (From Cabrera, M. L. and Sims, J. T., in *Beneficial Uses of Agricultural, Municipal, and Industrial By-Products*, J. Power, Ed., Soil Science Society of America, Madison, WI, in press. With permission.)

- *Animal manure management* — Farmers are encouraged to have their manure ana-lyzed to determine the content of PAN (see Equation 4.17) and then to apply the manure using well-calibrated spreading equipment at a rate that is consistent with a realistic yield goal and in a manner that maximizes crop N uptake. For corn this would be as close to planting as possible, while for winter wheat the best timing is in the early spring. Manure applications are not recommended for soybeans because research showed that manure use did not increase yields and did increase NO_3^- leaching below the root zone and the buildup of soil P to excessive levels.
- *Fertilizer N management* — Corn is the major crop that receives fertilizer N. Farmers who only use commercial fertilizer N normally apply a small amount of N at corn planting as a "starter" fertilizer and most of the fertilizer N in a single side-dress application about 6 weeks after planting. If irrigation is available farmers also use fertigation to provide small amounts of fertilizer N in several midsummer appli-cations. The amount of side-dress N applied is based on the results of a PSNT evaluated in >50 on-farm studies in Delaware. The PSNT provides a quantitative assessment of the soil N-supplying capacity and, if manure was applied, the amount of N that can be expected from the manure. Leaf chlorophyll meter (LCM) readings can be used in conjunction with fertigation to determine if additional N is required. Recent research has also shown that spring LCM readings are a reasonably accurate means to identify fertilizer N requirements for winter wheat. A corn stalk NO_3^- test at the end of the season is used to evaluate the overall success of the N management program (see Figure 4.16).

Improvements needed in nitrogen management

Despite the extensive research efforts on N management and the investment of state and federal funds into the demonstration of BMPs, N management remains a major challenge on the Delmarva Peninsula. The existence of annual, large-scale geographic N surpluses, in conjunction with the regulatory nature of the TMDL agreement, points to the need for even more intensive and innovative BMPs. Some options now being considered are as follows:

- *Alternatives to land application of animal manures* — Serious efforts are now under way to evaluate the potential use of poultry litters as "bioenergy" sources to gen-erate electricity and/or steam heat. Commercial ventures to pelletize manures and redistribute them to other regions for use as fertilizers, perhaps following enhance-ment with inorganic fertilizers, are now under construction. Composting manures to produce soil amendments that can be used not only by home owners but also by industries, government agencies (e.g., the state highway department), and in reclamation of strip mines in Pennsylvania and West Virginia is also being evalu-ated. The common goal of all these efforts is to reduce the N surplus on Delmarva by producing "value-added" uses for animal manures.
- *Increased educational efforts on N testing* — Even if the problem of surplus N is resolved, there will continue to be a need to improve N management on the Delmarva because of the sandy soils, shallow aquifers, and humid climate. The most successful practices to date (PSNT, LCM, stalk NO_3^-) have been accepted reasonably well by farmers, but an infrastructure to conduct these tests has not arisen. Farmers are not likely to be able to do the testing themselves because of the large land areas involved and the fact that many tests must be taken at the same time as other farming activities. Educational programs and efforts to involve the commercial fertilizer industry and crop consultants in expanded N-testing

programs are needed. Another option is to expand the use of new remote-sensing technologies that may be able to assess the N status in a field rapidly with images obtained by aircraft or satellites.

- *Soil and water conservation practices* — Increased use of winter annual cover crops, buffer strips, and constructed wetlands to recover residual NO_3-N by plant uptake or enhance denitrification are other options to minimize the movement of N from agricultural soils to surface waters. Research on innovative conservation practices is needed.

4.5   Soil nitrogen and environmental quality: management directions for the future

Efficient N management will become increasingly important in the future as the pressures to provide food and a safe environment for a growing world population increase. Projections today call for a global population of 10 billion by 2070, compared with today's 6 billion. In the face of this, two questions come immediately to mind. Can we manage N in a manner that will produce the food required to sustain population growth, without permanently damaging the environment? And, what do long-term studies show us about the impact of changing (increasing) N use on environmental quality? Four recent studies addressed these issues — examining them provides some insight into the future directions needed for environmentally and agricultural efficient use of N.

Nitrogen Fertilizer: Retrospect and Prospect *(Frink et al., 1999)*

World fertilizer N use increased from 1.3 Tg in 1930 to 80 Tg in 1988. Further, the ratio of N fertilizer used to that contained in harvested crops is estimated today as 1.5:1, with the most inefficient systems having ratios of as much as 2:1. This means that only about two thirds of the fertilizer N applied is harvested; the rest enters the soil N cycle where it can be conserved or lost to air or water. The driving forces for this growth in fertilizer N use, and the increased "leaks" of N to other sectors of the environment, were population growth and increased population wealth and thus a greater ability to purchase and use fertilizer N. However, in the past 10 years (1986 to 1995) the ratio of fertilizer N to crop N in the U.S. has *decreased* by nearly 2% per year. Today, in the corn belt of the U.S., ratios range from 1.2 to 1.5, compared with almost 2.0 in the early 1980s. Declining ratios reflect improved fertilizer N use efficiency, wider use of technologies that overcome yield-limiting factors (irrigation, pest control) and the development of higher yielding crop varieties (all of which increase crop N uptake). Three scenarios, shown in Table 4.12 illustrate the impact of altering the "fertilizer N use:crop N uptake" ratio, by improved conservation of N and raising crop yield potentials through scientific and technological advances, on our ability to feed the rapidly growing world population. Also illustrated is the impact on world ecology, as reflected by the reduced amount of cropland needed to feed 10 billion people in 2070 when we increase the efficiency of N use and crop N uptake, compared with what will happen if farming remains at the present level of efficiency. The projections for the future are thus both optimistic, if we assume that national and international policies promote the wider implementation of science and technology-based approaches to N management for agriculture, and pessimistic, should our agricultural practices stagnate at the efficiency level of the early 1990s.

Table 4.12 Comparison of Four Scenarios That Relate Nitrogen Management Efficiency and Advances in the Science and Technology of Crop Production to Fertilizer, Nitrogen Consumption, and the Amount of the World's Land Needed to Sustain the Population in 2070

Scenario	Fertilizer N use (Tg)	Ratio of fertilizer N use to crop N uptake (%)	Crop yield (kg/ha)	Crop N uptake (kg/ha)	Cropland required to sustain population (% of world land)
1990 conditions	79	150	1900	38	11
		Projection for 2070			
If farming stagnates at 1990 level of efficiency	284	150	1900	38	38
If crop yields increase slowly and N use efficiency improves	192	100	3800	77	19
If significant increases in crop yields and N use efficiency both occur	192	100	7600	155	10

Note: Projections based on an increase in world population from 5.3 in 1990 to 10 billion in 2070.

Source: Adapted from Frink et al. (1999).

Occurrence of Nitrate in Groundwater — A Review *(Spalding and Exner, 1993)*
Nitrate-N is the most widespread chemical contaminant in the world's aquifers, and
concentrations are increasing with time. Surveys in Europe show that more than 20% of
the population of France drink water exceeding the limits established by the European
Union (11.3 mg NO_3-N/L) and that from 5 to 10% of the waterworks in Denmark and
Germany supply drinking water exceeding this limit. In the Netherlands 25% of the well
fields used for drinking waters are contaminated with NO_3-N. Elevated NO_3-N concentrations in groundwater have also been documented in the Caribbean, Canada, Africa, the
Middle East, Australia, and New Zealand and attributed to fertilizers, manures, and high-
density housing with unsewered sanitation facilities.

The current situation in the U.S., while of concern, is not as serious on a national scale;
the situation is summarized in this study which evaluated >200,000 groundwater NO_3-N
measurements from several national surveys in the U.S. A highly skewed distribution was
reported, one that clearly indicates groundwater in many agricultural regions is *not* contaminated by NO_3-N. Beyond this, variations in the incidence of NO_3-N contamination in
groundwaters were often adequately explained by well-understood principles of N cycling
and N management. For instance, the most severe instances of contamination were found
in areas dominated by irrigated, row-crop agriculture with well-drained soils, permeable
vadose zones, and shallow aquifers (e.g., Delmarva, Minnesota, Arizona, California, Washington, and Nebraska). In contrast, in the southeastern U.S. low NO_3-N concentrations
were found in groundwaters because of a combination of high temperatures, abundant
rainfall, and riparian soils with high organic C, situations that promote denitrification
and/or plant N uptake. In large areas of the Midwest (the corn belt) a combination of
denitrification and the interception of NO_3-N by tile drains, with subsequent discharge to
surface waters (a separate environmental issue, see below), contributed to decreased
incidences of NO_3-N contamination in aquifers. This study emphasizes the importance of
a balanced perspective when addressing the issue of groundwater pollution by NO_3-N —
while it is clearly an important issue, it is usually geographically localized and often results
from activities occurring on only a small percentage of the overall land area.

Des Moines River Nitrate in Relation to Watershed Agricultural Practices: 1945 versus 1985 *(Keeney and DeLuca, 1993)*

True long-term studies of the effect of agricultural practices in a watershed on water quality
are rare. Studies such as this one, which compared the concentrations and flow of NO_3-N
in the Des Moines River (located in the heart of the U.S. corn belt) shortly after World War
II, when widespread use of N fertilizers was just beginning, with that of the mid-1980s,
when fertilizer use was at its peak, provide valuable insight for future planning. The issue
is of immediate importance to the city of Des Moines because it relies upon shallow wells
in the river's riparian zone for drinking water. Public health concerns forced the city to
construct and operate, at the cost of millions of dollars, a water purification facility to
prevent public drinking waters from exceeding the EPA drinking water standard (10 mg
NO_3-N/L). Analysis of the Des Moines River watershed indicated that today 78% of the
land is used for agricultural crop production, primarily corn and soybeans. Sources of N
to the river were estimated in 1945 and 1990. In both years the major N source was cropland,
although in 1945 most N originated from mineralization of soil organic matter (65%) while
in 1990 the dominant source was fertilizer N (46%) (Figure 4.22). Several important observations resulted from this study. First, the average *annual* NO_3-N concentration in the Des
Moines River never exceeded the EPA drinking water standard (10 mg NO_3-N/L) in any
year between 1980 and 1990. The mean value for that 10-year period was 5.6 mg NO_3-N/L
vs. 5.0 mg NO_3-N/L in 1945. On average, the NO_3-N concentration in the river exceeded

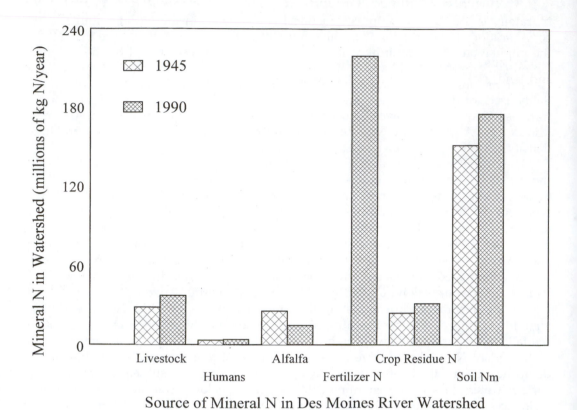

Figure 4.22 Sources of nitrogen to the Des Moines River in Iowa in 1945 and 1990. (Adapted from Keeney and DeLuca, 1993.)

the EPA standard twice per year. This suggested that either the causes of N loading had changed little in 45 years or that changes in sources have canceled each other out, giving no net change in inputs of NO_3-N to the river. Second, while the study supported the fact that agriculture is the major source of elevated NO_3-N concentrations in the Des Moines River, there was evidence that water quality began to decline early in the 20th century when farming began to be the dominant land use in the watershed. Third, the primary causes of N loading were identified as mineralization of soil organic N (Nm) and increased fertilizer and manure use combined with the widespread use of tile drains. These subsurface drains "short-circuit" natural water flow and reduce the likelihood of denitrification occurring in subsoils, groundwaters, or wetlands. Finally, improving water quality in the Des Moines River will require a fundamental restructuring of agriculture in the watershed, one that increases the use of practices such as filter and buffer strips at field margins, wetlands and prairie restoration, and significantly improved N management practices for fertilizers and livestock manures. Managing livestock manures has assumed even greater importance because of the recent explosive growth in the swine industry in Iowa (3.1, 4.1, and 12 million head in 1945, 1990, and 1999, respectively).

Agriculture and Nitrate Concentrations in Maryland's Community Water System Wells *(Lichtenberg and Shapiro, 1997)*

In many watersheds, groundwaters and surface waters are impacted by both agricultural and nonagricultural (stormwater runoff, septic systems) activities. In the U.S. more than

97% of the drinking water in rural areas is supplied by groundwaters; as rural areas urbanize the need for safe drinking waters increases proportionately with population growth. In this study, sources of NO_3-N contamination of rural community water system (CWS) wells in Maryland were examined and related to hydrologic factors in differing watersheds (e.g., depth to water table, confined vs. unconfined aquifer, fractured bedrock, and land use). Land-use factors that were associated with higher NO_3-N in CWS wells were corn and broiler chicken production, not surprising given the fact that corn requires high N inputs and broilers produce large amounts of N, often on a limited land base (see the environmental quality issues/events box on the Delmarva Peninsula). Conversely, soybean production, a leguminous crop that does not normally receive fertilizer or manure N, was negatively correlated with NO_3-N levels in CWS wells. The statistical model used in this study also suggested that (1) a 1% increase in the number of rural septic systems in a county would increase NO_3-N in CWS well water by 1% and (2) 26 septic systems are associated with as much NO_3-N leaching as one chicken house with a 10,000 bird capacity (most broiler houses have capacities of 20,000 to 25,000 birds). From a hydrologic perspective, CWS wells located in shallow, unconfined aquifers in limestone formations were identified as the most vulnerable to NO_3-N contamination. Several policy implications were identified as a result of this study. Public technical assistance and cost-sharing programs should focus on improving septic systems and extending public sewer lines and on more widespread implementation of best management practices for corn and broiler production.

Problems

4.1 List all the impacts that N agricultural N use can have on other sectors of the environment. Describe the pathways by which N in agricultural soils can move from the soil to air and water.

4.2 The drinking water standard for NO_3-N in the U.S. is 10 mg NO_3-N/L. The standard used in Europe is 45 mg NO_3^-/L. Show by calculation that these values are essentially the same.

4.3 Some areas in the U.S. have reported very high NO_3-N concentrations in groundwaters that are used for irrigation of crops. If irrigation water has a NO_3-N concentration of 28 mg NO_3-N/L and 2 cm of irrigation water per hectare is applied to a corn crop ten times during the growing season, what is the total amount of soluble NO_3-N added to the crop via irrigation, in kg/ha?

4.4 It has been estimated that 30% of the N entering a poultry farm in feed is lost to the atmosphere in NH_3 emissions from the poultry houses during the growth of the chickens. It requires about 10 lb of feed with an average N content of 2.8% to produce one broiler chicken. Assume that a poultry farm has two poultry houses, each with a capacity of 20,000 chickens per flock, and produces six flocks of chickens per year. How many kilograms of N are lost from this farm as NH_3 each year? If 75% of the NH_3 that was emitted from the total production facility was deposited on the 10 ha of cropland closest to the farm, how much lime (in Mg/ha) would be needed to neutralize the acidity that resulted from the nitrification of this NH_3 (assume that it takes 0.9 kg of agricultural-grade limestone to neutralize the acidity produced by 0.45 kg of NH_3).

4.5 One approach that has been shown to reduce NH_3 losses from animal wastes successfully is to add an acidifying material, such as "alum" $[(Al)_2(SO_4)_3]$, directly to the manure/litter. Explain why this approach would reduce NH_3 volatilization losses. Prior to recommending this practice for widespread use, what other factors would need to be considered?

4.6 Explain why there has been interest, from both crop production and water quality perspectives, in the development of chemical compounds that can inhibit nitrification (see Figure 4.19).

4.7 The long-term use of animal manures can increase the N-supplying capacity of soils and decrease the need for inputs of fertilizer N. A recent study (Gordillo and Cabrera, 1997) evaluated the mineralization of N in ten soils amended with poultry litters, using Equation 4.7. The authors reported the following mean values: $N_f =$ 11.0 g N/kg, $N_s = 13.8$ g N/kg, $k_f = 2.6$/day, and $k_s = 0.04$/day. Using Equation 4.7 and these values, calculate the amount of N that would be mineralized from a soil amended with poultry litter after 1, 7, 14, 21, 42, and 84 days. What are the implications of this mineralization pattern for poultry litter management?

4.8 If the Q_{10} value for the rate of N mineralization in the study in Problem 4.6 was 2.0, how much more N would be mineralized in a poultry litter amended soil in 7 days if the soil temperature was increased from 15°C (a cool, spring application) to 35°C (a hot, summer application)? (See Section 4.2.1 for definition and discussion of Q_{10} values.)

4.9 Many corn farmers in the eastern U.S. now use minimum- or no-tillage farming practices to reduce soil erosion, conserve soil moisture (crop residue acts as mulch to reduce evaporative losses of soil water), and build soil organic matter content. To save time these farmers may also spray a mixture of urea-ammonium NO_3^- fertilizer and preemergence herbicides onto the surface of the soil immediately after planting. Explain (a) how and why this practice would affect the potential for N losses by NH_3 volatilization; (b) what the implications are for corn production; and (c) how timing the application relative to expected rainfall could affect NH_3 loss and corn yields (see Table 4.3)?

4.10 Review Figure 4.6 and explain why corn grown in a field where the pharmaceutical waste was injected in this manner might exhibit N deficiency. How could you prevent N deficiency by changing the application method for the waste?

4.11 The fertilizer N recommendation for a wheat crop is 100 kg N/ha. How much of each of the following fertilizer materials would you need to apply, in kg/ha, to provide this amount of available N to the wheat: ammonium nitrate, ammonium sulfate, urea, and UAN solution (assume the density of a 30% UAN solution is 1.3 kg/L).

4.12 Explain the general relationship between N fertilizer application method and the efficiency of crop N recovery.

4.13 Assume you are working as a consultant for an organic farming operation that produces vegetables and wheat for bread making. The farm has access to a ready supply of poultry manure from a nearby farm and composted biosolids from a small municipality. The farmers ask you what application rate, in Mg/ha, of each material would be needed to provide 150 kg N/ha for their vegetable crops and 90 kg N/ha for their wheat. Using Equation 4.17 and the following information, determine the application rates required for (a) poultry manure (total N = 4.0%, NH_4-N = 1.0%, $k_m = 0.80$, $e_f = 0.80$ if incorporated within 1 day, 0.40 if incorporated within 1 week); (b) composted biosolids (total N = 2.0%, NH_4-N = 0.05%, $k_m = 0.15$, $e_f = 0.90$ if incorporated within 1 day, 0.60 if incorporated within 1 week). They also ask you to give them a general overview of any other factors to consider when deciding between the two organic sources of N. What is your response?

4.14 You are asked to evaluate the current approach used to apply N to corn in a watershed dominated by intensive dairy operations. Most farmers in the watershed apply the manure in late winter or early spring, about a month to 6 weeks before planting and then use a small amount of "starter" fertilizer at planting. You set up an "on-farm" demonstration where five rates of dairy manure are applied (0, 22, 44, 66, and 88 Mg/ha). The total N content of the dairy manure, as applied, was 1.4%, approximately 25% of which is plant available. At the end of the season you determine total N uptake at these N rates to be 90, 240, 270, 290, 310 kg N/ha. Calculate the nitrogen recovery efficiencies (NRE) for each approach using Equation 4.19, based on total N applied and PAN applied. Discuss (a) the possible fates of any "unrecovered" N and (b) any steps that could be taken to improve N use efficiency in this watershed.

4.15 The application of organic by-products such as animal manures, composts, and municipal biosolids to leguminous crops is an issue currently being debated in many watersheds. Describe the pros and cons of the use of organic by-products on crops such as alfalfa and soybeans. Do not confine your answer only to N, but consider all possible benefits and costs that are involved.

4.16 A 3-year study in Maryland (Angle et al., 1993) showed that NO_3 leaching losses from dairy manure and inorganic fertilizer were consistently lower in a no-tillage situation than when the soils were conventionally tilled (plowed and disked). Explain how each of the following could have contributed to lower NO_3^- leaching with no-tillage: (a) cover crop use in no-till but not conventional till; (b) tillage effects on N immobilization, denitrification, and NH_3 volatilization; and (c) improved soil moisture relations (more plant-available water in soils) with no-tillage.

4.17 To conduct the PSNT, a soil sample is collected to a depth of 30 cm, dried rapidly, extracted with water or a salt solution, and analyzed for soil NO_3-N. You extracted 5 g of soil with 100 mL of 2 M KCl and determined the concentration of NO_3-N in the extract was 1.4 mg NO_3-N/L. The critical level for the PSNT for corn is 25 mg/kg. Based on your analysis would you recommend side-dressing the corn with fertilizer N?

4.18 An engineering company initiates a project to revegetate a disturbed soil at a strip mine site with conservation grasses. The soil is highly acidic, deficient in nutrients, and has poor structural properties and a very low organic matter content. They wish to add a very high rate of organic matter to improve soil structure and increase soil water-holding capacity. They have access to an organic by-product from a papermill (C:N ratio = 60:1), poultry manure from a nearby farm (C:N ratio = 6:1), and yard waste (leaves, grass clippings) compost from a local municipality (C:N ratio = 20:1). Which material would you recommend they use, and why? What other steps will they need to take to establish and maintain a grass cover success-fully? If economics forced them to use one of the other sources, what steps should they take to counteract any problems they might encounter?

4.19 A new, nonagricultural approach to reuse municipal biosolids beneficially is "deep row tree farming." In this system, large trenches (~50 cm wide and 75 cm deep) are filled with biosolids, covered with soil, and planted with poplar trees for 6 years, then harvested for timber. One concern that had to be addressed in this system was the potential for NO_3^- leaching from the trenches. Explain how the environment in the trenches themselves (low oxygen, high moisture) and the types of plants being grown (perennials) affect the likelihood of NO_3^- leaching to shallow groundwaters.

4.20 Urea is sometimes used as a "de-icing" agent at airports. A small, rural airport using urea began to notice algal blooms and declining water quality in nearby ponds. A state agency constructed an "artificial" wetland between the airport and the ponds to improve water quality. Explain the processes by which this wetland removes N from airport runoff waters.

References

Alberts, E. E., W. E. Neibling, and W. C. Moldenhauer. 1981. Transport of sediment nitrogen and phosphorus in runoff through cornstalk residue strips, *Soil Sci. Soc. Am. J.*, 45, 1177–1184.

Angle, J. S., C. M. Gross, R. L. Hill, and M. S. McIntosh. 1993. Soil nitrate concentrations under corn as affected by tillage, manure, and fertilizer applications, *J. Environ. Qual.*, 22, 141–147.

Berry, J. T. and N. L. Hargett. 1986. 1986 Fertilizer Summary Data. National Fertilizer and Environmental Research Center, Tennessee Valley Authority, Muscle Shoals, AL, 132 pp.

Binford, G. D., A. M. Blackmer, and N. M. El-Hout. 1990. Tissue test for excess nitrogen during corn production, *Agron. J.*, 82, 124–129.

Bock, B. R. 1984. Efficient use of nitrogen in cropping systems, in *Nitrogen in Crop Production*, R. D. Hauck, Ed., American Society of Agronomy, Madison, WI, 273–294.

Boswell, F. C., J. J. Meisinger, and N. L. Case. 1985. Production, marketing, and use of nitrogen fertilizers, in *Fertilizer Technology and Use*, O. P. Engelstead, Ed., American Society of Agronomy, Madison, WI, 229–291.

Burton, A. J., J. B. Hart, and D. H. Urie. 1990. Nitrification in sludge-amended Michigan forest soils, *J. Environ. Qual.*, 19, 609–616.

Cabrera, M. L. and J. T. Sims. Beneficial uses of poultry by-products: challenges and opportunities, in *Beneficial Uses of Agricultural, Municipal, and Industrial By-products*, J. Power, Ed., Soil Science Society of America, Madison, WI (in press).

Chae, Y. M. and M. A. Tabatabai. 1986. Mineralization of nitrogen in soils amended with organic wastes, *J. Environ. Qual.*, 15, 193–198.

Deans, J. R., J. A. E. Molina, and C. E. Clapp. 1986. Models for predicting potentially mineralizable nitrogen and decomposition rate constants, *Soil Sci. Soc. Am. J.*, 50, 323–326.

Donovan, W. C. and T. J. Logan. 1983. Factors affecting ammonia volatilization from sewage sludge applied to soil in a laboratory study, *J. Environ. Qual.*, 12, 584–590.

Ellert, B. H. and J. R. Bettany. 1988. Comparison of kinetic models for describing net sulfur and nitrogen mineralization, *Soil Sci. Soc. Am. J.*, 52, 1692–1702.

Epstein, E., D. B. Keane, J. J. Meisinger, and J. O. Legg. 1978. Mineralization of nitrogen from sewage sludge and sludge compost, *J. Environ. Qual.*, 7, 217–221.

Fox, R. H., and L. D. Hoffman. 1981. The effect of N fertilizer source on grain yield, N uptake, soil pH, and lime requirement in no-till corn, *Agron. J.*, 73, 891–895.

Frink, C. R., P. E. Waggoner, and J. H. Ausubel. 1999. Nitrogen fertilizer: retrospect and prospect, *Proc. Natl. Acad. Sci. U.S.A.*, 96, 1175–1180.

Gartley, K. L. and J. T. Sims. 1993. Ammonia volatilization from poultry manure-amended soils, *Biol. Fertil. Soils*, 16, 5–10.

Gordillo, R. M. and M. L. Cabrera. 1997. Mineralizable nitrogen in broiler litter: II. Effect of selected soil characteristics, *J. Environ. Qual.*, 26, 1679–1686.

Hadas, A., B. Bar-Yosef, S. Davidov, and M. Sofer. 1983. Effect of pelleting, temperature, and soil type on mineral N release from poultry and dairy manures, *Soil Sci. Soc. Am. J.*, 47, 1129–1133.

Hamilton, P. A. and R. J. Shedlock. 1992. Are fertilizers and pesticides in the groundwater? A case study of the Delmarva Peninsula. U.S. Geological Survey Circular 1080. Denver, CO, 15 pp.

Hergert, G. W. 1987. Status of residual nitrate-nitrogen soil tests in the United States of America, in *Soil Testing: Sampling, Correlation, Calibration, and Interpretation*, J. R. Brown, Ed., SSSA Spec. Pub. 21, Soil Science Society of America, Madison, WI, 73–88.

Igo, E. C., J. T. Sims, and G. W. Malone. 1991. Advantages and disadvantages of manure analysis for nutrient management purposes, *Agron. Abstr.*, 154.

Jokela, W. E. 1992. Nitrogen fertilizer and dairy manure effects on corn yield and soil nitrate, *Soil Sci. Soc. Am. J.*, 56, 148–154.

Keeney, D. R. 1982. Nitrogen management for maximum efficiency and minimum pollution, in *Nitrogen in Agricultural Soils*, F. J. Stevenson, Ed., American Society of Agronomy, Madison, WI, 605–649.

Keeney, D. R. and T. H. DeLuca. 1993. Des Moines River nitrate in relation to watershed agricultural practices: 1945 versus 1985, *J. Environ. Qual.*, 22, 267–272.

Kelling, K. A., L. M. Walsh, D. R. Keeney, J. A. Ryan, and A. E. Peterson. 1977. A field study of the agricultural use of sewage sludge: II. Effect on soil N and P, *J. Environ. Qual.*, 6, 345–352.

Lichtenberg, E. and L. K. Shapiro. 1997. Agriculture and nitrate concentrations in Maryland's community water system wells, *J. Environ. Qual.*, 26, 145–153.

Magdoff, F. R., W. E. Jokela, R. H. Fox, and G. F. Griffin. 1990. A soil test for nitrogen availability in the Northeastern United States, *Commun. Soil Sci. Plant Anal.*, 21, 1103–1116.

McInnes, K. J., R. B. Ferguson, D. E. Kissel, and E. T. Kanemasu. 1986. Field measurements of ammonia loss from surface applications of urea solution to bare soil, *Agron. J.*, 78, 192–196.

Meisinger, J. J., V. A. Bandel, G. Stanford, and J. O. Legg. 1985. Nitrogen utilization of corn under minimal tillage and moldboard plow tillage. I. Four year results using labeled N fertilizer on an Atlantic coastal plain soil, *Agron. J.*, 77, 602–611.

Nielsen, E. G. and L. K. Lee. 1987. The magnitude and costs of groundwater contamination from agricultural chemicals: a national perspective, Agric. Econ. Rep. No. 576. U.S. Department of Agriculture, Washington, D.C., 38 pp.

Piekielek, W. P. and R. H. Fox. 1992. Use of a chlorophyll meter to predict sidedress nitrogen requirements for maize, *Agron. J.*, 84, 59–65.

Pitcairn, C. E. R., I. D. Leith, L. J. Sheppard, M. A. Sutton, D. Fowler, R. C. Munro, S. Tang, and D. Wilson. The relationship between nitrogen deposition, species composition, and foliar nitrogen concentrations in woodland flora in the vicinity of livestock farms, *Environ. Pollut.* (in press).

Rice, C. W., P. E. Sierzega, J. M. Tiedje, and L. W. Jacobs. 1988. Stimulated denitrification in the microenvironment of a degradable organic waste injected into soil, *Soil Sci. Soc. Am. J.*, 52, 102–108.

Roth, G. W. and R. H. Fox. 1990. Soil nitrate accumulations following nitrogen-fertilized corn in Pennsylvania, *J. Environ. Qual.*, 19, 243–248.

Sallade, Y. E. and J. T. Sims. 1992. Evaluation of thiosulfate as a nitrification inhibitor for manures and fertilizers, *Plant Soil*, 147, 283–291.

Schepers, J. S., M. G. Moravek, E. E. Alberts, and K. D. Frank. 1991. Maize production impacts on groundwater quality, *J. Environ. Qual.*, 20, 12–16.

Shirey, J. J. and A. J. Sexstone. 1989. Denitrification and nitrate reducing bacterial populations in abandoned and reclaimed mine soils, *FEMS Microb. Ecol.*, 62, 59–70.

Sims, J. T. 1987. Agronomic evaluation of poultry manure as a nitrogen source for conventional and no-tillage corn, *Agron. J.*, 79, 563–570.

Sims, J. T., D. W. Murphy, and T. S. Handwerker. 1992. Composting of poultry wastes: implications for dead poultry disposal and manure management, *J. Sust. Agric.*, 12, 67–82.

Spalding, R. F. and M. E. Exner. 1993. Occurrence of nitrate in groundwater — a review, *J. Environ. Qual.*, 22, 392–402.

Stanford, G. and E. Epstein. 1974. Nitrogen mineralization-water relations in soils, *Soil Sci. Soc. Am. Proc.*, 38, 103–106.

Stanford, G., M. H. Frere, and D. H. Schwaninger. 1973. Temperature coefficient of nitrogen mineralization in soils, *Soil Sci.*, 115, 321–323.

Touchton, J. T. and W. L. Hargrove. 1982. Nitrogen sources and methods of application for no-tillage corn production, *Agron. J.*, 74, 823–826.

Vigil, M. F. and D. E. Kissel. 1991. Equations for estimating the amount of nitrogen mineralized from crop residues, *Soil Sci. Soc. Am. J.*, 55, 757–761.

Supplementary reading

Andres, A. S. 1995. Nitrate loss via ground water flow, coastal Sussex County, Delaware, in *Animal Waste and the Land Water Interface*, K. Steele, Ed., Lewis Publishers, New York, 69–76.

Dahnke, W. C. and G. V. Johnson. 1990. Testing soils for available nitrogen, in *Soil Testing and Plant Analysis*, R. L. Westerman, Ed., Soil Science Society of America, Madison, WI, 127–140.

Fedkiw, J. 1991. Nitrate Occurrence in U.S. Waters (and Related Questions): A Reference Summary of Published Sources from an Agricultural Perspective. U. S. Department of Agriculture, Washington, D.C.

Follett, R. F., D. R. Keeney, and R. M. Cruse. 1991. *Managing Nitrogen for Groundwater Quality and Farm Profitability,* American Society of Agronomy, Madison, WI, 357 pp.

Greenwood, D. J. 1990. Production or productivity: the nitrate problem, *Ann. Appl. Biol.,* 117, 209–231.

Lanyon, L. E. 1995. Does nitrogen cycle? Changes in the spatial dynamics of nitrogen with industrial nitrogen fixation, *J. Prod. Agric.,* 8, 70–78.

Legg, J. O. and J. J. Meisinger. 1982. Soil nitrogen budgets, in *Nitrogen in Agricultural Soils,* F. J. Stevenson, Ed., Agron. Monogr. 22, American Society of Agronomy, Madison, WI, 503–564.

Magdoff, F. R., D. Ross, and J. Amadon. 1984. A soil test for nitrogen availability to corn, *Soil Sci. Soc. Am. J.,* 48, 1301–1304.

Minkara, M. Y., J. H. Wilhoit, C. W. Wood, and K. S. Yoon. 1995. Nitrate monitoring and GLEAMS simulation for poultry litter application to pine seedlings, *Trans. ASAE,* 38, 147–152.

National Research Council. 1995. *Nitrate and Nitrite in Drinking Water,* National Academy Press, Washington, D.C.

Power, J. F. and J. S. Schepers. 1989. Nitrate contamination of groundwater in North America, *Agric. Ecosyst. Environ.,* 226, 165–187.

Ritter, W. F. and A. E. M. Chirnside. 1987. Influence of agricultural management practices on nitrates in the water table aquifer. *Biol. Wastes,* 19, 165–178.

Sims, J. T. 1995. Organic wastes as alternative nitrogen sources, in *Nitrogen Fertilization in the Environment,* P. E. Bacon, Ed., Marcel Dekker, New York, 487–535.

Sims, J. T., N. Goggin, and J. McDermott. Nutrient management for water quality protection: integrating research into environmental policy, *Water Sci. Technol.* (in press).

Sommers, L. E. and P. M. Giordanoa. 1984. Use of nitrogen from agricultural, industrial and municipal wastes, in *Nitrogen in Crop Production,* R. D. Hauck, Ed., American Society of Agronomy, Madison, WI, 207–220.

chapter five

Soil phosphorus and environmental quality

Contents

5.1   Introduction: phosphorus and the environment

Phosphorus (P) is essential to all forms of life on earth and has no known direct toxic effects to humans or animals. Environmental concerns associated with P center on its stimulation of biological productivity in aquatic ecosystems. In most freshwater systems primary productivity (e.g., the growth of algae or aquatic plants) is limited by inadequate levels of P. External inputs of P from urban wastewater systems, surface runoff, or subsurface groundwater flow can remove this limitation and stimulate the growth of aquatic organisms to ecologically undesirable levels. Total P concentrations of >100 µg P/L (ppb) are regarded as unacceptably high in most surface waters and concentrations as low as 20 µg P/L can cause environmental problems in some waters.

Eutrophication is defined as "an increase in the fertility status of natural waters that causes accelerated growth of algae or water plants." Eutrophication is not caused by inputs of P alone, but by a complex interaction between nitrogen (N), P, environmental conditions (temperature, salinity, light) and the physical and hydrologic characteristics of different types of surface waters (streams vs. lakes vs. estuaries). For example, N limits primary productivity more in estuaries and other coastal waters than P, particularly in summer months.

The negative effects associated with eutrophication of surface waters are important from ecological, economic, and animal and human health perspectives. As nutrient inputs to surface waters gradually increase, the trophic state of the water body passes through four stages of eutrophication: *oligotrophic, mesotrophic, eutrophic,* and *hypereutrophic.* At each stage, progressive changes in the ecology of the water body occur that degrade habitat quality, reduce biodiversity, decrease economic value, and impair recreational uses (Figure 5.1a; also see Section 5.1.1).

If we are to manage soil P to prevent eutrophication and sustain agricultural productivity, we must address the following questions:

- What are the characteristics of a water body that control eutrophication? Is eutrophication limited most by P, or by N, or by both nutrients?
- What are the natural inputs of P into a water body, from groundwater discharge, erosion and runoff from natural areas, and atmospheric deposition?
- Do anthropogenic P inputs to a water body come from point- or nonpoint-source pollution? When and where do inputs of P from agricultural and urban sources impact water quality?
- How does the soil P cycle affect the availability of P for transport to water bodies? When does soluble P become important, as well as particulate P?
- What are the dominant transport mechanisms operational between the sources of P and the water body (e.g., erosion, surface/subsurface runoff, channel processes) and how do they vary with land use, management, and seasonal climatic patterns?
- What management practices can be used to reduce P loading to a water body from all sources, urban and rural, point and nonpoint? How can we develop an integrated, economically feasible approach to reduce P loading to water bodies?

5.1.1. *Eutrophication: the role of phosphorus*

As the sole environmental impact of P is its role in eutrophication, a clear understanding of this process is essential to the development of sound strategies for P management. In oligotrophic lakes, soluble and sediment-bound nutrient (N, P) concentrations are low and limit the growth of algae and other water plants. These lakes are normally deeper and have adequate oxygen (O_2) levels even in the summer when photosynthesis and temperatures favor maximum plant growth. They also have a high degree of species diversity for plants and animals. However, oligotrophic lakes, because of their low nutrient concentrations, also have low biological productivity and, if used for the economic production of fish, may require continuous fertilization to sustain a fishing industry. Alterations in the ecosystem (construction of wastewater treatment plants, conversion of land from forested to agricultural or urban use) can increase inputs of nutrients to oligotrophic lakes, especially N and P, stimulating the growth of algae and other water plants and initiating the eutrophication process (Figure 5.1a). For example, as illustrated in Figure 5.1b, converting land that is 90% forested to 90% agricultural, could result in a threefold increase in total P loss in runoff.

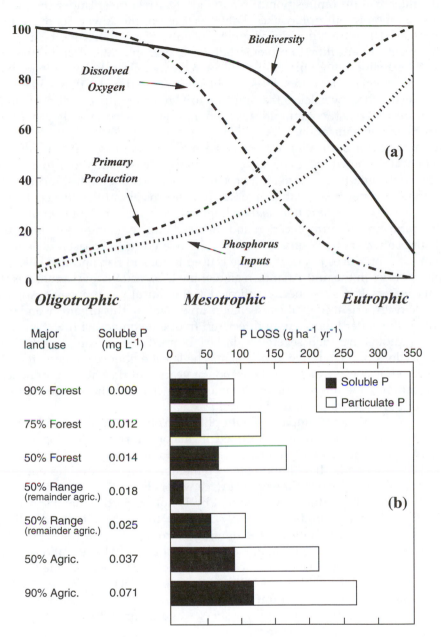

Figure 5.1 (a) Overview of the changes that occur when fresh waters become eutrophic (Adapted from Correll, 1998.) (b) Effect of land use on the loss of soluble and particulate P from soils. (Adapted from Sharpley and Halvorson, 1994.)

Eutrophic waters are widely regarded as undesirable, for many reasons (see Tables 2.3 and 2.4). Once eutrophic conditions are established, algal blooms and other ecologically damaging effects can occur, including low dissolved O_2 levels, excessive aquatic weed growth, increased sedimentation, and greater turbidity. Decreased oxygenation is the primary negative effect of eutrophication because low dissolved O_2 levels seriously limit the growth and diversity of aquatic biota and, under extreme conditions, cause fish kills.

The increased biomass resulting from eutrophication causes the depletion of O_2, especially during the microbial decomposition of plant and algal residues. Under the more turbid conditions common to eutrophic lakes, light penetration into lower depths of the water body is decreased, resulting in reduced growth of subsurface plants and benthic (bottom-living) organisms. In addition to ecological damage, eutrophication can increase the economic costs of maintaining surface waters for recreational and navigational purposes. Surface scums of algae, foul odors, insect problems, impeded water flow and boating due to aquatic weeds, shallower lakes that must be dredged to remove sediment, and disappearance of desirable fish communities are among the most commonly reported, undesirable effects of eutrophication.

In some cases eutrophication also causes ecological changes that may affect animal and human health. Some blue-green algae can directly release dangerous toxins, and algal blooms can contribute to the formation of dangerous trihalomethanes during the chlorination process in water treatment plants. Recent concerns about the effects of some toxic dinoflaggellates (e.g., *Pfiesteria piscicida*) on human health have heightened public awareness of the problems of eutrophication and the needs for reducing nonpoint-source pollution of surface waters by N and P. *Pfiesteria*, discovered in the coastal waters of North Carolina in 1988, produces toxins that have caused lesions in fish and fish kills in several rivers and estuaries in the mid-Atlantic region of the U.S., usually in warmer weather and in eutrophic waters. It has also been implicated as the causative organism in some human health problems experienced by the researchers investigating this organism and by people who work on the water who were in direct and frequent contact with waters containing *Pfiesteria*. Symptoms included severe headaches, blurred vision, sores, reddening of the eyes, memory loss, and cognitive impairment. The latest evidence suggests that *Pfiesteria* has been present in these waters for thousands of years, but that human influences, such as eutrophication, have altered the environment in a manner that causes more regular occurrence of the toxic stages of its life cycle.

The many environmental impacts of eutrophication on water quality, and now perhaps on human and animal health, have resulted in major efforts to develop management strategies to reduce nutrient inputs to surface waters and, where possible, to reclaim eutrophic lakes or ponds. It is important to remember that controlling eutrophication requires that both N and P enrichment of surface waters be minimized, although in most fresh waters biological productivity is limited by P, not N. Studies have shown that the ratio of N:P in the water body (commonly referred to as the "Redfield ratio") is an important indicator of which nutrient is limiting eutrophication. If the Redfield ratio is >16:1, P is most likely to be the limiting factor for algal growth; lower ratios indicate that N is of greater importance. The eutrophication threshold for most P-limited aquatic systems (~20 to 100 µg P/L) is also much lower than for N (500 to 1000 µg N/L). Water bodies with naturally low P concentrations will, therefore, be highly sensitive to external inputs of P from agricultural runoff, domestic sewage treatment systems, urban stormwater discharge, and industrial wastewaters.

Strategies to reduce P loading must consider the nature of the P source as well as the sensitivity of the water body to eutrophication. Even natural ecosystems, such as forests or grasslands, normally considered to be low in P fertility, may provide enough P from runoff and subsurface flow to alter the nutrient balance of a water body on the threshold of eutrophication. Controlling P inputs from these areas is generally not feasible, with the possible exception of commercial forestry operations where erosion control practices can reduce sediment loads to surface waters. In contrast, P discharge from a concentrated point source (e.g., sewage treatment plant) may be relatively easy to identify and control, but costly to correct due to the expenses associated with technological improvements to municipal or industrial wastewater treatment plants. Conversely, reducing P loading to

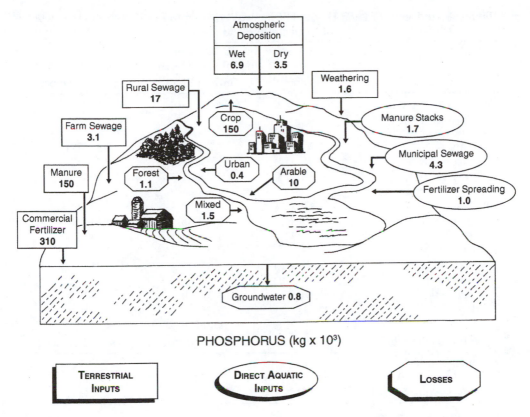

Figure 5.2   A P budget for agricultural and urban areas of the Lake Ringsjon (Sweden) watershed. (Adapted from Ryding et al., 1990.)

surface waters from nonpoint-sources, such as runoff from an urban or agricultural watershed, is often limited not by technological costs, but by the difficulties associated with implementing improved nutrient management practices over a large area with diverse agricultural (or rural and industrial) enterprises.

The relative magnitudes of these P inputs, and the complexities involved in reducing P (or N) loading to lakes and other water bodies are illustrated in Figure 5.2, which provides a broad view of the impact of civilization on a watershed scale P "budget" (P outputs – P inputs). In the case of the Lake Ringsjon watershed, the budget is decidedly negative, with P inputs from all sources, including municipal and rural sewage, animal manures, fertilizers, and runoff from urban and forested areas totaling ~512,000 kg/year, while P outputs (crops, groundwater) are ~150,000 kg/year. Although all of these P inputs do not directly enter the lake, as much of the P is stored in soils in slowly soluble forms, it seems clear that, in the long term, nutrient enrichment of this lake is due to the magnitude and intensity of the agricultural and urban activities in the watershed.

Another, somewhat different, example of the effects of land use on watershed-scale point and nonpoint-source P loading is given in Table 5.1 for Delaware's Inland Bays, an important coastal estuary dominanted by a nutrient-intensive poultry–grain agriculture and a large tourism industry. Agriculture is clearly the largest single source of P to the Inland Bays (~37% for the entire watershed), perhaps expected because it is the single largest land use (~45%). However, the large inputs of P from nonagricultural activities (34% from urban runoff, boating, septic tanks, and point sources combined) and natural inputs (29% from forests, wetlands, and rainfall) illustrate the need to

Table 5.1   Influence of Land Use on Phosphorus Loading to the Three Subwatersheds of Delaware's Inland Bays

Source of P	P Load to Inland Bays subwatershed (kg/year) (% total P load in subwatershed)		
	Rehoboth	Indian River	Little Assawoman
Point source	5,300 (25)	5,820 (16)	None
Nonpoint sources			
Groundwater discharge			
Urban	1,700 (8)	2,560 (7)	560 (7)
Agricultural	4,910 (24)	11,030 (30)	2,700 (34)
Forest	2,700 (13)	5,700 (16)	1,000 (13)
Erosion and runoff			
Urban	420 (2)	640 (2)	300 (4)
Agricultural	1,220 (6)	2,760 (8)	1,440 (18)
Forest	680 (3)	1,420 (4)	540 (7)
Other			
Rainfall	2,500 (12)	3,300 (9)	910 (12)
Boating	35 (<1)	47 (<1)	5 (<1)
Septic tanks	1,370 (7)	3,310 (9)	440 (6)
Total	20,835	36,587	7,895

Subwatershed (ha)	Land use (%)		
	Urban	Agriculture	Forest and wetlands
Rehoboth (16,500)	15	41	44
Indian River (39,000)	9	44	47
Little Assawoman (8,500)	11	54	35

Source: Adapted from Ritter (1992).

develop comprehensive management practices that reduce P export from all types of land use if we are to protect and improve water quality.

5.1.2   Environmental impacts of soil phosphorus

Although many point and nonpoint sources of P have the potential to induce eutrophication in surface waters, the focus of this chapter is the management of P to avoid water quality problems while sustaining agricultural productivity and environmental quality. Nonpoint-source pollution of P from urban soils will be considered to illustrate the different management options needed for areas where erosion, stormwater runoff, and intensive horticultural operations are the major sources of P.

The total quantity of P in a lake or other surface water body will be controlled by the balance between the inputs from external sources (agriculture, urban runoff, point sources) and the outputs as water drains from the lake via rivers, streams, or other watercourses. A net increase in P will increase the likelihood of eutrophication, but the cycling of P between soluble, organic, and sediment-bound forms within any aqueous ecosystem will regulate the bioavailability of P and thus the extent of eutrophication that occurs. Therefore, insofar as soil P is concerned, its importance in eutrophication will be regulated by the chemical, biological, and physical reactions that control P solubility, and the transport processes that move all forms of P to and within water bodies. An overview of these complex phenomena is given in Figure 5.3, which summarizes the movement of P in soils, the major transport mechanisms to a water body, and the major transformations that P undergoes in an aquatic system. In brief, most P in agricultural (or urban) soils is found as either insoluble precipitates of calcium (Ca), iron (Fe), and/or aluminum (Al) or as a

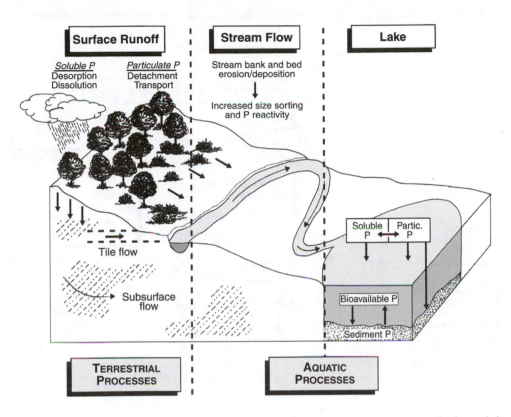

Figure 5.3 Phosphorus transport and fate in terrestrial and aquatic ecosystems. (Adapted from Sharpley and Halvorson, 1994.)

constituent of a wide range of organic compounds. Water moving across or through soils dissolves soluble P and removes sediments enriched with P, usually the lighter, finer-sized particles such as clays and organic matter, initiating the transport process. The soluble or particulate P can then either enter a flowing water body (stream, river) where it can be deposited as sediment or be carried into standing waters (lakes, ponds, reservoirs). Phosphorus can also leach downward in the soil, perhaps to a tile drainage system or to shallow groundwaters, where subsurface transport can then discharge the P into surface waters. Once in the lake, the soluble P is immediately available for uptake by aquatic organisms such as algae, while much of the particulate P is removed from the lake by deposition as sediment. It is important to note, however, that P bound to sediments in streams, lakes, or coastal waters may become available for biological uptake at a later date. This can occur when P uptake by aquatic plants depletes soluble P, when sediments are reduced under anoxic conditions, or when climatic conditions cause lake turbulence and resuspension of sediments. Clearly, an in-depth understanding of the chemistry, biology, and physics of P in the soil and aquatic environment is needed to design management practices that will reduce P impacts on water quality.

5.2 The soil phosphorus cycle

Adequate P levels in soils are essential to produce agricultural crops, revegetate disturbed lands, and to grow plants for aesthetic or recreational purposes. Phosphorus fertilization is thus a vital component of modern agriculture. Phosphorus, found in all

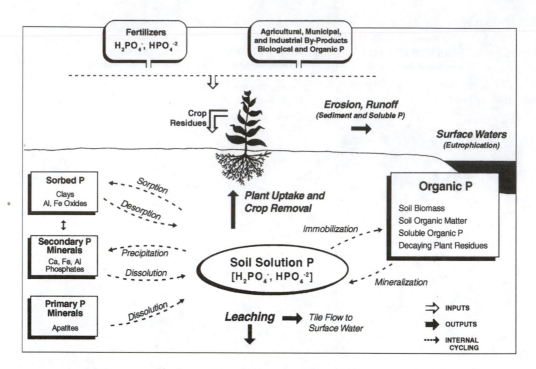

Figure 5.4 The soil P cycle. An overview of the physical, chemical, and microbiological processes controlling the availability of P to plants and P transport in runoff or leaching waters. (Adapted from Gachon, 1969.)

terrestrial environments, primarily originates from the weathering of soil minerals and other more stable geological materials. As P is solubilized in soils by the chemical and physical processes of weathering, it is accumulated by plants and animals, reverts to stable inorganic forms in the landscape, or is eroded from soils and deposited as sediments in fresh waters, estuaries, or oceans. Urban areas or large animal production operations convert biologically accumulated P (food or feed) into human or animal wastes (biosolids, manures) and recycle this P into the ecosphere with varying degrees of efficiency. Soil factors that control the rate of conversion of P between the inorganic and organic forms regulate the short- and long-term fates of P in the environment. The soil P cycle, illustrated in Figure 5.4, consists of many complex chemical and microbiological reactions. The overall goals of agronomic and environmental management programs for soil P are to maximize the economic efficiency of plant growth, while minimizing losses of P to surface waters and groundwaters. It is important, therefore, to understand the role of each reaction in the soil P cycle in controlling the availability of soil P for plant uptake or loss in erosion, surface runoff, leaching and lateral flow, and artificial drainage waters.

5.2.1 Inorganic soil phosphorus

Total P levels in soils range from 50 to 1500 mg/kg, with 50 to 70% found in the inorganic form in mineral soils. The principal minerals that weather and release P into more soluble forms differ between soils, primarily as a function of time and soil development (Figure 5.5a). In soils that are unweathered or moderately weathered, the dominant minerals are the *apatites*, Ca phosphates with the general chemical formula $Ca_{10}(PO_4)_6X_2$, where X represents anions such as F^-, Cl^-, OH^-, or CO_3^{2-}. Although over 200 forms of mineral P are known to occur in nature, fluorapatites are the most common form both as a mineable ore and in most arable soils. In areas of intense weathering Ca and other basic minerals are eventually leached from the soil, the pH of the soil solution decreases, and Fe and Al solubilization occurs. Precipitates of Fe, Al, and P then form and become the major mineral sources of P in highly weathered soils. Amorphous (noncrystalline) oxides of Fe and Al are also common in these soils and act as "sinks" for P, through a variety of chemical reactions collectively referred to as *phosphorus fixation*.

As P is dissolved or desorbed from soil minerals and colloids, it enters the soil solution, where its ionic form is largely controlled by soil pH (Figure 5.5b). In the pH ranges commonly found in agricultural soils (pH 4.0 to 9.0), P will be found as a monovalent ($H_2PO_4^-$) or divalent (HPO_4^{2-}) anion, both of which are readily available for plant uptake.

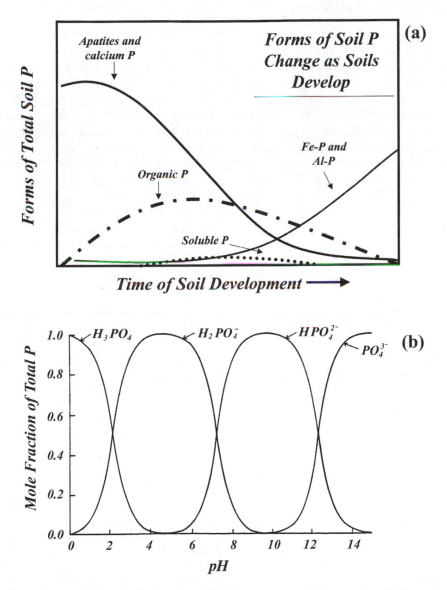

Figure 5.5   Changes in the form of soil P. (a) As affected by time and soil development for total P. (From Foth, H. D. and Ellis, B. G., *Soil Fertility*, 2nd ed., Lewis Publishers, Boca Raton, FL, 1997. With permission.) (b) As affected by soil pH for soluble P.

One common characteristic of P in soils at all stages of weathering is the low solubility of P-bearing minerals. In most agricultural soils the soil solution concentration of P ranges from <0.01 to 1 mg P/L (ppm), although concentrations as high as 6 to 8 mg P/L have been measured in recently fertilized soils. Soluble P concentrations required for normal plant growth vary somewhat between species and with yield potential. A value of 0.2 mg P/L is commonly reported as the desired soil solution P concentration needed to optimize plant growth, but P concentrations as low as 0.03 mg P/L have been adequate to produce high yields of many agronomic crops. Interestingly, a soluble P concentration of 0.03 mg P/L is also sometimes used as the threshold for eutrophication in fresh waters, suggesting a commonality in P nutritional requirements between terrestrial and aquatic plants.

5.2.2 Organic soil phosphorus

Phosphorus dissolved from minerals enters the soil solution where biological accumulation by plants and the soil microbial biomass occurs. Plants vary widely in their P contents, with most agricultural crops containing from 0.1 to 0.5% P, much less than the concentrations of other major essential elements such as N and potassium (K) (2.0 to 4.0%). In agricultural situations some of the P accumulated by plants is removed from the soil as harvested grain or forage, with the remainder returned to the soil as crop residues (Table 5.2). In urban horticulture and native ecosystems, where "harvesting" does not occur, all of the P accumulated is eventually recycled back into the soil as plants die or lose vegetative matter in response to normal biological development. One major environmental concern in urban areas, however, is the conversion of biologically incorporated P in private or industrial landscapes into waste organic matter (leaves and lawn clippings) that must then be disposed of by municipalities through landfills or large-scale composting operations. Similarly, human wastes represent a transformation of biologically accumulated P from food chain crops to waste organic matter (biosolids). An agricultural analogy to this

Table 5.2 Phosphorus Concentrations and Removal in the Harvested Portion of Some Major Agricultural Crops

Crop category	P concentration (%)	Crop yield	P removal (kg P/ha)
Grains			
Corn	0.28	9.4 Mg/ha (150 bu/ac)	26
Rice	0.24	4.0 Mg/ha (80 bu/ac)	10
Soybeans	0.58	2.7 Mg/ha (40 bu/ac)	12
Wheat	0.45	2.7 Mg/ha (40 bu/ac)	12
Forages			
Alfalfa	0.25	13.4 Mg/ha (6 tons/ac)	34
Bermudagrass	0.32	22.4 Mg/ha (10 tons/ac)	71
Corn silage	0.06	67.2 Mg/ha (30 tons/ac)	39
Red clover	0.25	9.0 Mg/ha (4 tons/ac)	22
Tall fescue	0.44	9.0 Mg/ha (4 tons/ac)	39
Specialty			
Beans, snap	0.06	26.9 Mg/ha (12 tons/ac)	15
Celery	0.14	2470 crates/ha (1000 crates/ac)	90
Onions	0.05	40.3 Mg/ha (18 tons/ac)	19
Potatoes	0.04	448 quintals (400 cwt.)	17
Sugar cane	0.02	224 Mg/ha (100 tons/ac)	48
Sweet corn	0.07	6.7 Mg/ha (3 tons/ac)	5
Tomatoes	0.04	67.2 Mg/ha (30 tons/ac)	27

Source: Adapted from Pierzynski and Logan (1993).

Table 5.3 Forms of Phosphorus in Soils Amended with Organic Wastes

	Total P (mg/kg)	Organic P (mg/kg)	Inorganic P (mg/kg)
Manure-amended soils (Sims, 1992)			
Four Delaware soils	1467	281 (19%)	1166 (81%)
Seven Pennsylvania soils	2240	427 (19%)	1813 (81%)
Pullman clay loam (Sharpley et al., 1984)			
Untreated soil	353	202 (57%)	151 (43%)
Soil + fertilizer P	457	231 (51%)	226 (49%)
Soil + feedlot waste at 67 Mg/ha/year	996	323 (32%)	673 (68%)
Greenfield sandy loam (Chang et al., 1983)			
Untreated soil	579	60 (10%)	519 (90%)
Soil + biosolids at 45 Mg/ha/year	1433	122 (9%)	1211 (91%)

situation is concentrated animal-based agriculture (feedlots, dairy, swine, and poultry operations) where harvested organic P (grains, forages, and silage) is transformed into animal manure organic and inorganic P (as much as 60% of the P in some animal manures is inorganic P). With time, much of the organic P added in biosolids and manures is degraded by soil microorganisms and converted to soluble and inorganic forms of P (Table 5.3). Because organic wastes, in both urban and agricultural systems, are often continually applied to the same soils within a limited geographic area, the buildup of soil P to levels of environmental concern often occurs. The development of economic means for the beneficial reuse of organic wastes as soil amendments without enhancing P losses to surface waters is perhaps the most significant environmental issue related to soil P management we face today (see Section 5.5).

Common forms of organic P found in soils include inositol phosphates, phospholipids, phosphoglycerides, phosphate sugars, and nucleic acids. Microbial decomposition of organic P results in the release of soluble organic P that, with time, is normally converted into stable inorganic forms of P. Studies have shown, however, that as much as 50% of the P transported in runoff can be soluble organic P. The extent and rate of conversion of organic P into soluble or stable inorganic forms is highly dependent upon the nature of the original organic material, as well as environmental factors such as pH, temperature, and soil moisture. Fresh plant residues may quickly release P into the soil solution where more stable forms of organic matter, such as soil humus, animal manures, municipal biosolids, or composts, generally act as long-term, slow-release sources of P. Mineralization, the conversion of organic P to inorganic P, usually occurs rapidly if the carbon (C):P ratio of the organic matter is <200:1, while immobilization, the incorporation of P into microbial biomass, occurs if C:P ratios are >300:1. At this time, it is unclear whether plants can directly absorb soluble organic P or whether further enzymatic action is required to cleave phosphate molecules from organic compounds prior to uptake by plant roots.

5.2.3 Phosphorus additions to soils

The low solubility of P in many soils indicates that the soil solution must be replenished frequently and rapidly with soluble P if adequate plant growth is to occur. In soils with high fertility levels the soil alone can normally provide all the P required by plants. The need for additional P can be determined by *soil testing* (see Section 3.2). Agronomic soil testing involves comparison of available soil P, as estimated by a chemical extracting

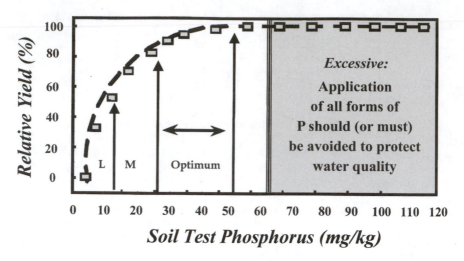

Figure 5.6   The relationships between soil test P (Mehlich 1, 0.05 *M* HCl + 0.0125 *M* H$_2$SO$_4$), crop yield, and the potential for environmental problems due to excessive soil P (L = Low, M = Medium).

solution, with *critical values* determined from long-term research (Figure 5.6). As the soil test P value of a field becomes progressively lower, the likelihood of an economically profitable crop yield response to P fertilization increases. Similarly, in fields that have soil test values above the critical level, addition of P fertilizers will, in all likelihood, not result in crop yield increases. In some areas of the world there is concern that long-term overapplication of P to soils, usually in animal manures, has not only resulted in soil test P values well above the agronomic critical level, but has also "saturated" the P sorption capacity of soils (see Section 5.5.1). This results in soil particles that are not only highly enriched in P, but that maintain higher equilibrium concentrations of P in the soil solution. As soils become increasingly saturated with P, the risk of nonpoint-source pollution of surface waters and shallow groundwaters by erosion, runoff, and leaching increases.

Application of fertilizers, animal manures, or other P sources to fields with soil test values well above the critical level is unnecessary and increases the chance of soluble or sediment-bound P losses to surface waters. One exception has been the use of small amounts of "starter" fertilizer placed near the seed at planting. Starter fertilizers have frequently been shown to produce more vigorous early plant growth, give greater resistance to early season pest and weather stresses, and occasionally, but not always, to increase crop yields (Figure 5.7). Plant response to starter fertilizers, even in high P soils, can be explained by the lower rates of mineralization of organic P and diffusion of soluble P early in the growing season when soil temperatures are cooler. As soils warm, microbial and chemical reactions are more rapid and the supply of P to plant roots increases to acceptable levels.

Many soils, however, due either to low P parent material, low organic matter contents, or intensive cropping, do not contain adequate P to achieve the desired level of plant growth. Fertilization with inorganic P sources, or the use of animal manures, municipal biosolids, composts, or other types of by-products, is then necessary. Commercial P fertilizers are produced by industrial processes that react phosphate ores (e.g., fluorapatite, or "rock phosphate") with sulfuric acid (ordinary superphosphate, 9% P), phosphoric acid (triple superphosphate, 20 to 22% P), or phosphoric acid with ammonia (monoammonium phosphate, 21% P; diammonium phosphate, 23% P). Rock phosphate (13 to 17% P) can also be used directly as a P fertilizer, although its low solubility often

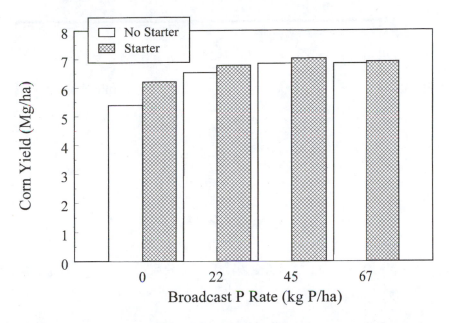

Figure 5.7 The effect of starter fertilizer on corn yield. Results of a 20-year study in Iowa comparing the use of starter P applied with differing rates of broadcast P at 3-year intervals. (Adapted from Young et al., 1985.)

means high application rates are needed to provide sufficient available P for optimum yields. The situations where rock phosphate is most effective include highly weathered, acid soils with low Ca contents, soil conditions that promote the dissolution of apatite minerals (rock phosphate). The P content of the most commonly used inorganic and organic sources of P is provided in Table 5.4.

Animal manures, municipal biosolids, or composts applied to agricultural lands as part of waste disposal operations contain relatively low concentrations of P (0.1 to 3.0%). However, the large and frequent applications normally used with these organic wastes, along with the relatively small amounts of P actually removed from the soil in harvested grain or forage, not only provide for crop P requirements, but also, as mentioned above, can build soil P to extremely high levels within a few years (Figure 5.8). Unfortunately, because of transportation and handling costs, most animal-based agricultural operations have few alternatives to land application of manures within a short distance from the area where the animals are grown. This often results in the long-term buildup of soil P to excessive levels as typical manure application rates usually add P at rates beyond plant P requirements. If these soils are near a sensitive water body, the potential environmental impacts of increasing soil P values above agronomic optimum concentrations can become a socioeconomic and political issue (see Section 5.5 and the environmental quality issues/events box, Improving agricultural phosphorus management in the Chesapeake Bay Watershed). Similarly, for municipalities, the alternatives to land application of organic wastes (landfills, ocean dumping, and incineration) are increasingly limited, costly, and may have more serious environmental impacts than well-managed land application programs.

Inorganic wastes from industrial operations, such as basic slag (3 to 5% P), a steel manufacturing by-product, or coal ash (<1% P), a by-product of the electric power industry, are usually inexpensive sources of P. While the solubility and thus plant availability of P in industrial wastes is less than fertilizers and animal manures, land application is an economically desirable alternative to landfill disposal of these wastes.

Table 5.4 Major Inorganic and Organic Sources of Phosphorus for Crop Production

P Source and chemical composition	P (%)	P_2O_5 (%)	Other nutrients
Commercial fertilizers			
Ordinary superphosphate [$Ca(H_2PO_4)_2 + CaSO_4$]	7–10	16–23	Ca, S (8–10%)
Triple superphosphate [$Ca(H_2PO_4)_2$]	19–23	44–52	Ca
Monoammonium phosphate (MAP) [$NH_4H_2PO_4$]	26	61	N (12%)
Diammonium phosphate (DAP) [$(NH_4)_2HPO_4$]	23	53	N (21%)
Urea-ammonium phosphate [$CO(NH_2)_2 + NH_4H_2PO_4$]	12	28	N (28%)
Ammonium polyphosphates (liquid) [$(NH_4)_3HP_2O_7$]	15	34	N (11%)
Rock phosphates			
U.S. (Florida) $Ca_{10}F_2(PO_4)_6 \cdot XCaCO_3$ (varies between mineral deposits)	14	33	Major impurities: Al, Fe, Si, F, CO_3^{2-}
Brazil	15	35	
Morocco	14	33	
Former U.S.S.R.	17	39	
Organic P sources			
Beef manure	0.9	2.1	N, K, S, Ca, Mg, and trace elements
Dairy manure	0.6	1.4	
Poultry manure	1.8	4.1	
Swine manure	1.5	3.5	
Aerobically digested sludge	3.3	7.6	
Anaerobically digested sludge	3.6	8.3	
Composted sludge	1.3	3.0	

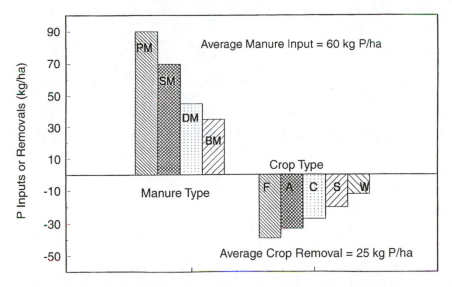

Figure 5.8 Comparison of typical inputs of P from animal manures and crop removal in harvested grain or forage (PM = poultry manure, SM = swine manure, DM = dairy manure, BM = beef manure; F = fescue, A = alfalfa, C = corn, S = soybeans, and W = wheat).

Management programs that use these materials as soil amendments must also consider the effects on soil pH, soluble salts, the presence of potentially phytotoxic elements, e.g., boron (B), sodium (Na), and the long-term fate of nonessential heavy metals found in these by-products.

5.3 Phosphorus transformations in soils

The ultimate goal of soil P management is to maintain the concentration of soil solution P at a value adequate for plant growth, while minimizing the movement of soluble P, organic P, and sediment-bound P to sensitive water bodies. Soil, plant, and water management practices, such as fertilization techniques and timing, crop rotation, tillage, irrigation, drainage, terracing, grassed waterways, and buffer strips, all act to maximize the efficiency of P recovery from fertilizers and wastes. These best management practices (BMPs) represent the practical applications of the basic knowledge gained from our understanding of the soil P cycle. For crop production purposes these practices focus on ensuring a rapid replenishment of the soil solution to optimize P nutrition and on minimizing soil loss to prevent long-term degradation of the soil resource and thus soil productivity. Although similar in many respects to agricultural practices, environmental management of soil P must be more intensive and has less room for error due to the low levels of P required to induce eutrophication. In either case, basic knowledge of the chemical, microbiological, and physical transformations of soil P is necessary to develop innovative approaches that accomplish both goals.

The major soil inorganic P transformations are the fixation of P in sparingly soluble forms by *adsorption* and *precipitation* reactions and the solubilization of P by *desorption* reactions and mineral *dissolution*. Soil organic P transformations are primarily *mineralization-immobilization* reactions mediated by soil microorganisms and P uptake by plant roots alone or in association with mycorrhizal fungi.

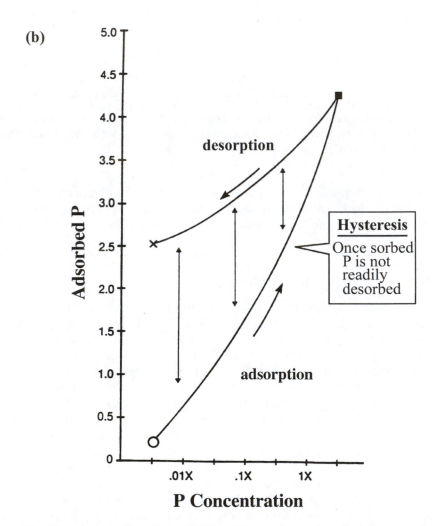

Figure 5.9 Phosphorus adsorption and desorption processes in soils (a) proposed mechanisms of adsorption of P by soil constituents, M represents a metal (Fe, Al) ion in a clay or oxide. (b) Idealized adsorption–desorption curve for P, illustrating hysteresis. When P is added to the soil, as in fertilization, and the concentration of P in the soil solution (x) increases (e.g., from 0.01 to 1 times), adsorbed P increases as well. However, there is a the lack of complete reversibility in desorption of P from the solid phase, referred to as hysteresis, when the concentration of P in the soil solution decreases from 1 to 0.01 times by plant uptake or leaching. (From Fixen, P. B. and Grove, J. H., in *Soil Testing and Plant Analysis*, R. L. Westerman, Ed., American Society of Agronomy, Madison, WI, 1990. With permission.)

5.3.1 Adsorption and desorption of soil phosphorus

Adsorption refers to the removal of ionic P ($H_2PO_4^-$, HPO_4^{2-}) from solution by a chemical reaction with the solid phase of the soil. It is a general term that refers to the formation of a chemical bond between the phosphate anion and a soil colloid, but does not specify the mechanism of P retention, although several bonding mechanisms have been proposed (Figure 5.9a). The primary reactive phases in soils and sediments responsible for adsorption reactions are clays, oxides or hydroxides of Fe and Al, Ca carbonates, and organic matter. Adsorbed P is considered to be slowly available and capable of gradually replenishing the soil solution in response to plant uptake, or losses of soluble P in runoff and leaching waters. The term *labile* soil P refers to that portion of adsorbed P that will be readily available for plant uptake and can be easily extracted, along with soluble P, by a chemical soil test.

Desorption refers to the release of P from the solid phase into the soil solution. Desorption occurs when plant uptake, runoff, or leaching deplete soluble P concentrations to very low levels, or in aquatic systems when sediment-bound P interacts with natural, low P waters. Only a small fraction of adsorbed P in most soils is readily desorbable. Much of the P added in fertilizers or manures is rapidly "fixed," and not easily desorbed, a phenomenon referred to as *adsorption hysteresis* (Figure 5.9b). Hysteresis reflects the transformation of soluble P into very unavailable forms by processes such as precipitation of P as insoluble compounds, occlusion of P by other precipitates (e.g., Al/Fe oxides), and slow diffusion of P into soil solids. Studies have shown that while the rate of P desorption is initially quite rapid it often decreases markedly within a short time. From an environmental perspective this suggests that short, intense rainfall events may quickly deplete the soluble and readily desorbable fractions of soil P and redistribute this P into runoff.

Soils and sediments differ widely in their capacity to adsorb and desorb P. In general, highly weathered soils will adsorb more P due to their greater clay and Al/Fe oxide contents. As shown in Figure 5.10, the amount of P adsorption that must occur to attain a P concentration in solution adequate for plant growth (0.2 mg/L) would be ~2500 mg P/kg for a Hawaiian Inceptisol (70% clay) compared with ~50 mg P/kg for a Peruvian Ultisol (6% clay). Adsorption also varies with soil depth and is affected by cultural operations that alter soil P levels, soil pH, and organic matter content, such as fertilization, liming, manuring, or reduced tillage operations. The importance of subsoils in P adsorption and leaching is shown in Figure 5.11 for two Atlantic Coastal Plain soils. The Evesboro soil, with sandy surface and subsoil horizons, has much less adsorption capacity in the soil profile than the Matawan sandy loam, which has a subsoil horizon with 60% clay. The higher soil test P (STP) values in the Evesboro subsoil reflect the greater mobility of P in profiles with little accumulation of clay. Tillage practices that affect the vertical distribution of P in the soil profile can also alter the adsorption and desorption of P. No-tillage agriculture, where P is not incorporated by plowing, often results in stratification of P in the soil, with extremely high P values in the top few centimeters. This can result not only in a lower P adsorption capacity in this depth (0 to 2 cm), but also in the presence of very high levels of easily extractable P relative to the remainder of the topsoil horizon (6 to 18 cm), a situation that could contribute to high levels of soluble P in runoff (Figure 5.12).

In general, soils that are low in P, acidic, and high in clay or Fe/Al oxides, particularly amorphous oxides, have the greatest P adsorption capacities. In calcareous soils, $CaCO_3$ and Fe oxides are the main factors controlling P adsorption. Sandy soils have low P adsorption capacities and are susceptible to leaching of P, as are soils with very high

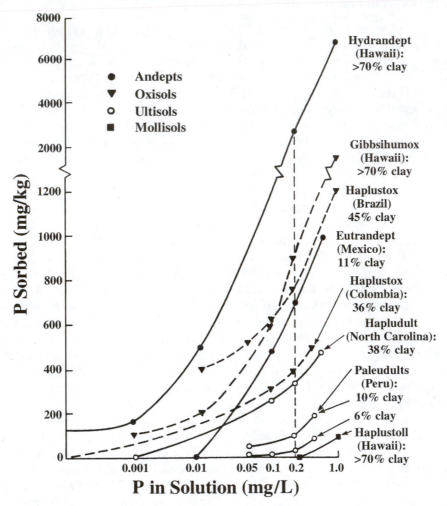

Figure 5.10 Differences in P adsorption capacity of soils from four soil groups. Vertical dashed line represents the amount of P that must be adsorbed to attain a level in solution adequate for the growth of most crops (0.2 mg/L) (From Sanchez, P. A. and Uehara, G., in *The Role of Phosphorus in Agriculture*, F. E. Khasawneh, Ed., American Society of Agronomy, Madison, WI, 1980. With permission.)

organic matter contents (peats, heavily manured soils) where soluble organic matter may enhance P mobility by coating colloidal surfaces responsible for P adsorption. Organic P complexes also leach more rapidly and to greater depths than inorganic soluble P, raising questions about the subsurface movement of P into surface waters by lateral groundwater flow, particularly in artificially drained soils.

    Adsorption of P by soils can be quantified by the use of mathematical equations that relate the amount of P retained by a solid phase (P_{ads}) to the concentration of P in solution at equilibrium (P_{eq}). Adsorption data are usually obtained by reacting soils or sediments with solutions ranging in initial P concentration from 0 to 20 mg P/L. Plots of the quantity of P adsorbed vs. the equilibrium P concentration are referred to as "adsorption isotherms" (Figures 5.11 and 5.12). The most common approaches to quantify P adsorption in soils are the Langmuir and Freundlich equations:

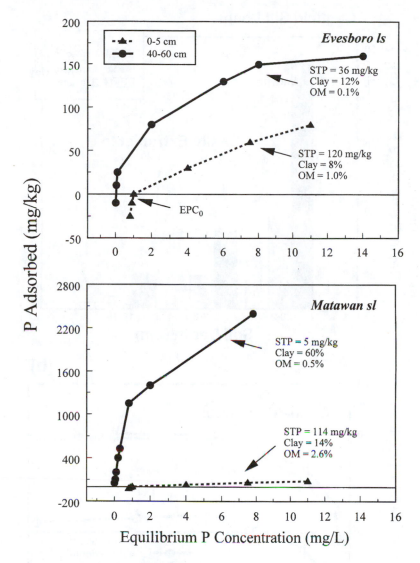

Figure 5.11 Phosphorus adsorption isotherms for two depths of two Atlantic Coastal Plain soils. The 0 to 5 cm depth represents the zone of maximum interaction between runoff and soil P, and the 40 to 60 cm depth represents the potential for subsoils to retain P against leaching. STP refers to soil test P value prior to adsorption (soil test = Mehlich 1, 0.05 M HCl + 0.0125 M H$_2$SO$_4$). EPC$_0$ is equilibrium concentration of P in solution when the rate of P adsorption and desorption are the same. (Adapted from Mozaffari and Sims, 1994.)

Langmuir equation:

$$P_{ads} = \frac{kP_{eq}b}{[1 + kP_{eq}]} \qquad (5.1)$$

The linearized version of the Langmuir equation (Equation 5.2) is often used to determine k and b mathematically, parameters that are presumed to characterize the energy of

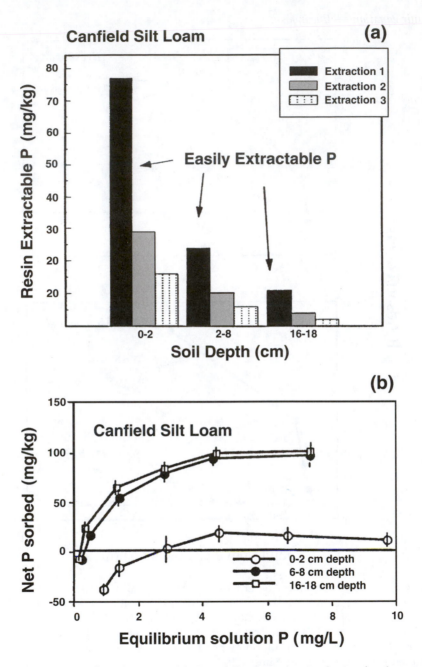

Figure 5.12 Comparison of easily extractable P and P adsorption isotherms for the upper (a) and lower (b) portions of the surface horizon of a Canfield silt loam used for the production of no-till corn. (From Guertal, E. A., *Soil Sci. Soc. Am. J.*, 55, 410, 1991. With permission.)

bonding of P with the solid phase (k) and to estimate the maximum amount of P that can be adsorbed (b). Units for P_{ads} and b are mg P/kg soil, for P_{eq} are mg P/L, and for k are L/kg.

In this approach, a graph of $[P_{eq}/P_{ads}]$ against P_{eq} is prepared and, if the graph is linear, used to determine b from the slope (slope = $1/b$) and k from the intercept of the line (intercept = $1/kb$).

Langmuir equation — linear form:

$$\frac{P_{eq}}{P_{ads}} = \frac{1}{kb} + \frac{P_{eq}}{b}$$ (5.2)

Freundlich equation:

$$P_{ads} = kP_{eq}^{1/n}$$ (5.3)

For the Freundlich equation, k and n are empirical constants that vary according to soil properties. Although these two equations do not provide mechanistic information about the processes of P adsorption, they are useful in comparing adsorption capacities between soils or changes in P adsorption due to varying soil properties conditions (e.g., pH, organic matter, flooding, and draining) or variations in long-term management practices (manuring, tillage). They are also used to approximate the quantity of P that must be adsorbed by various soils to raise the P concentration in the soil solution at equilibrium to a desired, or maximum, value.

Another useful parameter that can be obtained from P adsorption isotherms is the EPC_0 or equilibrium P concentration at zero sorption. The EPC_0 value represents the P concentration maintained in solution by a solid phase (soil, sediment) when the rates of P adsorption and desorption are the same. EPC_0 values are used to characterize the potential of surface soils, stream sediments, and streambank materials in contact with runoff or stream flow to remove P from or release P into flowing waters. Soil particles or sediments with high EPC_0 values that contact low P waters (rainfall, subsurface discharge) will desorb P into the waters; conversely, if the solid phases have low EPC_0 values, they can reduce the P concentration of stream flow or runoff and thus decrease the potential for downstream eutrophication. The high EPC_0 values for top few centimeters of the Evesboro soil shown in Figure 5.11 (~1 mg/L) and the Canfield soil shown in Figure 5.12 (~3 mg/L) indicate that both of these high P soils should readily desorb soluble P into runoff waters.

5.3.2 Precipitation of soil phosphorus

Precipitation can be defined as the formation of discrete, insoluble compounds in soils, and can be viewed as the reverse of mineral dissolution. The most common forms of precipitated P include the products of reactions between soluble, ionic P and Ca, Al, and Fe. In neutral or calcareous soils, where the pH is higher and soluble Ca is the dominant cation, the addition of soluble P initially results in the formation of dicalcium phosphate dihydrate ($CaHPO_4 \cdot 2H_2O$), which, with time, slowly reverts to other more stable Ca phosphates such as octacalcium phosphate ($Ca_4H(PO_4)_3 \cdot 2.5H_2O$), and in the long term to apatite ($Ca_{10}(PO_4)_6F_2$). The chemistry of Ca phosphates is reasonably well understood and has been studied by the use of mineral stability diagrams that, based on chemical equilibria, can predict the mineral that controls soil P solubility. Examining the effect of cultural practices or soil properties on changes in these stability diagrams can help to understand the long-term cycling of P in soils. In acidic soils, where soluble Al and Fe are the major soluble cations, Fe and Al phosphates are the dominant precipitates. The presence of amorphous or partially crystalline Fe and Al oxides in these soils that can "occlude" P as they crystallize, makes identification of discrete solid phases of Fe-P and Al-P difficult. Dissolution of precipitates in calcareous and acidic soils is highly pH dependent. The solubility and thus availability of P to plants (and runoff or leaching waters) is generally greatest under slightly acidic conditions (pH 6.0 to 6.5). Application of the principle of chemical precipitation to remediation of eutrophic lakes has been successful and commonly involves addition of alum (Al sulfate; $Al_2(SO_4)_3$) to the epilimnion (upper portion of the water column) to precipitate P as Al phosphate. In addition to direct precipitation of soluble P, the alum forms a layer of Al hydroxide on the lake bottom that is highly reactive for P subsequently desorbed from lake sediments.

Another example of the successful use of precipitation reactions to mitigate the environmental impact of P has been the treatment of poultry litters with alum. In much of the southeastern U.S., poultry litters are applied to pastureland as a fertilizer. In this situation the application of an animal waste not only contributes to the long-term buildup of soil P to excessive levels, but also represents a direct source of soluble P loss to runoff, because the litters are not incorporated with the soil, but remain on the surface of the pasture. About 20% of the total P in most poultry litters is water soluble, hence rainfall events can release soluble P directly from the litters and transport this P to surface waters in runoff, where it is immediately available to algae. Research has shown that alum, which is also used to remove soluble P from municipal wastewaters, could be applied to the litters in the poultry houses after the poultry were removed for processing. The alum precipitated the P in the litters in an insoluble form (as Al-P), thus decreasing the risk of P loss in runoff (Figure 5.13). Other benefits resulted from this BMP. The acidifying nature of alum decreased ammonia levels in the poultry houses, improving poultry performance and profitability and decreasing ammonia emissions (an air quality benefit), and also decreased runoff losses of soluble organic carbon and some trace elements.

5.3.3 Mineralization and immobilization of soil phosphorus

Organic P in soils, or P added to soils in organic wastes or crop residues, represents an important source of P for plant growth, as well as a potential source of soluble P that may be lost in runoff or leaching. Mineralization refers to the microbially mediated decomposition of organic compounds, resulting in the release of inorganic forms of nutrients into the soil solution. Immobilization, conversely, is defined as the conversion of mineral

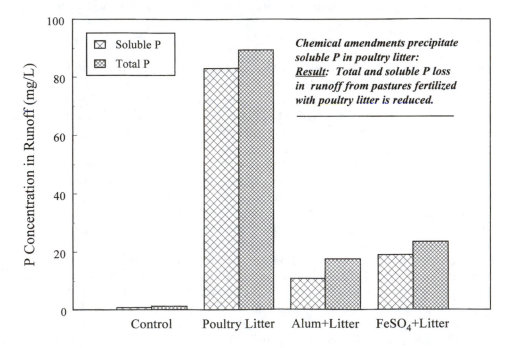

Figure 5.13 Effect of aluminum sulfate [alum: $Al_2(SO_4)_3$] treatment of poultry litters on total and soluble P concentrations in runoff from fescue pastures. (Adapted from Shreve et al., 1995.)

elements, such as P, by soil microorganisms into biochemical compounds essential for microbial metabolism. Mineralization of organic P is largely controlled by the relative amount of C in the organic substrate, which acts as the energy source for the decomposing microorganisms. High C:P ratios provide substantial energy and stimulate microbial growth that consumes all available P; low C:P ratios can result in excess soluble P beyond microbial needs that is then available for plant uptake.

Environmental conditions, as would be expected, play a key role in the rate and extent of decomposition of organic P compounds. Optimum conditions for mineralization of organic P are similar to those for soil organic N (see Chapter 4). Once mineralized from organic matter, P reverts rather quickly to inorganic forms such as adsorbed or precipitated P. In many instances frequent, long-term additions of animal manures and other organic wastes to soils have been shown to primarily increase total inorganic P in soils, not total organic P (Table 5.3).

5.4   Phosphorus transport in the environment

All forms of P, soluble, adsorbed, precipitated, and organic, have been shown to be susceptible to transport from soils to water bodies. The fact that the greatest reservoir of P in the earth's environment is ocean sediments is proof that "leakage" of P from geologic sources is a continuous process. Reducing P transport helps to avoid the undesirable effects of eutrophication and the waste of finite natural resources that occurs when fertilizers produced from geologic deposits of P, or reclaimed waste organic P (e.g., animal manures, municipal biosolids), are used as soil amendments to improve plant growth and yield.

Transport of soil P occurs primarily via surface flow (erosion and runoff), although the background levels of P entering streams and lakes via groundwater base flow certainly reflects the impacts of land use (see Table 5.1). Leaching and subsurface lateral flow of P can also be a concern in soils with high degrees of P saturation and artificial drainage systems (tiles, ditches). These artificial drainage systems "short-circuit" the movement of water and dissolved solutes that would normally percolate slowly through the soil to groundwaters, and eventually discharge, through base flow, to surface waters. Considerable adsorption or precipitation of P in soil profiles and by geologic materials in aquifers could occur during the slow, long-term process of deep leaching and groundwater discharge. Tiles and drainage ditches, by design, result in a much more rapid discharge of percolating waters and shallow groundwaters to surface waters. Natural short-circuiting of water flow can also occur in soils, through a process referred to as macropore, preferential, or bypass flow. *Macropore flow* is defined as the rapid leaching of water and solutes through natural channels, fissures, and structural voids in the soil in a manner that bypasses the soil matrix. Flow of this sort can rapidly transport P from surface soils to subsurface horizons, tile lines, and shallow water tables. Leaching of P through macropores is greater in soils with higher "cracking potential" (soils that shrink or swell due to changes in soil moisture because of a high concentration of 2:1 expanding clay minerals) or soils with well-established root channels, such as pastures and forages. Any form of water flow that bypasses the soil matrix (e.g., tiles, macropores) is of environmental concern because it decreases the likelihood that P will be retained by soils and thus increases the potential for soluble P to enter groundwaters and surface waters.

5.4.1   Transport of phosphorus by surface flow

Water flowing across the soil surface can dissolve and transport soluble P or erode and transport particulate P. Soluble P can be either inorganic or organic, while particulate P generally consists of finer-sized soils particles (e.g., clays) and lighter organic matter. The small quantity of soluble or readily desorbable P in most soil environments is due to the low solubility of P, the considerable adsorption capacities of clays for P, and the high concentration of P in soil organic matter. This results in the majority of total P transport occurring as particulate P. This is particularly true where runoff contains high quantities of suspended solids, as often occurs during storm flow or snowmelt events. However, if natural or artificial filtration processes are operational (e.g., forests, wetlands, grassed waterways, crop residues in reduced tillage systems) to remove sediment-bound P, then the transport of soluble P becomes more important. Similarly, if soils are excessive in P, such as often occurs in areas where organic wastes are frequently applied to soils, the amount of soluble P increases. In both situations practices that not only control soil erosion but reduce the transport of soluble P must then be considered.

Virtually all soluble P transported via runoff is biologically available, but particulate P that enters streams and other surface waters must first undergo some type of solubilization reaction (e.g., desorption, change in oxidation–reduction conditions) before becoming available for aquatic biota. Complicating the matter further is the fact that during the transport process itself, soluble and particulate P can undergo reactions with soils bordering fields and lakes, flowing waters, and streambed and stream bank sediments. These reactions can dramatically alter the potential for the P originally lost in runoff to induce eutrophication downstream. Soil management practices seek to reduce P losses in runoff by decreasing excessive soluble soil P and reducing erosive loss of fine-sized particles and organic matter. Nutrient management programs that focus on improved management of fertilizers and organic wastes and the use of conservation tillage practices and erosion control mechanisms (e.g., terraces, buffer strips, grassed waterways) are examples of strategies used to minimize P loss in runoff (see Section 5.5).

The mechanisms involved in soluble P transport are straightforward, and include an initial desorption or dissolution of P bound by soil particles, followed by water movement from the source soil to a stream or river that later intercepts a sensitive water body. Soluble organic P that is not adsorbed by soil particles may also be carried by surface or subsurface runoff. Of importance from a soil management point of view is the fact that most of the soluble P entering runoff originates from the interaction of flowing water with the uppermost surface (0 to 2 cm) of the soil. Any management practice that minimizes soluble P concentrations at this depth (e.g., fertilizer or manure incorporation), or enhances infiltration of water into the soil, where soluble P can be adsorbed by soil constituents, will decrease soluble P losses to the environment. In a similar manner, the natural filtration capacity of nonagricultural areas bordering fields, or the high sorption capacities of stream and stream bank sediments, can reduce P transport by adsorbing P from runoff waters, assuming adequate time of contact between P enriched waters and sediments.

Some examples of methods used to reduce P losses in runoff are provided in Table 5.5, which shows the value of reduced tillage practices (chisel plow or no-till vs. conventional), crop residue strips, grassed waterways, and soil amendments in trapping sediment-bound P and reducing soluble P transport. The value of altering the properties of a P source to reduce P solubility and thus direct loss in runoff from an applied organic waste was discussed earlier and illustrated in Figure 5.13. The examples in Table 5.5 also illustrate the importance of considering all aspects of runoff control practices to maximize the overall efficiency of a management system. Chisel plowing, for example, may be just as effective in controlling erosion and runoff as complete no-tillage in certain soils. Chisel plowing also improves crop yield and nutrient uptake by breaking apart compaction zones and improving root growth; it also provides a method to at least partially incorporate organic wastes. Narrower crop residue strips may be just as effective as wider strips in reducing P loss, if adequate soil cover is present. Multiple cutting practices for grassed

Table 5.5   Examples of the Methods Used to Control Phosphorus Loss in Runoff

Tillage practices for corn (Andraski et al., 1985)
Total P loss in runoff (mg/m²)

Tillage method	1980 September	1981 June	1981 July	1982 October	1983 June	1983 July
Conventional	133	8	230	220	175	22
Chisel plow	21	<1	20	37	67	11
No-till	39	<1	20	24	21	5

Cornstalk residue management (Alberts et al., 1981)
Available P (g/h/m of width)

Strip width (m)	% Cover	Entering	Leaving	Reduction in P loss (%)
1.8	27	0.98	0.77	21
1.8	50	1.39	0.70	50
2.7	50	2.63	0.48	82
4.6	50	1.80	0.51	72

Wastewater renovation by a reed canarygrass filter strip (Payer and Weil, 1987):
P Removal efficiency from wastewater (%)

Cover crop management	Dissolved P	Total P
Multiple cuttings, harvest crop residue	32	70
Multiple cuttings, leave crop residue	20	62
Cut once, leave crop residue	4	50

Use of by-products to decrease soluble and soil test (Mehlich 3) P (Peters and Basta, 1996):

	Dickson silt loam		Keokuk very fine sandy loam	
	Soluble P (mg/kg)	STP (mg/kg)	Soluble P (mg/kg)	STP (mg/kg)
Control (no by-product)	14	550	4	300
Water treatment residual #1	2	425	1	225
Water treatment residual #2	1	375	1	220
Cement kiln dust	4	375	2	180
Bauxite red mud	4	480	3	230

waterways increased the efficiency of P removal in an overland flow wastewater renovation system. However, this study also found that equipment damage to the grassed waterway caused by the extra cuttings produced ruts and channels that could contribute to increased runoff. Drinking water treatment residuals (WTR: an Al/Fe-based by-product of the treatment of drinking waters to remove suspended inorganic solids, organics, and color) can precipitate or adsorb soluble P and increase soil P sorption capacity when added to soils in buffer strips or grassed waterways. While this practice may decrease soluble P and increase the capacity of the amended soil to remove P from runoff waters, it must be balanced against the need to maintain an adequate P supply for vegetation in these areas. In each of the examples in Table 5.5, the need to reduce P loss must be balanced against other aspects of the system (e.g., crop growth in severely compacted soils, loss of cropland by conversion to filter strips, sustainability of grassed waterway) to optimize the efficiency of the management practice.

The transport and subsequent reactions of particulate P are considerably more complex. Particulate P originates not only from soil erosion, but from the beds and banks of streams or drainage areas that carry water from a field to a surface water body. As with all erosion, the loss of particulate P represents the detachment of soil particles as a result of the energy contained in falling or flowing water. The nature of the rainfall or runoff event will determine the energy level of the water, while the texture and structure of the soil will affect the amount of energy required to detach and transport soil particles. As smaller, lighter particles require less energy to dislodge and carry, it is not surprising that clays and organic matter, soil constituents with relatively low bulk densities, are preferentially transported in runoff. Once soil solids enter a watercourse, they eventually settle from the water as sediment. Resuspension of these sediments during intense rainfall events or rapid stream flow, along with detachment of soil particles from stream banks, can increase particulate P transport at a later date. In general, more intensive land use practices such as urban construction activities, surface mining, and conventionally tilled agricultural soils will result in the highest soil erosion and thus the greatest transport of particulate P. In natural ecosystems and areas where soil conservation practices decrease erosion of soil solids, the role of stream bank erosion and stream sediments in the transport of particulate P is of more importance.

The particles transported in runoff are normally higher in P, organic matter, and other nutrients, relative to the soil from which they originated. This enrichment occurs due to the greater sorption capacity of the finer-sized clay particles and the relatively high concentrations of P in organic matter, relative to silt and sand particles (Table 5.6).

Table 5.6 Average Enrichment Ratios for Six Western Soils

Soil property	Enrichment ratio[a]
Organic C	2.00
Clay (%)	1.56
Surface area	1.32
Bioavailable P	1.49
Total P	1.46
EPC_0 (mg/L)	−1.80[b]

[a] Enrichment ratio = $\dfrac{[\text{Value of soil property in runoff sediment}]}{[\text{Value for soil property in source soil before runoff}]}$

[b] EPC_0 = equilibrium P concentration at zero sorption. Negative enrichment ratio indicates that source soil maintained a higher P value in solution at equilibrium than runoff sediment.

Source: Adapted from Sharpley (1985).

The "enrichment ratio" concept was developed to identify watersheds where more intensive management practices are required to control nutrient enrichment of surface waters. The enrichment ratio is calculated as the ratio of the concentration of P in the particulate matter found in runoff to that in the source soil. Research has shown that enrichment ratios for total and available P are closely related to the amount of soil lost from a watershed. This provides a quantitative basis for estimating the impact of land use on the loss of soluble, bioavailable, and particulate P and thus productivity of soils and surface waters in a watershed. Recent advances in research have resulted in models that have been remarkably successful in predicting P loss from readily obtainable parameters such as soil loss from erosion, soil test P, and soil properties associated with the range of values for soil series found in the landscape (e.g., clay content, organic matter, soil total P) (Figure 5.14).

Once particulate P has entered a lake, many of the processes operating in the soil P cycle continue, albeit at different rates. The size of the particles along with the relative temperatures of the inflowing water and the lake influence the initial distribution of

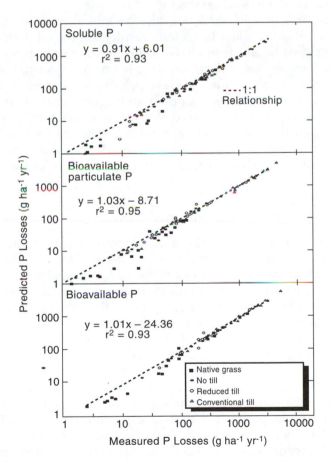

Figure 5.14 Comparison of predicted and measured losses of P. Soluble P was predicted from soil loss, soil test P, and variables related to runoff. Bioavailable P was predicted from soil total or bioavailable P, sediment concentration in runoff, and enrichment ratios estimated from soil loss. (Adapted from Sharpley and Halvorson, 1994.)

the sediments within the lake. One of the major causes of eutrophication is the high degree of turbidity associated with fine-sized particles that settle very slowly to the lake bottom. The turbid nature of the upper water column reduces light penetration, decreasing the photic zone. As the finer-sized particles are usually higher in P than the quickly settling coarser particles, the overall initial effect is to stimulate biological productivity in the *epilimnion* (upper layer of the water column). Of long-term importance, however, is the later desorption of P from sediments that, with time, have settled to the bottom of the lake. The P adsorbed by clays, Fe and Al oxides, and $CaCO_3$ are the main sources of potentially desorbable P.

The physical and chemical processes operational in surface waters greatly influence the desorption process. Changes in temperature, or turbulence induced by storms, can resuspend bottom sediments in the water column and enhance the desorption of P in photic zones where algal growth occurs. The O_2 status of the water body can also affect desorption or dissolution of P from sediments. Studies have shown, for instance, that, under anaerobic conditions, P desorption is enhanced and that when lake O_2 levels increased P sorption increased as well. Similar results were obtained with sediments from agricultural ditch drainage systems where Fe-P was shown to dissolve under reducing conditions and release P into solution (Figure 5.15).

The residence time of water in a lake also greatly affects the importance of P in sediments. When inflowing water has a long residence time (at least several months), the P inputs from streams and rivers have sufficient time to accumulate as sediments on lake bottoms. If water flow through the lake is rapid, however, P retained by finer-sized particles may be removed from the lake prior to settling from the water column. Indeed, one common strategy used in urban and agricultural situations to minimize nutrient enrichment of lakes and reservoirs is to construct settling ponds, impoundments, or constructed wetlands with long residence times that act as filtration systems for particulate P in runoff. Subsequent removal or chemical "sealing" of the enriched sediments may still be needed eventually, but can be anticipated and planned for maximum efficiency.

5.4.2   *Phosphorus leaching and subsurface flow*

The natural process of groundwater discharge into surface waters contributes P that can play a role in eutrophication. In most instances the input of P by subsurface flow represents the effect of decades of land use and must be viewed as a baseline contribution that can only be altered by long-term improvements in management of all urban and agricultural operations in a watershed. One possible exception is the subsurface discharge from artificially drained fields, where corrective measures can be taken at either the source or at the point of discharge. Improved nutrient management programs and artificial wetlands are examples of strategies that can be used to reduce nutrient loads in drainage waters from urban and agricultural areas.

In most cases the concentrations of P in subsurface flow have been found to be quite low and well below the eutrophication threshold. This reflects the considerable sorption capacity of soils for P, and particularly of P-deficient subsoil horizons. Because of this, P leaching, unlike NO_3^- leaching, is rarely viewed as an important environmental issue. Exceptions include organic soils with fluctuating water tables, soils where macropore flow is significant, and heavily manured, sandy soils with shallow water tables. The role of organic matter in P leaching, mentioned above, is not well understood, but is believed to be key. Artificial drainage, commonly used in organic and poorly drained soils, normally increases infiltration and percolation of water, increasing the likelihood of P leaching, but decreasing P losses in runoff. If heavy applications of animal manure are combined with

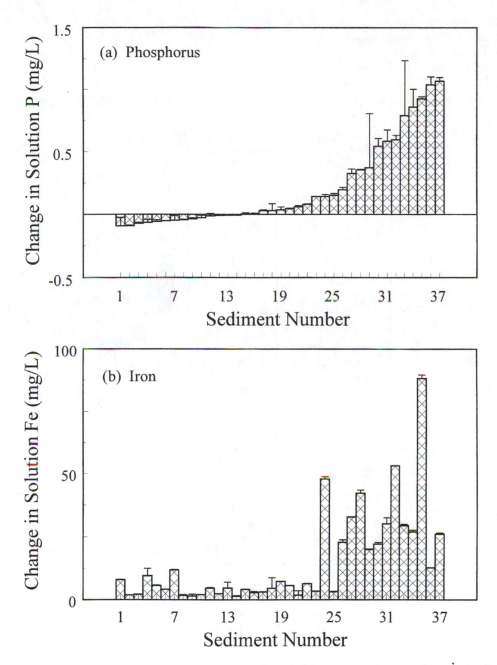

Figure 5.15 Effect of reducing conditions on Fe and P solubility in 37 sediments from agricultural drainage ways. Vertical bars are the change in Fe or P concentration following incubation of sediments under anoxic conditions. The concurrent release of P and Fe from sediments under reducing conditions indicates that Fe-P compounds in sediments may be a source of P to overlying waters under anoxic conditions (From Sallade, Y. E. and Sims, J. T., *J. Environ. Qual.*, 26, 1579, 1997. With permission.)

artificial drainage of soils, P leaching and loss via subsurface pathways can be significant, as shown in Figure 5.16 where the application of 200 Mg/ha of dairy manure sharply increased dissolved P concentrations in tile drainage. Improved aeration in organic soils

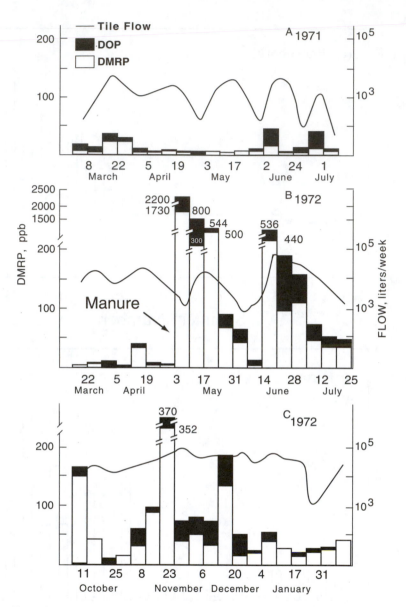

Figure 5.16 Effect of dairy manure application (200 Mg/ha) and tile flow on the loss of dissolved organic P (DOP) or dissolved soluble P (DMRP) in tile drainage during a 2-year period. (From Hergert et al., *J. Environ. Qual.*, 10, 338, 1981. With permission.)

due to artificial drainage has also been shown to stimulate P mineralization and loss in drainage waters via leaching and shallow groundwater discharge.

The most common situations where P leaching from mineral soils have been observed are in municipal and agricultural wastewater treatment systems and sandy soils where animal manure disposal operations have been necessitated by shortages of suitable land for proper manure management. Examples of this are shown in Figure 5.17 for sandy soils used for a wastewater irrigation system and for agricultural crops in an area dominated by a highly concentrated poultry industry. In both cases, P movement into subsoils has clearly occurred. In the case of the wastewater irrigation system, groundwater P values

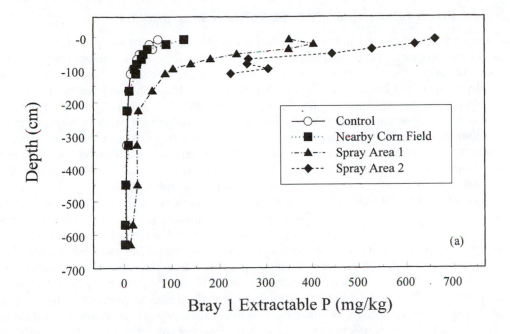

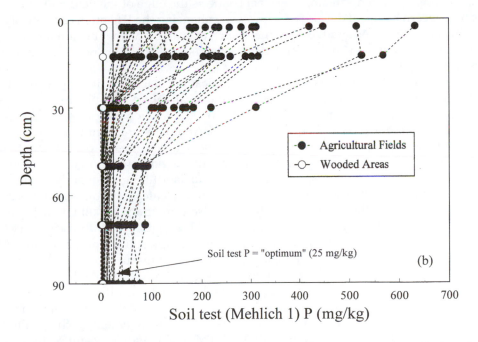

Figure 5.17 Phosphorus leaching studies illustrating distribution of soil test P with depth; soil tests: Bray P (0.03 M NH$_4$F + 0.025 M HCl; M1-P (Mehlich 1, 0.05 M HCl + 0.0125 M H$_2$SO$_4$). (a) A wastewater irrigation system compared with a commercially fertilized corn field. (Adapted from Adriano et al., 1975.) (b) 34 agricultural soils from Delaware where animal manures are regularly applied. (Adapted from Mozaffari and Sims, 1994.)

in the areas irrigated with wastewaters averaged 0.75 mg/L, relative to a local environmental standard of 0.05 mg/L.

5.5   *Environmental management of soil phosphorus*

Management programs to minimize the environmental impacts of soil P, while sustaining agricultural productivity, must be multidisciplinary in scope. As eutrophication is the primary negative environmental effect of P, the strategies to control P pollution are conceptually simple and begin with the development of a comprehensive nutrient management plan (CNMP; see Chapter 9 for details on CNMPs). The general goal of a CNMP is to assess thoroughly and quantitatively the nutrient management practices for a farm (or other operation), including nutrient budgets (inputs vs. outputs and nutrient balance), soil nutrient status, sources and properties of all nutrients used, methods and timing of nutrient applications, soil conservation practices, and the availability and use of monitoring techniques that can assess and track the efficiency of nutrient use.

Once a CNMP is in place, specific management practices for the efficient use of P can be implemented. In general, soil P levels should be maintained at levels that are adequate, but not excessive for plant growth, and runoff of soluble and particulate P should be minimized by proper soil and water management. In water bodies where eutrophication has already occurred, remediation involves practices that decrease the solubility and transport of soil P and the desorption/dissolution of P already present in the water body. Most processes designed to reduce P transport are designed to prevent "edge of field losses" (e.g., reduced tillage, buffer strips, grassed waterways). To date, no systematic approach that reduce P impacts on surface waters by somehow "trapping" P in the channel (e.g ., the drainage ditch, stream, river) that connects the edge of the field with the surface water of concern has been developed. However, enhancing the P fixation capacity of buffer strips by amending them with by-products, such as water treatment residuals (Table 5.5) and the construction of artificial wetlands to act as "sinks" for P, are new ideas now being investigated in some states and countries.

Unfortunately, in many cases there are significant obstacles to the implementation of soil P management practices. Foremost are the availability of human and economic resources needed to design and implement improved management programs that reduce soil loss (erosion) in urban and agricultural areas. Another significant limitation, particularly in urban areas, or areas where animal-based agriculture is predominant, is the concentration of P from other geographic areas into organic (or inorganic) wastes. The increasing costs of waste transport, treatment, and landfilling frequently result in the application of these wastes to soils in a manner that creates environmentally excessive levels of soil P. Given the formidable nature of these limitations to soil P management, it is apparent that a systematic, long-term approach will be required. Multidisciplinary teams with expertise in aquatic ecology and chemistry, hydrology, soil science, crop production, urban and industrial waste management, engineering, and resource economics must identify the areas requiring immediate attention, as well as broadly applied, long-term solutions. Strategies required for urban and agricultural situations although different, must be coordinated, given the interdependence of farm and municipal populations. Computer-based water quality models have been developed for P management, such as EPIC (Erosion-Productivity-Impact-Calculator) and FHANTM (Field Hydology and Nutrient Transport Model). These models require a large database with many parameters, but can accurately predict changes in soil P levels, crop yield responses, or P losses in erosion and runoff (Figure 5.18).

Simpler approaches that can be used by advisory agencies (Cooperative Extension, USDA Natural Resources Conservation Service) or crop consultants are also available. One example of a field-oriented approach to predicting the potential environmental impact of soil P is the P site index system that uses readily obtainable site characteristics to obtain an index of site vulnerability to P loss (Figure 5.19). The P index was developed by a

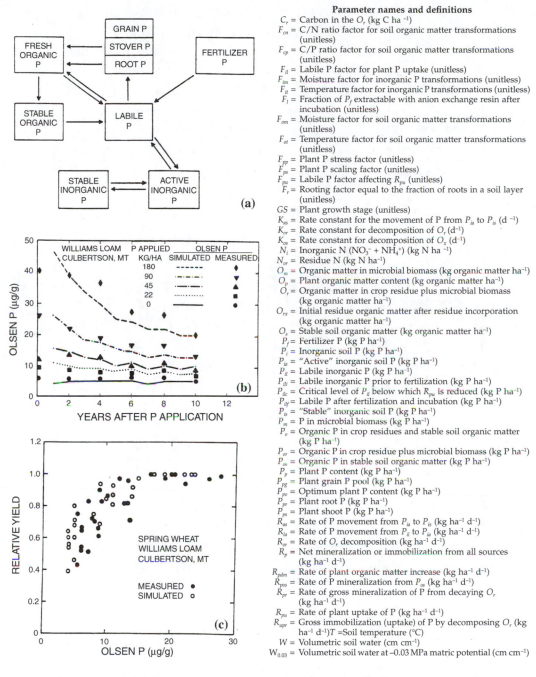

Parameter names and definitions

C_r = Carbon in the O_r (kg C ha^{-1})
F_{cn} = C/N ratio factor for soil organic matter transformations (unitless)
F_{cp} = C/P ratio factor for soil organic matter transformations (unitless)
F_{il} = Labile P factor for plant P uptake (unitless)
F_{im} = Moisture factor for inorganic P transformations (unitless)
F_{it} = Temperature factor for inorganic P transformations (unitless)
F_l = Fraction of P_f extractable with anion exchange resin after incubation (unitless)
F_{om} = Moisture factor for soil organic matter transformations (unitless)
F_{ot} = Temperature factor for soil organic matter transformations (unitless)
F_{pp} = Plant P stress factor (unitless)
F_{ps} = Plant P scaling factor (unitless)
F_{pu} = Labile P factor affecting R_{pu} (unitless)
F_r = Rooting factor equal to the fraction of roots in a soil layer (unitless)
GS = Plant growth stage (unitless)
K_{as} = Rate constant for the movement of P from P_{ia} to P_{is} (d^{-1})
K_{or} = Rate constant for decomposition of O_r (d^{-1})
K_{os} = Rate constant for decomposition of O_x (d^{-1})
N_i = Inorganic N (NO$_3^-$ + NH$_4^+$) (kg N ha^{-1})
N_{or} = Residue N (kg N ha^{-1})
O_m = Organic matter in microbial biomass (kg organic matter ha^{-1})
O_p = Plant organic matter content (kg organic matter ha^{-1})
O_r = Organic matter in crop residue plus microbial biomass (kg organic matter ha^{-1})
O_{rx} = Initial residue organic matter after residue incorporation (kg organic matter ha^{-1})
O_s = Stable soil organic matter (kg organic matter ha^{-1})
P_f = Fertilizer P (kg P ha^{-1})
P_i = Inorganic soil P (kg P ha^{-1})
P_{ia} = "Active" inorganic soil P (kg P ha^{-1})
P_{il} = Labile inorganic P (kg P ha^{-1})
P_{ilb} = Labile inorganic P prior to fertilization (kg P ha^{-1})
P_{ilc} = Critical level of P_{il} below which R_{pu} is reduced (kg P ha^{-1})
P_{ilf} = Labile P after fertilization and incubation (kg P ha^{-1})
P_{is} = "Stable" inorganic soil P (kg P ha^{-1})
P_m = P in microbial biomass (kg P ha^{-1})
P_o = Organic P in crop residues and stable soil organic matter (kg P ha^{-1})
P_{or} = Organic P in crop residue plus microbial biomass (kg P ha^{-1})
P_{os} = Organic P in stable soil organic matter (kg P ha^{-1})
P_p = Plant P content (kg P ha^{-1})
P_{pg} = Plant grain P pool (kg P ha^{-1})
P_{po} = Optimum plant P content (kg P ha^{-1})
P_{pr} = Plant root P (kg P ha^{-1})
P_{ps} = Plant shoot P (kg P ha^{-1})
R_{as} = Rate of P movement from P_{ia} to P_{is} (kg ha^{-1} d^{-1})
R_{la} = Rate of P movement from P_{il} to P_{ia} (kg ha^{-1} d^{-1})
R_{or} = Rate of O_r decomposition (kg ha^{-1} d^{-1})
R_p = Net mineralization or immobilization from all sources (kg ha^{-1} d^{-1})
R_{pdm} = Rate of plant organic matter increase (kg ha^{-1} d^{-1})
R_{pos} = Rate of P mineralization from P_{os} (kg ha^{-1} d^{-1})
R_{pr} = Rate of gross mineralization of P from decaying O_r (kg ha^{-1} d^{-1})
R_{pu} = Rate of plant uptake of P (kg ha^{-1} d^{-1})
R_{upr} = Gross immobilization (uptake) of P by decomposing O_r (kg ha^{-1} d^{-1}) T =Soil temperature (°C)
W = Volumetric soil water (cm cm^{-1})
$W_{0.03}$ = Volumetric soil water at –0.03 MPa matric potential (cm cm^{-1})

Figure 5.18 Illustration of the parameters required and the use of the EPIC model for P cycling in soils. Figures represent (a) the major components of the model and simulations conducted using the model to (b) predict changes in soil test P with time and (c) crop yield response to soil test P. (From Jones, C. A., *Soil Sci. Soc. Am. J.*, 48, 810, 1984. With permission.)

national committee of soil scientists from U.S. universities and government agencies in the early 1990s and was later verified using field runoff data. As originally proposed, the P index integrates agronomic soil test P with other criteria that quantify soil erosion and runoff, and P fertilizer and/or organic P source application rate, method, and timing in

SITE CHARACTERISTIC (*Weighting Factor*)	PHOSPHORUS LOSS RATING (*VALUE*)				
	None (0)	Low (1)	Medium (2)	High (4)	Very High (8)
Soil Erosion (*1.5*)	N/A	< 11 Mg/ha	11–22 Mg/ha	23–34 Mg/ha	> 34 Mg/ha
Irrigation Erosion (*1.5*)	N/A	Infrequent irrigation on well-drained soils	Moderate irrigation on soils with slope < 5%	Frequent irrigation on soils with slopes of 2-5%	Frequent irrigation on soils with slopes > 5%
Soil Runoff Class (*0.5*)	N/A	Very Low or Low	Medium	High	Very High
Soil Test P (*1.0*)	N/A	Low	Medium	High	Excessive
P Fertilizer Rate (kg P/ha) (*0.75*)	N/A	< 15	15–45	46–75	> 75
P Fertilizer Application Method (*0.5*)	N/A	Placed with planter deeper than 5 cm	Incorporate immediately before crop	Incorporate > 3 months before crop or surface applied > 3 months before crop	Surface applied > 3 months before crop
Organic P Source Application Rate (kg P/ha) (*1.0*)	N/A	< 15	15–45	46–75	> 75
Organic P Source Application Method (*1.0*)	N/A	Injected deeper than 5 cm	Incorporate immediately before crop	Incorporate > 3 months before crop or surface applied < 3 months before crop	Surface applied to pasture or > 3 months before crop

Phosphorus Index for Site	Generalized Interpretation of Phosphorus Index for Site
< 8	**LOW** potential for P movement from the site. If farming practices are maintained at the current level there is a low probability of an adverse impact to surface waters from P losses at this site.
8–14	**MEDIUM** potential for P movement from the site. The chance for an adverse impact to surface waters exists. Some remedial action should be taken to lessen the probability of P loss.
15–32	**HIGH** potential for P movement from the site and for an adverse impact on surface waters to occur unless remedial action is taken. Soil and water conservation as well as P management practices are necessary to reduce the risk of P movement and water quality degradation.
> 32	**VERY HIGH** potential for P movement from the site and for an adverse impact on surface waters. Remedial action is required to reduce the risk of P loss. All necessary soil and water conservation practices, plus a P management plan must be put in place to avoid the potential for water quality degradation.

Figure 5.19 A P site index system used to rate sites for their potential to deliver P to sensitive water bodies. (Adapted from Lemunyon and Gilbert, 1993; Sims, 1996.)

a simple, weighted matrix system to identify soils, landforms, and management practices with the potential for unfavorable impacts on water bodies because of P losses from agricultural soils. Specifically, the original version of the P index uses eight characteristics to obtain an overall rating for a site. Each characteristic is assigned an *interpretive rating* with a corresponding *numerical value* — Low (1), Medium (2), High (4), or Very High (8) — based on the relationship between the characteristic and the potential for P loss from a site. Suggested ranges appropriate to each rating for a site characteristic are then assigned. Each of the characteristics in the P index has also been given a *weighting factor* which reflects its relative importance to P loss. For instance, erosion (weighting factor = 1.5) is generally more important to P loss than P fertilizer application method (weighting factor = 0.5). At present, the weighting factors are based on the professional judgment of the scientists that developed the P index; they are not derived directly from field research with the P index. From the outset, the intent of the scientists developing the P index was that individual states or regions should modify the ratings, values, and weighting factors as appropriate to local conditions, based on soil properties, hydrology, climate, and agricultural management practices.

As others have worked with the P index concept, modifications have been considered, or adopted, that have improved its usefulness as a field-scale risk assessment tool; however, soil P testing has remained an integral part of all modified versions. Suggested modifications, for example, include adding *catchment factors* (condition of receiving water, and ratio of land:water) that reflect the sensitivity of the impacted water body to P inputs and *farm factors* such as proximity of the farm to receiving waters. It has also been suggested that the P index should not be additive, but should be derived by multiplication of site factors. This avoids the situation where a very high soil test P value in an area with little or no runoff or erosion results in a very high P index value; similarly, it allows for sites with low soil test P values and extremely high erosion/runoff potentials and/or very bad management practices to be identified as high-risk areas in need of more intensive conservation practices. Other factors such as soil drainage class, soil texture and cracking potential, and information on mean high water tables and/or depth of tile drain lines may be needed in regions where P leaching and subsurface lateral flow to streams and artificial drainage ditches are important pathways for P transport. Data from routine or special soil tests, such as pH, organic C, extractable Al, Fe, Ca, and magnesium (Mg), and tests for water soluble P or the degree of soil P saturation, may also be useful in the identification of soils with higher potentials for P loss.

5.5.1 *Management of soil phosphorus: soil testing and nutrient budgets*

Agricultural operations include agronomic crop production, animal production, vegetables and specialty crops, and commercial horticultural operations such as greenhouses, nurseries, and turf farms. Adequate P fertility is necessary for maximum economic plant production in all these operations, but, in many situations, poor nutrient management has resulted in excessive use of P fertilizers or manures, with resultant losses of soluble P or P-enriched sediments. Phosphorus management for agricultural purposes requires the development and implementation of a nutrient management program that focuses not only on profitability, but also on minimizing environmental impacts of P. Sustainable agricultural practices also recognize that P is a finite natural resource, one that requires considerable effort and expense to mine, process, transport, and apply. Hence, it should be used as efficiently as possible and not wasted.

In general, key aspects of an efficient P management program for agriculture include realistic assessments of plant P requirements, nutrient budgeting, use of soil testing and plant analysis to monitor soil P levels, timely and efficient P fertilization practices, cultural operations that reduce soil loss, and a sound approach to organic waste management.

Although P is an essential plant nutrient with numerous key biochemical functions, the quantity required by plants is much less than the other macronutrients (N, K). Typical concentrations of P in crops range from 0.1 to 0.5%, and P fertilizer recommendations usually range from 10 to 100 kg P/ha. Balancing fertilizer or manure inputs of P with crop removal is important to prevent the accumulation of P in soils to values in excess of those needed for optimum yields. This can be difficult when manures are used, as illustrated in Figure 5.8, because the amount of P removed in the harvested portion of most agronomic crops is rather small compared to manure P inputs.

The nature of the crop rotation and characteristics of the farming operation (e.g., number of animals per acre of cropland) will ultimately determine the long-term soil nutrient "budget." The two scenarios illustrated in Figure 5.20 show the differences in complexity of a P management for a cash grain farm as compared to a livestock farm. Clearly, it is much more difficult to avoid a P "surplus," and buildup of soil P, in a livestock operation than in one that only uses commercial fertilizer. This is because the inputs of P in animal feed commonly exceed the P outputs in animal products. A decision can be made not to purchase fertilizer P when soils are in the agronomically "optimum" range but a similar reduction cannot be made in feed inputs of P unless the number of animals on the farm are reduced proportionally.

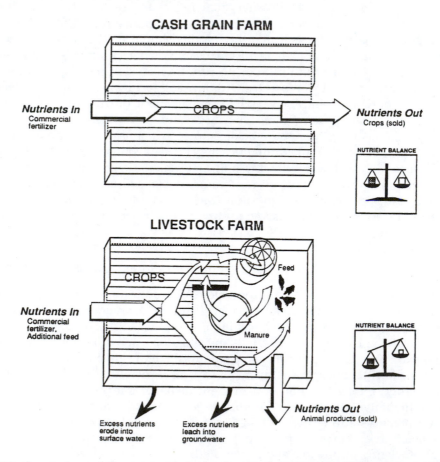

Figure 5.20 Comparison of the nutrient balances for a cash grain and a livestock farm, illustrating the more complex nature of the nutrient cycle for a livestock operation. (Adapted from Weidner, 1988.)

Table 5.7 Phosphorus Mass Balance Analyses for Poultry-Based Agriculture in Delaware

		Statewide P budget[a]		
Crop	Hectares	Annual P requirement[b] (kg/crop)	P Source	Annual amount of available P (kg/source)
Corn	61,000	530,000	Poultry manure	5,200,000
Soybeans	91,000	1,120,000	Fertilizer sales	3,500,000
Wheat	28,000	350,000	Other animal manures	210,000
Barley	12,000	150,000	Municipal biosolids	415,000
Other	34,000	425,000		
Total	226,000	2,575,000	Total	9,325,000

	Mass balance analyses for P at several scales in Delaware		
Nutrient	Poultry–grain farm[c] (kg/year)	Inland Bays watershed (kg/year)	Sussex County (kg/year)
Inputs			
Feed	12,500	2,175,000	8,300,000
Fertilizers	1,500	420,000	1,800,000
Animals	100	15,000	45,000
Total inputs	14,100	2,610,000	10,145,000
Outputs			
Harvested crops	1,800	550,000	1,700,000
Animal products	3,700	600,000	2,300,000
Total outputs	5,500	1,150,000	4,000,000
Nutrient balance			
kg/year	+ 8,600	+ 1,460,000	+ 6,145,000
kg/ha/year	+ 34	+ 52	+ 70

[a] Statewide P balance: An estimated excess of 6,750,000 kg/state/year (equivalent to 30 kg P/ha/year).

[b] Based on current fertilizer recommendations and soil test summaries indicating 65% of the cropland in the state is optimum or excessive in P and thus requires little or no fertilizer P.

[c] Poultry–grain farm assumed to have 100 ha cropland and 88,000 broiler chicken capacity.

Nutrient budgets must be viewed from state, regional, and national perspectives as well, especially in areas dominated by animal-based agriculture, or in urbanized regions where land application of municipal biosolids is an accepted practice. Formidable problems exist in certain areas, both at the farm and state level, as illustrated in Table 5.7 for Delaware, site of a highly concentrated poultry operation, and for the U.S. as a whole in Figure 5.21.

Soil testing and plant analysis have traditionally been used to identify P deficiency and the likelihood of crop response to P fertilization. In many areas of the U.S., as shown in Figure 5.22, P deficiencies are common and soil testing plays a vital role in ensuring maximum economic crop yields. In other areas, however, the routine use of P fertilizers and organic wastes has built soil test P levels to such high values that it may be years before economic response to P fertilization occurs (with the possible exception of P in "starter" fertilizers). An example of the long "draw-down" period required for high P soils is given in Figure 5.23, where ~15 years were required to reduce soil test P from

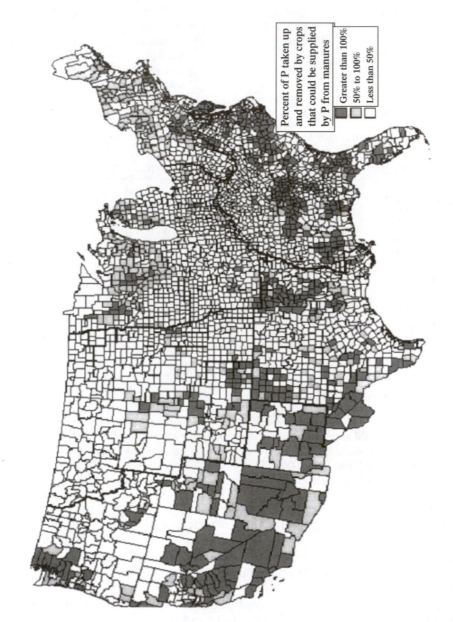

Figure 5.21 County-by-county analysis of the availability of animal manure P, relative to crop P requirements, in the U.S. Calculations assume manure is only applied to harvested cropland, nonlegumes, and pasture. (Adapted from Lander et al., 1998.)

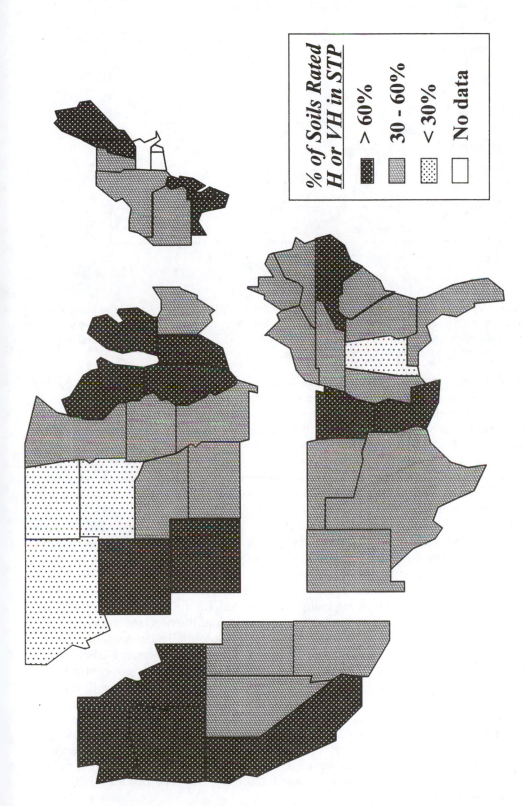

Figure 5.22 Soil test P (STP) summaries for the U.S., for agricultural cropland only. H and VH represent high and very high, respectively. (Adapted from Potash and Phosphate Institute, 1998.)

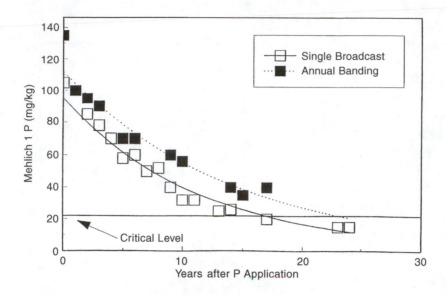

Figure 5.23 Decrease in soil test (Mehlich 1, 0.05 *M* HCl + 0.0125 *M* H_2SO_4) extractable P in a Portsmouth soil cropped to a corn–soybean rotation for 26 years. Initial soil test P levels (T = 0 years) resulted from a single broadcast application of 324 kg P/ha or eight annual banded applications of P at 60 kg P/ha. (Adapted from McCollum, 1991.)

~100 mg P/kg to a yield-limiting level (~20 to 25 mg P/kg). It has been argued recently that the increasing number of areas in the U.S. where soil P levels are "excessive" is an indication that the global P cycle has been seriously disrupted. In the U.S., for example, P fertilizers are produced from rock phosphate mined in Florida, transported via railways to croplands in the Midwest where the P is converted to grain that is subsequently used to produce animals in areas such as the Delmarva Peninsula. Much of the P entering the animal production facility is then converted to manures that are continuously land applied to nearby cropland because it is not economically feasible to redistribute this waste P to other areas where it is needed for crop production. The P cycle thus "breaks down," P accumulates in soils, and eventually is transported via surface and subsurface flow to groundwaters and surface waters. Recently, public concern about the environmental impacts of P has stimulated an effort to resolve these complex problems, particularly in watersheds sensitive to eutrophication (see environmental quality issues/events box, Improving agricultural phosphorus management in the Chesapeake Bay Watershed, p. 199).

   New soil testing procedures are being investigated for these P-enriched areas to identify the long-term P-supplying capacity of the soils, and the need for more intensive soil management strategies due to the high levels of potentially bioavailable and/or soluble P present in the soils. For example, until recently many soil-testing laboratories, due to equipment or economic limitations, did not measure the actual concentration of P in soils beyond some "high" value where crop response was no longer likely. The advent of environmental concerns about soil P and new instrumentation have made routine measurement of actual soil P values possible and, in many cases, have identified farming operations with levels far in excess of the amount of P required for normal plant growth. As an example, a survey conducted of 70 farms in southeastern Pennsylvania, an area dominated by animal-based agriculture, found Bray P_1 soil test values ranged from 36 to 411 mg P/kg, and averaged 130 mg P/kg. The optimum value for Bray P_1 is 30 mg P/kg.

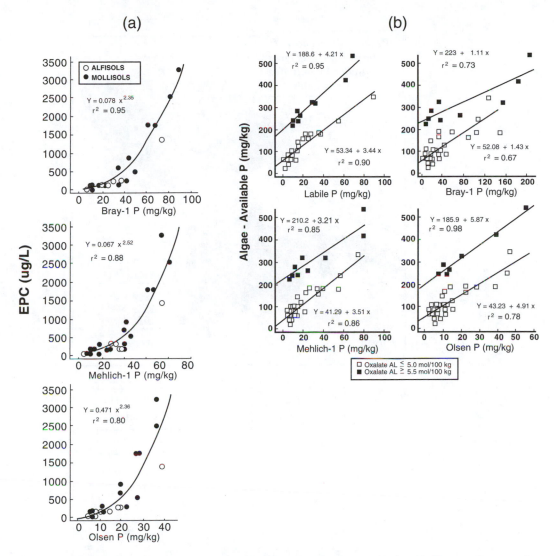

Figure 5.24 Examples of the use of routine soil tests (Olsen P, Mehlich-1 P, Bray-1 P) to predict parameters useful for the assessment of the environmental impact of soil P. (a) EPC_0 and (b) algal-available P. (From Wolf, A. M., *J. Environ. Qual.*, 14, 341, 1985. With permission.)

Other monitoring approaches include use of soil tests to assess the degree of soil P "saturation" (DPS), water-soluble P, the amount of "bioavailable" or "algal available" P (Figure 5.24), to determine, through subsoil testing, if P leaching has occurred, and quick tests for P sorption capacity. Many of these tests would be not conducted routinely, but would be part of a more intensive testing procedure used when routine soil tests identified areas with soil P levels excessive enough to warrant further investigation. For example, the degree of P saturation is now used in the Netherlands and Belgium to identify soils where P leaching to shallow groundwaters is a risk (Figure 5.25a). Upper limits for DPS of 25 to 40% have been proposed. Research in the U.S. found increases in water-soluble and desorbable P above these DPS values, suggesting a greater risk of P loss to surface waters and shallow groundwaters (Figure 5.25b).

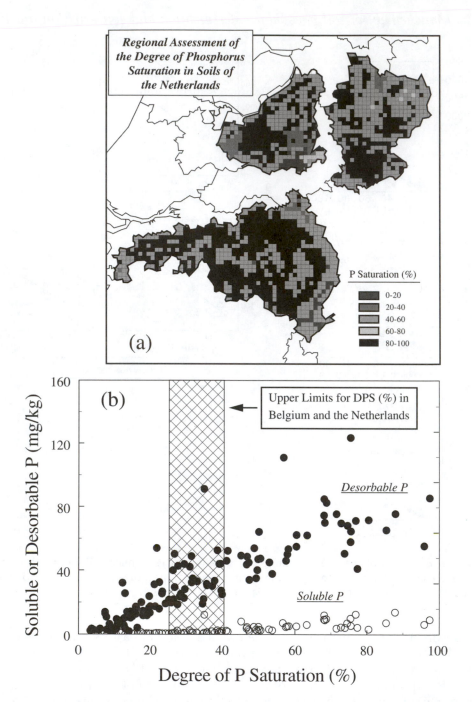

Figure 5.25 (a) Regional assessment of the degree of P saturation in the southern Netherlands. Areas with darker shadings are progressively more saturated with P in the soil profile to the depth of the mean high water table. (From Breeuswma et al., in *Animal Waste and Land-Water Interface*, K. Steele, Ed., Lewis Publishers, Boca Raton, FL, 1995. With permission.) (b) Relationship between the degree of P saturation in 125 Delaware soils and easily desorbable and soluble P. (Adapted from Pautler and Sims, 1998.)

5.5.2   *Management of soil phosphorus: fertilization and conservation practices*

Once the need, or lack of need, for P has been ascertained, the next step is to select the most efficient P source and application technique that together maximize crop P recovery. From an environmental perspective, the source of P generally has little influence on P loss via runoff or leaching, especially if the P source is incorporated into the soil. One possible exception to this is when chemical amendments are used to "stabilize" P in animal wastes that are later applied to the surface of pastures and no-till soils. In this case, source properties will be an important consideration because chemical stabilization reduces the amount of soluble P in the manure. This then decreases the risk of P loss when runoff waters interact directly with manure lying on the soil surface.

In general, the most efficient means to avoid P loss is to incorporate fertilizers and animal manures with the soil, where sorption and precipitation reactions will reduce P solubility and interaction with runoff waters. Fertilizer placement techniques, such as subsurface banding near the seed or seedling, are generally superior to broadcast applications, because they reduce the distance required for P diffusion to plant roots, as well as the extent of contact of fertilizer P with soil particles. This increases crop recovery of applied P. Application techniques that rely on soil tillage to incorporate P in the soil are sometimes necessary, particularly if soil P levels are low and higher rates of fertilizers or organic P sources are needed. The effects of conventional (moldboard plowing, disking) and conservation tillage ("no-till," "minimum-till," "chisel plowing," etc.) practices on P losses in erosion and runoff have been intensively investigated. As was shown in Table 5.5, the presence of crop residues on the soil surface in conservation tillage systems can appreciably reduce the loss of P. Other erosion control practices such as contour plowing, terracing, buffer strips, and grassed waterways can also reduce erosion, runoff, and thus P loss (Table 5.8). However, in many conservation tillage situations, unless fertilizer P is banded or manure P is injected these P sources are placed directly on the soil surface without incorporation. In some situations, a cultural operation such as conservation tillage that is designed to control one environmental issue (soil erosion, sediment-bound P loss) may create another (soluble P loss) if P levels in the upper few centimeters of the soil surface become excessive due to long-term overfertilization or manuring or if P is dissolved and lost directly from the surface-applied manure. Nutrient management programs that maintain soil P at reasonable levels, reduced tillage practices that partially incorporate fertilizers and animal manures (e.g., chisel plowing, Figure 5.26), or chemical stabilization of soluble P in manures (Figure 5.13) can help avoid the problem of excessive soluble P loss in runoff from soils.

Conservation tillage practices are particularly difficult to reconcile with animal waste applications where equipment to incorporate manures without tillage is frequently not available. Applying solid or liquid manures directly to the surface of untilled soils produces a stratification of P in the topsoil horizon, with very high levels present in the erosion-prone uppermost few centimeters. It also increases the likelihood of direct losses of soluble P and of P-enriched, highly erodible organic matter. Complicating the issue further is the fact that manure applications (and P fertilization) are often done after the growing season, when farmers have more time available, or when frozen ground makes it easier to move equipment through the fields. Fall, winter, and early spring surface applications of inorganic or organic sources of P greatly enhance the likelihood of P losses during the spring when rainfall and snowmelt normally results in the greatest percentage of annual rainfall occurring. Similar situations occur in agricultural and urban soils that are rarely plowed, such as pastures and turf. In these situations, where particulate P loss is low because of the filtering action of the grass, the best management strategy is to avoid unnecessary applications of P that will result in high soluble P losses.

Table 5.8 Overview of Soil and Water Conservation Practices That Can Be Used
to Reduce Phosphorus Losses in Erosion and Runoff[a]

Practice	Soil and water conservation benefits
Conservation tillage	Conservation tillage is a tillage system that leaves at least 30% of the soil surface covered by plant residue after planting to reduce erosion and runoff. A variety of tillage systems fall under conservation tillage, including stubble mulch tillage, minimum tillage, reduced tillage, and no-till. Conservation tillage may also make the soil surface porous, cloddy, rough, or ridged, which increases infiltration and reduces runoff.
Contour cultivation	Contour cultivation follows the contour lines of sloping fields. Contour ridges produced by tillage and planting form barriers that slow or stop downhill water movement and decrease runoff and erosion.
Deep chiseling	Deep chiseling or deep plowing can break up pans at plow depth or slightly below to increase water infiltration and reduce runoff and erosion.
Vegetative soil cover	Rainfall breaks up soil aggregates, allowing soil particles to move with runoff, and compacts and puddles the soil surface, decreasing water infiltration and increasing runoff volume. Crops can intercept up to 45% of rainfall, dissipating raindrop energy and reducing the ability of the rainfall to initiate runoff and erosion. After planting cash crops, cover crops, or green manures, it is essential to achieve canopy closure quickly to minimize soil exposure. Adequate soil fertility, optimum seed bed preparation and planting conditions, such as soil moisture and temperature, and pest control will promote seed germination, plant growth, and canopy closure. Plant populations, row spacing, and plant spacing that achieve equidistant plant separation also maximize soil coverage.
Crop rotation	Crop rotation enables a grower to plant a new crop soon after the preceding crop is harvested and minimize soil exposure time. Rotations can also help break insect, disease, or weed cycles that could impede plant growth, increase time to canopy closure, and reduce plant cover.
Companion crops	With companion crops such as small grain/forage crops the small grain starts first and provides quick plant cover. Harvesting the small grain leaves residue and the growing forage crop to protect the soil. Later, the forage crop provides excellent plant cover.
Cover crops	Cover crops grown during cold or dry seasons unfavorable for cash crops protect the soil by filling time and space gaps when cash crops leave the soil bare. Cover crops should germinate easily, grow quickly, provide sufficient plant cover, and be hardy against weather and pests.
Green manures	Green manures act as cover crops and when plowed under can add nutrients to soil to increase fertility for succeeding crop growth and soil organic matter content to promote the soil structure, permeability, and aeration needed to maximize rainfall infiltration and minimize runoff and erosion.
Crop residues	Crop residues act as mulch to increase soil coverage. Although plowing the residue under decreases soil coverage, it can promote residue decomposition and benefits similar to green manures.
Strip cropping	Strip cropping divides fields into long, narrow segments that help control runoff and erosion while growing crops in rotation. More heavily vegetated strips slow runoff and catch soil eroded from more exposed strips. Strips planted on the contour of slopes (contour strip cropping) are especially effective in controlling runoff and erosion.
Terraces	Terraces divide a slope so that runoff water is intercepted and carried to a protective outlet. Terraces help to decrease erosion by shortening slope length, slowing runoff velocity, and trapping sediments.

continued

Table 5.8 **(continued)**   Overview of Soil and Water Conservation Practices That Can Be Used to Reduce Phosphorus Losses in Erosion and Runoff[a]

| Vegetated waterways | Runoff water concentrates in waterways that when bare or unstable are extremely erodable. Vegetated waterways are natural or constructed channels that, when properly established and maintained, transport runoff water at a nonerosive velocity from fields, prevent gully formation, and greatly decrease erosion. |
| Buffer strips | Vegetated buffer strips keep soil from being carried into streams, ponds, or drainage ditches that need protection. Buffer strips primarily control water pollution. Erosion reduction may only be secondary. |

Source: Adapted from Sims and Vadas (1997) and Troeh et al. (1991).

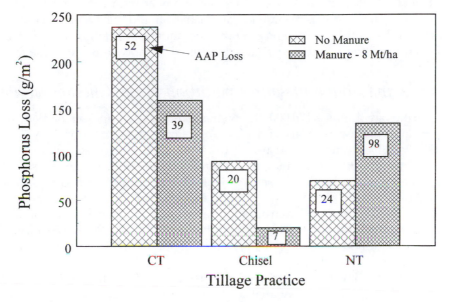

Figure 5.26 Effects of tillage practices and manure use on the loss of total P and algal-available P (AAP = values inside bars) in runoff from cornfields (CT = Conventional; Chisel = Chisel plow; NT = No-tillage). (Adapted from Mueller et al., 1984.)

Environmental quality issues/events

Improving agricultural phosphorus management in the Chesapeake Bay Watershed

Background

The Chesapeake Bay is one of the most important estuarine ecosystems in the world. In addition to its priceless value as a natural ecosystem, the bay is vital to the economies of the surrounding states, which rely upon its waters for fishing, recreation, and tourism. Unfortunately, years of point- and nonpoint-source pollution by nutrients, sediments, and toxics have degraded water quality in the bay to the point that it is now regarded as seriously polluted. In 1998, the Chesapeake Bay Foundation (CBF) rated the health of the bay as "27 out of 100," using an index derived from parameters related to habitat quality and quantity, water pollution, and the status of fish and shellfish populations. The CBF was cautiously optimistic, however, stating that the coordinated efforts of public agencies, private groups, and individuals have begun to restore the health of the bay. Reductions have been measured in the two major sources of pollution to the bay — nutrients and sediments — and increases have been observed in submerged aquatic vegetation and some fish populations. A key factor in the effort to restore the bay was the establishment of the Chesapeake Bay Program, which has supported research and education efforts and regulatory programs in the watershed since 1983. Partners in the Chesapeake Bay Program include the EPA, state governments (Maryland, Pennsylvania, Virginia), the District of Columbia, and a wide range of advisory groups. While it is unlikely that the bay will ever return to the pristine status of the 1600s, a health index of 70 is regarded as attainable. Should this occur, marked improvements in fish and shellfish populations, submerged aquatic vegetation, and water quality are anticipated, resulting in a more resilient, healthier Chesapeake Bay ecosystem.

Agriculture and nonpoint-source pollution of the Chesapeake Bay

The Chesapeake Bay drains a 170,000 km² watershed that encompasses parts of six major eastern U.S. states (Delaware, Maryland, New York, Pennsylvania, Virginia, West Virginia) and the District of Columbia. Land use in the watershed is diverse. About 60% of the watershed is forested, 30% is used for agriculture (cropland and pastures), and 10% for a variety of urban or suburban land uses. Agriculture has been identified as a major nonpoint-source of nutrients to the bay — studies suggest that >50% of the P and > 33% of the N entering the bay originate from agricultural sources. In some areas of the watershed these values are much higher. For instance in the lower Eastern Shore of Maryland (site of a highly concentrated poultry industry) ~70% of the N and 82% of the P inputs are from agriculture.

The Chesapeake Bay Program recognized from its inception in 1983 that reducing nutrient inputs to the bay was essential to its restoration goals. The initial target was a 40% reduction in the amount of N and P entering the bay by the year 2000. Indications today are that, while substantial progress has been made toward this goal, continued and more intensive efforts will be needed to reduce nutrient inputs to desired levels. Some important steps taken to reduce nutrient loading have included banning phosphate detergents in some states (Maryland, Virginia) and the District of Columbia, more efficient nutrient removal by wastewater treatment plants, expanded efforts to restore forested riparian buffers, and a concerted effort to develop and implement BMPs for agriculture. The main agricultural BMPs to date have been nutrient management planning, runoff controls for fields and barnyards, better and more animal manure storage facilities, planting cover crops, and restoring forested buffers and wetlands that had been cleared for crop production. Initially, these BMPs were directed more toward N than P. However, as discussed below, interest in the development of BMPs that can reduce agricultural P loading to the bay has intensified in the past few years.

Innovations in phosphorus management in the Chesapeake Bay Watershed

It was recognized from the outset of the Chesapeake Bay Program that reductions in P loading would be necessary to restore the health of the bay — hence, the improvements in wastewater treatment plants and the ban on detergents containing phosphate. Nutrient management planning, however, has been unable to address fully the issue of agricultural

P loading. Extensive reviews of the soil P status in the watershed found that ~70% of the soil samples from agricultural cropland were rated as "optimum" or "excessive" in P; in these situations little or no additional P is required either as fertilizers or animal manures (Figure 5.27). This creates serious problems for animal-based agriculture, such as the

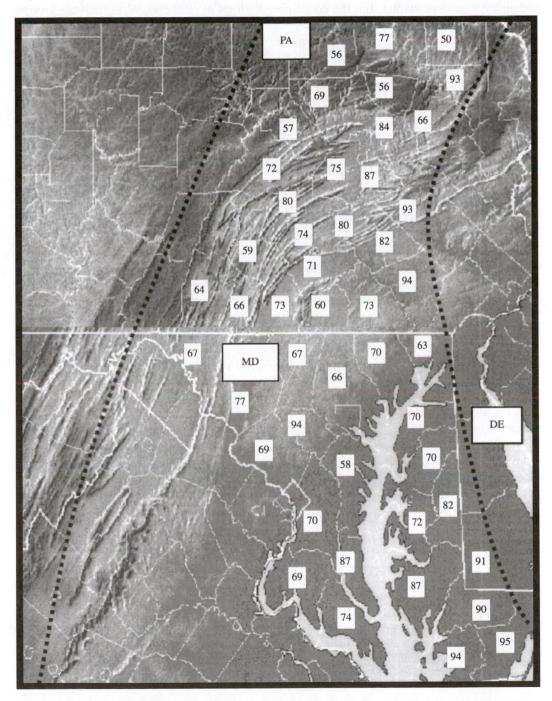

Figure 5.27 Soil test summaries for Delaware, Maryland, and Pennsylvania counties in the Chesapeake Bay Watershed. Values are percentage of samples rated as "optimum" or "excessive" in soil test P. Dashed lines indicate approximate boundaries of the watershed. (Adapted from Sims, 2000.)

geographically intensive poultry industry on the Delmarva Peninsula, where farmers currently have few, if any, options to land application of animal manures containing P. Along with the recognition that significant regional surpluses of P were present in the watershed (see Table 5.7), came the concern that minimizing soil erosion (the longstanding approach to reduce P inputs to surface waters) would not be enough to reduce agricultural P loads to the bay. Now there is concern that many soils in the watershed are sufficiently saturated with P that significant amounts of soluble P are lost to tributaries of the bay in runoff and shallow lateral flow. Given this, what steps are being taken today to improve agricultural P management in the Chesapeake Bay Watershed?

CNMPs: A much more comprehensive approach is now being taken to nutrient management planning, now required by law in three bay states (Maryland, Pennsylvania, Virginia) and by the USDA-NRCS and EPA Unified Strategy for Animal Feeding Operations (see Section 5.5). Comprehensive Nutrient Management Plans (CNMPs) address not only field-scale nutrient recommendations, but also farm-scale nutrient budgets, animal waste storage and treatment, manipulation of animal diets, redistribution of nutrients from areas of excess, and wider implementation of soil and water conservation practices.

Source reductions and animal waste treatment: Fertilizer P inputs can be reduced by the simple economic decision not to buy fertilizers when soils have optimum (or higher) soil test P values. Reducing P inputs via animal manures is more difficult. Strategies now being implemented include using phytase enzymes to increase the digestibility of phytate P in the grain used for poultry and swine diets, thus decreasing the need to supplement the feed with inorganic sources of P (monogastric animals such as poultry and swine lack the enzyme needed to digest phytate P). In a similar approach, new corn varieties with lower phytic acid contents ("high available P" corn) are now being introduced to the region. These dietary manipulations have been shown to reduce the amount of P excreted by poultry, and thus the overall P input to the watershed, by 20 to 50%. Wastewater treatment technologies are now being evaluated for use with animal wastes, such as the addition of alum to poultry litters to decrease the solubility of P (see Section 5.4 and Figure 5.13).

Bioenergy: The use of poultry litters as a renewable energy source, to produce electricity and/or heat for local industries, has attracted interest. Power plants in England, which now safely burn ~500,000 tons of poultry litter per year, are one model for this option. Evaluation of air quality impacts and the development a market for the resulting ash are two key issues that must be resolved.

Redistribution and alternate land uses: The potential to move surplus animal manure P to other states or counties for use by farmers without animals is being tested. Use of poultry litter as a nutrient source for managed pine forests and for reclamation of mine spoils is also under evaluation.

Value-added products: Finally, joint ventures to produce "pelletized" fertilizers from poultry litters have been initiated by poultry companies and fertilizer manufacturers. Pelletizing these materials will produce a uniform product that is easier and more economic to transport and that can be enriched with commercial fertilizers to give specific nutrient ratios for differing cropping systems.

Problems

5.1 Explain why soil conservation and nutrient management practices for P have traditionally focused on reducing P losses by soil erosion, while those for N have been directed toward reducing NO_3^- leaching.

5.2 Assume you are asked to construct a "nutrient budget" for a 500 ha dairy farm (500 cows, 300 heifers). The major crops grown are corn silage and alfalfa. All dairy manure produced on the farm is applied to the cropland where most (>75%) of the soils are now rated as either "optimum" or "excessive" in soil test P. Based on soil test results and crop yields, the total farmwide crop P requirement is determined to be 4000 kg P/farm. You determine from farm records that each year the farm produces 1.4 million L of liquid animal manure with a total P content of 0.06 kg per 1000 L and 3000 Mg of solid manure, with a total P concentration of 0.2%. Calculate the total P surplus on the farm in lb P_2O_5/acre and in kg P/ha.

5.3 One of the most widely used practices for P management is soil testing. However, the many different soil P tests and units used to express the results are often confusing to farmers and others involved in interpreting test results. You receive the results of a soil test and find that the soil test P value, using the Mehlich 1 test ($0.05\ M$ HCl + $0.0125\ M H_2SO_4$), is 94 mg P/L.

   a. Convert these results to mg P/kg; lbs P_2O_5/acre; kg P/ha.

   b. You are told that laboratory comparison studies have developed the following "conversion" equation between the Mehlich 1 and Mehlich 3 soil P tests:

   Mehlich 3-P (mg P/kg) = [1.5 × Mehlich 1-P (mg P/kg)] + 35

   An upper limit for soil test P (Mehlich 3 test) of 75 mg P/kg has been suggested in some states to identify soils where no P should be applied based on surface water quality concerns. Is the sample you received above or below this limit?

5.4 In some European countries the degree of P saturation (DPS) is used to determine whether fertilizer or manure P can be applied to soils. For instance, in the Netherlands if an agricultural soil is >25% saturated with P to the depth of the mean high water table, P inputs are limited or prohibited. Assume you extracted 1 g of topsoil with 40 mL of acid ammonium oxalate solution to determine DPS by the following equation:

   DPS (%) = 100 × [(Oxalate P, mmol/kg) ÷ α (Oxalate Fe + Al, mmol/kg)]

   The concentrations of Al, Fe, and P in the oxalate extract are 10.8, 7.4, and 4.5 mg/L, respectively. What is the DPS value in the topsoil (assume α = 0.5)?

5.5 In a mine spoil reclamation project you have the option of using rock phosphate, triple superphosphate, or composted municipal biosolids to build soil P fertility to an "optimum" level for the conservation grasses and trees to be grown. If the amount of P recommended to raise the soil from a "very low" to "optimum" value was 90 kg P/ha, how much of each material, in lbs per acre, would you need to apply, based on the P composition values given in Table 5.4? Assume you decided to use the composted biosolids because it was the most cost-effective material. What other factors would you need to consider before applying the composts?

5.6 Explain the difference between the following soil processes relevant to both plant availability and environmental impacts of soil P: (a) desorption and dissolution; (b) erosion, runoff, and leaching.

5.7 Explain why "starter" fertilizers are of value in crop production, even on soils that are considered to have "optimum" soil P levels.

5.8 Adsorption isotherms are useful for comparing the P sorption capacities of soils with differing physical and chemical properties or management histories. You are provided with the following data from adsorption isotherm experiments with two soils:

Initial P concentration (mg P/L)	Final P concentration at equilibrium (P_{eq}, mg P/L)	
	Soil 1	Soil 2
0	0.5	0.02
1	0.8	0.05
2	1.2	0.10
4	2.6	0.2
10	7.75	1.2
20	17.7	5.2
40	37.1	22.0

Calculate the amount of P adsorbed (P_{ads}) for both soils. Then plot (a) P_{ads} vs. P_{eq} for each soil (adsorption isotherm) and (b) the linearized version of the Langmuir equation (Equation 5.2) and calculate k and b. Contrast the suitability of the soils for (i) use in a wastewater irrigation system where disposal of high P wastewaters is desired and (ii) growth of alfalfa, a crop with a high P requirement (assume both soils have a low soil test P value).

5.9 Phosphorus in soil extracts is commonly and easily measured by colorimetric procedures. Soil P is "extracted" by shaking the soil and chemical extracting solution at certain soil:solution ratios for specific times, followed by filtration and colorimetric analysis. A key aspect of all colorimetric procedures is the use of a standard curve that relates known concentrations of the element of interest to the absorbance of light by the colored solution that results when the element reacts with added chemical reagents. In the case of P the chemical reagents that are added (ammonium molybdate) react with orthophosphate, forming a blue color that increases in intensity in proportion to the concentration of P in solution. This technique is commonly referred to as the Murphy–Riley method. Assume you are asked to determine the amount of water-soluble and plant-available P in the soil from a cornfield located near a surface water sensitive to eutrophication where there is considerable concern about overfertilization of soils with P and losses of P in runoff. You extract the soil with both deionized water (soil:solution ratio = 1:10) and the Mehlich 1 soil test (soil:solution ratio = 1:4). You then prepare a series of P standards and measure the absorbance of light by each standard solution and by the two samples and obtain the following results:

Standard Curve and Sample Absorbance Results

P concentration (mg/L)	Absorbance	
Standard curve		
0	0	
0.2	0.12	
0.4	0.24	
0.6	0.35	
0.8	0.47	
1.0	0.62	
Soil sample	Water soluble:	0.13
	Mehlich 1:	0.88

a. Plot the standard curve and calculate the amount of water soluble and plant available P in the soil in mg P/kg, in lb P/acre, in lb P_2O_5/acre, and in kg P/ha.

b. If the "optimum" soil test P value for corn production is 25 mg P/kg (Mehlich 1 soil test), would you recommend that P be applied to this field?

5.10 The owner of a commercial turf farm near the Chesapeake Bay contacts you with the following analysis from a municipal biosolids compost sample: [$N-P_2O_5-K_2O$ = 2.8-3.2-4.0]. A local waste management firm has told the owner that this compost can provide all the N, P, and K required to fertilize much of the soil on the turf farm. However, the grower is concerned, because of the proximity of the farm to the Chesapeake Bay, that using the compost may result in overapplication of P to this soil. Analyses of the soil at the site indicate that application of an 8-8-8 fertilizer ($N-P_2O_5-K_2O$) at 400 kg/ha is required. What rate of biosolids compost, in Mg/ha, would be needed to equal the plant available N added in 8-8-8 fertilizer (assume 20% of the total N in the compost is plant available)? How much P would be added in the biosolids compost at this rate and what effect would this have on the soil test P levels on the farm?

5.11 Some more complex soil testing procedures can be used to "fractionate" the total P in soils into Al-P, Fe-P, and Ca-P. This is done by sequentially extracting the same soil sample with stronger or more selective chemical reagents. Assume you conduct a fractionation procedure for soil P and obtain the following results:

Form of soil P	Extracting solution	Soil:solution ratio	P concentration in solution (mg/L)
Al-P	0.1 N NaOH	1 g and 40 mL	21.4
Fe-P	0.5 M citrate-dithionite bicarbonate	1 g and 25 mL	23.7
Ca-P	1 N HCl	1 g and 50 mL	3.2

a. Calculate the amount of P in each fraction in mg/kg.

b. Calculate the %P in each fraction, using the "sum of fractions" to represent total P.

5.12 Constructing artificial wetlands has been suggested as a means to "trap" soluble and sediment-bound P moving from agricultural soils to surface waters. What are the advantages and disadvantages of this "BMP" for water quality protection?

5.13 Explain the physical and chemical principles by which buffer strips and grassed waterways operate to prevent nonpoint-source pollution of surface waters by soil P.

5.14 Explain the value and limitations of (a) computer-based models such as EPIC (Figure 5.18) and (b) the P site index (Figure 5.19) in our ongoing efforts to improve the management of agricultural P.

References

Adriano, D. C., L. T. Novak, A. E. Erickson, A. R. Wolcott, and B. G. Ellis, 1975. Effect of long term land disposal by spray irrigation of food processing wastes on some chemical properties of the soil and subsurface water, *J. Environ. Qual.*, 4, 242–248.

Alberts, E. E., W. E. Neibling, and W. C. Moldenhauer. 1981. Transport of sediment nitrogen and phosphorus in runoff through cornstalk residue strips, *Soil Sci. Soc. Am. J.*, 45, 1177–1184.

Andraski, B. J., D. H. Mueller, and T. C. Daniel. 1985. Phosphorus losses in runoff as affected by tillage, *Soil Sci. Soc. Am. J.*, 49, 1523–1527.

Breeuswma, A., J. G. A. Reijerink, and O. F. Schoumans. 1995. Impact of manure on accumulation and leaching of phosphate in areas of intensive livestock farming, in *Animal Waste and the Land-Water Interface*, K. Steele, Ed., Lewis Publishers, Boca Raton, FL, 239–251.

Chang, A. C., A. L. Page, F. H. Sutherland, and E. Grgurevic. 1983. Fractionation of phosphorus in sludge-affected soils, *J. Environ. Qual.*, 12, 286–290.

Correll, D. L. 1998. The role of phosphorus in the eutrophication of receiving waters: a review, *J. Environ. Qual.*, 27, 261–266.

Fixen, P. E. and J. H. Grove. 1990. Testing soils for phosphorus, in *Soil Testing and Plant Analysis*, R. L. Westerman, Ed., American Society of Agronomy, Madison, WI, 141–180.

Foth, H. D. and B. G. Ellis. 1997. *Soil Fertility*, 2nd ed., Lewis Publishers, Boca Raton, FL.

Gachon, L. 1969. Les methodes d'appreciation de la fertilite phosphorique des sols, *Bull. Assoc. Fr. Etude Sol*, 4, 17.

Guertal, E. A., D. J. Eckert, S. J. Traina, and T. J. Logan. 1991. Differential phosphorus retention in soil profiles under no-till crop production, *Soil Sci. Soc. Am. J.*, 55, 410–413.

Hergert, G. W., D. R. Bouldin, S. D. Klausner, and P. J. Zwerman. 1981. Phosphorus concentration–water flow interactions in tile effluent from manured land, *J. Environ. Qual.*, 10, 338–344.

Jones, C. A., C. V. Cole, A. N. Sharpley, and J. R. Williams. 1984a. A simplified soil and plant phosphorus model: I. Documentation, *Soil Sci. Soc. Am. J.*, 48, 800–805.

Jones, C. A., A. N. Sharpley, and J. R. Williams. 1984b. A simplified soil and plant phosphorus model: II. Testing, *Soil Sci. Soc. Am. J.*, 48, 810–813.

Lander, C. H., D. Moffitt, and K. Alt. 1998. Nutrients available from livestock manure relative to crop growth requirements, Res. Assess. Strat. Planning Paper 98-1, USDA Natural Resources Conservation Service, Washington, D.C.

Lemunyon, J. L. and R. G. Gilbert. 1993. Concept and need for a phosphorus assessment tool, *J. Prod. Agric.*, 6, 483–486.

McCollum, R. E. 1991. Buildup and decline in soil phosphorus: 30-year trends on a typic umbraquult, *Agron. J.*, 83, 77–85.

Mozaffari, P. M. and J. T. Sims. 1994. Phosphorus availability and sorption in an Atlantic Coastal Plain watershed dominated by animal-based agriculture, *Soil Sci.*, 157, 97–107.

Mueller, D. H., R. C. Wendt, and T. C. Daniel. 1984. Phosphorus losses as affected by tillage and manure application, *Soil Sci. Soc. Am. J.*, 48, 901–905.

Pautler, M. C. and J. T. Sims. 1998. Integrating environmental soil P tests into nutrient management plans for intensive animal agriculture, in *Proc. 16th World Congress of Soil Science*, Montpelier, France, August 20–26, 484.

Payer, F. S. and R. R. Weil. 1987. Phosphorus renovation of wastewater by overland flow application, *J. Environ. Qual.*, 16, 391–397.

Peters, J. M. and N. T. Basta. 1996. Reduction of excessive bioavailable phosphorus in soils by using municipal and industrial wastes, *J. Environ. Qual.*, 25, 1236–1241.

Pierzynski, G. M. and T. J. Logan. 1993. Cropping patterns and their effects on phosphorus soil test levels, *J. Prod. Agric.*, 6, 513–520.

Potash and Phosphate Institute. 1998. Soil test summaries: phosphorus, potassium and pH, *Better Crops*, 82, 16–19.

Ritter, W. F. 1992. Delaware's Inland Bays: a case study, *J. Environ. Sci. Health*, 27, 63–88.

Ryding, S. O., M. Enell, and R. E. White. 1990. Swedish agricultural nonpoint source pollution: a summary of research and findings, *Lake Reserv. Manage.*, 6, 207–217.

Sallade, Y. E. and J. T. Sims. 1997. Phosphorus transformations in the sediments of Delaware's agricultural drainageways: II. Effect of reducing conditions on phosphorus release, *J. Environ. Qual.*, 26, 1579–1588.

Sanchez, P. A. and G. Uehara. 1980. Management considerations for acid soils with high phosphorus fixation capacity, in *The Role of Phosphorus in Agriculture*, F. E. Khasawneh, Ed., American Society of Agronomy, Madison, WI, 471–514.

Sharpley, A. N. 1985. The selective erosion of plant nutrients in runoff, *Soil Sci. Soc. Am. J.*, 49, 1527–1534.

Sharpley, A. N. and A. D. Halvorson. 1994. The management of soil phosphorus availability and its transport in agricultural runoff, in *Soil Processes and Water Quality*, R. Lal, Ed., Advances in Soil Science, Lewis Publishers, Boca Raton, FL, 1–84.

Sharpley, A. N., S. J. Smith, B. A. Stewart, and A. C. Mathers. 1984. Forms of phosphorus in soil receiving cattle feedlot waste, *J. Environ. Qual.*, 13, 211–215.

Shreve, B. R., P. A. Moore, Jr., T. C. Daniel, D. R. Edwards, and D. M. Miller. 1995. Reduction of phosphorus in runoff from field-applied poultry litter using chemical amendments, *J. Environ. Qual.*, 24, 106–111.

Sims, J. T. 1992. Environmental management of phosphorus in agricultural and municipal wastes, in Future Directions for Agricultural Phosphorus Research, Bulletin Y-224, National Fertilizer and Environmental Research Center, Muscle Shoals, AL, 59–65.

Sims, J. T. 1996. The Phosphorus Index: A Phosphorus Management Strategy for Delaware's Agricultural Soils, Fact Sheet ST-05, College of Agricultural Sciences and Cooperative Extension, University of Delaware, Newark.

Sims, J. T., 2000. The role of soil testing in environmental risk assessment for phosphorus, in *Agriculture and Phosphorus Management: the Chesapeake Bay*, A. N. Sharpley, Ed., Lewis Publishers, Boca Raton, FL.

Sims, J. T. and P. A. Vadas. 1997. Nutrient Management Strategies for the Profitable, Environmentally Sound Use of Phosphorus, Fact Sheet ST-08, College of Agricultural Sciences and Cooperative Extension, University of Delaware, Newark.

Troch, F. R., J. A. Hobbs, and R. L. Donahue. 1991. *Soil and Water Conservation*, Prentice-Hall, Englewood Cliffs, NJ.

Weidner, K. 1988. Murky water, *Pa State Agriculture*, Spring/Summer, 2, 19.

Wolf, A. M., D. E. Baker, H. B. Pionke, and H. M. Kunishi. 1985. Soil tests for estimating labile, soluble, and algae-available P in agricultural soils, *J. Environ. Qual.*, 14, 341–348.

Young, R. D., D. G. Westfall, and G. W. Colliver. 1985. Production, marketing, and use of phosphorus fertilizers, in *Fertilizer Technology and Use*, O. P. Engelstead, Ed., American Society of Agronomy, Madison, WI.

Supplementary reading

Burkholder, J. M., M. A. Mallin, H. B. Glasgow, Jr., L. M. Larsen, M. R. McIver, G. C. Shank, N. Deamer-Melia, D. S. Briley, J. Springer, B. W. Touchette, and E. K. Hannon. 1997. Impacts to a coastal river and estuary from rupture of a large swine waste holding facility, *J. Environ. Qual.*, 26, 1451–1466.

Foy, R. H. and P. J. A. Withers. 1995. The contribution of agricultural phosphorus to eutrophication, Proc. No. 365, The Fertilizer Society, Petersborough, U.K.

Haygarth, P. M. and S. C. Jarvis. Transfer of phosphorus from agricultural soils, *Adv. Agron.* (in press).

Journal of Environmental Quality, 1998. Vol. 27: Special Issue: Agricultural phosphorus and eutrophication.

Mason, C. F. 1991. *Biology of Freshwater Pollution*, 2nd ed. John Wiley & Sons, New York.

Moore, P. A., Jr., T. C. Daniel, and D. R. Edwards. Reducing phosphorus runoff and inhibiting ammonia loss from poultry manure with aluminum sulfate, *J. Environ. Qual.* (in press).

Parry, R. 1998. Agricultural phosphorus and water quality: a U.S. Environmental Protection Agency perspective, *J. Environ. Qual.*, 27, 258–261.

Sharpley, A. N., S. C. Chapra, R. Wedepohl, J. T. Sims, T. C. Daniel, and K. R. Reddy. 1994. Managing agricultural phosphorus for protection of surface waters: issues and options, *J. Environ. Qual.*, 23, 437–451.

Sibbesen, E. and A. N. Sharpley. 1997. Setting and justifying upper critical limits for phosphorus in soils, in *Phosphorus Loss from Soil to Water*, H. Tunney et al., Eds., CAB International, London, 151–176.

Sims, J. T., R. R. Simard, and B. C. Joern. 1998. Phosphorus losses in agricultural drainage: historical perspective and current research, *J. Environ. Qual.*, 27, 277–293.

Sims, J. T., A. C. Edwards, O. F. Schoumans, and R. R. Simard. Integrating soil phosphorus testing into environmentally-based agricultural management practices, *J. Environ. Qual.* (in press).

Soil Use and Management. 1999. Vol. 14: Special Issue on Phosphorus, Agriculture, and Water Quality, CAB International, Oxon, U.K.

Taylor, A. W. and V. J. Kilmer. 1980. Agricultural phosphorus in the environment, in *The Role of Phosphorus in Agriculture*, F. E. Khasawneh, E. C. Sample, and E. J. Kamprath, Eds., American Society of Agronomy, Madison, WI, 545–590.

Tunney, H., O. T. Carton, P. C. Brookes, and A. E. Johnson. 1997. *Phosphorus Loss from Soil to Water*, CAB International, Harpenden, U.K., 468 pp.

chapter six

Soil sulfur and environmental quality

Contents

6.1   Introduction

Sulfur (S) exists in the atmosphere, hydrosphere, biosphere, and lithosphere as various chemical species with different oxidation states. Environmental impacts are primarily the result of atmospheric pollution and acidic deposition from gaseous S emissions (see Chapter 10), and acid soils, acid mine drainage (AMD), and groundwater contamination from aqueous S species. Because S is a secondary macronutrient required for plant, animal, and human subsistence, both toxicities and deficiencies can occur depending on soil conditions, cultural practices, and climatic influences. To manage S properly, we must understand its role in the environment. First, we know, with respect to our concern of S in the environment, that acidic deposition and acid-producing soils and subsurface materials are the most formidable problems that must be addressed. Second, S is an essential plant nutrient that has received greater attention recently because of the reduction in S-containing materials (e.g., low-S fertilizers, use of non-S pesticides, and reduced atmospheric S deposition) added to crop- and pasturelands. These reductions increase the probability of S deficiencies in crops. Third, it is essential that we appreciate the connection among S and the other biogeochemical cycles, including the essential elements — carbon (C), hydrogen (H), oxygen (O), nitrogen (N), phosphorus (P) — and micronutrients. This chapter will discuss the role of S in agricultural, forest, mine land, and salt-affected and aquatic ecosystems, with special emphasis on environmental concerns, biogeochemistry, inorganic and organic forms, retention, transformations, nutrition, and management.

6.2   Importance of sulfur

Sulfur is an essential nutrient that is required for all biological systems including plants, animals, and humans. Although both S deficiencies and toxicities have been extensively reported for plants and animals, only cases involving plant S deficiencies appear to be of significance because S-deficient areas have been reported throughout the world. Sulfate (SO_4^{2-}) is the predominant form of soil S taken up (e.g., absorbed) by plants, but is not the predominant form in soils, which explains at least partially why certain areas may exhibit plant S deficiencies despite what may seem to be adequate S levels in soils. Low S parent materials, high rainfall, intense leaching, and extremely weathered soils, along with low S atmospheric deposition, are common conditions in S-deficient areas.

Sulfur occurs naturally in several inorganic and organic forms that are part of the biogeochemistry of the global S cycle (see Chapter 9 for discussion on biogeochemical cycles). The major S pools are part of the lithosphere, similar to that of P. However, the global S cycle is much like that of N, with the major N and S pools being the lithosphere. Although most of the S that cycles through the atmosphere is a result of human activity, the atmospheric S pool is small and the short mean residence time (MRT) of S compounds actually results in more S cycling through the atmosphere than N (i.e., S flux is greater than N flux). In soils, S transformations are predominately controlled by biochemical processes, which also play an important role in the availability of S to plants.

6.2.1   Sulfur nutrition

Like N, S is an essential element in several amino acids that are part of plant, animal, and human proteins; S is also an important constituent in vitamins and hormones. Sulfur plays an active role in plant structural composition and metabolic processes, and is crucial to animals because S performs several functions in structural, metabolic, and regulator processes. Sulfur has often been overlooked as an essential plant nutrient primarily because, until recently, S deficiencies were not considered to be very common. The importance of S in plant nutrition and the determination of S availability is discussed further in Sections 6.5.1 and 6.5.2.

For animals, S-containing amino acids are integral components of many biochemical constituents including proteins, enzymes, and some hormones. Organic S-containing biochemical compounds are important to the production of animal bones, tendons, cartilage, skin, and heart valves. The major structural protein in animals is collagen, which contains S amino acids. Enzymatic S plays an essential role in enzyme activity and function. Sulfur deficiencies in animals may be manifested as decreased or slow weight gains, lethargy, reduced milk, egg, or wool production, as well as several other symptoms, and if S deficiency is prolonged, possibly death. In addition, the form of S supplied to animals is critical because nonruminants are unable to utilize inorganic S compounds. Ruminants, such as sheep and cattle, are capable of metabolizing inorganic forms of S through the activity of rumen microorganisms. A constant supply of S is required for proper growth in plants and animals, but especially for animals because S is not stored in their bodies.

Sulfur toxicity is also a potential problem for plants and animals. For plants, S toxicity may be exhibited through retarded growth, interveinal chlorosis that extends along leaf margins, or premature senescence of leaves. However, except for citrus, most crops appear to be unaffected by high SO_4^{2-} levels, and for those plants that do display symptoms, it is generally the associated cation that is detrimental. The toxic effect of gaseous sulfur dioxide (SO_2) and hydrogen sulfide (H_2S) compounds on plants is discussed further in Section 6.5. As for animals, S toxicity is primarily a function of the S species. Ruminants have suffered from S toxicity when ammonium sulfate [$(NH_4)_2SO_4$] or gypsum ($CaSO_4$) were used as nonprotein N or calcium (Ca) sources, respectively. Nonruminants suffer from S toxicity when they consume excessive amounts of S-containing amino acids, especially methionine, in their diet.

6.2.2   Environmental impacts of sulfur

Environmental concerns related to S are listed in Table 6.1. Acidic deposition results from the oxidation of S and N in the atmosphere and will be discussed in greater detail in Chapter 10. Acid soils and acid mine drainage (AMD) occur when reduced S compounds and minerals are oxidized, forming sulfuric acid (H_2SO_4). Examples of AMD are most prevalent (78%) at inactive or abandoned mine sites, which continue to be of greatest environmental concern and liability for the mining industries. In the U.S., AMD and other toxins from abandoned sites have polluted 75,000 ha of impoundments and lakes as well as over 20,000 km of streams and rivers. The U.S. Bureau of Mines has estimated the cost of remediating the impacted waterways at $30 to $75 billion. In addition, geothermal activity can also influence the surrounding soil and plant ecosystems by releasing significant amounts of gaseous S into the atmosphere. Sulfate is highly soluble and is known to accumulate in groundwaters of arid and semiarid environments. A drinking water standard of 250 mg SO_4/L was established because SO_4^{2-} was found to act as a laxative in some humans and animals.

Burning of fossil fuels, petroleum refining, and processing of metals such as copper (Cu), lead (Pb), zinc (Zn), and nickel (Ni) have resulted in enormous amounts of S emitted into the atmosphere every year (Table 6.2). Estimates for natural S emission rates range from 80 to 143 Tg S/year (Tg = 10^{12} g), suggesting humans are responsible for roughly 40 to 60% of the S entering the atmosphere. Natural sources of S include volcanic activity and biogenic gases from terrestrial and ocean ecosystems as well as

Table 6.1 Potential Environmental Impacts Due to Sulfur

Environmental concern	Description of problem
Acidic deposition	Sulfuric acid formation in the atmosphere results in wet deposition (rain, snow, mist, fog) and dry deposition (particulates) that may be detrimental to vegetation, surface waters, buildings and structures, and humans; see Table 4.1 for environmental impacts due to N and Chapter 10 for further discussion on acidic deposition
Acid sulfate soils	Caused by the release of H_2SO_4 into the soil solution through the oxidation of sulfidic materials that are commonly associated with coastal regions and lignite coal mining operations
Acid mine drainage	Oxidation of reduced forms of S from mining activities produces H_2SO_4 that can impact the soils or mine spoils and surface waters in the surrounding environments
Geothermal activity	Geysers and other geothermal releases can bring significant amounts of gaseous and soluble S compounds to the earth's surface; vegetation in the surrounding area can be killed by S gases or the extremely acidic soils that form; soil pH values as low as 0.9 have been reported in Yellowstone National Park, WY
Groundwater contamination	High SO_4 concentrations can render groundwaters unsafe for human and livestock consumption
Sulfur toxicity to animals	Both ruminant and nonruminant animals can suffer from S toxicity; ruminants are susceptible to excess S in the forms of $(NH_4)_2SO_4$ and gypsum $(CaSO_4)$ when these are used as nonprotein N or Ca feed sources, respectively; nonruminants are susceptible to excessive S amino acids, especially methionine, in their diet; livestock are also sensitive to H_2S (Goodrich and Garrett, 1986)

Table 6.2 Estimations of Global S Emissions from Various Industrial Sources

Source	S Emitted (Tg/year)[a]	Source	S Emitted (Tg/year)
Coal	51	Cr Smelting	7.5
Lignite (brown coal and peat)	17.5	Cu Refining	1.4
Coke (briquettes)	1.3	Pb Smelting	0.8
Total coal	**69.8**	Zn Smelting	0.5
		Ni Smelting	0.8
Petroleum refining	3.2	**Total metallurgical operations**	**11.0**
Motor gasoline	0.4		
Kerosene	0.1	Sulfuric acid manufacturing	1.3
Fuel and diesel	1.7	Incineration	0.3
Petroleum coke	0.2	Pulp and paper	0.3
Residual fuel oils	18.0	Total miscellaneous	1.9
Total petroleum	**23.6**		
		Total global S emissions	**106.3**

[a] $Tg = 10^{12}$ g.

Source: Krupa, S. V. and Tabatabai, M. A., in *Sulfur in Agriculture*, Agron. Monogr. No. 27, M. A. Tabatabai, Ed., American Society of Agronomy, Madison, WI, 1986.

wetland, swamp, and other anaerobic environments. Coal-fired, electric-power-generating plants are the main anthropogenic source of S emissions, which was a concern that prompted the 1990 Clean Air Act. The eastern U.S. has seen a decline in S emissions because of the Clean Air Act regulations.

Industrial regions are the predominant source of anthropogenic S released into the atmosphere. A major environmental concern associated with atmospheric S release is acidic deposition (wet and dry deposition containing high levels of H_2SO_4), which has caused problems with vegetative growth, aquatic ecosystem acidification, human health and structural deterioration that many nations throughout the world are now attempting to solve (see Chapter 10 for further discussion on acid deposition). Precipitation in affected areas can have a pH below 4, which, compared with rainfall in many nonindustrial areas with pH values of 6 to 7, suggests a 100- to 1000-fold increase in H^+ ions in acidic precipitation.

The amount of S that falls annually in rainfall varies depending on outputs from natural and anthropogenic sources. Because plants require S (9 to 39 kg/ha each year for average crop yields), some regions benefit from the S added in rainfall. In some areas, atmospheric deposition of S may be sufficient to sustain adequate yields. For example, S deposition in the U.S. has been estimated for several urban (nonindustrial) and rural regions to be 1 to 37 kg S/ha in the southeast, 4 to 42 kg S/ha in the Midwest, and 5 to 17 kg S/ha in the north-central states. Sulfur deposition in urban and rural regions of other counties is comparable to those listed above; 2 to 16 kg S/ha in Canada, 13 to 27 kg S/ha in China, and 10 to 20 kg S/ha in the former Soviet Union.

In addition to acidic deposition, AMD is another environmental concern due to the mining of minerals such as coal, Cu, Zn, uranium (U), and others. After exposure of overburden materials during the mining process, oxidation and leaching can produce drainage waters that are acidic and contain harmful concentrations of dissolved minerals. In addition to potential environmental problems due to acid production, several elements such as aluminum (Al), arsenic (As), cadmium (Cd), Cu, iron (Fe), manganese (Mn), Ni, Pb, sodium (Na), Zn, may be released into the environment at high enough levels to be toxic to plants and soil and aquatic organisms (Barton, 1978). Further discussion on AMD can be found in Section 6.7.2.

Environmental quality issues/events
Sulfur origins and contents of major U.S. coal deposits

The S content of coal is related to its parent material (i.e., plant matter, and the sedimentary environment in which the coal formed). Low-S coal ($\leq$1% S) is the result of incorporation of plant S into the accumulating peat that ultimately forms coal within a freshwater setting. Medium-S (>1 to <3% S) and high-S ($\geq$3% S) coals are the consequence of two major sources of S, namely, that derived from parent plant material and SO_4-S in seawater that inundated the peat accumulation areas. Some high-S coals can contain a significant amount of S such as bituminous coal from Texas, which has been reported to have an S content as high as 8.9%. Geologic materials overlying coal deposits also contain S levels that parallel the abundance of that present in the underlying coal. Acid mine drainage, therefore, is a greater concern in areas where high-S coals exist. Additional information related to AMD will be presented in Section 6.7.2.

The major forms of S in coal are pyritic-S, organic-S, and SO_4-S, with pyritic and organic forms generally dominant. Pyrite is the sulfide (reduced-inorganic-S) form and marcasite, pyrrhotite, sphalerite, galena, and chalcopyrite are secondary sulfide sources. Several S-containing organic compounds are found in coal, including thiols, organic sulfides and

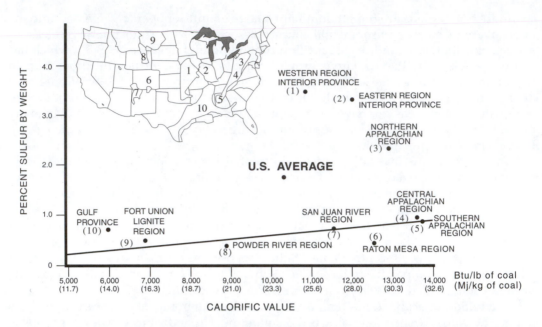

Figure 6.1   Average percent S content and heat energy (BTU/lb or MJ/kg) of major U.S. coal regions. The line represents the EPA compliance value of 1.2 lb SO_2 per million BTU output. (Information from U.S. Geological Survey Open-File Report No. 96-92.)

disulfides, and thiophene and its derivations. Some of the organic-S compounds may have been formed during the early stages of coal formation (humification process) when plant matter is decomposed by microbial activity. Only small amounts of SO_4-S minerals are present in coal, including gypsum ($CaSO_4$), barite ($BaSO_4$), and other Fe and Na SO_4-S minerals.

The high-S bituminous coals mined in Appalachian and north-central regions of the U.S. (see Chapter 8 for further discussion on coal) have high heat energy (MJ/kg), but, due to their higher S contents (average 1 to 4% S contents, Figure 6.1), burning these coals results in the production of environmentally significant amounts of SO_2. Low-S subbituminous and lignite coals in the Rocky Mountain region have lower heat energy, but are mined extensively (more than 300 Mg from Wyoming in 1998) because these materials are less harmful to the environment and are usually easier and cheaper to mine (i.e., low overburden-to-coal ratio). Sulfur dioxide produced from the combustion of coal reacts with water in the atmosphere to produce sulfurous (H_2SO_3) and sulfuric (H_2SO_4) acids according to the following simplified reactions:

$$SO_2 + H_2O \rightleftharpoons H_2SO_3 + 1/2O_2 \rightleftharpoons H_2SO_4 \qquad (6.1)$$

—*Source:* Chou (1989)

6.3   Global sulfur cycle

Sulfur exists in several inorganic and organic forms within the atmosphere, hydrosphere, and soil environments. Soil S exists as aqueous species such as SO_4^{2-}, solid forms that consist of mineral matter and organic constituents of plant, animal, and microbial origin, and gaseous species. Oxidation states for S range from –2 to +6 (see Table 6.3 for examples

Table 6.3 Forms of Inorganic Sulfur in the Environment

Mean oxidation state	Category	Compound	Formula
-2	Sulfides	Sulfide ion	S^{2-}
		Bisulfide ion	HS^-
		Hydrogen sulfide	H_2S
		Carbonyl sulfide	COS
-1	Polysulfides	Disulfide	S_2^{2-}
		Pyrite	FeS_2
0	Elemental	Sulfur	S^0
+2[a]	Thiosulfate	Thiosulfate ion	$S_2O_3^{2-}$
		Bithiosulfate ion	$HS_2O_3^-$
		Thiosulfuric acid	$H_2S_2O_3$
+4	Sulfites	Sulfite ion	SO_3^{2-}
		Hydrogen sulfite	HSO_3^-
		Sulfur dioxide	SO_2
+6	Sulfates	Sulfate ion	SO_4^{2-}
		Bisulfate ion	HSO_4^-
		Sulfuric acid	H_2SO_4

[a] Sulfur oxidation states for thiosulfates are -2 and +6 with an average of +2.

Source: Stevenson, F. J. and Cole, M. A., in *Cycles of Soil: Carbon, Nitrogen, Phosphorus, Sulfur, Micronutrients,* 2nd ed., John Wiley & Sons, New York, 1999.

of oxidation states for inorganic and gaseous S species). As S cycles through atmosphere, hydrosphere, and soil ecosystems, several transformations can occur — due to biochemical and chemical processes — and form S species of different oxidation states. For example, carbonyl sulfide (COS), the most abundant atmospheric S gas, is oxidized to SO_4^{2-} or taken up by plants and metabolized. Sulfate entering soils is utilized by plants and microorganisms where SO_4^{2-} is reduced via metabolic pathways to amino acids, proteins, and other S-containing biochemicals. Decomposition of organic matter releases oxidized and reduced S forms depending on environmental conditions, in which oxygen plays a dominant role. Several pathways have been proposed for the transformation of organic S to volatile S compounds including the following:

$$\text{organic S}$$
$$\swarrow \quad \downarrow \quad \searrow \qquad\qquad (6.2)$$
$$SO_4^{2-} \rightarrow \text{cysteine} \rightarrow \text{methionine} \rightarrow \text{volatile S compounds}$$

Physical processes, such as movement of S as dust, wet and dry deposition, or precipitation–dissolution reactions involving S evaporites (e.g., S minerals that form during desiccation), are also responsible for the transfer of S from one ecosystem to another.

The largest of the earth's S pools (or reservoir) is the lithosphere (metamorphic > sedimentary > igneous rock), followed by the hydrosphere (primarily oceans and seas), biosphere, and last the atmosphere, which contains only a minor amount of S (Figure 6.2). By far the greatest S pool exists in sedimentary pyrite, shale, and evaporite materials. In soils, S concentrations range from a low of approximately 0.002% in coarse-textured, highly weathered, leached soils in humid areas to greater than 5% in gypsiferous soils of arid and semiarid areas, and in coastal soils developed in tidal marshes. Soil S is predominately composed of inorganic and organic forms, and the ratio of the two varies with soil properties (pH, moisture status, organic matter and clay contents), soil depth, and climatic conditions. Soil organic matter consists of approximately 0.5% S.

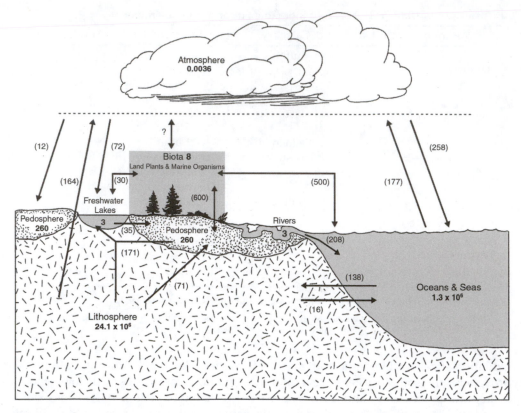

Figure 6.2   Environmental S pools and transformations (fluxes). Bold numbers represent pool concentrations (Gt = 10^{15} g or 1000 Tg) and flux rates are shown in parentheses (Tg/year). (Data from Trudinger, 1986.) An S balance (inputs = outputs) is presumed between the biosphere and atmosphere.

In the S cycle, transfer of S (the flux) from one pool (or reservoir) to another occurs at a rate proportional to the amount entering and exiting a pool (Figure 6.2). Although only small amounts of S are transferred from one S pool to another, humans have had an immense influence on the annual atmospheric S flux. As with the increase in C-based materials entering the atmosphere since the beginning of the industrial era (see Chapter 10), so too are S-based materials increasing. Over the past 100 years, the rate of S entering the atmosphere by fossil fuel emissions has doubled. Over time, the amount of oxidized S is increasing, some of which is returned to earth's surface as wet and dry deposition (i.e., acidic deposition). Further discussion on atmospheric deposition of S and N can be found in Chapter 10; however, it should be emphasized that S deposition in the U.S. is actually lower now than during the 1980s.

Example problem 6.1

The atmospheric S mean residence time (MRT) characterizes the transfer of S through the atmosphere and is calculated based on the size of the atmospheric S pool and the flux rate for wet and dry S deposition. The atmospheric S MRT is based on the assumption that the atmospheric S pool is representative of all locations, and that the flux rate is universal. Clearly, this is not the case in industrial areas and under variable climatic conditions. However, the concept is useful for examining the general characteristics of atmospheric S inputs and outputs.

$$\text{atmospheric S MRT} = \text{atmospheric S pool} / \text{atmospheric S flux rate}$$
$$= (3.6 \text{ Tg S}) / (342 \text{ Tg S/year}) = 0.01 \text{ year} \approx 4 \text{ days} \qquad (6.3)$$

The primary S forms in wet and dry deposition are SO_4^{2-} and SO_2. Specific atmospheric S compounds have MRTs that range from approximately 1 day for dimethyl sulfide $(CH_3)_2S$ and H_2S to 160 days for COS, and are based on their concentrations, chemical reactivity, and chemical properties.

The soil S cycle is similar in many respects to N and P soil cycles in that each is affected by biochemical and chemical processes. The S cycle — emphasizing soil, plant, and animal S transformations — is shown in Figure 6.3. Four dominant S forms are found in soils, including elemental S, sulfides, sulfates, and organic S compounds. Additions of soluble S to soil results from mineral weathering, atmospheric deposition, fertilization (and other soil amendments), and decomposition of plant, animal, and soil organic matter. Losses of S occur during leaching, surface runoff, and crop removal. Transformations that take place between the various S forms are mediated primarily by biochemical processes that will be discussed in more detail later in this chapter.

6.3.1   Inorganic sulfur in soils

Several forms of inorganic S exists in soils (Table 6.3). In well-drained soils, inorganic S occurs primarily as SO_4^{2-} in a dissolved, adsorbed, or solid state. Soils containing Al and Fe hydrous oxides are capable of adsorbing SO_4^{2-} and preventing losses due to leaching.

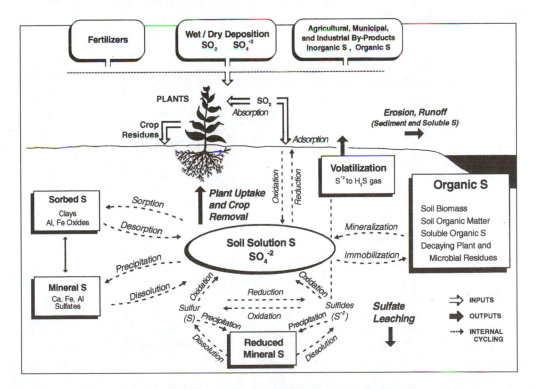

Figure 6.3   Environmental S additions, losses, and transformations.

Gypsum ($CaSO_4 \cdot 2H_2O$) is found primarily in calcareous soils of arid and semiarid regions and may also occur as a coprecipitated mineral form associated with $CaCO_3$ (limestone). Other less common solid mineral forms of SO_4^{2-} include barite ($BaSO_4$), celestite ($SrSO_4$), jarosite [$KFe_3(SO_4)_2(OH)_6$], and coquinbite [$Fe_2(SO_4)_3 \cdot 5H_2O$]. In anaerobic soils such as wetlands, tidal marshes, and poorly drained soils, reduced forms of S are common and can include FeS, FeS_2, H_2S, and S^0.

Gaseous forms of S are also present in soils and range from trace concentrations in well-drained, aerobic soils to high concentrations in waterlogged, anaerobic soils. Soil S is less volatile than soil C and N compounds, whereas P has essentially no gaseous forms in nature. Gases formed during organic matter decay that may be emitted from soils include carbon bisulfide (CS_2), COS, methyl mercaptan (CH_3SH), dimethyl sulfide [$(CH_3)_2S$], dimethyl disulfide [$(CH_3)_2S_2$], and H_2S, all of which have characteristic, distinctive odors. Sulfur emissions from anaerobic soils and wetlands (i.e., swamps) are approximately 30 Tg S/year.

Elemental S and sulfides are oxidized to H_2SO_4 when exposed to aerobic conditions. Oxidation of reduced S occurs when subsurface geological materials are brought to the earth's surface during mining activities or on draining tidal marshes. Under these conditions the solution pH level may be sustained as low as 2 or less until the reduced S is oxidized, neutralized, or otherwise removed from the soil/geologic environments.

Speciation of solution inorganic S is controlled by redox conditions (as measured by pe, e^- activity or oxidation–reduction condition) and pH (H^+ activity) relationships as shown in Figure 6.4. Sulfate is the dominant species found at high pe levels and all pH levels. At low to intermediate pe levels, sulfide species predominate, with H_2S at low pH (<7) and HS^- at high pH (>7) levels. Elemental S can also exist at moderate pe levels and acidic pH values. Some S-containing amino acids are also stable at pe + pH levels within the SO_4^{2-} range, thus suggesting organic S compounds may persist in aerobic soil environments.

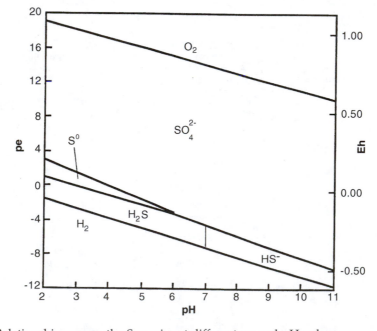

Figure 6.4   Relationship among the S species at different pe and pH values.

6.3.2   Organic sulfur in soils

Organic S is the dominant form of S in most soils and sediments. The distribution of S in several types of soils is listed in Table 6.4, which clearly shows that organic S is the predominant form of soil S. Generally, concentrations of inorganic S will only exceed organic S in low-organic-matter, calcareous soils typical of arid and semiarid regions. Further examination of the inorganic and organic S compounds can be obtained by S fractionation methods as discussed in the next section.

The content of organic S in soil and aquatic environments varies considerably, depending primarily on the amount of organic matter present. In soils, organic matter accumulates at the soil surface as plant materials decay; subsurface horizons can also accumulate organic S as shown in Table 6.5. Note in Table 6.5 that total and organic S in Soil 1 decreases with depth while inorganic S stays relatively constant, whereas in Soil 2 there is an accumulation of all three forms of S that occurs at the 30 to 60 cm depth. Soil 2 is typical of Spodosols (Podzols), which through translocation of organic matter into subsurface horizons have accumulated organic S. The increase in the inorganic S is related to increased SO_4^{2-} adsorption by Fe and Al hydrous oxides that also accumulate in the subsurface horizons of Spodosols.

Organic S compounds in soils and aquatic environments are derived from several plant, animal, and microbial sources. Sulfur-containing amino acids are generally the most prevalent of the known organic S compounds, accounting for 10 to 30% of the total S in various soils. These amino acids are readily decomposed and may only exist for short periods of time before they are utilized by microorganisms. In addition, S-containing amino acids are often bound to mineral and organic matter, making it difficult to extract amino acids from soils. Sulfur-containing polysaccharides and lipids are also present in soils and may comprise approximately 5% of the total S.

Table 6.4   Distribution of Inorganic and Organic Sulfur in Different Soils

Type of soil	Location	Total S (mg/kg)	Inorganic S (%)	Organic S (%)
Agriculture	Iowa[a]	78–452	1–3	97–99
	Brazil	43–398	5–23	77–95
	West Indies	110–510	2–10	90–98
	Alberta[b]	80–700	8–15	85–92
	Queensland[c]	11–725	2–18	82–98
	New South Wales[c]	38–545	4–13	87–96
	New Zealand	240–1360	2–9	91–98
Forest[d]	New Hampshire[a]	452–1563	1–8	92–99
	New York[a]	68–2003	1–18	82–99
	Alberta[b]	364–1593	2–10	90–98
	Illinois[a]	112–555	2–10	90–98
	Germany	74–328	7–28	72–93
Surface	Iowa[a]	55–618	1–5	95–99
Organic	England	7405	5	95
Acid	Scotland	300–800	2–10	90–98
Calcareous	Scotland	460–1790	21–89	11–79
Volcanic ash	Hawaii	180–2200	6–50	50–94

[a] U.S.

[b] Canada.

[c] Australia.

[d] Ranges represent variation in a single profile.

Sources: Freeney (1986) and Mitchell et al. (1992).

Table 6.5 Relationship of Total, Inorganic, and Organic Sulfur (mg/kg)
by Depth in Two Soil Profiles

Depth (cm)	Total S Soil 1	Total S Soil 2	Inorganic S Soil 1	Inorganic S Soil 2	Organic S Soil 1	Organic S Soil 2
0–10	436	205	7	8	429	197
10–20	389	93	6	6	383	88
20–30	342	118	6	7	336	111
30–45	266	134	6	18	260	116
45–60	230	134	8	29	222	105
60–75	188	115	7	22	181	93
75–90	162	99	5	10	147	89

Note: Soil 1 is a Mollisol with decreasing organic matter with depth, and Soil 2 is a Spodosol with organic matter, Fe, and Al accumulations in the B horizons.

Source: Williams, C. H., in *Handbook of Sulphur in Australian Agriculture*, K. D. McLachlan, Ed., CSIRO, Melbourne, Australia, 1974.

Municipal biosolids (sewage sludge) and animal manures contain several forms of organic S. Although these materials are perceived as waste products that require disposal, several benefits can be realized when utilizing biosolids and manures in a land application program (see Chapter 9 for more information on management of waste materials). In addition to the organic matter, N, and P contained in biosolids and manures, both inorganic and organic forms of S are generally part of these materials. For example, some biosolids have been found to contain elemental S and sulfides, and organic S such as S-containing amino acids, alkyl benzene sulfonates, sulfonated acidic polysaccharides, and ester sulfates (a significant portion of which is derived from detergents).

6.3.3 Sulfur fractionation scheme

Due to the heterogeneous nature of organic matter (see Chapter 3), soil and aquatic organic S compounds are generally characterized by fractionation methods. The S components identified in an S fractionation method are illustrated in Figure 6.5, which shows S can be separated into five basic fractions. Both inorganic S (SO_4^{2-} and nonsulfate S, primarily sulfides) and organic S (C-bonded and ester sulfate) that have been mentioned in the previous sections can be determined by these methods. Examples of the concentrations and proportions of the S fractions in municipal biosolids, soils, and lake sediments and waters are listed in Table 6.6. The S fractionation method can also be used to evaluate soil and aquatic S dynamics including transformations and flux rates.

The nature and content of soil S can be quite variable. Using the S fractionation method, total S is determined on a subsample of the soil. Water-soluble and phosphate-extractable S are also determined on subsamples, and the difference between the two is considered specifically adsorbed SO_4^{2-} (SO_4^{2-} adsorption will be discussed in Section 6.4.1). Inorganic nonsulfate S is evaluated directly by Zn-HCl digestion. Organic S is comprised of both ester sulfate (–C–O–S–) and C-bonded (–C–S–) forms with the former determined along with inorganic S through hydriodic acid (HI) digestion. Carbon-bonded S is obtained by difference (total S minus HI-digested S) because there is no method available that can accurately and completely account for all the C-bonded S in a sample.

The distribution of inorganic and organic S in several soils was shown in Table 6.4. Organic S, which consists of ester sulfate (Figure 6.6a) and C-bonded S (Figure 6.6b), is the dominant form of S in most soils. There appears to be a difference in the content of

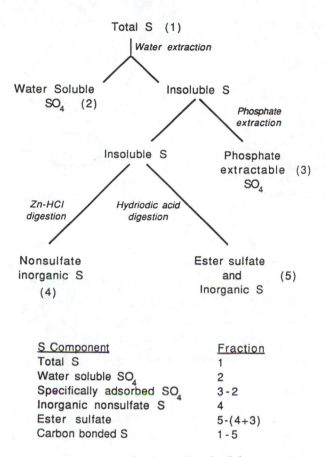

S Component	Fraction
Total S	1
Water soluble SO$_4$	2
Specifically adsorbed SO$_4$	3-2
Inorganic nonsulfate S	4
Ester sulfate	5-(4+3)
Carbon bonded S	1-5

Figure 6.5 Schematic representing the methods used in the S fractionation procedure.

Table 6.6 Average Sulfur Constituents (mg/kg) in Solid or Aqueous Samples from Different Ecosystems

Sample	Total S	Extractable SO$_4$	Inorganic nonsulfate S	Ester S	C-Bonded S
Municipal biosolids (aerobic digestion)	11,000	1,180 (11)	1,480 (13)	4,040 (37)	4,300 (39)
Soil (Oi horizon)	1,605	15 (1)	20 (1)	230 (14)	1,340 (84)
Soil (Bhs horizon)	530	23 (4)	22 (4)	115 (22)	370 (70)
Lake sediment	7,320	1,130 (15)	60 (1)	1,450 (20)	4,680 (64)
Lake water	2.05	1.7 (83)	0.00 (0)	0.22 (11)	0.13 (6)

Notes: Values in parentheses are the percent of total S in that S fraction. For lake water, values reported as µg/L.
Source: Landers, D. H. et al., *J. Environ. Anal. Chem.*, 14, 245, 1983.

organic S forms in agricultural vs. forested soils. The percentage of total S in agricultural soils that is composed of ester sulfates ranges between 35 to 60%, whereas in forested soils the range is lower, from 20 to 30%. The amount of C-bonded S in forested soils typically averages 50 to 80% of the total S. In agricultural soils, higher percentages of ester sulfates, relative to forest soils, have been attributed to the removal of C-bonded S with cropping.

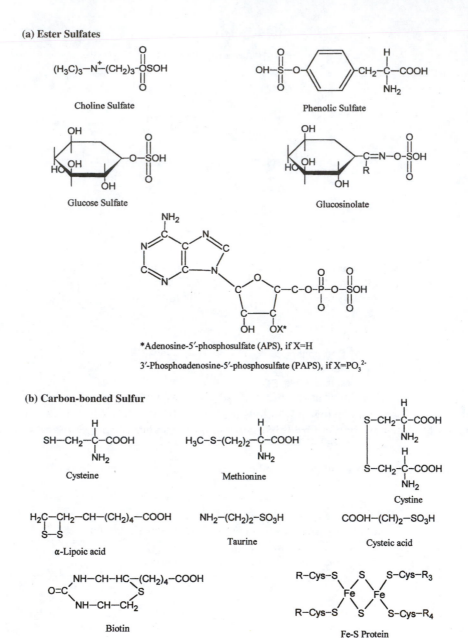

Figure 6.6 Different organic S compounds that have been identified in soils: (a) ester sulfate compounds and (b) C-bonded S compounds.

6.4 *Sulfur retention and transformations in soils*

Several processes that occur in soil and aquatic environments affect S movement and transformations. The most important processes are adsorption–desorption of SO_4^{2-}, inorganic S oxidation–reduction, and organic S transformations. It was pointed out in previous sections that organic S comprises the greatest proportion of total S in most soils. Organic S transformations, therefore, are of particular importance when considering soil S cycling and reactions.

6.4.1 Adsorption and desorption of soil sulfate

Adsorption of SO_4^{2-} by soils results in the retention of S that otherwise would have been leached from the soil profile. Because SO_4^{2-} is the major form of S that is taken up by plants, the retention of SO_4^{2-} may enhance the ability of a soil to supply S to plants in the future. The amount of water-soluble and exchangeable SO_4^{2-} can be low in some soils, and is dependent on factors such as soil type, climate, and management practices. The mineralogical composition of a soil can also have a significant impact on its ability to retain adsorbed SO_4^{2-}. For example, a Tennessee forest soil had an adsorbed SO_4^{2-} pool 15 times greater than the total S stored in the aboveground biomass.

Sites for SO_4^{2-} adsorption include Fe and Al hydrous oxides, edges of aluminosilicate clays, and possibly metal–organic matter complexes. In some soils, coatings of Fe and Al hydrous oxides on soil particles (e.g., spodic B horizons) contribute to the majority of SO_4^{2-} adsorption sites. Sulfate adsorption is expected to be greater on amorphous or poorly crystalline Fe and Al hydrous oxides because these materials have greater surface area than other crystalline minerals. The number of edge sites on aluminosilicate clays determines the amount of SO_4^{2-} adsorbed. Some studies have shown that 1:1 clays adsorb greater amounts of SO_4^{2-} than do 2:1 clays; SO_4^{2-} is also adsorbed by metal–organic complexes.

Sulfate adsorption is influenced by pH, amount and crystallinity of Fe and Al hydrous oxides, and the presence of organic matter. Maximum adsorption of SO_4^{2-} occurs at low pH, typically around pH 3 to 4.5, which is due to a net positive surface charge that develops on Fe and Al hydrous oxides. The increasing number of positively charged sites with decreasing pH enhances SO_4^{2-} adsorption; however, Fe and Al hydrous oxides are not stable at extremely low pH values and their dissolution will reduce the number of adsorption sites available. Organic matter can effectively hinder SO_4^{2-} adsorption by coating mineral and hydrous oxide surfaces or competing with SO_4^{2-} for adsorption sites (see Example Problem 6.2).

Desorption of SO_4^{2-} may occur if soils have received high S inputs, have a relatively low SO_4^{2-} adsorption capacity, and/or SO_4^{2-} is displaced by competing anions. The amount of readily desorbable SO_4^{2-} decreases with time, which has been attributed to the formation of more tightly bound (specifically adsorbed) SO_4^{2-} or a conversion of the SO_4^{2-} to organic forms. Desorption of SO_4^{2-} is enhanced by using a phosphate-extracting solution that readily displaces SO_4^{2-} present in specifically bound forms.

Sulfate adsorption can be analyzed by using Langmuir and Freundlich mathematical models (see Chapters 5 and 8 for a discussion of these models in P and organic chemical adsorption studies), and a recently proposed model called the initial mass (IM) isotherm. The IM isotherm was shown to fit forest soil SO_4^{2-} adsorption data better than Langmuir and Freundlich models, which were unable to provide a linear relationship or be transformed for use in the linearization process (Figure 6.7). The IM isotherm is written as

$$SO_4^{2-} \text{ (adsorbed or released)} = mX_i - b \qquad (6.4)$$

where

m	=	slope
X_i	=	initial amount of solution SO_4^{2-} with respect to the soil mass
b	=	intercept

Because native SO_4^{2-} levels are high in some soils, there are cases where SO_4^{2-} release occurs. The reserve SO_4^{2-} soil pool (RSP) can be calculated from the IM isotherm as

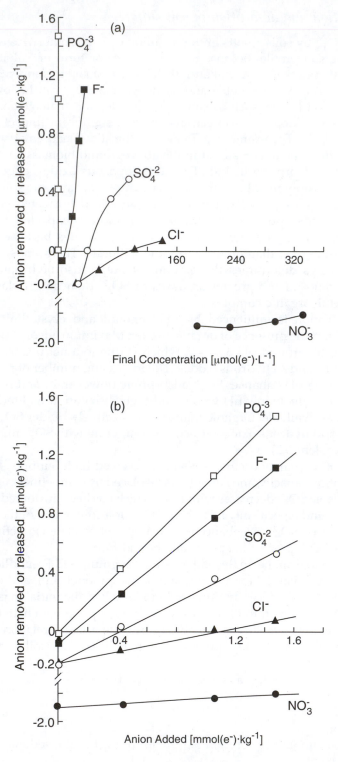

Figure 6.7 Examples of adsorption studies evaluating PO_4^{3-}, F^-, SO_4^{2-}, Cl^-, and NO_3^- retention by a sandy loam forest soil using (a) final solute concentration and (b) initial mass of anions added. Positive ordinate (*y*-axis) values indicate anion adsorption, whereas negative values represent anion release. (Adapted from Nodvin et al., *Soil Sci.*, 142, 27, 1986. With permission.)

$$RSP = \frac{b}{1 - m} \tag{6.5}$$

The RSP is a measure of the labile soil SO_4^{2-}, and is an estimate of biologically active SO_4^{2-} in a soil.

Example problem 6.2

Initial mass isotherms were evaluated in a study examining dissolved organic carbon (DOC) and SO_4^{2-} sorption by Spodosol mineral horizons from Maine. By using a Tunbridge Bh soil horizon, the hypothesis that DOC interferes with the sorption of SO_4^{2-} was tested by conducting SO_4^{2-} adsorption studies in the presence or absence of DOC. First, adsorption studies were performed with SO_4^{2-}-only solutions; then similar concentrations of SO_4^{2-} were prepared with 67 mg C/L as DOC in each solution (Table 6.7). Adsorption was calculated as follows:

$$(SO_4^{2-}{}_{initial} - SO_4^{2-}{}_{final}) \times (\text{volume of solution/soil wt}) \tag{6.6}$$
$$= \text{mass of } SO_4^{2-} \text{ adsorbed/mass of soil}$$

where $SO_4^{2-}{}_{initial}$ and $SO_4^{2-}{}_{final}$ represent the amount of SO_4^{2-} added and that which is determined in the soil solution after reacting for a specified time, respectively.

Table 6.7 Sulfate Adsorption by a Tunbridge Bh with and without DOC

Initial solution SO_4^{2-} (mg/kg)	Adsorbed SO_4^{2-} (mg/kg)	m	$-b$	RSP	r^2
Sulfate only adsorption study					
0	−32	0.48	32	62	0.998
60	−7				
120	24				
240	90				
480	195				
Sulfate plus DOC adsorption study					
0	−63	0.45	61	111	1.000
60	−34				
120	−5				
240	45				
480	153				

Source: Vance, G. F. and David, M. B., *Soil Sci.*, 154, 136, 1992.

Results of this study indicate DOC caused a reduction (i.e., lower b and higher RSP values) in SO_4^{2-} adsorption due to competition for the adsorption sites; note that the slope (m) was not greatly influenced by DOC addition (Figure 6.8). The greater affinity of DOC for the adsorption sites may have been due to anion exchange, specific adsorption, and/or physical interactions between the soil and DOC.

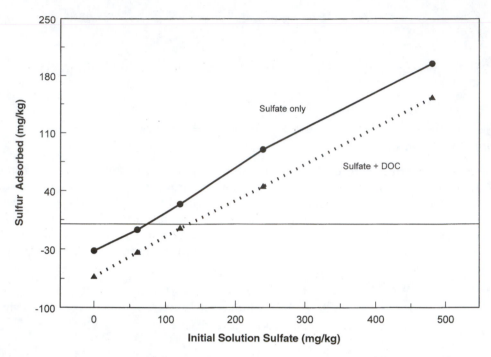

Figure 6.8 Sulfate adsorption isotherms illustrating the influence of DOC on the amount of SO_4^{2-} adsorbed by a Spodosol (Tunbridge) Bh horizon. (Data from Vance and David, 1992.)

6.4.2 Inorganic sulfur oxidation–reduction reactions

Oxidation–reduction (redox) reactions involving inorganic S are primarily mediated by soil microorganisms. The role of S redox reactions in soils were shown in Figure 6.3, which portrays several pathways available for the oxidation or reduction of S. As mentioned earlier, SO_4^{2-} is the dominant inorganic S form in aerobic soils. Sulfides are more important in anaerobic soils found in wetlands, swamps, and tidal marshes. Therefore, reduction reactions involving SO_4^{2-} and oxidation reactions involving S^{2-} are considered here.

Sulfate reduction occurs by assimilatory or dissimilatory reduction reactions. The assimilation reduction process occurs when microorganisms utilize SO_4^{2-} in the assimilation of cellular constituents. In converting inorganic S to organic S, the S is immobilized (see next section for a discussion of mineralization and immobilization processes). The dissimilatory reduction process is unique to a specific genera of bacteria (e.g., *Desulfovibrio*) that reduce SO_4^{2-} to sulfides. Dissimilatory SO_4^{2-} reduction occurs in anaerobic environments such as waterlogged soils, swamps, stagnant waters, and tidal marshes.

Conditions under which dissimilatory S-reducing microorganisms are capable of surviving are quite varied. They can live in soils having an extensive pH range and high salt content such as found in saline lakes, evaporation beds, deep-sea sediments, and oil wells. A low redox potential (E_h) of less than 100 mV (pe = 1.69) is required for their growth; under higher redox potentials, dissimilatory S-reducing microorganisms can persist in a dormant state. Further discussion of redox reactions is presented in Chapter 9.

Environmental problems associated with SO_4^{2-} reduction are varied. In municipal biosolid treatment systems, the production of H_2S leads to excessive corrosion of stone, concrete, and pumps as well as other components of the biosolid distribution system. Hydrogen sulfide also has an odor similar to that of rotten eggs. The presence of reduced forms of S in fossil fuels, sediments, and anaerobic soils can cause a number of problems, discussed in the next section, as S oxidizes back to SO_4^{2-}.

Oxidation of reduced inorganic S compounds occurs under aerobic conditions and follows several pathways. Although S oxidation is considered to be primarily biochemically driven, chemical oxidation of sulfides (S^{2-}), sulfites (SO_3^{2-}), and thiosulfates occurs readily in nature; however, the rate of chemical oxidation is slower than biochemical oxidation. Oxidation of reduced S is also an acidifying process, which is a cause for concern in certain situations. Examples of S oxidation that result in the formation of acidity are

$$H_2S + 2O_2 \rightleftharpoons H_2SO_4 \rightleftharpoons 2H^+ + SO_4^{2-} \tag{6.7}$$

$$2S + 3O_2 + 2H_2O \rightleftharpoons 2H_2SO_4 \rightleftharpoons 4H^+ + 2SO_4^{2-} \tag{6.8}$$

There are several types of autotrophic and heterotrophic microorganisms capable of oxidizing S. Bacteria of the *Thiobacillus* genus can survive within a broad range of soil and environmental conditions; some *Thiobacillus* species are capable of living in soils with pH levels ranging from 1.5 to 9. *Sulfolobus* species are another group of S-oxidizing organisms that live in S-rich geothermal hot springs. They survive in waters that range in pH from 2 to 5 and temperatures of 60 to 80°C. Another form of S-oxidizing microorganisms that survives in hot springs is the *Thermothrix*, but unlike *Sulfolobus*, *Thermothrix* grows in near-neutral pH waters.

In neutral and alkaline pH soils, the primary S oxidizers are often a group of heterotrophic microorganisms. The heterotrophic microorganisms of interest include several genus of bacteria (*Arthrobacter*, *Bacillus*, *Micrococcus*, *Mycobacterium*, and *Pseudomonas*), as well as some actinomycetes and fungi that are capable of oxidizing inorganic S compounds. These microorganisms are believed to oxidize inorganic S only as a consequence of other metabolic processes.

Factors that influence S oxidation in soils include soil type, pH, temperature, moisture, and organic matter. Oxidation of inorganic, reduced forms of S is an acidifying process that can result in acidic soils that are toxic to plants, animals, and microorganisms. Acid mine drainage occurs when reduced forms of S are oxidized during mining processes. Oxidation of pyrite, a common S-containing mineral found in reduced subsurface environments, will be discussed in Section 6.7.2.

6.4.3 Organic sulfur transformations

The transformation of organic S to inorganic S and volatile gases are processes mediated by soil microorganisms. Release of inorganic S during the decomposition of organic matter is important for supplying adequate S for plant growth. Soil factors that influence the growth and activity of microorganisms (e.g., pH, temperature, soil moisture, organic matter) will also affect the rate of organic S transformations.

Transformation of organic S to inorganic S, and of inorganic S to organic S, are processes that describe S mineralization and immobilization, respectively, as shown in the following equation:

$$\text{Organic S} \underset{\text{immobilization}}{\overset{\text{mineralization}}{\rightleftharpoons}} \text{Inorganic S} \tag{6.9}$$

During the mineralization of S, inorganic S — primarily SO_4^{2-} — is released as microorganisms decompose organic matter, utilizing some of the SO_4^{2-} to synthesize cell constituents and releasing some inorganic SO_4^{2-} to the soil solution. Immobilization of S is a process by which inorganic S is assimilated by microorganisms when low-S organic energy sources (i.e., organic matter) are added to soils. Mineralization–immobilization processes were also discussed for N and P in Chapters 4 and 5.

Rates of S mineralization are not proportional to the total amount of S in organic matter due to the variety of S-containing organic compounds in soils that have different decomposition rates, the type of plant and animal residues that affect mineralization–immobilization rates and release, and the formation of S-containing precipitates, e.g., $CaSO_4$, $Al_2(SO_4)_3$, that can influence the amount of plant-available S. The C:S ratio of these materials is extremely important because, in general, net mineralization occurs when the C:S ratio is less than 200:1, net immobilization occurs when the C:S ratio is greater than 400:1, and a steady state results when the C:S ratios are between 200:1 and 400:1.

6.5   Sulfur effects on plants

Sulfur is one of the secondary plant nutrients — S, Ca, magnesium (Mg) — but it has been considered by many as the fourth most important plant nutrient after N, P, and potassium (K). Sulfur deficiencies have become increasingly common as the use of fertilizers that contained S (e.g., normal superphosphate, ammonium sulfate) has been replaced by use of low-S fertilizers (e.g., triple superphosphate, ammonium phosphates); as S-containing pesticides have been converted to organic-based pesticides; and as atmospheric S inputs have been reduced — all of which have lowered the amount of S added to soils. Since the development of new high-grade fertilizers with low-S impurities, S deficiencies have become more evident in many parts of the world. Higher crop yields have also put a greater demand on soil S reservoirs to supply extra amounts of available S for increased plant growth. Reports of S deficiencies and increased crop yields due to S fertilization have been reported in several regions throughout the U.S. and in other parts of the world. Generally, S deficiencies are more likely in the Southern Hemisphere than the Northern Hemisphere, and are also a greater problem in the Tropics than in temperate environments.

Gaseous atmospheric S compounds may have a detrimental, beneficial, or neutral effect on plants. The concentrations of ambient gaseous S forms, SO_2 and H_2S, are low and can be beneficial; however, in the vicinity of high discharge areas, the concentration of these ambient gases may be harmful to plants. Plants absorb SO_2 primarily through the stomata and are capable of oxidizing the SO_2 to sulfites and sulfates. When plants are exposed to levels of SO_2 that exceed the ability of the plant to oxidize sulfite to sulfate, sulfite can accumulate in the plant to toxic levels. Visual toxicity symptoms on affected leaves can be mistaken for other types of injury; noting the close proximity of a SO_2-generating source will help in diagnosing whether S toxicity has occurred. Short-term exposure of plants to H_2S generally causes little, if any, damage; however, continuous exposure to low H_2S concentrations, such as in the air near geothermal sites, may affect plants by causing leaf lesions, defoliation, and decreased plant growth.

6.5.1   Sulfur in plant nutrition

Sulfur requirements vary with plant species, cultivars, and stages of development. Several S-containing compounds are present in plants, but the majority of the S is part of the amino acids cysteine, cystine, and methionine, which are found largely in proteins. Additional S compounds include vitamins, biotin, thiamine, B_1, and other coenzymes such as lipoic acid and coenzyme A (CoA). Sulfur deficiencies not only reduce yields due to improper plant nutrition, but also lower the quality of certain crop products (e.g., digestibility of forages and baking quality of flours).

Figure 6.9 illustrates the various S pathways that are involved in the synthesis of organic S compounds, starting with SO_4^{2-}. Once S (primarily SO_4^{2-}) enters the plant, it is activated through the formation of adenosine-5'-phosphate (APS); APS is a precursor to 3'-phospho-APS (PAPS), which is an intermediate in the formation of ester sulfates. The activated SO_4^{2-} in APS is reduced, and the reduced S is incorporated into either Fe-S proteins or as cysteine, which acts as a precursor to several organic S compounds (thiamin, lipoic acid, biotin, CoA, glutathione, and S-substituted cysteines). The majority of the two S-containing amino acids — cysteine and methionine — are incorporated into various proteins.

A number of important plant metabolic functions depend upon S-containing compounds. Methionine and cysteine are essential components to many plant enzymes and their structure and function are, in part, due to S interactions. The Fe-S protein — ferredoxin — is involved in redox reactions, most notably as a stable redox compound in the photosynthesis process. Nitrogen metabolism involving nitrate (NO_3^-) reduction, N_2 fixation, and NH_4^+ assimilation has been directly or indirectly related to S. Sulfur deficiencies can impair symbiotic N_2 fixation, resulting in plant N deficiencies.

Sulfur deficiency is usually manifested through visual symptoms of light green to yellow leaves that appear first along the veins of young leaves. Legumes are particularly susceptible to S deficiencies and are often diagnosed as having N-deficiency problems because of the similarity in the symptoms. Alfalfa and canola, because of their high S requirements, can be very susceptible to S deficiency. In corn, S deficiencies are generally noted by yellowing of the newer leaves, whereas N deficiencies show up on the older leaves. Sometimes emerging plants will show symptoms of S deficiency (e.g., more uniform chlorosis than that which is caused by N deficiencies), but these symptoms diminish as roots extend into the subsoil and absorb SO_4^{2-} that has been adsorbed by clays and metal oxides.

6.5.2 Availability index of sulfur

Several methods are available to assess the status of S in soils and plants. However, there are no universal techniques for evaluating S availability in soils, or diagnosing S deficiencies in plants. It is well known that total S in soils is not a reliable test of S availability because

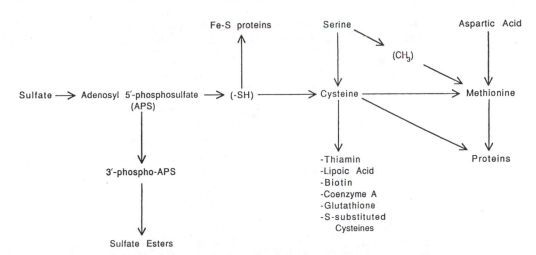

Figure 6.9 Pathways involving the transformation of SO_4^{2-} through intermediates to different S-containing organic compounds. The reduction of SO_4^{2-} in APS is denoted –SH. (Adapted from Thompson et al., 1986.)

much of the S exists in forms unavailable to plants. For soils, S availability is generally estimated by extracting the soil with water or solutions containing a salt, dilute acid, or phosphate (see Chapter 3). Depending on the nature of the extracting solution, information can be obtained on the soil content of (1) labile SO_4^{2-}; (2) labile and adsorbed SO_4^{2-}; or (3) labile and adsorbed SO_4^{2-} and portions of organic S. In addition to extraction methods, incubation techniques and microbial assays have also been used to estimate plant-available S.

Some commonly used techniques for diagnosing S deficiencies in plants include total S, SO_4-S, ratio of SO_4-S to total S, N:S ratio, and the Diagnosis and Recommendation Integrated System (DRIS). Total S levels in plants are generally an indication of soil S availability because plants acquire most of their S from soil solution SO_4^{2-}. However, several factors must be considered when interpreting plant total-S data because this varies with plant parts, age of plant and tissue, and interactions with other elements. Low, sufficient, and high total-S levels for a number of forages and agronomic crops are listed in Table 6.8. The amount of plant SO_4-S can be used to estimate when S is deficient or in excess. A low plant SO_4-S content would indicate SO_4^{2-} has been incorporated into organic S compounds, whereas high SO_4-S contents suggest the plant is accumulating S. The ratio of SO_4-S to total S is relatively independent of plant growth stage and may provide a better estimate of plant S requirements; the critical ratio is about 0.1. Total N:S ratios have also been used to estimate the status of plant S, but the ratio by itself does not indicate whether N or S is high or low.

In the DRIS method, several comparisons are made among indexes for different elemental ratios against established norm values. Three advantages to using the DRIS method are (1) analyses are independent of plant age and tissue; (2) nutrients are ranked in order of the most to the least limiting; and (3) nutrient balance is stressed. The DRIS method is a holistic approach to plant diagnostics, and the information derived from this method can be used to evaluate deficiencies of most of the macronutrients and some of the micronutrients. Further discussion on the DRIS systems can be found in Walworth and Sumner (1987), Westerman (1990), and Jones et al. (1991).

6.6   Management of sulfur in agriculture

As with N, P, K, Ca, Mg, and the micronutrients, S must also be managed to achieve a proper balance between deficiency and excess. Typically, plant S needs are similar to P

Table 6.8   Total Sulfur Contents in Various Forages and Agronomic Crops of Varying Levels of Sulfur Nutrition

Crops	S (%)		
	Low	Sufficient	High
Alfalfa	<0.25	0.25–0.50	>0.50
Barley	<0.15	0.15–0.40	>0.40
Clover, white	<0.25	0.25–0.50	>0.40
Corn	<0.25	0.25–0.80	>0.80
Cotton	<0.25	0.25–0.80	>0.80
Grass, brome	<0.17	0.17–0.30	>0.30
Oat	<0.15	0.15–0.40	>0.40
Peanuts	<0.20	0.20–0.35	>0.35
Soybeans	<0.20	0.20–0.40	>0.40
Sugarcane	<0.14	>0.14	—
Wheat, spring	<0.15	0.15–0.40	>0.40

Source: Jones, J. B., Jr. et al., *Plant Analysis Handbook: A Practical Sampling, Preparation, Analysis, and Interpretation Guide*, Micro-Macro Publishing, Athens, GA, 1991.

requirements, with average yields for grasses and cereal grains removing less than 15 kg S/ha, forage crops, sugar beets, cabbage, and cotton 15 to 35 kg S/ha, while some plants remove as much as 90 kg S/ha. A management program must provide sufficient amounts of all plant nutrients in order to obtain economic yields while giving due consideration to potential environmental impacts. Unlike N and P, the agricultural role in S pollution of air and water is minimal, or nonexistent. Therefore, management of S in agriculture focuses more on supplying adequate amounts of S for crop production than on S impacts on the environment. North American soils generally contain readily soluble SO_4^{2-} levels of approximately 20 mg/L in the soil solution, which usually is sufficient for most crops. When readily soluble SO_4^{2-} concentrations are less than 10 mg/L, S supplementation may be needed for proper plant growth. Coarse-textured, S-deficient soils will have SO_4^{2-} concentrations that are often less than 5 mg/L.

Example problem 6.3

A 10-g soil sample is extracted with 0.025 L of calcium phosphate $(Ca(H_2PO_4)_2 \cdot H_2O)$ by shaking for 2 h at room temperature. The sample is filtered (Whatman No. 42 filter paper) and the supernatant (filtered solution) analyzed for SO_4-S. Results suggest there are 2 mg SO_4-S/L in the supernatant solution. Does this soil require additional S to produce an average yield of cotton (20 kg/ha)? An approximate relationship that is commonly used in determining requirements on a kilogram per hectare basis is to assume that a hectare furrow slice (18 cm depth) is about 2,250,000 kg/ha, assuming a soil density of 1.3 g/cm³. From our analysis and the hectare furrow slice relationship, the amount of S needed is calculated as

$$\frac{2 \text{ mg } SO_4\text{-S}}{\text{L solution}} \times \frac{0.025 \text{ L solution}}{0.010 \text{ kg soil}} = \frac{5 \text{ mg } SO_4\text{-S}}{\text{kg soil}} \qquad (6.10)$$

$$\frac{5 \text{ mg } SO_4\text{-S}}{\text{kg soil}} \times \frac{2,250,000 \text{ kg soil}}{\text{ha}} = \frac{11,250,000 \text{ mg } SO_4\text{-S}}{\text{ha}} \quad \text{or} \quad \frac{11.2 \text{ kg S}}{\text{ha}} \qquad (6.11)$$

Based on these results, an additional 8.8 kg S/ha would be required.

Management of S for agricultural purposes is dependent on the properties of S and the factors that govern its availability. Figure 6.10 describes the various S inputs and S outputs that are generally part of an agricultural production setting. Inputs of S include crop residues and animal manures, S in rainfall, fertilizers, mineral weathering, and organic matter mineralization to enhance the soil S supply. Outputs or losses of S include immobilization, volatilization, crop removal, leaching, and erosion. Preventing erosion, returning crop residues to the soil, and adding other sources of S (i.e., municipal bio-solids and composts) will increase the amount of S in soils, in addition to supplying other essential plant nutrients.

No universal management plan can be designed for S because each situation warrants a close examination of the role S plays in both economic and environmental issues and controls. For example, considering our current state of affairs with regard to acidic deposition, and its immense publicity, it is not surprising that any problem potentially associated with acid rain is going to warrant government action. The recent enactment of the Clean Air Act (1990) has specifically designed regulations that call for a reduction in the

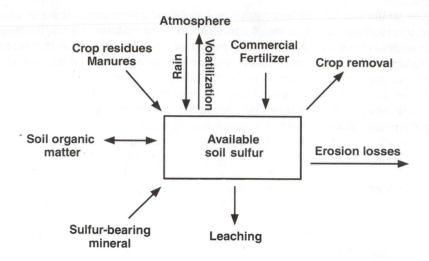

Figure 6.10   Example of S inputs and outputs in an agricultural ecosystem.

amount of S that can be released to the atmosphere. Although acidic deposition has been shown to be detrimental, especially within areas surrounding the S-emitting source, atmospheric S contributions to crop S requirements have been a clear benefit to areas with lower-S soils.

6.6.1 Management of sulfur in cultivated cropping systems

Cultivation results in an initial mineralization of organic matter, releasing nutrients such as N, P, and S. Loss of S, as well as C, N, and P, is related to the intensity of cultivation, soil properties, and cropping practices, but usually ranges from 20 to 40% total S compared with uncultivated fields. As an approximate measure, about 1 to 3% of the soil organic matter will be mineralized each growing season. The amount of SO_4^{2-} released in the mineralization processes will depend on the nature of the organic matter and microbial assimilation. Predicting S needs is therefore difficult because most of the S in soils exists in organic forms that must undergo mineralization for the S to be released. Factors that affect the amount of plant-available S include temperature, pH, and moisture, all of which directly influence the growth and activity of microorganisms. If crop production uses less S than the amount of S mineralized, a net annual release of nutrients in excess of plant needs could result in a potential loss of S through erosion or leaching, and decrease availability of S to future crops. Therefore, it is important to use management practices that balance S needs.

There are many types of S fertilizers, several of which contain other essential nutrients. Examples include elemental S, K_2SO_4, $(NH_4)_2S_2O_3$, $CaSO_4 \cdot 2H_2O$, and $MgSO_4 \cdot 7H_2O$. The choice of one S-containing fertilizer over another should be determined based on cost when comparing fertilizers on an equivalent S basis. Factors that should be taken into consideration in S management programs include type of fertilizer S, timing of application, method of application, and placement. Both the physical and chemical characteristics of the fertilizer and soil will determine the best method and timing of application. Municipal biosolids and animal manure can also be used as a source of S. For example, the S content (dry weight basis) of biosolids ranges from 0.6 to 1.5%.

Criteria that can be used to determine the economic benefits of S fertilizer use include economic optimum S fertilizer rate, minimum S fertilizer rate, S fertilizer rate to maximize return on investment, minimum cost per unit of yield, and discounting future returns. All but the last criterion are based on present-year conditions such as cost of S fertilizer and additional expenses, expected yield, and anticipated return on investment. Discounting for future returns relies on additional benefits due to residual effects. These residuals may include S carryover, mineralization of other nutrients, or prevention of S deficiencies in perennial crops. In a long-term S management program, the latter strategy may be the most beneficial; however, on a short-term basis, the other economic criteria may result in increased profit margins.

6.6.2 Management of sulfur in grazed systems

Grazed pastures and rangelands are different from cultivated croplands because most of the S is cycled through plants and returned to the soil as plant residues or animal excreta. The major S loss in grazed systems occurs when the animals are removed from the pasture or rangelands, or by fire, erosion, and disturbance events. Grazing animals can influence their environment by foraging, soil surface deformation (e.g., hoof action), selective eating habits, spreading undesirable weed species, and excreta impacts. Grazing can also reduce plant photosynthetic area, which results in low C assimilation and translocation to plant roots. Plant residue S is often retained in immobilized soil organic matter, with as much as 80% or more of S released from plant litter incorporated into soil organic S. Animals, however, enhance the process of plant decomposition and reduce the turnover rate of S cycled from soil through plants back to the soil. Much of the excreta, and in particular urine, provides a plant-available source of SO_4^{2-}.

Domestic animals such as sheep and cattle produce S-containing fecal excreta that is proportional to the dry matter consumed; approximately 1 g S is excreted through fecal material per 100 g of dry matter consumed. The amount of S in urine varies based on the grazing environment, ranging from 6 to 90% with a value around 50 to 60% for pastures that contain sufficient S levels. Characterization of S forms in sheep indicated fecal matter contained 87 to 94% C-bonded S (C–S), 4 to 5% ester-SO_4^{2-}, and 1 to 4% SO_4^{2-}, whereas urine contents are 10 to 14% C–S, 5 to 8% ester-SO_4^{2-}, and 80 to 85% SO_4^{2-}. Enhanced rates of S cycling are primarily attributed to urine due to high SO_4^{2-} levels that are more readily utilized by plants and microorganisms.

Application of S to grazed ecosystems is needed if natural S sources such as atmospheric deposition, mineral weathering, or irrigation water, if used, are insufficient for plant growth requirements. Arid or semiarid environments with soils that are coarse textured, have low soil S reserves and have been intensively grazed are particularly susceptible to S deficiencies. Sulfur requirements are characterized by local soil and climatic conditions as well as land-use practices. Added S is usually incorporated into organic matter with most of the S converted into C–S forms. Irrigation waters in some parts of the world may add sufficient levels of S to pastures and rangelands to support plant growth.

6.7 Problem soils and surface waters

Salt-affected soils can have a wide pH range (Figure 6.11) and vary from acid-sulfate soils at low pH to Na-affected soils at high pH. Low-pH soils result from biotic and abiotic reactions involving reduced-S compounds. Gypsiferous soils (e.g., sodic soils) contain $CaSO_4$, have low to high pH values ranging from 4 to 9, and are generally found in arid and semiarid environments. High-pH soils are the consequence of Na^+ salts contributing to alkaline hydrolysis reactions that result in poor soil physical conditions,

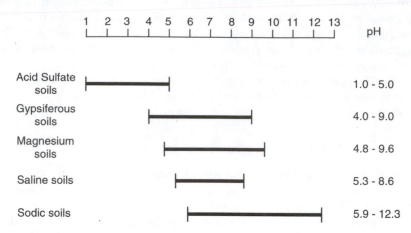

Figure 6.11   Range in pH values for different salt-affected soils. (Redrawn from Szabolcs, 1989.)

inhibition to plant growth, and often inadequate water availability. Sulfur-containing minerals and compounds are important in the formation and reclamation of the low- and high-pH soils, respectively. In addition, the oxidation and mobilization of products from reduced-S-containing soils and mining materials can have a destructive impact on surface water quality.

6.7.1   Acid soils and mine spoils

When material containing reduced-S compounds is exposed to oxygen, high concentrations of H_2SO_4 can develop and form acid soils. Two examples where this might occur are: (1) draining of coastal plains or tidal marsh sediments containing pyritic materials and (2) exposing pyritic overburden materials to the atmosphere during surface mining activities. Reclamation of these areas requires the neutralization of both current (e.g., active, exchangeable, and nonexchangeable forms) and potential acidities that result when oxidation of reduced S occurs. Neutralization of the potential acidity is an important part of a reclamation program because over time reduced-S minerals will oxidize and contribute to the current acidity in soils. In most cases, the neutralization of potential acidity requires greater reclamation efforts to neutralize than active acidity. Acid-sulfate soils, commonly referred to as cat clays, develop in drained coastal floodplains located mainly in temperate and tropical regions. If drained, the reduced-S compounds in these soils can oxidize to produce H_2SO_4, which in turn can dissolve other minerals. Both acid sulfate soils and mine spoils can contain large amounts of pyrite (FeS_2) that, upon oxidation, can result in soil pH levels below 2.

Neutralization of some of the acidity produced during the oxidation of reduced-S compounds occurs when silicate minerals dissolve; however, during this process high levels of potentially toxic metals such as Al, Cd, Cu, Fe, Mn, Ni, Pb, and Zn may be released. Micronutrient deficiencies involving molybdenum (Mo) and boron (B) may occur with some vegetation due to the low solubility of these nutrients in acid soils. The survival and function of certain soil microorganisms (e.g., rhizobia, mycorrhizae, and others) may be impacted due to acidity factors. Reclamation of acid soils and mine spoils requires addition of liming materials to neutralize all current and future acidity and the consideration of acid-tolerant crops. Soil tests should be done every 1 to 2 years, and the use of finely ground liming materials is recommended for faster acid neutralization. Reclamation efforts can turn unproductive soils into lands useful for crop production or rangeland use, in addition to alleviating a potential source of environmental pollution.

6.7.2 Acid mine drainage

Acid mine drainage occurs when certain sulfide minerals are exposed to oxidizing conditions and the resulting products are transported to surface waters such as streams, rivers, and lakes. Much of the AMD worldwide is associated with surface and underground mining activities, but acid drainage can occur under natural conditions or where sulfides in geologic materials are exposed by surface disturbances and subsurface excavations. The product of pyrite oxidation is an acidic, Fe- and SO_4^{2-}-rich water. Detailed information on AMD control and treatment is provided by Skousen and Ziemkiewicz (1996) and Brady et al. (1998).

The primary iron sulfides that produce AMD are pyrite and marcasite (FeS_2), with secondary sources being chalcopyrite ($CuFeS_2$), covellite (CuS), and arsenopyrite ($FeAsS$) (Table 6.9). Acidity levels, metal composition, and metal concentrations in AMD depend on the type and amount of sulfide mineral and the presence or absence of alkaline materials. In areas where sulfides are present, the oxidation of Fe disulfides and subsequent conversion to acid occur through several reactions. The following chemical equations describe the processes involved in the oxidation and acidification reactions associated with reduced S transformations.

$$FeS_2 + \frac{7}{2}O_2 + H_2O \rightleftharpoons Fe^{2+} + 2SO_4^{2-} + 2H^+ \qquad (6.12a)$$

$$Fe^{2+} + \frac{1}{4}O_2 + H^+ \rightleftharpoons Fe^{3+} + \frac{1}{2}H_2O \qquad (6.12b)$$

$$Fe^{3+} + 3H_2O \rightleftharpoons Fe(OH)_3 + 3H^+ \qquad (6.12c)$$

$$FeS_2 + 14Fe^{3+} + 8H_2O \rightleftharpoons 15Fe^{2+} + 2SO_4^{2-} + 16H^+ \qquad (6.12d)$$

Initially, Fe sulfide is solubilized (Equation 6.12a) and ferrous iron (Fe^{2+}, reduced Fe), sulfate (SO_4^{2-}, oxidized S), and acid are formed. Ferrous iron is then oxidized (Equation 6.12b) to form ferric iron (Fe^{3+}, oxidized Fe). Ferric iron may be hydrolyzed and form ferric hydroxide, $Fe(OH)_3$, and H^+ acidity (Equation 6.12c), or directly attack pyrite and act as a catalyst in generating more Fe^{2+}, SO_4^{2-}, and acidity (Equation 6.12d). Under many conditions, Equation 6.12b is the rate-limiting step in pyrite oxidation due to the slow conversion of Fe^{2+} to Fe^{3+} at pH values below 5 under abiotic conditions. However, Fe-oxidizing bacteria, principally *Thiobacillus*, greatly accelerate the oxidation of Fe^{2+}; bacterial activity is believed to be crucial for generation of most AMD.

Table 6.9 Important Metal Sulfides in Geologic Subsurface Environments

FeS_2	–	Pyrite	MoS_2	–	Molybdenite
FeS_2	–	Marcasite	NiS	–	Millerite
Fe_xS_x	–	Pyrrhotite	PbS	–	Galena
Cu_2S	–	Chalcocite	ZnS	–	Sphalerite
CuS	–	Covellite	$FeAsS$	–	Arsenopyrite
$CuFeS_2$	–	Chalcopyrite			

If any of the reactions described above are limited, the generation of AMD would be slowed or possibly cease altogether. Removal of O_2 and/or H_2O from the system (two of the three principal reactants) would suppress pyrite oxidation. Almost complete absence of O_2 occurs in nature where pyrite forms; under these conditions, the pyrite remains almost completely unreacted. The rate of pyrite oxidation depends on numerous variables, such as reactive surface area, form of pyritic S, O_2 concentrations, solution pH, catalytic agents, flushing frequencies, and presence of *Thiobacillus* bacteria. The possibility of identifying and quantifying the effects of these and other controlling factors with all the various rock types in a field setting is unlikely. However, when pyritic material is fractured and exposed to oxidizing conditions, such as during mining operations or other major land disturbances, pyrite solubilization and leaching can result in reaction products (Fe and other metals, SO_4^{2-}, and acid) migrating into groundwater and surface water sources.

The drainage quality (acid or alkaline) emanating from underground mines or backfills of surface mines is dependent on the acid (sulfide) and alkaline (carbonate) minerals contained in the disturbed geologic material. Alkaline mine drainage is water that has a pH of 6.0 or above, contains alkalinity, but may still have dissolved metals that can create acidity by reactions described in Equation 6.12b and c. In general, sulfide-rich and carbonate-poor materials are expected to produce acidic drainage. In contrast, alkaline-rich materials, even with significant sulfide concentrations, often produce alkaline drainage waters.

The natural base content (alkali and alkaline earth cations, commonly present as carbonates or exchangeable cations on clays) of overburden materials is important in evaluating the future neutralization potential (NP) of the materials. The amount of alkaline material in unweathered overburden may be sufficient to overcome the acid-producing potential of reduced-S materials. Of the many types of alkaline compounds present in rocks, carbonates (specifically calcite ($CaCO_3$) and dolomite ($CaMg(CO_3)_2$)) are the primary alkaline compounds that occur in sufficient quantities to be effective impediments to AMD generation. Higher alkalinities also help control bacteria and restrict the solubility of ferric iron, which are both known to accelerate AMD generation.

6.7.3   Sodic soils

Soils of the arid and semiarid regions of the world can accumulate appreciable amounts of S salts. However, sodic soils have an exchangeable sodium percentage (ESP) of 15% or greater and low soluble salt concentrations (<4 dS/m) that results in a greater influence by Na ions. The equation that is used to calculate ESP is as follows:

$$ESP = \frac{[\text{exchangeable Na}]}{CEC} \times 100 \qquad (6.13)$$

where CEC = cation exchange capacity.

Under these conditions the soil generally has a pH of 8.5 to 10 and excess Na causes dispersion of clays, a reduction in soil pore diameter, destruction of soil structure, and decreased soil water permeability. In addition to the major influence of Na on soil physical properties, Na can also be toxic to certain plants when other cation concentrations are low.

Reclamation of sodic soils can be expensive if there is little or no water infiltration and permeability is very slow. The exchangeable Na must be replaced by other cations, preferably Ca, which is divalent and readily replaces Na. Gypsum ($CaSO_4$) is very soluble and once dissolved can supply the necessary Ca needed to displace Na. An application of several Mg/ha of gypsum may be required to supply enough Ca. The amount of gypsum required (GR) for reclamation can be approximated from the ESP and CEC of the sodic soil.

Example problem 6.4*

If a sodic-affected area has an ESP of 25% and a CEC of 20 cmol/kg, and the goal is to lower the ESP to 5%, the gypsum requirement (GR) would be calculated as follows:

$$GR = (Na_x) \cdot 4.5 \text{ Mg/ha to a depth of 30 cm} \qquad (6.14)$$

where Na_x represents the cmol Na/kg to be replaced (i.e., 20% of 20 cmol/kg). The 4.5 value is a constant that takes into consideration the equivalent weights for Na and Ca exchange, an assumed bulk density of 1.35 g/cm^3, and a 70% efficiency factor. Therefore, the GR for the example given is determined as

$$GR = 4 \times 4.5 = 18 \text{ Mg/ha} \qquad (6.15)$$

A total of 18 Mg of gypsum/ha will be required to replace the 4 cmol Na/kg. For best results, the gypsum should be incorporated into the surface and subsurface soil and adequate water, if available, should be applied after gypsum incorporation to leach out the Na that is displaced by Ca ions.

Materials containing reduced-S forms can also be used to reclaim sodic soils. Upon incorporation into sodic soils, reduced S will readily oxidize and produce sulfuric acid that lowers soil pH. Amounts of different S-containing materials required to reclaim sodic soils, using gypsum as the standard, are 0.18 for elemental S, 0.57 for sulfuric acid, 0.75 for lime-sulfur, and 1.62 for iron sulfate. Thus, for every Mg of gypsum calculated from the GR equation, 0.18 Mg of elemental S or 1.62 Mg of iron sulfate would be needed.

For salt-affected croplands that utilize irrigation, a leaching requirement or leaching fraction needs to be determined to control salt and Na problems. The leaching requirement is determined based on the amount of water needed to produce sufficient drainage, and is related to the salinity of both the irrigation and drainage waters. The leaching requirement (LR) is calculated as:

$$LR = \frac{\text{volume of drainage water}}{\text{volume of surface-applied water}} = \frac{\text{irrigation water salinity}}{\text{drainage water salinity}} \qquad (6.16)$$

An acceptable salinity level is based on irrigation water costs, crop product value, and the sensitivity of the crop to salt. Salt-tolerant crops do not require as much leaching and can survive under higher salt conditions. For example, halophytes are plants that have adapted to salt-affected conditions; however, halophytes are normally nonagricultural crop species.

---

* Based on Donahue et al. (1983).

Environmental quality issues/events
Acid mine drainage in the Appalachian coal region: extent and controls

In 1995, the EPA compiled a database of Maryland, Ohio, Pennsylvania, Virginia, and West Virginia streams with fisheries impacted by AMD (Figure 6.12). The EPA database defined two levels of stream impacts: (1) streams with severe impacts were characterized as having "no fish," and (2) streams with less severe impacts were described as containing "some fish" but where AMD had reduced the number of species or reduced their productivity. The EPA study indicated that over 8230 km of streams were impacted by AMD in the five-state area, with Pennsylvania containing over 60%, or 5212 km, of the streams listed in the survey.

Most of the AMD in the Appalachian coal region (estimates are from 80 to 90%) comes from abandoned mines where no individual or company is responsible for the poor quality drainage. Therefore, these streams will remain impacted by AMD unless some state or federal agency, or a watershed organization, takes measures to improve the stream quality. While federal and state agencies or watershed organizations cannot afford chemical treatment (see below), passive treatment systems and land reclamation activities can reduce the acidity and metal load into streams from abandoned surface mines or underground mines. Significant reductions in acid loads to streams were found after land reclamation, and the acid load reductions were due both to reductions in water flow from the site and reductions in acid concentration in the water.

Passive systems, including wetlands, anoxic limestone drains, vertical flow wetlands (SAPS), and open limestone channels, have been constructed on many sites and have demonstrated measurable decreases in acid load to streams. These systems are low-maintenance and do not require continual addition of chemicals; they remove metals from AMD by oxidation and precipitation processes, microbial reduction reactions, and adsorption–exchange reactions. Longevity of system effectiveness is not entirely known for all systems, but passive treatment is an option for many stream restoration projects. On current mining operations, operators must mine in a responsible manner to restrict the production of AMD on their site. If AMD is produced during or after mining, they must develop a system for treating the water to be in compliance with water quality discharge limits.

In the Appalachian coal region, recommended reclamation procedures have included segregating and placing acid-producing materials above the water table and then treating, compacting, and covering the materials to reduce surface water infiltration. Another possible alternative is to remine areas where AMD comes from underground mines. An evaluation of ten remined sites in Pennsylvania and West Virginia showed that all were reclaimed to current standards, thereby eliminating mine portals and previously constructed highwalls, covering refuse, and revegetating the entire area. All sites also had improved water quality and some completely eliminated on-site AMD. At a surface remining operation of an underground mine in Preston County, WV, alkaline overburden (15,000 Mg/ha) from an adjacent surface mine was used for neutralizing acid materials on the remined site. Water quality from an acid-producing deep mine prior to remining had an average pH of 3.7 and 75 mg/L acidity as $CaCO_3$, but after remining and reclamation, the pH was greater than 7.0 with no acidity.

Costs have been developed for five AMD treatment chemicals under four sets of flow (gpm) and acidity concentration (mg/L as $CaCO_3$) conditions in an effort to help coal

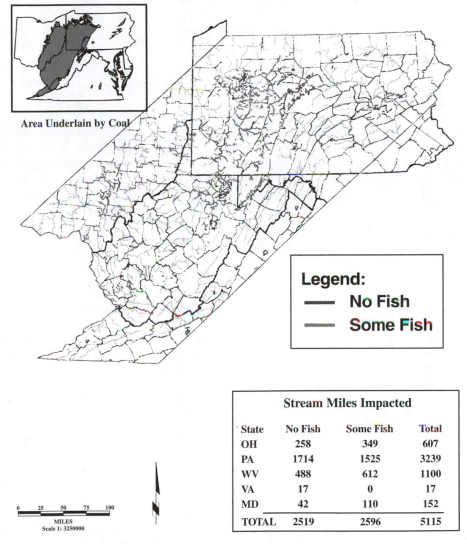

Streams with Fisheries Impacted
by Acid Mine Drainage
in MD, OH, PA, VA, WV
(Based on EPA Fisheries Survey - 1995)

Area Underlain by Coal

Legend:
— No Fish
— Some Fish

Stream Miles Impacted			
State	No Fish	Some Fish	Total
OH	258	349	607
PA	1714	1525	3239
WV	488	612	1100
VA	17	0	17
MD	42	110	152
TOTAL	2519	2596	5115

0   25   50   75   100

MILES
Scale 1: 3250000

Figure 6.12   Fish habitats impacted by AMD within Maryland, Ohio, Pennsylvania, Virginia, and West Virginia. (Information based on a 1995 EPA fisheries survey).

operators choose the most appropriate chemical for treating AMD (Table 6.10). These conditions are: (1) 189 L/min flow and 100 mg/L acidity; (2) 3780 L/min flow and 100 mg/L acidity; (3) 945 L/min flow and 500 mg/L acidity; (4) 3780 L/min flow and 2500 mg/L acidity, which represent a sufficiently wide range for valid comparison for treatment systems. The net present value (NPV) is the value of the total treatment system plus annual operating and chemical expenses over the specified duration of treatment. A rate of 6% per year was

Table 6.10   Costs in 1996 of Five Chemicals to Treat AMD in West Virginia

Chemical	Flow and acidity conditions			
	189[a] 100[b]	3780[a] 100[b]	945[a] 500[b]	3780[a] 2500[b]
Soda ash				
Reagent costs	$3,731	$44,000	$58,300	$1,166,000
Repair costs	0	0	0	0
Annual labor	14,040	14,040	14,040	14,040
Installation costs	229	229	229	229
Salvage value	0	0	0	0
Net present value	75,052	244,679	245,774	4,911,804
Annualized cost	**$17,817**	**$58,086**	**$58,346**	**$1,166,046**
Ammonia				
Reagent costs	$2,543	$22,440	$28,050	$561,000
Repair costs	495	495	495	495
Tank rental	480	1,200	1,200	1,200
Annual labor	7,020	7,020	7,020	7,020
Electricity	600	600	600	600
Installation costs	1,936	6,357	6,357	6,357
Salvage value	0	0	0	0
Net present value	48,547	139,117	162,749	2,407,725
Annualized cost	**$11,525**	**$33,026**	**$38,636**	**$571,586**
Caustic soda (20% liquid)				
Reagent costs	$5,174	$79,341	$99,176	$1,983,520
Repair costs	0	0	0	0
Annual labor	7,020	7,020	7,020	7,020
Installation costs	283	5,478	5,478	5,478
Salvage value	0	0	0	0
Net present value	51,601	368,398	451,950	8,389,433
Annualized cost	**$12,250**	**$87,457**	**$107,292**	**$1,991,636**
Pebble quicklime				
Reagent costs	$1,478	$9,856	$12,320	$246,400
Repair costs	500	2,500	2,500	10,000
Annual labor	6,500	11,200	11,200	11,200
Electricity	0	0	0	0
Installation costs	16,000	80,000	80,000	120,000
Salvage value	0	5,000	5,000	20,000
Net present value	49,192	162,412	172,790	1,127,220
Annualized cost	**$11,678**	**$38,556**	**$41,020**	**$267,600**
Hydrated lime				
Reagent costs	$814	$9,768	$12,210	$244,200
Repair costs	1,000	3,100	3,500	10,500
Annual labor	6,500	11,232	11,232	11,232
Electricity	3,500	11,000	11,000	11,000
Installation costs	58,400	102,000	106,000	200,000
Salvage value	5,750	6,500	7,500	25,000
Net present value	94,120	228,310	242,809	1,313,970
Annualized cost	**$22,344**	**$54,200**	**$57,642**	**$311,932**

Note: Prices for the reagents, equipment, and labor are based on actual costs to mining operators in West Virginia. The analysis is based on a 5-year operation period and includes chemical reagent costs, installation and maintenance of equipment, and annual operating costs.

[a] Flow, L/min.

[b] Acidity, mg/L as $CaCO_3$.

used to devalue the dollar during future years of the treatment period. The annualized cost was obtained by converting the total system cost (NPV) to an equivalent annual cost so that each system could be compared equally on an annual basis. The information suggests soda ash has the highest labor requirements (10 h/week) because the dispensers must be filled by hand and inspected frequently. Caustic soda has the highest reagent cost per unit (i.e., moles) of acid-neutralizing capacity and soda ash has the second highest; however, soda ash is much less efficient in treating water than caustic soda. Hydrated lime treatment systems have the highest installation costs of the five technologies because of the need to construct a lime treatment plant and the need for installing a pond aerator, although the cost of hydrated lime is low. The combination of high installation costs and low reagent cost make hydrated lime systems particularly appropriate for long-term treatment of high-flow (>375 L/min), high-acidity situations.

For a 5-year treatment period, ammonia had the lowest annualized costs for the low-flow/low-acid situation. Pebble quicklime is similar to ammonia in cost, and caustic is third. Soda ash is fourth because of its high labor and reagent costs, and hydrated lime is last because of its high installation costs. With the intermediate flow and acid cases, NH_3 is the most cost effective, with pebble quicklime second. Hydrated lime and soda ash were next. Caustic soda is the most expensive alternative with these intermediate conditions. In the highest flow/acidity category, pebble quicklime and hydrated lime are clearly the least costly treatment systems, with an annualized cost of $260,000 less than NH_3, the next best alternative. The use of soda ash and caustic soda is prohibitively expensive at high flow and high acidity.

(Based on information supplied by J.G. Skousen, West Virginia University)

Problems

6.1 Describe the potential environmental S impacts that could be expected to be associated with (a) an industrial region with factories that produce gaseous and solution S waste products that require disposal; (b) surface-mining operations that stockpile overburden materials, high in reduced-S minerals, close to a river; and (c) excessive S fertilization of coarse-textured soils that are overirrigated for the production of forages used for nonruminant animal feed.

6.2 Describe how animal toxicities occur when different types of S-containing feed supplements are used.

6.3 A coal-fired power plant must produce 250,000 MJ/h to supply enough steam to operate its electric generators to meet customer needs. How much coal would be required if the power plant used coal from the Northern Appalachian Region, San Juan River Region, or the Powder River Region? What additional operational concerns (e.g., price, quality, location) should be considered?

6.4 Calculate the difference in the quality (mol H^+/L) of rainwater in an unpolluted area with a precipitation pH of 6.8 vs. an industrial region with a precipitation pH of 4.3.

6.5 What are the mean resident times (MRTs) for S transfer through the biosphere, pedosphere, freshwater lakes, and oceans and seas?

6.6 What percentage of S released into the atmosphere is related to industrial sources (see Table 6.2 and Figure 6.2)?

6.7 What forms of S are present in soils? Which S forms are (a) the most important for plant nutrition; (b) dominant in surface horizons; and (c) found in subsoils of arid climates?

6.8 Describe the soil factors that are important in the adsorption of S. Determine the initial mass isotherm parameters (i.e., m, b, and RSP) for the following soil SO_4^{2-} adsorption study.

Initial SO_4^{2-} (mg/L)	Equilibrium SO_4^{2-} (mg/L)	Soil = 5 g	Solution = 50 ml
0	−2		
5	5.5		
10	9		
20	16		
40	30		
80	58		

6.9 Why are S mineralization–immobilization processes important and what influence does the C:S ratio have on organic N and P mineralization and immobilization?

6.10 Why do plants require S? Describe the symptoms of plant S deficiencies. Discuss how plant S analysis can be used to determine if supplemental S might be needed for optimum plant growth (see Table 6.8).

6.11 A soil test indicates that 4.5 mg SO_4-S/kg is potentially available in a field that will be planted in alfalfa. This crop requires 22 kg S/ha for adequate yields. What amount of fertilizer S should be recommended?

6.12 Design a S management program that includes irrigation, manure, fertilizers, and crop requirements.

6.13 Describe some of the problems that occur when hydric (wet) soils that contain reduced S are drained.

6.14 Calculate the difference in cost per liter for treating acid mine drainage that has a flow of 3780 L/min and acidities of 100 or 2500 mg/L using the information in Table 6.10. Determine the difference for each of the reagents — soda ash, ammonia, caustic soda, pebble quicklime, and hydrated lime — using the 1996 data.

6.15 A bentonite (Na-saturated 2:1 clay) mine land contains an area with sodic soils that have an ESP of 40% and a CEC of 30 cmol/kg. Determine the gypsum requirement (GR) for lowering the ESP to 5%. If a local source of gypsum is available, how much of this material should be applied to a 10-ha area on the bentonite mine site?

References

Barton, P. 1978. The acid mine drainage, in *Sulfur in the Environment*, J. O. Nriagu, Ed., John Wiley & Sons, New York, 313–358.

Brady, K. B. C., M. W. Smith, and J. Schueck, Technical Eds., 1998. Coal Mine Drainage Prediction and Pollution Prevention in Pennsylvania. The Pennsylvania Department of Environmental Protection, Harrisburg.

Chou, C. L. 1989. Geochemistry of sulfur in coal, in *Geochemistry of Sulfur in Fossil Fuels*, W. L. Orr and C. M. White, Eds., ACS Symposium Series No. 429, American Chemical Society, Washington, D.C., 30–52.

Donahue, R. L., R. W. Miller, and J. C. Shickluna. 1983. *Soils: An Introduction to Soils and Plant Growth*, Prentice-Hall, Englewood Cliffs, NJ.

Freeney, J. R. 1986. Forms and reactions of organic sulfur compounds in soils, in *Sulfur in Agriculture*, M. A. Tabatabai, Ed., Agronomy Monograph No. 27, American Society of Agronomy, Madison, WI, 207–232.

Goodrich, R. D. and J. E. Garrett. 1986. Sulfur in livestock nutrition, in *Sulfur in Agriculture*, M. A. Tabatabai, Ed., Agronomy Monograph No. 27, American Society of Agronomy, Madison, WI, 617–634.

Jones, J. B., Jr., B. Wolf, and H. A. Mills. 1991. *Plant Analysis Handbook: A Practical Sampling, Preparation, Analysis, and Interpretation Guide*, Micro-Macro Publishing, Athens, GA.

Krupa, S. V. and M. A. Tabatabai. 1986. Measurement of sulfur in the atmosphere and in natural waters, in *Sulfur in Agriculture*, M. A. Tabatabai, Ed., Agronomy Monograph No. 27, American Society of Agronomy, Madison, WI, 251–278.

Landers, D. H., M. B. David, and M. J. Mitchell. 1983. Analysis of organic and inorganic sulfur constituents in sediments, soils and water, *Int. J. Environ. Anal. Chem.*, 14, 245–256.

Mitchell, M. J., M. B. David, and R. B. Harrison. 1992. Sulphur dynamics of forest ecosystems, in *Sulfur Cycling in Terrestrial Systems and Wetlands*, R. W. Howarth and J. W. B. Steward, Eds., John Wiley & Sons, New York.

Nodvin, S. C., C. T. Driscoll, and G. E. Likens. 1986. Simple partitioning of anions and dissolved organic carbon in a forest soil. *Soil Sci.*, 142, 27–35.

Skousen, J. G. and P. F. Ziemkiewicz. 1995. Acid Mine Drainage: Control and Treatment, National Research Center for Coal and Energy, Mine Land Reclamation Division, West Virginia State University, Morgantown.

Stevenson, F. J. and M. A. Cole. 1999. The sulfur cycle, in *Cycles of Soil: Carbon, Nitrogen, Phosphorus, Sulfur, Micronutrients*, 2nd ed., John Wiley & Sons, New York, chap. 10.

Szabolcs, I. 1989. *Salt-Affected Soils*, CRC Press, Boca Raton, FL.

Thompson, J. F., I. K. Smith, and J. T. Madison, Sulfur metabolism in plants, in *Sulfur in Agriculture*, M. A. Tabatabai, Ed., Agronomy Monograph No. 27, American Society of Agronomy, Madison WI, 1986, 57.

Tisdale, S. L., W. L. Nelson, and J. D. Beaton. 1985. *Soil Fertility and Fertilizers*, Macmillan, New York.

Trudinger, P. A. 1986. Chemistry of the sulfur cycle, in *Sulfur in Agriculture*, M.A. Tabatabai, Ed., Agronomy Monograph No. 27, American Society of Agronomy, Madison, WI, 1–22.

Vance, G. F. and M. B. David. 1992. Adsorption of dissolved organic carbon and sulfate by Spodosol mineral horizons, *Soil Sci.*, 154, 136–144

Walworth, J. L. and M. E. Sumner. 1987. The Diagnosis and Recommendation Integrated System, *Adv. Agron.*, 40, 194–288.

Westerman, R. L., Ed. 1990. *Soil Testing and Plant Analysis*, SSSA Book Series No. 3, Soil Science Society of America, Madison, WI.

Williams, C. H. 1974. The chemical nature of sulphur in some New South Wales soils, in *Handbook of Sulphur in Australian Agriculture*, K. D. McLachlan, Ed., CSIRO, Melbourne, Australia.

Supplementary reading

Adams, F., Ed. 1984. *Soil Acidity and Liming*, Agronomy Series No. 12, Soil Science Society of America, Madison, WI.

Brimblecombe, P. and A. Y. Lein, Eds. 1989. *Evolution of the Global Biogeochemical Sulphur Cycle*, John Wiley & Sons, New York.

Maynard, D. G., Ed. 1998. *Sulfur in the Environment*, Marcel Dekker, New York

Nriagu, J. O., Ed. 1978. *Sulfur in the Environment: Part II, Ecological Impacts*, John Wiley & Sons, New York.

Paul, E. A. and F. E. Clark. 1996. *Soil Microbiology and Biochemistry*, 2nd ed., Academic Press, San Diego, CA.

Schlesinger, W. H. 1997. *Biogeochemistry: An Analysis of Global Change*, 2nd ed., Academic Press, New York.

Tabatabai, M. A., Ed. 1986. *Sulfur in Agriculture*, Agronomy Monograph Series No. 27, American Society of Agronomy, Madison, WI.

chapter seven

Trace elements

Contents

7.1   Introduction

Trace elements are elements that are normally present at relatively low concentrations in soils, plants, or natural waters, and which may or may not be essential for the growth and development of plants, animals, or humans. With such a broad definition, trace elements obviously include a large number of elements with widely ranging chemical characteristics and effects on various organisms. Figure 7.1 is the periodic table of elements with those that are not considered trace elements shaded. Shaded elements (1) are present in relatively high concentrations in plants, soils, and natural waters and therefore do not fit the definition of trace elements; (2) have existed only as radioactive isotopes that have since decayed to daughter products; (3) do not occur naturally in the environment; or (4) are present only as inert gases. There is some debate whether certain elements should be classified as trace elements when present in high concentrations in soils or the earth's crust, but in low concentrations in plants. Titanium (Ti), iron (Fe), and aluminum (Al) are three examples. They are not discussed in detail in this chapter, but they would be considered trace elements by our definition. Most discussions of trace elements do not

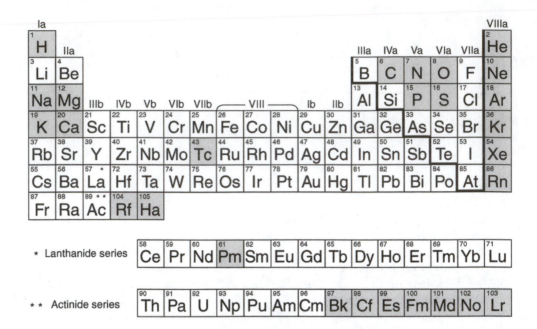

Figure 7.1 Periodic table of the elements. Shaded elements are not considered trace elements.

include the radionuclides, many of which would also be trace elements by our definition and, thus, will be discussed in this chapter.

A variety of other terms are often used to describe trace elements or subsets of the elements that we have defined as trace elements. *Micronutrients* and *heavy metals* are two such terms. The use of micronutrients is avoided here because the term implies that the elements in question are essential for the growth and development of some organism, and many trace elements are not. Zinc (Zn) is essential for many organisms and would be a micronutrient while lead (Pb) is not essential and would not be a micronutrient. Both are trace elements. Heavy metals is another term that is avoided here because it generally refers to only metallic elements with an atomic weight greater than that of Fe (55.8 g/mol) or a density greater than 5.0 g/cm³, which also excludes many trace elements. For example, chromium (Cr) is a metal with an atomic weight less than Fe and arsenic (As) and selenium (Se) are metalloids rather than trace metals. All of these elements are important trace elements that are not heavy metals, although they are often mistakenly referred to as heavy metals. Other terms that have been used to describe trace elements include trace metals, microelements, minor elements, and trace inorganics.

Some 78 elements are considered trace elements based on Figure 7.1. It would not be possible, or even necessary, to discuss all of these in detail in this book. A simple means of reducing the number of elements considered would be to limit the discussion to those elements that are of concern because they are either essential or potentially toxic to humans, animals, or plants. By this method we might consider As, boron (B), beryllium (Be), cadmium (Cd), cobalt (Co), Cr, copper (Cu), fluorine (F), Fe, mercury (Hg), iodine (I), manganese (Mn), molybdenum (Mo), nickel (Ni), Pb, Se, tin (Sn), vanadium (V), and Zn. A description of the behavior of all of these elements in soils is not possible here, although some will be used as examples to illustrate general principles.

Another convenient way to categorize soil trace elements is by their expected chemical form in soils and soil solutions. *Cationic metals* are metallic elements that occur predominantly in the soil solution as cations. Examples are silver (Ag^+), Cd^{2+}, Co^{2+}, Cr^{3+}, Cu^{2+}, Hg^{2+}, Ni^{2+}, Pb^{2+}, and Zn^{2+}. *Oxyanions* are elements that are combined with oxygen in molecules with an overall negative charge. Examples are arsenate (AsO_4^{3-}), borhydrate ($B(OH)_4^-$), chromate (CrO_4^{2-}), molybdate (MoO_4^{2-}), selenite (SeO_3^{2-}), and selenate (SeO_4^{2-}). The *halides* are members of group VIIA in the periodic table and are present as anions in the soil solution. The halides are flouride (F^-), chloride (Cl^-), bromide (Br^-), and iodide (I^-). The categories are not mutually exclusive, however, as some elements can occur in more than one category.

As will be shown, the chemical form in the soil dictates the behavior of the element in the soil, including factors such as leachability, risk to human health, and remediation strategies. In previous chapters we have discussed three important nutrient oxyanions, namely, nitrates (NO_3^-), orthophosphates (e.g., $H_2PO_4^-$), and sulfates (SO_4^{2-}). It is reasonable to expect the trace element oxyanions and the oxyanions of the nutrients to exhibit similar characteristics for properties such as sorption to soil solids, changes in oxidation state, formation of gaseous compounds, and microbial interactions. Similarly, we would expect the cationic metals to interact with the cation exchange sites of the soil the same as any other cation, although they are at a competitive disadvantage because of the relatively high concentrations of calcium (Ca^{2+}), magnesium (Mg^{2+}), sodium (Na^+), and potassium (K^+). Cationic metals do have more of a tendency to form both inorganic and organic complexes than the major exchangeable cations.

7.2   Sources of trace elements for the terrestrial environment

Soil trace elements have both natural and anthropogenic sources. Soil parent materials will contain trace elements; therefore, all soils are expected to have small quantities of the elements we have defined as trace elements. The issue of what constitutes typical trace element concentrations in soils arises occasionally. Usually this is in reference to a potential contamination problem and one wants to know if the results from a soil analysis reflect a contaminated situation. Unfortunately, there is not a straightforward answer to this issue as the natural variability in soil trace element concentrations can be quite high. Table 7.1 presents some normal soil concentrations for selected trace elements and some geochemically anomalous concentrations. A soil with a total Pb concentration of 600 mg/kg, for example, could represent a situation in which a soil with an original Pb concentration within the normal range was contaminated or a geochemically anomalous situation. Additional information is required to make the correct determination.

Table 7.1   Selected Trace Element Concentrations in Soils at Normal and Geochemically Anomalous Levels

Element	"Normal" range (mg/kg)	Metal-rich range (mg/kg)
As	<5 to 40	up to 2500
Cd	<1 to 2	up to 30
Cu	2 to 60	up to 2000
Mo	<1 to 5	10 to 100
Ni	2 to 100	up to 8000
Pb	10 to 150	10,000 or more
Se	<1 to 2	up to 500
Zn	25 to 200	10,000 or more

Source: Bowie, S. H. U. and Thornton, I., Eds., *Environmental Geochemistry and Health*, Kluwer Academic, Hingham, MA, 1985.

Uses for trace elements are quite numerous, and, consequently, there are a variety of ways trace elements can enter the terrestrial environment and become an environmental concern. Historically, the mining and smelting of trace elements has created soil contamination problems of the greatest magnitude. Figure 7.2 is a scene in southwest Missouri showing some of the environmental problems associated with Pb and Zn mining. The material in the background is chat, a waste rock that can have Zn and Pb concentrations as high as 20,000 mg/kg (2%). Fine particles selectively eroded from the chat piles have the potential of contaminating nearby soils and becoming sediments in surface waters, which greatly enlarges the area impacted by the original mining activity. In addition, these chat piles are a source of dust that can lead to Pb exposure to people across a wide area. Mining activities such as this produce primary contaminants consisting of waste rock, tailings, and slag. Secondary contamination occurs in groundwater beneath open pits and ponds, sediments in river channels and reservoirs, floodplain soils impacted by contaminated sediment, and soils affected by smelter emissions. River sediments reworked from floodplains and groundwater from contaminated reservoir sediments are tertiary contaminants. Substantial areas have been impacted by such activities in Arizona, Colorado, Kansas, Missouri, Montana, Oklahoma, Pennsylvania, Utah, and Wyoming in the U.S. Many other such areas exist around the world. Mining and smelting activities have caused widespread ecological damage because relatively large areas have been impacted and the trace element concentrations in soils can be quite high.

The use of motor vehicles has been a source of trace elements for the terrestrial environment. Motor vehicles emit Cd from diesel fuel; Zn and Cd from tire attrition; Ni, Cr, V, tungsten (W), and Mo from attrition of steel; and Pb from gasoline. The use of Pb in gasoline was phased out in the U.S. beginning in the 1970s, but it is still used in gasoline marketed in much of the world. The concentrations of certain trace elements in soils along roads that have been in use for a long time will undoubtedly be elevated. This problem is most severe in older urban areas. The use of motor vehicles has also led to trace element–contaminated soils indirectly through practices such as automobile battery recycling and the use of automobile salvage yards.

Figure 7.2   Abandoned Pb and Zn mining site in southwest Missouri. (Photograph credit: Gary M. Pierzynski.)

Paints are a major source of Pb in the environment. Similar to Pb in gasoline, Pb in paint was banned over 20 years ago (1978) in the U.S. However, homes and other structures built prior to the ban still have painted surfaces. As the paint on the exterior surfaces of buildings weathers, the Pb ends up in the soil. Similarly, paint from interior surfaces can become a component of house dust (as does the soil from outside the building), and produce a significant exposure pathway for people living in that environment. Paint chips can also be ingested directly by children. Obviously, older homes that are not well maintained pose the greatest risk.

It is logical to conclude that urban soils will generally have higher soil trace element concentrations than rural areas (except mining areas) due to their proximity to more trace element sources. Figure 7.3 shows a general decline in Cu, Ni, and Pb concentrations in forest floor and soil samples with distance from the center of New York City, illustrating the decline in trace element concentrations with distance from a generalized source.

Land application of waste products, either as a beneficial reuse or as a disposal strategy, can lead to trace element contamination of soils. Biosolids, the solid material remaining after municipal wastewater is treated, may enrich soils with trace elements particularly if the municipal wastewater treatment plants receive industrial as well as domestic inputs. This subject has been the source of considerable debate within the scientific community in recent years. At issue is the degree to which enrichment should be allowed, and this will be examined further in our discussion of risk assessment in Chapter 12. Similarly, animal manures may contain elevated levels of trace elements such as As and Cu that are present in livestock rations for various reasons.

Products sold as inorganic fertilizers may be enriched in certain trace elements. Cadmium and other nonessential elements may exist in the materials that were used to make the fertilizers and end up in the finished product. Rock phosphate is the primary source of phosphorus (P) for fertilizer and can be a carrier for unwanted trace elements. These impurities are an integral part of the ore. In some unusual cases industrial by-products containing Zn, as well as some other nonessential trace elements, were processed into a Zn-containing fertilizer (Zn as a micronutrient in this case). This generated considerable public outcry since the initial by-products were classified as hazardous waste and the materials were then legally applied to land as fertilizers. While there are no widespread reports of damages from these practices, the public is demanding closer scrutiny of all fertilizer materials. Table 7.2 provides averages and ranges for As, Cd, and Pb concentrations in commercial inorganic fertilizers. In general, the phosphate fertilizers have higher Cd concentrations than the other materials. Trace element concentrations in blended and micronutrient fertilizers are highly variable.

A variety of other trace element uses and sources for contamination of the terrestrial environment are of less significance. The mining and use of materials such as coal or iron ore often lead to trace element–contaminated soils as well. In Wyoming, surface coal mining exposes Se-containing overburden that would not otherwise interact with the environment. Metal recycling facilities, particularly for metals like Cu and brass, have been known to emit significant amounts of trace elements. Disposal of batteries containing Pb, Ni, Cd, and other trace elements is another example.

Widespread trace element contamination of soils has occurred in some areas of the world. Our knowledge of the behavior of these elements indicates that soils are difficult to decontaminate once the contamination has occurred. Thus, society is increasing efforts to prevent trace element contamination. Simultaneously, efforts are under way in some areas to mitigate the negative effects of existing contamination as our knowledge of the human health and ecological risks increases. Soil remediation will be discussed at length in Chapter 11.

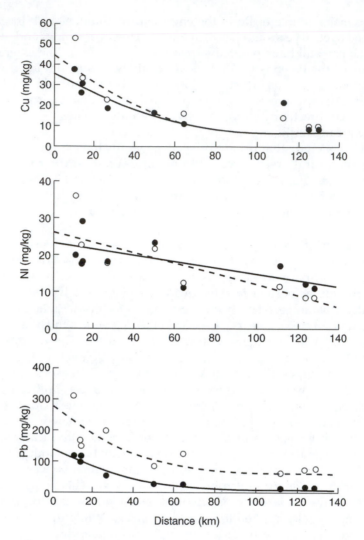

Figure 7.3 Forest floor and soil trace element concentrations as a function of distance from Central Park, Manhattan, New York City. Open circles represent forest floor and closed circles represent soil values. (From Pouyat, R. V. and McDonnell, M. J., *Water Air Soil Pollut.*, 57/58, 797, 1991. With permission.)

7.3   Adverse effects

It is difficult to summarize all adverse effects due to trace elements because of the large number of elements and the fact that each element can cause a number of different problems. Human health effects are generally of greatest concern. As will be shown, an astonishing number of people are affected by trace element toxicities. Impacts on our ability to produce food are also of great concern. Here the impacts of the trace elements on food quality as well as plant and animal productivity are important. Of growing interest in recent years are the ecological impacts of trace elements in the terrestrial and aquatic environments. All of the effects can be placed into four broad categories, which will be discussed individually: human health, animal health, effects on plants, and effects on aquatic ecosystems. Ecological impacts are divided among the final three categories.

Table 7.2   Average and Range of Arsenic, Cadmium, and Lead Concentrations (mg/kg) in Selected Inorganic Commercial Fertilizers and Soil Amendments Sold in California

Material	Fertilizer grade	As	Cd	Pb
Soil amendments and liming materials				
Limestone	—	2.2 (0.3–5.1)	6.2 (4.0–8.1)	33.9 (3.2–52.0)
Gypsum	—	1.3 (0.2–3.4)	0.9 (0–2.5)	4.4 (0.8–9.2)
Phosphate fertilizers				
Triple superphosphate	0-20-0[a]	4.8 (0.5–14.0)	110 (0–163)	2.0 (0–4.0)
Monoammonium phosphate	11-23-0	8.1 (1.5–16.0)	74.1 (0–166)	2.6 (0–9.0)
Diammonium phosphate	18-20-0	4.9 (0.3–8.5)	38.1 (5.0–125)	4.0 (0–15.6)
Soft phosphate rock[b]	Unknown	13.0	130	13.0
Other fertilizers				
Muriate of potash	0-0-50	0.4 (0.1–0.6)	0.8 (0–1.6)	3.3 (2.5–4.0)
Urea[b]	46-0-0	0	0	0
Blended[b]	11-5-0	11.2	95.0	3600
Zinc sulfate	—	2.6 (0–29.5)	32.3 (0–233)	203 (0–1430)

[a] % N, P, and K by weight.

[b] One sample only.

Source: California Department of Food and Agriculture, 1998.

Any trace element can have an adverse effect on any organism if the dose is high enough. Some of the trace elements are essential for humans, animals, or plants, which further complicates issues. Generally, exposure of organisms to high doses of trace elements is uncommon and, therefore, not considered an environmental problem. Table 7.3 indicates the species that are most commonly at risk due to exposure to elevated doses of 13 important trace elements. Note, for example, that Pb is of concern for humans, animals, aquatic organisms, and birds, but rarely induces phytotoxicities as compared with B for which phytotoxicities are of primary concern.

7.3.1   Human health

The primary routes of exposure of humans to trace elements in soil are through food chain transfer and by direct ingestion of soil particles (Figure 1.1). Documented cases

Table 7.3   A Summary of Species Most Commonly Affected by Toxicities of Selected Trace Elements

Element	Species adversely affected				
	Humans	Animals	Aquatic organisms	Birds	Plants
Cd	*	*	*	*	*
As, Pb, Hg, Cr, Se	*	*	*	*	
Cu, Ni, Zn			*		*
Mo, F, Co	*				
B					*

Source: A. L. Page, personal communication, 1992.

Table 7.4   Estimated Magnitude of the Extent of Trace Element Poisonings

Element	Global emissions (1000 Mg/year)			People affected	Comments
	Air	Water	Soil		
Pb	332	138	796	>1 billion	Blood Pb > 20 μg/dL
Cd	7.6	9.4	22	500,000	Producing renal dysfunction
Hg	3.6	4.6	8.3	80,000	Certified Hg poisonings
As	18.8	41	82	>100,000	Skin disorder and H_2O As >2 μg/L

Source: Nriagu, J. O., *Environ. Pollut.,* 50, 139, 1988.

of acute trace element poisonings in humans due to elevated soil trace element concentrations are rare. Documented cases of chronic adverse health effects due to trace element exposure through food chain transfer or direct ingestion of soil are more numerous. The number of people affected by trace element poisoning worldwide is quite astounding, as shown in Table 7.4. Note that soil is the primary recipient of global emissions of trace elements.

Cadmium and Pb can be used to illustrate the two general exposure routes for humans. Each can have profound health effects on humans, but means of exposure are quite different due to their chemical behavior in the soil environment and the concentration ranges at which they occur in contaminated soils. Cadmium is readily taken up by plants (a high bioavailability for plant uptake) and food chain transfer is the primary route of exposure. An often cited example deals with Japanese farmers suffering from Cd toxicity after long-term exposure to Cd-enriched rice. Two symptoms of Cd toxicity are renal dysfunction and *itai-itai* disease. *Itai* is Japanese for "ouch," and the disease is characterized by Cd-induced bone loss that produces localized, severe pain in the joints of victims. The Cd-enriched rice had been grown in paddies polluted by Pb and Zn mining and smelting operations. There is a geochemical association between Zn and Cd, and therefore Zn-contaminated soils are often Cd contaminated as well. Table 7.5 indicates that rice grown in areas having >10% morbidity had Cd concentrations as much as 14 times higher than rice grown in nonpolluted areas. This situation was exasperated by the general poor nutrition of the population, which made people more susceptible to Cd toxicity. Food chain transfer of Cd is often the limiting exposure pathway (negative effects at the lowest soil Cd concentration) for risk assessments conducted for trace element–contaminated soils.

For trace element–contaminated soils, Pb concentrations are generally much higher than Cd concentrations (see Table 7.1). Despite the higher soil Pb concentrations, plant uptake of Pb is very limited (a low bioavailability for plant uptake). In order for substantial

Table 7.5   Cadmium Concentrations (mg/kg) in Rice by Prevalence of *Itai-Itai* Disease

Rice type	Area	
	High endemic[a]	Nonendemic
Nonglutinous		
Polished	0.52	0.048
Unpolished	0.54	0.079
Glutinous		
Polished	1.03	0.071
Unpolished	1.12	0.150

[a] >10% morbidity.

Source: Tsuchiya, K., in *Cadmium Studies in Japan: A Review,* K. Tsuchiya, Ed., Elsevier, New York, 1979.

exposure to occur the soil itself has to be ingested, and because soil Pb concentrations can be fairly high this can result in a physiologically significant dose of Pb to the recipient. It is also recognized that the bioavailability of the Pb in the soil for the human receptor can vary considerably. That is, two soils with equal total Pb concentrations ingested in equivalent amounts can produce differing effects (see Figure 1.2). Lead has a number of physiological effects on the human body and infants and children are at much greater risk than adults. One such effect is impairment of mental development that causes children with chronic exposure to Pb to score lower on IQ tests than children not exposed. Blood Pb concentrations are often used as a screening test for Pb exposure in children and the levels have been steadily falling in the U.S. because Pb has been eliminated from gasoline and paint. People living in older urban areas will typically have elevated Pb exposures due primarily to the use of leaded house paints. The current guideline for children is to have the blood Pb concentration less than 10 µg/dL, but in areas with Pb-contaminated soils as many as 25% of the children will exceed this guideline.

Two other trace elements that are somewhat notorious for human health effects are Hg and As. The phrase "mad as a hatter" is derived from individuals who used mercuric nitrate ($Hg(NO_3)_2$) in the making of felt hats and treatment of furs. The symptoms of this subacute Hg poisoning included tremors, vertigo, irritability, moodiness, and depression. Liquid (elemental) Hg evaporates at room temperature and inhalation of Hg vapor can lead to significant exposure from this form of Hg. This was once significant in certain occupations such as mining where Hg amalgams were used to concentrate metals such as Au. This is also the reason attempts are made to vacuum up Hg droplets after a Hg spill. The most extensive exposure route for humans today is through methylmercury ($Hg(CH_3)_2$) in seafood. A highly publicized case of this occurred in Japan following the dumping of Hg into Minamata Bay and the subsequent consumption of fish from the bay by local residents. This acute form of Hg poisoning, since called *Minamata disease*, manifested itself in mental retardation, cerebral palsy, and fetus mortality in the children of the exposed population. While outbreaks as serious as that in Japan have not occurred since, health advisories against consumption of seafood because of Hg are still common today. The methylated forms of Hg are formed readily in the anaerobic environment found in sediments.

Human health effects from As vary depending on the dose received and the chemical form of As. General symptoms include weakness, muscular aches, hearing loss in children, and possibly skin cancer. Exposure most often occurs from drinking water, although soils that have received arsenical pesticides, such as orchards and areas where vegetables were produced, can be highly contaminated. In these situations, the direct consumption of soil is the primary route of exposure, similar to Pb. A serious problem exists in western India and Bangladesh today where an estimated 200,000 people have As-induced skin lesions. This is a result of consumption of As-contaminated drinking water. New wells were drilled into high As sediments to supply water for irrigation to improve agricultural productivity, and the water was also used for personal consumption.

Of increasing interest recently is the possibility of subclinical effects due to exposure to slightly elevated amounts of trace elements. Subclinical effects represent symptoms or diseases not easily diagnosed by medical examination. Attention deficit disorder, hyperactivity, or low IQ test scores are three relevant examples. Epidemiological studies are required to detect statistically significant trends. This is an area needing much study.

7.3.2 Animal health

Animal health effects include domestic animals and wildlife. Toxicities of Mo in livestock (molybdenosis) and Se in livestock and wildlife (selenosis) are two relatively common trace

element animal health problems. Ruminant animals are the most susceptible to molybdenosis, a Mo-induced Cu deficiency. Some soils have naturally occurring high Mo concentrations and can produce forages that can induce molybdenosis in grazing animals. Soils
can be naturally enriched in Mo or become enriched due to poor quality irrigation water or
Mo-rich biosolids additions. Selenium problems are much more widespread. Outcroppings
of seleniferous sedimentary rocks occur in certain regions throughout the U.S. and elsewhere.
In semiarid to arid regions the associated soils may produce plants with Se concentrations
that are harmful to animals if consumed regularly. Coal mining activities may also bring
seleniferous materials to the surface. The Kesterson Wildlife Refuge area of central California
represents a situation where agricultural activities have led to some serious outbreaks of Se
poisoning in wildfowl. Irrigated cropland produced drainage water with a high Se concentration. As this water flowed into the Kesterson area, it was further concentrated by evaporation and produced widespread Se toxicity problems for aquatic life and waterfowl.

7.3.3 Phytotoxicities

Phytotoxicity refers to reduced yields or death of plants by substances in the soil. Symptoms for trace element–induced phytotoxicities can include stunting, chlorosis, necrosis,
and death of the plant. Trace elements most often associated with phytotoxicities are B,
Cu, Ni, and Zn. Phytotoxicity problems are of concern for two primary reasons. First, the
reduction in soil quality induced by elevated trace element concentrations can reduce both
the quantity and the quality of the food produced from that soil. An example is illustrated
in Figure 7.4. Here the relative yield of three crops declines as the $Mg(NO_3)_2$-extractable

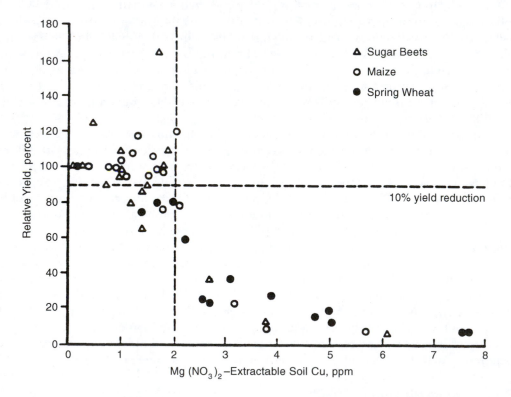

Figure 7.4 Relationships between $Mg(NO_3)_2$-extractable Cu in soil and crop yield. (From Lexmond,
T. M. and deHaan, F. A. M., in *Proc. Int. Seminar on Soil Environment and Fertility Management in
Intensive Agriculture (Tokyo)*, Soc. Sci. Soil Manure, Tokyo, 1977. With permission.)

soil Cu concentration increases. When the extractable Cu concentration exceeds that shown by the vertical dashed line, relative yield is predicted to decline by >10%. In this situation, Cu was being studied because of potential phytotoxcities from Cu additions due to swine manure applications. Manure-amended soils had not, in general, reached the critical level of 2 mg/kg of $Mg(NO_3)_2$-extractable soil Cu. The second reason relates to areas where vegetation is sparse because of trace element phytoxicities so that wind and water erosion can occur uninhibited. These conditions often exist around sites where metal mining or smelting once took place (Figure 7.2). In the case of Figure 7.2, elevated soil Zn caused phytotoxicity problems but was not a direct human health threat. The lack of vegetative cover allowed further dispersal of the Cd and Pb present at these sites thereby increasing human exposure indirectly. Direct inhalation of windblown dusts from such areas can also be an exposure route for humans and animals.

7.3.4 Aquatic environments

The effects of elevated concentrations of trace elements in aquatic environments are extremely difficult to assess, due in part to the mobility of some of the species and the difficulty in separating out the effects of contaminated water from contaminated sediment. Soil erosion is the primary mechanism by which trace elements are transferred from the terrestrial to the aquatic environment. The enrichment ratios for most trace elements are ≥1. Aquatic species can be divided into groups of plants (phytoplankton and benthic), invertebrates, and fish. The species can be characterized according to their diversity, productivity, and density. Diversity refers to the number of species present in a given area, productivity refers to the weight or size of individual organisms, and density refers to the number of a given species present in a given area. The effects of trace elements are generally to reduce the diversity, productivity, and density of aquatic organisms.

7.4 Trace element cycles in soils

Detailed soil nitrogen (N), P, and sulfur (S) cycles have been presented in Chapters 4, 5, and 6, respectively. Since the trace element category includes a large number of elements, it would be impractical to present individual soil trace element cycles. Figure 7.5 presents a generalized soil trace element cycle that illustrates processes and transformations that are important for most trace elements. Trace element inputs to the soil cycle were discussed in Section 7.2. Most of these processes should be familiar to the reader based on previous discussions of the soil N, P, and S cycles. The influence of these processes on bioavailability will be discussed in the next section and the use of our knowledge of these processes for soil remediation will be discussed in Chapter 11.

The soil solution is the focal point of any soil element cycle. The soil solution is where plant roots and other soil organisms interact with essential and nonessential elements in the soil. Plant uptake of trace elements occurs from the soil solution and its composition reflects the interactions between soluble species and the solid phases. For example, strongly adsorbed species or those of sparingly soluble compounds will be present in the soil solution at very low concentrations, and vice versa.

The primary trace element species of interest in the soil solution are the cationic metals, oxyanions, and halides, as described earlier. These species also form a variety of soluble complexes with other elements and with dissolved organic carbon (C) constituents. Sometimes these soluble complexes are a combination of two ions, such as $CdCl^+$, or are a combination of a trace element and elements that are normally present in much higher concentrations in soils, such as $CaSeO_4^0$. Techniques are available to speciate the total amount of a trace element in the soil solution into its constituents (e.g., $Cu_T = Cu^{2+} +$

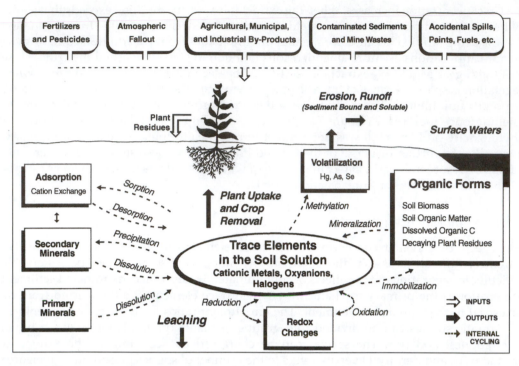

Figure 7.5   Generalized soil trace element cycle.

$CuOH^+ + CuCO_3^0 + ...$). However, the chemistry and environmental significance of this speciation is beyond the scope of this book.

Most trace elements undergo some sort of sorption processes with the solid phases in the soil. *Sorption* is a generic term that refers to the tendency of the soil to retain substances (see Chapter 3). Substances that are strongly retained are relatively immobile in soils. Adsorption generally refers to the formation of a chemical bond between soluble substances and the soil solids. With trace elements this behavior can be exhibited by the oxyanions. Molybdate (MoO_4^{2-}), for example, is adsorbed by Fe and Al oxides in a fashion similar to orthophosphate. Two important oxyanions, NO_3^- and orthophosphate, which have been discussed already in this book, have very different adsorption characteristics. Nitrate undergoes little if any adsorption, whereas orthophosphates are strongly adsorbed. The behavior of most of the trace element oxyanions is intermediate relative to nitrates and orthophosphates. The reason that selenosis in livestock is more prevalent in arid or semiarid regions is because SeO_4^{2-} is not strongly adsorbed by the soil, is not leached from the soil, and is readily available for plant uptake. In more humid environments, SeO_3^{2-} is a predominant species present and it is strongly adsorbed by the soil, making it less available for plant uptake. Molybdate is less prone to leaching than SeO_4^{2-} but is still more mobile than orthophosphates.

Cation exchange refers to the electrostatic attraction between cations in the soil solution and the negatively charged soil solids, both clay minerals and soil organic matter. The cationic trace elements must compete for exchange sites against the predominant cations in soils (Ca^{2+}, Mg^{2+}, Na^+, and K^+) and generally do so effectively. There are differences between the cationic metals, as was illustrated earlier for Cd and Pb and their availability for plant uptake.

In uncontaminated soils trace elements do not generally form secondary minerals because the concentrations are low and undersaturated with respect to the solubilities of possible solid phases. In contaminated soils, however, trace elements may form secondary minerals. When secondary minerals do form, the possibility exists that the solid phase will control the concentration (more accurately, the activity) of that element in the soil solution, which can greatly influence the trace element bioavailability. Examples include wulfenite ($PbMoO_4$), hydroxypyromorphite ($Pb_5(PO_4)_3OH$), and franklinite ($ZnFe_2O_4$).

In most soils the dissolution of primary minerals is of minor importance because the quantities of such minerals are quite low. In areas where trace element–rich deposits are near the surface, the dissolution of primary minerals can be a significant source of trace element enrichment for soils. We have already described an example with Se. Other elements where this might be a concern include Cr, Cu, Ni, Pb, and Zn.

Redox changes are important for a number of trace elements including As, Cr, Hg, Mn, and Se. The chemistry behind the redox changes is beyond the scope of this text, but the reader should be aware that such changes can greatly impact the fate and transport of the trace element in the environment as well as the toxicological characteristics. Chromium, for example, exists as both Cr^{3+} and CrO_4^{2-} (with Cr(VI)). The hexavalent Cr is much more toxic and mobile in the environment compared with the trivalent form. The Kesterson Wildlife Refuge situation was a result of the oxidation of reduced forms of Se to SeO_4^{2-}, which is more mobile and which allowed Se to leach from irrigated soils and eventually reach the refuge. There are indirect effects of redox as well. Most of the cationic metals do not undergo changes in oxidation state under conditions found in soils but can form insoluble sulfides under reducing conditions (see Figure 6.3). This is one mechanism by which wetlands can accumulate trace elements.

Organic forms can be significant for certain trace elements. The amounts present in soil biomass are generally small and of little importance. Certain trace elements form strong complexes with functional groups in soil organic matter or dissolved organic C, which greatly influences the interactions with the soil solution. The reaction given below illustrates this process for Cu, which is generally strongly associated with the organic C fraction in soils. These cation exchange type of reactions are obviously of importance for the cationic metals. Oxyanions can also interact with soil organic matter by forming bonds with other cations that are already associated with the organic C fraction. Mercury forms a variety of methylated compounds under reducing conditions that are readily accumulated by aquatic species. The consumption of seafood is a major route of exposure for Hg to humans, as discussed earlier.

(7.1)

Volatilization is only of importance for a few trace elements, notably Hg, As, and Se. Elemental Hg can form a vapor and be lost from soils or sediments as a gas. Mercury, As, and Se can be methylated by soil microorganisms and some of these methylated forms are gases. Dimethylselenide, for example, can be formed from Se-contaminated soils and this process has been explored as a means of remediating Se-contaminated soils and sediments.

Runoff and erosion losses of trace elements can be a major transport mechanism in the environment. Soil erosion occurs when soil particles are detached and then transported across the landscape. Vegetation and plant residues greatly reduce soil erosion because they absorb the impact of raindrops, thereby reducing detachment of soil particles, and by decreasing the flow of water across the soil surface, which inhibits the transport process.

If phytotoxic conditions have prevented or limited plant growth, then the transport of trace elements in runoff becomes more prevalent (Figure 7.2). Aquatic ecosystems are sensitive to trace element contamination so significant negative effects can occur with only small inputs of trace elements.

In situations where plant residues are not removed from the landscape, soil trace elements are continually recycled through plant uptake and decomposition of plant residues. Where biomass removal occurs (e.g., agricultural or turf grass scenarios), significant amounts of trace elements can be removed from the soil through plant uptake. In general, plants do not accumulate high concentrations of trace elements so removal rates from plant uptake are considerably less than those for plant nutrients such as N or P. Trace elements in plants represent an exposure route for any organism that consumes the plant tissue, so the concentration issue is not trivial. Certain types of plants are known to *hyperaccumulate* some trace elements. There are Se hyperaccumulators that are native to rangeland in arid and semiarid regions. When these are present, grazing animals can greatly increase their risk for selenosis when these plants are consumed rather than other species with much lower Se concentrations. Scientists have proposed using hyperaccumulators of Zn, Pb, and other trace elements as a means of removing trace elements from soils, as will be discussed in Chapter 11.

7.5   Bioavailability of trace elements in soils

As was described in Chapter 1, bioavailability is related to the possibility that a substance can cause an effect, either positive or negative, on an organism. This concept is of vital importance for trace elements for a variety of reasons, although we must take a broad view of the subject and carefully identify both the organism of interest and the effect. Once the organism and the effect are identified, the factors influencing bioavailability can be studied. Small changes in bioavailability can dramatically influence the effect. Two examples will be used, one involving plants as the organism of interest and the other involving humans. Of course, any of a number of organisms could be used including soil invertebrates, domestic or wild animals, or even birds.

There is a long history of research dealing with the bioavailability of soil trace elements to plants. Early work focused on deficiencies of trace elements that were essential for plant growth, including B, Cu, Fe, Mn, Mo, and Zn. When these elements are not present in soil in sufficient available quantities for plants, dramatic increases in growth can be realized by increasing the bioavailability of the trace element. Iron, typically present in relatively high concentrations in soils, can have very limited availability for plant uptake when the soil pH is high. At the opposite end of the spectrum are phytotoxicities where trace element bioavailability becomes too high.

Total trace element concentrations in a soil are, in fact, poor predictors of trace element bioavailability. The bioavailable fraction is some subfraction of the total trace element fraction that best predicts the response. Here, this response generally refers to plant uptake of the trace element. It must be recognized, however, that trace element additions to soils, which will increase total trace element concentrations, will likely increase trace element bioavailability at least to some extent. In this situation, a relationship may exist between total and bioavailable trace element concentrations. In the aforementioned Fe example, however, the direct application of Fe salts to high-pH soils is a notoriously poor way to correct Fe deficiencies because the soluble Fe added precipitates as insoluble secondary minerals of low plant availability.

Trace element bioavailability to plants is strongly related to the concentration and speciation of the element in the soil solution because this is where the plants get the trace elements that they take up. Typically, plants can only take up one or two forms of trace elements from the soil solution. Hence, one key to understanding bioavailability to plants is to understand the factors that influence the soil solution concentration of the soluble species that are taken up by the plant.

Copper can be used as an example to illustrate some of the processes involved in bioavailability to plants along with our soil trace element cycle. One could easily measure the total Cu concentration in the soil solution (not to be confused with the total Cu concentration in the soil) with atomic absorption spectrophotometry. As described earlier, the concentration would be denoted Cu_T and is a summation of various soluble Cu species. As Figure 7.6 implies, it is the free Cu^{2+} concentration (more precisely, the Cu^{2+} activity) that determines the Cu bioavailability. Figure 7.6a shows the relationship between Cu_T and the Cu concentration in snapbean plants. The filled circles represent Cu additions to the soil from biosolids and the open circles represent Cu additions from $CuSO_4$. Figure 7.6b shows the relationship between pCu (negative logarithm of the Cu^{2+} activity; as pCu increases, the Cu^{2+} activity decreases), as measured with an ion-selective electrode, and the Cu concentration in snapbean plants. Qualitatively, one can see that pCu is a better predictor of Cu concentrations in the plants than Cu_T.

Turning to our soil trace element cycle, we see that the free Cu^{2+} ion can undergo adsorption reactions, precipitate as some Cu-containing solid phase, interact with organic matter, form other soluble species, or change oxidation state if the oxidation/reduction (redox) conditions of the soil change. If any of these processes change the Cu^{2+} concentration, the Cu bioavailability to plants will change. For example, if one were to add certain clay minerals to a soil, the cation exchange capacity of the soil would increase. This increase in cation exchange sites would likely reduce the Cu^{2+} concentration in the soil solution.

The solubility and, as a consequence, the bioavailability of trace elements can be greatly influenced by soil pH. Generally, the bioavailability of the cationic metals will increase as soil pH decreases. Soil pH effects on the bioavailability of oxyanions are more variable. Arsenic, Mo, Se, and some forms of Cr often become more available as pH increases.

There are a number of reasons the bioavailability of trace elements is affected by changes in soil pH. Protons or hydroxyl ions can compete for adsorption or complexation sites and, therefore, a change in the activity of H^+ or OH^- can alter the partitioning of trace elements between the soil solution and the soil solids. A change in the soil solution pH can change the suite of trace element–containing soluble species. For example, at pH < 6 Pb^{2+} predominates while at pH 6 to 11 the $PbOH^+$ form predominates. Trace element solid phases can also have pH-dependent solubilities. In situations where these solid phases are controlling the solubility of a trace element, the bioavailability of the element will be influenced by changes in pH. A simple example is given below where the mineral tenorite (CuO) is dissolved by the addition of hydrogen ions. The equilibrium expression is also given along with an algebraic expression describing the changes in the logarithm of the Cu^{2+} activity with changes in pH. In deriving the last expression we have assumed that the activities of H_2O and the solid mineral CuO are equal to one, taken the logarithm of the equilibrium expression, and converted log (H^+) to pH.

$$CuO + 2H^+ \rightleftharpoons Cu^{2+} + H_2O \qquad (7.2)$$

$$\frac{(Cu^{2+})(H_2O)}{(CuO)(H^+)^2} = 10^{7.66} \qquad (7.3)$$

$$\log(Cu^{2+}) = 7.66 - 2\,pH \qquad (7.4)$$

So as the pH decreases, the Cu^{2+} activity will increase, and vice versa.

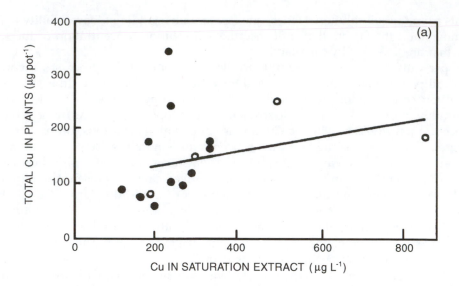

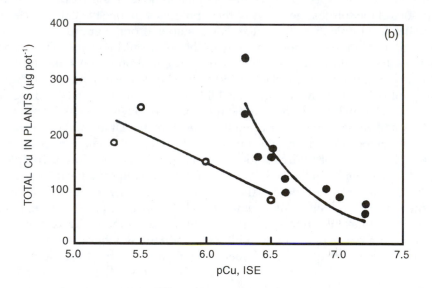

Figure 7.6 Relationship between total Cu in snapbean plants and Cu_T (a) or pCu (b) in the soil solution. Open circles represent Cu additions to the soil from $CuSO_4$ and filled circles represent Cu additions from biosolids. (From Minnich, M. M. et al., *Soil Sci. Soc. Am. J.*, 51, 573, 1987. With permission.)

Table 7.6 illustrates the influence of soil pH on Cd, Cu, Mo, and Ni concentrations in alfalfa tissue. The bioavailability of Cd, Cu, and Ni decreases with increasing pH, whereas the opposite is true for Mo.

Our discussion thus far has focused on plants and the soil solution. It is not hard to imagine that the bioavailability of trace elements to other organisms, such as earthworms or other soil invertebrates that are in intimate contact with the soil solution, would also be influenced by similar factors.

For the bioavailability of trace elements to humans a completely different environment must be considered. Risk assessments for some trace elements, such as As and Pb, typically

Table 7.6 Concentrations (mg/kg) of Selected Elements in Alfalfa Tissue as Influenced by Soil pH

pH	Cd	Cu	Ni	Mo
6.0	0.8	17.7	1.9	193
7.0	0.6	16.8	0.8	342
7.7	0.4	16.0	0.8	370

Source: Pierzynski (1985).

identify direct ingestion of soil as the primary route of exposure. Young children (6 months to 6 years old) are usually the most susceptible to this exposure pathway because they and their toys are often in contact with the floor or the ground and they have a tendency to place their hands in their mouths. Most children ingest approximately 30 mg of soil per day but some can consume several grams of soil per day. Children with an abnormal craving for nonfood substances are called *pica* children and are at greatest risk for exposure to trace elements in soils.

Once soil is consumed by the child, it enters the digestive system and the behavior of the trace elements in that environment becomes important for considering bioavailability of the elements. The stomach is very acidic under fasting conditions with a pH as low as 1.5. The pH rises to approximately 7.0 when food is consumed and as the food moves into the upper intestines. In addition, the digestive fluids contain salts and enzymes that are not normally found in a soil environment. In such a harsh environment, the possibility exists that trace element solid phases may dissolve and release the trace elements into the digestive fluids where they are available for absorption through the intestinal walls. Bioavailability is then related to the likelihood that the trace element will desorb or dissolve in the digestive system. This tendency may be either a kinetically controlled process or an equilibrium-controlled process. Kinetics may be important because the soil particles may not come to equilibrium with the digestive fluids in the time that they are present in the digestive system. In this case, a trace element that is released faster may have a higher bioavailability than one that is released more slowly. If desorption or dissolution is controlled by equilibrium conditions, then trace element bioavailability will increase with increasing solubility. As will be shown in Chapter 11, converting soil Pb to insoluble forms is one way to reduce bioavailability.

Table 7.7 shows the relative bioavailability of Pb from various Pb minerals and contaminated soils as determined by pig-feeding studies. Here the pigs have been fed the Pb-contaminated materials in their regular diet and the amount of Pb that ends up in the blood, bone, liver, or kidney is used to estimate bioavailability compared with doses of Pb from Pb acetate, a soluble form of Pb, that give similar responses (hence, the name relative bioavailability). In this case galena (PbS), a form of Pb that is often mined, has a very low relative bioavailability while the relative bioavailability of Pb from different contaminated soils varies considerably. Lead bioavailability from leaded paint is quite high. Data such as these can be related to potential human response to Pb in contaminated soils, as will be shown in Chapter 12. Grazing animals are also exposed to soil trace elements because the animals consume considerable amounts of soil that adhere to vegetation or plant roots. If they are ruminant animals, then the behavior of the trace elements in the ruminant digestive system (which is quite different from the mongastric system) would need to be considered.

7.6 Radionuclides

The radionuclides include a wide variety of radioactive isotopes of elements that are the result of naturally occurring isotopes and our efforts to produce electricity with nuclear power, to develop and use nuclear weapons, and to use radionuclides in research and for

Table 7.7   Relative Bioavailability, as Compared to Lead Acetate, of Lead in Galena (PbS), Paint, and Various Lead Contaminated Soils and Mine Waste Materials as Determined by Swine Feeding Studies

Pb Source	Relative bioavailability as determined by				Ref.
	Blood	Bone	Liver	Kidney	
Galena (PbS)	0.01	0.01	0.00	0.01	1
Leaded paint	0.82	0.63	0.85	0.70	2
Leadville, CO (soil)	0.71	0.62	0.92	0.92	3
Lake Fork Trailer Park, CO (soil)	0.87	0.84	0.96	1.24	3
Arkansas Valley, CO (mine waste)	0.20	0.18	0.11	0.10	3
Oregon Gulch (mine waste)	0.06	0.004	0.05	0.04	3
Butte, MT (soil)	0.22	0.13	0.09	0.13	4
Midvale, UT (mine waste)	0.20	0.09	0.08	0.08	5

Sources:

1. Casteel et al., 1998a.
2. Casteel et al., 1998b.
3. Casteel et al., 1998c.
4. Casteel et al., 1998d.
5. Casteel et al., 1998e.

medical purposes. Radionuclides might include various isotopes of americium (Am), cerium (Ce), Co, cesium (Cs), Fe, I, krypton (Kr), plutonium (Pu), radium (Ra), radon (Rn), ruthenium (Ru), thorium (Th), uranium (U), and Zn that are normally present in plants and soils at low concentrations and as such would be considered trace elements. In addition, there are elements such as barium (Ba), C, H, P, and S that are typically present in high concentrations in soils or plants that would not be considered trace elements and that have radioactive isotopes. Regardless, since many of the concepts and concerns about radionuclides are unique, they will be discussed separately — despite the fact that most are trace elements — rather than trying to include these unique aspects into the previous sections. The environmental chemistry of the radionuclides is similar to that already discussed for trace elements. Radionuclides occur as cationic metals, oxyanions, and halides and the fate and transport processes will be the same. Bioavailability issues will also be similar with the addition of the effects of radiation on organisms.

A brief explanation of some of the concepts of radionuclides is necessary for an understanding of the sources of the radionuclides and the potential environmental problems. Two parameters that are used to describe the nucleus of an atom are the *atomic number* and the *mass number*. The atomic number is the number of protons in the nucleus and this determines the identity of the element. The mass number is the sum of the protons and neutrons in the nucleus. A given element will have a set atomic number but different atoms can have varying mass numbers, meaning that the number of neutrons in the nucleus varies. This collection of nuclei is called the *isotopes* for that element, defined as having the same atomic number but varying mass numbers. The isotopes of an element are described by the system: ^{N}X where X is the elemental symbol and N is the mass number. For example, P, with an atomic number of 15, has the known isotopes of ^{28}P, ^{29}P, ^{30}P, ^{31}P, ^{32}P, ^{33}P, and ^{34}P. The number of isotopes per element varies from H with 3 to Cs and xenon (Xe) with 36 each. Isotopes that do not undergo radioactive decay are called *stable isotopes*, and those that do are called *radionuclides*.

It would seem possible that an element could have a large number of isotopes because, at first glance, it would appear that any number of neutrons could be associated with a fixed number of protons. However, relatively few combinations are stable enough to persist in nature and the range of neutron to proton ratios for known isotopes is relatively narrow (0.5 to 3.0). This leads to a discussion of *radioactive decay*. Radioactive decay occurs when the ratio of neutrons to protons in the nucleus of an atom is such that the nucleus is unstable. Under these conditions the isotopes will undergo nuclear changes sooner or later such that the resulting nucleus is in a position of greater stability than the initial condition. So radioactive decay can be viewed as an adjustment of the neutron-to-proton ratio from a less stable condition to a more stable condition for the nucleus of an atom. There are numerous modes of decay that can generally be described as the conversion of neutrons to protons, the conversion of protons to neutrons, the combination of an electron and a proton to form a neutron, or the emission of an alpha particle which consists of two protons and two neutrons (elements of higher atomic number only). In each mode the neutron-to-proton ratio is changed by gaining or losing protons or neutrons, or by the loss of two protons and two neutrons. These changes are accompanied by the loss of other subatomic particles and energy in various forms. The emission of the subatomic particles and the energy are what makes the radionuclides both useful and dangerous.

A measure of the rate of radioactive decay of an element is its *half-life*. The half-life is the length of time required for half of the atoms to undergo radioactive decay. Each radioactive isotope has its own half-life, and the values can range from milliseconds to millions of years. The rate of radioactive decay is described with the unit of becquerel (Bq), which is defined as one decay per second. This unit is named after Henri Becquerel, a French scientist who discovered radioactivity in 1896. The rate of radioactive decay is also described with the unit of curie (Ci), defined as the rate of decay of one gram of pure radium (^{226}Ra), or 3.7×10^{10} disintegrations per second, and named after Marie Curie, who was a leader in the study of radioactive elements. Logically, 1 Ci = 3.7×10^{10} Bq. To put this in perspective, an operating nuclear power plant would contain approximately 5.5×10^{20} Bq (1.5×10^{10} Ci); relatively concentrated radionuclides stored for research purposes would contain approximately 3.7×10^8 Bq (10 mCi); diluted radionuclides used in biological experiments would contain 3.7×10^5 Bq (10 μCi); and the naturally occurring radioactivity in a 70-kg person from ^{40}K would be approximately 4440 Bq (0.12 μCi). As an example, if one had 4.0×10^6 Bq of ^{90}Sr (half-life = 29 years), there would be 1.0×10^6 Bq after 58 years (two half-lives).

All elements that occur naturally today have at least one stable isotope or one isotope that has a very long half-life. A *stable isotope* is an isotope that does not undergo radioactive decay. Many elements have several stable isotopes. Phosphorus, used as an example previously, has seven isotopes but only one stable isotope (^{31}P). The remaining isotopes all have relatively short half-lives (<25 days) that have long since decayed so the *natural abundance* of ^{31}P is 100%. Iron, on the other hand, has ten isotopes, four of which are stable. The isotopes are ^{54}Fe, ^{56}Fe, ^{57}Fe, and ^{58}Fe with natural abundances of 5.8, 91.8, 2.1, and 0.3%, respectively.

7.6.1   Occurrence of radionuclides

Radioactive elements are present for a variety of reasons. A number of *primordial isotopes* were present when the earth was formed. These isotopes obviously have long half-lives compared to the age of the earth. Some of these isotopes decay to nonradioactive daughter products while others, such as the isotopes of U, decay to a series of radioactive daughter products. The decay scheme for ^{238}U, for example, has 14 radioactive daughter products with half-lives ranging from microseconds to thousands of years before the stable isotope ^{206}Pb is produced (Figure 7.7). The inert gas ^{222}Rn (radon gas) is one of the daughter products that has recently received attention as a hazard in the home when it builds up in the indoor air and is inhaled; this will be discussed under environmental effects in Section 7.6.3. The Rn comes from the decay of U daughter products that are present in small amounts in the soil and rocks underneath the house.

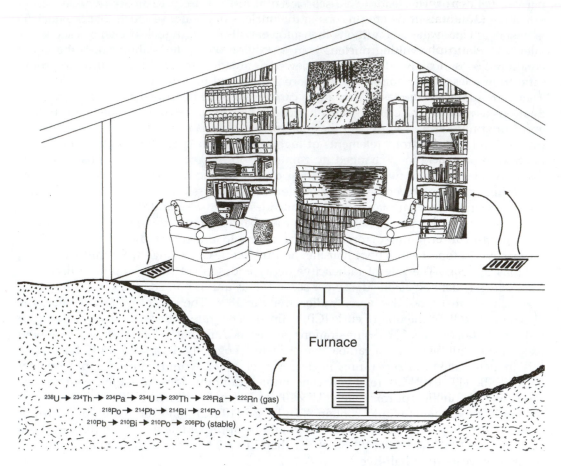

Figure 7.7 Production of Rn from the radioactive decay of ^{238}U and possible entry pathways for a home basement. (Drawing by Sarah Blair.)

A fairly constant input of radionuclides from the atmosphere originates from the interaction of cosmic rays with air and other atmospheric constituents. One such important isotope is ^{14}C, which is often used for determining the age of organisms that have been dead for long periods of time. The atmosphere contains a fairly uniform amount of cosmogenic ^{14}C that is incorporated into living tissue — the uptake of carbon dioxide (CO_2) during photosynthesis, for example. When the organism dies, ^{14}C assimilation stops and the age of the organism can be estimated based on the initial amount of ^{14}C assumed to be present and the activity of ^{14}C remaining in the sample. This works if the organism is not too much older than the 5760-year half-life of ^{14}C .

The infrastructure associated with the nuclear power industry also concentrates and generates a significant quantity of radioactive materials. The mining of U for nuclear fuel poses a host of environmental and occupational problems ranging from worker safety to potential ecosystem exposure to radioactivity from dust and wastewater. Conditions inside the reactor are such that a variety of radionuclides are produced as daughter and fission products. *Fission products* are the result of splitting of nuclei to produce elements of lower atomic weight. Nearly 40 radioactive isotopes are present in an operating light-water nuclear reactor, although many of these have short half-lives and decay quickly when the reactor is shut down. When these daughter and fission products are contained, they pose no threat to the environment, but when accidentally released, as occurred in the Chernobyl nuclear accident in the former Soviet Union in 1986, they present significant risks to exposed individuals. Similarly, the infrastructure associated with the production, use, and testing of nuclear weapons concentrates and generates significant quantities of radioactive materials. Atmospheric fallout from aboveground testing of nuclear weapons has delivered low doses of radioactivity worldwide, and many of the facilities used in the development of the weapons are contaminated with radionuclides.

Finally, the use of radionuclides for research or medical purposes produces large quantities of radioactive materials. In research, radioactive isotopes are used to trace the path of elements through chemical reactions and biochemical processes. In medicine, radioactive isotopes are used for diagnostic and therapeutic purposes. Generally, the use of radionuclides in this capacity does not pose a significant threat to the environment, although the disposal of the resulting low-level radioactive waste has been a major issue in society.

Table 7.8 presents general information on a number of environmentally important isotopes. These isotopes are either naturally occurring and contribute to background radiation or have been released into the environment from anthropogenic activities and have been a concern from a human health perspective.

7.6.2   Characterizing radiation dose

As there are numerous modes of decay for radioactive elements, there are also different types of nuclear radiation that can be produced. A single nucleus can decay and release more than one type of radiation. The most important types of radiation from an environmental standpoint are alpha (α) particles, gamma (γ) rays, X rays, and positively or negatively charged beta (β) particles. Alpha particles are helium nuclei containing two neutrons and two protons that are emitted when some isotopes with a high atomic number decay. They are large particles that travel at a relatively low velocity. Gamma and X rays are electromagnetic radiation of varying energies that travel at the speed of light. The emission of γ and X rays does not change the number of neutrons or protons in a nucleus. Rather, these are generated when atoms release energy as a result of radioactive decay. Beta particles are small charged particles that travel at velocities near the speed of light.

A complete description of the methods for detecting radiation produced from radioactive decay is beyond the scope of this book. The methods depend on the type and energy of the radiation. Survey-type instruments, such as the Geiger counter, can be used to detect gross counts of nuclear disintegrations. More sophisticated methods are used in research applications or for environmental monitoring for the quantification of radioactivity from individual isotopes.

Table 7.8   Selected Environmentally Significant Radionuclides

Isotope	Half-life	Principal mode of decay	Sources/concerns
^{14}C	5760 years	beta	Cosmogenic, nuclear reactors, and explosions
^{60}Co	5.27 years	beta	Nuclear reactors and explosions
^{134}Cs	2.06 years	beta	Nuclear reactors and explosions
^{137}Cs	30.1 years	beta	Nuclear reactors and explosions
^{3}H	12.3 years	beta	Nuclear reactors and explosions
^{131}I	8.1 days	beta	Nuclear reactors and explosions
^{40}K	1.3×10^9 years	beta	Ubiquitous, contributes to background radiation
^{222}Rn	3.8 days	alpha	Inert gas that can build up in homes; contributes to background radiation
^{90}Sr	29 years	beta	Nuclear reactors and explosions
^{235}U	7.1×10^8 years	alpha	Used in nuclear weapons and reactors, source of Rn
^{238}U	4.5×10^6 years	alpha	Used in nuclear weapons and reactors, source of Rn

From an environmental standpoint we are interested in the effects of the nuclear radiation on organisms and how exposure occurs. In this case it is justified to focus on the human health aspects because some of the health effects, such as cancer, have long latency periods and are generally not a concern for other organisms. Acute radiation effects are quite rare and are associated with serious industrial accidents or nuclear war. Low-level chronic exposure over long periods of time is far more common. In both cases, the ability of the radiation to strip electrons from atoms and form ions in living tissue leads to much of the damage that occurs and is the basis of the name *ionizing radiation*. The ionization ruptures chemical bonds and disturbs normal cell functions. The number of cells that are disrupted, their functions, and the biochemical processes that are affected determine the overall effect on the whole organism.

The absorbed dose of radiation by living tissue is described with the units of *gray* (Gy) or *radiation absorbed dose* (rad). The Gy is the preferred SI unit. These units characterize the amount of energy that is deposited in living tissue as radiation passes through; 1 Gy is equivalent to 10,000 ergs of energy deposited per gram of tissue and 1 rad is equivalent to 100 ergs of energy per gram (1 Gy = 100 rads). The effect of that absorption varies depending on the type of tissue and the energy or source of the radiation. For this reason, the rad or Gy is not used to characterize radiation effects. To do so, these additional factors are taken into account and the equivalent dose is used. The *equivalent dose* has the units of *sievert* (Sv) or *roentgen equivalent man* (rem), which reflects the amount of energy dissipated in the tissue and the potential biological damage from that energy. The Sv is the preferred SI unit; 1 Sv is equivalent to 100 rem. For small doses the mSv is often used, where 1000 mSv = 1 Sv. The advantage of the equivalent dose is that a given dose will have the same effect regardless of the radiation source.

Whole-body exposure to 250 mSv in a short period of time can have measurable effects on white blood cell count, and 4000 mSv will result in death to half of the exposed population. Natural background radiation exposures are approximately 2 mSv/year from outdoor activities and 5 mSv/year from living in a brick house. A dental or chest X-ray film supplies 10 mSv per film. Allowable exposure levels for the whole body, head, and trunk are on the order of 50 mSv/year or 12.5 mSv for each 13-week period with higher exposures allowed if confined to the extremities. These levels are for workers who are occupationally exposed and should not be viewed as exposure guidelines for the general public. There are additional concerns for internal exposure. Internal exposure includes inhalation and ingestion of radionuclides in air, food, and water or absorption through the skin. In this case, all of the energy from the isotopes may be dissipated within the body. In addition, some radionuclides are selectively concentrated within the body. Examples include I in the thyroid gland and Ca, strontium (Sr), and P in the bones. From an environmental standpoint, internal exposure through food chain transfer or inhalation pathways is generally of greatest concern.

7.6.3   Environmental effects

Significant environmental problems from radionuclides are the result of mining and processing of U ores and fuels for weapons and nuclear power plants, nuclear weapons detonations and testing, nuclear power plant accidents, and the occurrence of radon gas in homes. Some additional risk exists in storage facilities for nuclear wastes (including spent nuclear reactor fuels), but no significant problems have occurred in recent history. With the exception of radon gas, the locations of most of the contaminated sites are known and there is little exposure to the general public from radiation beyond normal background levels.

Any contamination resulting from U will involve a variety of radioactive isotopes. As is shown in Figure 7.7, the decay scheme for ^{238}U has 14 radioactive daughter products. This isotope represents approximately 99.3% of the U in U ore. The balance of the U in the ore is generally ^{235}U, which produces nine radioactive daughter products before the stable isotope ^{207}Pb is formed. In addition, nuclear reactors and nuclear detonations produce a variety of other radioactive isotopes with varying half-lives. Some of greater significance include ^{137}Cs, ^{90}Sr, ^{131}I, and ^{129}I. The Cs and Sr isotopes have long half-lives and are the primary contaminants remaining after nuclear explosions and reactor accidents involving the core material. The I isotopes have short half-lives, but are concentrated in the thyroid gland.

Uranium mining activities are relatively small compared with mining activities for metals or coal so the potential area of impact is restricted. Mine worker exposure is the primary concern. Much of our knowledge related to radon exposure comes from data on U mine workers. Facilities involved with the processing of the ores, particularly the production of highly enriched materials for use in weapons and as reactor fuel, have created a number of highly contaminated sites. Early work with radioisotopes proceeded without much knowledge of the health effects from radiation exposure or an appreciation for the fate and transport of radionuclides in the environment. Radioactive wastes were often disposed in open pits with resulting contamination to soil, sediment, and groundwater environments. Major facilities are located at the Savanah River Site in South Carolina, the Oak Ridge Site in Tennessee, Rocky Flats in Colorado, and the Fernald Feed Materials Production Center in Ohio. The elements of concern include radioactive isotopes of U, Ra, Rn, Th, Pu, and Cs as well as ^3H, ^{90}Sr, and ^{60}Co. Public access to these areas is restricted and public exposure to radiation hazards is essentially nonexistent. The Department of Energy is spending millions of dollars in cleanup efforts. Most of the remediation thus far has been either *in situ* stabilization or excavation and storage at approved waste disposal sites.

The most severe nuclear reactor accident occurred on April 16, 1986 at the Chernobyl nuclear power plant in the Ukraine. The combination of some safety systems for one reactor that had been turned off during an experiment and a number of safety violations allowed an explosion to occur that punctured the roof of the reactor building and exposed the reactor core to the atmosphere. Ensuing fires allowed radioactive elements to escape from the reactor until May 6 when liquid nitrogen was used to cool the reactor core. Approximately 3.5% of the reactor inventory was released as 16 different radioactive isotopes. From a human health perspective, the isotopes ^{131}I, ^{134}Cs, ^{137}Cs, and ^{90}Sr were the most important. The Soviet republics of Ukraine and Byelorussia were the most affected. Approximately 135,000 people were evacuated from within 30 km of the reactor. Outside of the former Soviet Union, much of the fallout occurred north of the Chernobyl site in the Nordic countries. Sweden received approximately 5% of the fallout. Areas in Sweden received fallout of ^{137}Cs 100 times higher than that during atmospheric testing of nuclear weapons.

Of the 203 people diagnosed with acute radiation sickness, 31 died within a few weeks of the accident. People in the immediate vicinity received substantial quantities of ^{131}I, which accumulates in the thyroid gland and is expected to increase the incidence of thyroid cancer. Attempts were made to provide people with KI supplements to prevent uptake of ^{131}I by the gland, but they did not occur quickly enough to prevent all of the problems. Equivalent doses to people within 30 km of the reactor ranged from 4 to 3000 mSv.

Outside the Soviet Union, scientists focused on the exposure of the general public to ^{137}Cs. Exposure could occur by external radiation from exposure of contaminated soils or other materials, inhalation of contaminated dust, or by consumption of contaminated food. Food chain transfer was considered to be a significant long-term threat and considerable efforts were made to minimize this risk. Tillage was recommended for soils to bury and dilute the radionuclides. Forage crops that were in the field during the fallout were not fed to animals. When the forages were cut, a higher cutting height than normal was recommended. Animals were kept out of pastures that had received fallout for some time afterward. Potassium fertilization was recommended because high K levels helped prevent ^{137}Cs uptake by plants. Overall, the highest radiation doses estimated for people outside the Soviet Union for the year after the accident were approximately 7 mSv, which is more than the typical background dose but less than what is believed to be the minimum dose to have health effects. The true health effects will not be known for a long time.

Several minor (by comparison to Chernobyl) nuclear reactor accidents have occurred in the U.S. The most recent was the Three Mile Island Reactor near Harrisburg, PA in 1979 and is the only accident involving a reactor producing electricity for commercial purposes. Approximately 120,000 L of cooling water was released accidentally during an incident in which a partial reactor core meltdown occurred. Most of the radioactivity that was released was in the form of radioactive Kr and Xe gases so no widespread contamination of soils or vegetation occurred. Some ^{131}I was released and exposure was detectable in a few individuals. However, the maximum exposure to any individual was estimated at 1 mSv, considerably less than the annual dose from background radiation.

The occurrence of radioactive Rn gas in homes has received considerable attention in the last 10 years. Of our average annual dose of radiation, approximately 80% comes from natural sources and of that approximately 55% comes from Rn. Next to smoking, Rn is believed to be a significant contributor to lung cancer. As shown in Figure 7.7, ^{222}Rn is a daughter product from the decay of ^{238}U. The decay of ^{235}U also produces Rn as ^{219}Rn. Uranium is a minor constituent of the soil and rocks nearly everywhere but is more prevalent is some areas. When Rn is produced it enters the basement or crawl space of homes, where it accumulates. The health effects of Rn are not actually caused by Rn itself. Radon is a relatively inert gas and nearly as much is exhaled as is inhaled so there is little accumulation in the lungs. The problems arise because both isotopes of Rn have short half-lives and decay to one of the isotopes of polonium (Po). Radon-222 decays to ^{215}Po, and ^{219}Rn decays to ^{216}Po. The Po atoms have an electrostatic charge that allows them to adsorb to dust particles in the air, which facilitates their transport around the home. Radon itself is a heavy gas that would remain close to the floor if left undisturbed. Levels of Rn that can be found in homes vary tremendously from location to location. Testing of individual homes is the only way to be certain if a problem exists and to what extent it occurs. If a problem is discovered, the remedy generally involves supplying ventilation for the basement or crawl space so that the radionuclides are removed before they can be inhaled. For new construction, plastic barriers can prevent the movement of Rn into the home in the first place.

Environmental quality issues/events
Lead and zinc mining in the Tri-State Mining Region

The Tri-State Mining Region encompasses southeast Kansas, southwest Missouri, and northeast Oklahoma in the U.S. Lead and Zn were mined extensively in this area for nearly 100 years beginning in the mid-1800s. The process of mining the ores and producing the final products of elemental Pb or Zn involved three steps: (1) the actual mining of the ores (sphalerite, ZnS, and galena, PbS, were the primary ores); (2) concentration of the ores; and (3) smelting of the concentrated ores. Early mining was generally done by excavating the ore from shallow pits, followed by crude concentrating and smelting techniques. Thousands of small operations dotted the landscape in the mining region. As mining techniques improved and the surface deposits were played out, shaft mining became the dominant method. Concentrating techniques also improved and fewer but larger smelters were used to process the ores. Most mining had ended by the 1950s with some smelting operations continuing into the 1970s.

The environmental problems from the mining are widespread and complex. Ecological and human health risk assessments have indicated that Cd, Zn, and Pb are the primary contaminants of concern. In the areas where mining took place, the landscape is dominated by abandoned pits, shafts, smelters, and other equipment used in processing the ores, reflecting the range of technology that was used in the mining efforts, and representing direct safety threats to persons venturing onto the sites. A variety of waste materials are also evident. Chat was the first by-product produced from processing ore. It still has high concentrations of Cd, Pb, and Zn and is present in large quantities at different locations. Chat has found use as a construction material and consequently has been transported all over the region. Smelter slags are present in much lower quantities but are highly enriched in metals. Smelters that were in operation most recently had tall smokestacks that emitted metal-enriched dust that was spread over large areas. One such smelter in Joplin, MO was surrounded by a residential neighborhood and created contaminated soils requiring remediation over 3 km from the stack. Since potential health effects of the contaminants of concern were not known until relatively recently, homes and schools have been built directly on or adjacent to mine wastes. These areas, plus those impacted by smelter emissions, demonstrate the potential intimate contact between the contaminants of concern and the local population.

Depending on the location, mine shafts and pits often intersected shallow aquifers that were used as drinking water sources for individual homes. Surveys of water from these wells found that many exceeded the maximum contaminant levels (MCLs) for both Cd and Pb. Water samples from deeper aquifers had acceptable Cd and Pb concentrations.

Epidemiological studies have suggested that people living in or near the affected areas have higher incidences of chronic kidney disease, heart disease, skin cancer, and anemia compared with people living in nearby control areas. Surveys have found 10 to 20% of children 6 to 72 months old in some communities in the area have blood Pb concentrations higher than the maximum health guideline of 10 μg/dL. Human health risk assessments clearly demonstrate consumption of Cd above suggested guidelines (reference doses) for both adults and children. Most of the increased risk came from consumption of locally grown food with dermal contact with soil, consumption of drinking water derived from shallow wells, and incidental ingestion of soil also being significant contributors. Data such as these provide evidence that genuine human health problems exist and should be addressed.

The lack of vegetative growth on the mine spoils is direct evidence of ecological effects from the mining activities. This problem can be attributed to poor soil physical properties, low soil fertility, and to Zn phytotoxicity. The areas of exposed soil and mine wastes further exasperate problems because erosion moves metal-rich materials into sensitive aquatic environments and increases human exposure through windblown dust. Ecological risk assessments for terrestrial environments also indicate adverse effects on earthworms and other soil decomposers. Ecological risk assessments for aquatic habitats consistently show water and sediment contaminant concentrations exceeding toxicity benchmarks with observable impacts on aquatic invertebrate and fish communities.

Remediation efforts have been under way for some time, but are proceeding slowly. The region is divided into various subsites that are handled separately. Some of these subsites were placed on the National Priority List (NPL) as early as 1985 while others may still need to be listed. The region covers portions of three states and two EPA regions, which creates some administrative challenges for prioritizing and funding the cleanup efforts. In general, the initial remediation efforts in a subsite involve addressing acute human health problems followed by chronic human health problems and then possibly ecological problems. For example, to minimize Pb exposure in Joplin, MO, an emergency removal order allowed contaminated soil to be removed from around schools, playgrounds, child-care centers, and other areas where children congregate. This was followed by soil removal from around homes with children known to have blood Pb concentrations in excess of 10 μg/dL, soil removal from around homes with children, and then finally from around homes having soil exceeding 800 mg/kg Pb. In addition to soil removal, people were educated on how to reduce Pb exposure by proper cleaning of their homes, frequent hand washing, and maintaining proper vegetative cover in their yards. Homes using water from the shallow aquifer were supplied with an alternative water source, either bottled or through a new public water system. Ecological problems have not received a great deal of attention yet; however, areas around Galena, KS were revegetated and recontoured in an attempt to stabilize large areas immediately around the city that were completely void of vegetation.

Preliminary results suggest treatment of acute health problems has been somewhat successful. Blood Pb concentrations of children in Picher, OK that were initially >10 μg/dL did decrease after their homes were remediated. Additional follow-up surveys will be required to determine definitively if the remediation efforts have been successful. Such surveys are required 5 years after the remediation is complete. While to some the remediation efforts seem to be taking too long, people must realize that it took over 100 years to create the problems and they will not be overcome with just a few years of effort.

Problems

7.1 A poultry litter contains 600 mg/kg Cu, 500 mg/kg Zn, 35 mg/kg As, and 5 mg/kg Se. If the material is applied at 5 Mg/ha (dry-weight basis), calculate the application rates for Cu, Zn, As, and Se in kg/ha.

7.2 Using the highest reported concentrations for As, Cd, and Pb given in Table 7.2, calculate the application rates for As, Cd, and Pb from the use of triple superphosphate, monoammonium phosphate, and diammonium phosphate supplying 40 kg P/ha.

7.3 Using the As application rate in Problems 7.1 and 7.2, determine how many applications of triple superphosphate would be required to equal the As application from one application of poultry litter.

7.4 The top 30 cm of soil is contaminated and has 950 mg/kg Pb. Calculate the mass of Pb (in kg) in 1 ha of soil to the 30 cm depth. Assume a bulk density of the soil of 1.3 g/cm^3.

7.5 List and describe at least five anthropogenic sources of trace elements for soils.

7.6 With 25 mL of a salt solution, 2 g of soil are extracted. The concentration of Cd in the extract is 0.1 mg Cd/L. In a separate analysis, 1 g of soil is completely digested in a small volume of strong acid. The digest solution is diluted to 50 mL and the concentration of Cd in the solution is 0.5 mg/L. Calculate the extractable and total Cd concentration of the soil in mg Cd/kg.

7.7 A soil contaminated with ^{137}Cs received 200 kBq/m^2 in nuclear fallout. Calculate the radioactivity 90 years later in kBq/m^2.

7.8 After the Chernobyl accident, plant uptake of ^{137}Cs was characterized by a transfer factor (TF) where:

$$TF = \frac{\text{activity in plant dry matter (Bq/kg)}}{\text{activity deposited (Bq/m}^2)}$$

An area in Sweden received an average fallout of 33 kBq/m². Grass grown in a pasture contained 680 Bq/kg prior to deep tillage (plowing) and 125 Bq/kg afterward. Calculate the TF before and after plowing to determine the effectiveness of this treatment for reducing ^{137}Cs uptake. (Data calculated from Rosen, 1996.)

7.9 Describe the major adverse effects from high levels of trace elements in the environment. Be sure to link the effect with the element that causes the effect.

7.10 List five trace elements that occur as cations and five that occur as oxyanions in the soil solution. Use the elemental symbols and provide the valence for the cations. Use the molecular formula and provide the valence for the oxyanions.

7.11 Define bioavailability as it relates to trace elements in soils. Describe how bioavailability influences the potential adverse effects from trace elements in soils.

7.12 Describe how factors that influence the bioavailability of trace elements in soils to plants might be similar or different from those that influence the bioavailability of trace elements in soils to humans after the soil has been ingested.

References

Bowie, S. H. U. and I. Thornton, Eds. 1985. *Environmental Geochemistry and Health*, Kluwer Academic, Hingham, MA.

California Department of Food and Agriculture. 1998. Development of risk-based concentrations for arsenic, cadmium, and lead in inorganic commercial fertilizers, prepared for the California Department of Food and Agriculture by Foster Wheeler Environmental Corporation, Sacramento, CA.

Casteel, S. W., L. D. Brown, M. E. Dunsmore, C. P. Weis, G. M. Henningsen, E. Hoffman, W. J. Brattin, and T. L. Hammon. 1998a. Bioavailability of Lead in Unweathered Galena Rich Soil, U.S. EPA Region VIII, Denver, CO.

Casteel, S. W., L. D. Brown, M. E. Dunsmore, C. P. Weis, G. M. Henningsen, E. Hoffman, W. J. Brattin, and T. L. Hammon. 1998b. Bioavailability of Lead in Paint, U.S. EPA Region VIII, Denver, CO.

Casteel, S. W., L. D. Brown, M. E. Dunsmore, C. P. Weis, G. M. Henningsen, E. Hoffman, W. J. Brattin, and T. L. Hammon. 1998c. Bioavailability of Lead in Soil and Mine Waste from the California Gulch NPL Site, Leadville, CO, U.S. EPA Region VIII, Denver, CO.

Casteel, S. W., C. P. Weis, G. M. Henningsen, E. Hoffman, W. J. Brattin, and T. L. Hammon. 1998d. Bioavailability of Lead in a Soil Sample from the Butte NPL Site, Butte, MT, U.S. EPA Region VIII, Denver, CO.

Casteel, S. W., C. P. Weis, G. M. Henningsen, E. Hoffman, W. J. Brattin, and T. L. Hammon. 1998e. Bioavailability of Lead in a Slag Sample from the Midvale Slag NPL Site, Midvale, UT, U.S. EPA Region VIII, Denver, CO.

Lexmond, T. M. and F. A. M. deHaan. 1977. In *Proc. Int. Seminar on Soil Environment and Fertility Management in Intensive Agriculture (Tokyo)*, Soc. Sci. Soil Manure, Tokyo, Japan, 383–393.

Minnich, M. M., M. B. McBride, and R. L. Chaney. 1987. Copper activity in soil solution: II. Relation to copper accumulation in young snapbean plants, *Soil Sci. Soc. Am. J.*, 51, 573–578.

Nriagu, J. O. 1988. A silent epidemic of environmental metal poisoning, *Environ. Pollut.*, 50, 139–161.

Pierzynski, G. M. 1985. Agronomic Considerations for the Application of a Molybdenum-Rich Sewage Sludge to an Agricultural Soil, M.S. thesis, Michigan State University, East Lansing.

Pouyat, R. V. and M. J. McDonnell. 1991. Heavy metal accumulations in forest soils along an urban-rural gradient in southeastern New York, USA, *Water, Air, and Soil Pollut.*, 57/58, 797–807.

Rosen, K. 1996. Transfer of radiocaesium in sensitive agricultural environments after the Chernobyl fallout in Sweden. II. Marginal and semi-natural areas in the county of Jamtland, *Sci. Total Environ.*, 182, 135–145.

Tsuchiya, K. 1979. Environmental pollution and health effects, in *Cadmium Studies in Japan: A Review*, K. Tsuchiya, Ed., Elsevier/North-Holland Biomedical Press, New York.

Supplementary reading

Three excellent references for trace elements in soils are Adriano, D.C., 1986. *Trace Elements in the Terrestrial Environment*, Springer-Verlag, New York; Kabata-Pendias, A. and H. Pendias, 1992. *Trace Elements in Soils and Plants*, CRC Press, Boca Raton, FL; and Adriano, D.C., Ed., 1992. *Biogeochemistry of Trace Metals*, CRC Press, Boca Raton FL.

Eisenbud, M. 1987. *Enviromental Radioactivity*, Academic Press, New York, 475 pp.

Evangelou, V. P. 1998. *Environmental Soil and Water Chemistry, Principles and Applications*, John Wiley & Sons, New York.

Jacobs, L. W. 1989. *Selenium in Agriculture and the Environment*, Soil Science Society of America Spec. Pub. No. 23, American Society of Agronomy, Madison, WI.

chapter eight

Organic chemicals in the environment

Contents

8.1   Introduction

Organic chemicals are compounds that contain carbon (C) and usually at least one C–hydrogen (H) bond; C-based substances such as carbon monoxide (CO), carbon dioxide (CO_2), and carbonates (CO_3^{2-}) are inorganic compounds. Over the last century millions of new organic chemicals have been developed. Thousands find their way into our daily lives even though many can be toxic if used or disposed of improperly. Therefore, it is imperative to understand the nature of organic chemicals in the environment, i.e., their impacts on different ecosystems and organisms, factors related to their mobility and fate, and evaluation of analytical techniques for determining the type and quantities of various synthetic and indigenous organic substances.

The focus of this chapter is organic chemicals in atmosphere, soil, and aquatic environments. Specific areas that will be covered include adverse effects of organic chemicals on organisms and human health; fate of organic chemicals in the environment; analytical methods for evaluating organic chemicals in the environment; and integrated and alternative pest management.

8.2   Organic chemicals

The production of synthetic organic chemicals has increased rapidly since the turn of the century, with growth in the chemical industry resulting in new and improved materials that have affected the way we live. However, with the good comes the bad; increased production (over 60 million Mg of C-based materials are manufactured yearly in the U.S.) and utilization of synthetic organic compounds such as pesticides, lubricants, solvents, fuels and propellants has increased the number of incidents where organic chemicals have accidently entered the atmosphere, hydrosphere, and soil environment, along with uptake by plant and microbial biota. More than 70,000 synthetic chemicals are used daily, most of which are organic chemicals, and human activity has contributed to the release of enormous amounts of these organic chemicals into the environment every day. Petroleum products are the dominant source of hydrocarbons released into the atmosphere due to human activities, with nearly 50% related to gasoline emissions.

Organic chemicals are also present in surface waters and groundwaters. Groundwaters, once thought to be pristine, have been found to contain a broad spectrum of organic chemicals that are both indigenous (natural materials) and anthropogenic (derived from human activities). A study of groundwaters commissioned by the U.S. Congress identified approximately 175 different natural and anthropogenic organic chemicals. Many of these natural organic chemicals, as well as microorganisms present in surface waters and groundwaters, will react with chlorine (Cl), which is commonly used in drinking water treatment, to produce trihalomethanes, a group of volatile halogenated organic compounds.

8.2.1   Sources of organic chemicals

The introduction of organic chemicals into the environment can occur by design, accident, or neglect. A variety of sources can contribute to the release of organic chemicals; industrial sites typically manufacture, store, and distribute the greatest amount of organic chemicals and therefore have a higher probability of contaminating the environment. Several industrial sites with the potential for contaminating surface waters, groundwaters, and soil environments are listed in Table 8.1. Other sources that may be a cause for concern include land application of wastes, feedlot operations, improper pesticide use, landfills, oil or gas operations, to name a few. Although the quality of groundwater is influenced by natural processes, as well as human activities, organic chemical contamination is generally a

Table 8.1 Some Examples of Industrial Sites with the Potential
for Contaminating the Environment

Acid/alkali plants	Pharmaceutical, perfumes, cosmetics, toiletries
Asbestos	Polymers and coatings
Chemical and allied products	Railway yards
Explosives and munitions	Scrap yards
Gas works and storage	Shipping docks
Metal treatment and finishing	Smelting and refining
Mining and extractive industries	Sewage treatment
Oil production	Tanning
Oil storage	Waste disposal
Paints	Wood preservation
Pesticide manufacture	

consequence of improper management by humans. As an example, creosote from a railroad tie operation has resulted in contamination of surface water, groundwater, soil and subsurface materials that, as described in the environmental quality issues/events box: Restoration of a contaminated wood-preservation site (p. 301), can be difficult to clean up.

8.2.2 Categories of organic chemicals

Organic chemicals can be broadly classified as aliphatic (straight-chain), cyclic (closed-ring), or combinations of aliphatic and cyclic, depending on their structural configuration. Many of these compounds also possess oxygen (O), nitrogen (N), phosphorus (P), sulfur (S) and halides (e.g., fluorine (F), chlorine (Cl), iodine (I)), atoms that impart specific qualities to these substances that can make them extremely useful; however, some organic chemicals can be highly toxic and hazardous to humans and the environment. Several industrial organic chemicals can be categorized as solvents, preservatives, petroleum products, refrigerants, explosives, polymers, and pesticides, which are some of the products of most concern.

Hydrocarbons, for instance, are comprised of C and H linked together to form alkanes (single-bonded C), alkenes (double-bonded C), alkynes (triple-bonded C), alicyclic (single-bonded, cyclic C), and aryl (aromatic C) compounds. These types of compounds are generally considered to be of most environmental interest due to the great number of organic chemicals with these structures and their potential toxicity. Alkanes are the simplest of the hydrocarbons and contain C linked to C and/or H such as methane (CH_4), ethane (C_2H_6), propane (C_3H_8), with other unbranched alkane formulas of C_xH_{2x+2}, where x represents the number of C atoms. Alkenes contain at least one double-bonded pair of C atoms and include compounds such as ethene (C_2H_4) and propene ($CH_2=CH-CH_3$), as well as chlorinated substances such as trichloroethylene ($CCl_2=CHCl$) and tetrachloroethene ($CCl_2=CCl_2$). Alkyne compounds have one or more C-pair of triple bonds ($C\equiv C$). Of these, acetylene (CHCH) has the simplest structure, is highly reactive, and when burned in an oxygen stream produces temperatures in excess of 3000°C (oxyacetylene torch). Alicyclic hydrocarbons contain single-bonded, C-to-C structures such as cyclopropane (C_3H_6) and cyclohexane (C_6H_{12}). Aromatic compounds are a large and important group of hydrocarbons comprising ring structures that contain double-bonded C pairs, which have a significant influence on the stability of these compounds. The simplest aromatic member is benzene (C_6H_6), one of the most common and stable organic structures.

Organic chemicals, particularly the hydrocarbons, have many uses. Alkanes with one to four C atoms are gases; those alkanes with five or more C atoms are liquids or solids under ordinary conditions. Methane, a natural gas of environmental concern due to its increased levels in the atmosphere, has been implicated in global climate change (see Chapter 11). The chlorinated alkenes trichloroethene (TCE) and tetrachloroethene (PCE) are used primarily as solvents. PCE is used extensively in the dry-cleaning industry to remove grease, oil, and other stains from clothing. TCE is a liquid under typical environmental conditions, slightly volatile, with a high density (1.46 mg/L), and has become one of the most common environmental contaminants. Because TCE has been identified in one third of U.S. groundwater supplies and is the most common contaminant at Superfund sites, there has been increasing interest in restricting the use of TCE and remediating soils and groundwaters contaminated by TCE; release of TCE in the atmosphere has also contributed to photochemical smog. Petroleum products contain the aromatic hydrocarbons benzene, toluene, ethylbenzene, and xylenes (*o-*, *m-*, and *p*-xylenes), otherwise known as BTEX, alkanes, and polynuclear aromatic hydrocarbons (PAHs) that are some of the most common contaminants of soils, sediments, and waters. Recently, the EPA has identified 90,000 leaking underground storage tanks (LUST) that have resulted in environmental contamination of groundwaters with gasoline. The EPA has identified BTEX compounds among the nine most common pollutants in relation to frequency of occurrence. PAHs are derived primarily from fossil fuels and significant amounts are distributed in air, water, and food resources. Some of these materials are carcinogens and mutagens (see Chapter 12 for more information on these types of substances). Another group of organic chemicals that have polluted the environment are the polychlorinated biphenyls (PCBs), which were used as insulating, cooling, and hydraulic fluids and have been detected in tissues of fish and marine mammals; the U.S. Food and Drug Administration (FDA) has set an advisory limit of 2.0 mg/kg PCBs in commercially traded fish.

Pesticides are classified according to the target organisms they are designed to control (Table 8.2). Of the target organisms, weeds by far cause the greatest economic loss due to their interference with crop production. It is not surprising, therefore, that herbicides are the most commonly used pesticides, comprising over 60% of the pesticides applied in the U.S. every year. Insects are the next greatest problem, with insecticide use totaling approximately 25% of the pesticides used annually in the U.S. Fungicides are the third most used pesticides, totaling about 6% of U.S. pesticides product consumption.

In the U.S. alone there are approximately 1/2 million Mg of pesticides used annually, most of which are used in crop production to control weeds, insects, and diseases. An estimated 25,000 Mg of pesticides are also used in nonagricultural situations such as pest control in lawns, flowerbeds, golf courses, forests, and in and along waterway, utility, and rail easements. In an EPA survey of pesticides in groundwater in the U.S., 67 different pesticides were found in 33 states (a total of 35 states participated in the survey); however,

Table 8.2 Categories of Pesticides Used to Control Unwanted Pests

Pesticide	Control
Herbicide	Prevents the growth of weeds in agricultural crops, lawns, golf courses, etc. or is applied directly to weeds that are established
Insecticide	Kills or controls harmful or undesirable insects that live on plants, in animals, or in buildings
Fungicides	Protects plants from infestations by diseases; usually used prior to conditions that are favorable to disease development
Bactericides	Controls bacteria that can cause damage to fruit and develop galls on plants
Nematicides	Protects young plant roots from microscopic soil worms that infect plants and feed upon their roots
Acaricides	Controls mites and spiders that can infest or damage agricultural crops or ornamental plants
Rodenticides	Kills mice and rats living in homes and other buildings and prevents infestations or losses of food products in storage

only 17 pesticides in 17 states were detected at levels greater than the EPA Health Advisory (HA) level. The percentage of groundwaters exceeding the HA level for any particular pesticide ranged from 0 to 10%.

Environmental quality issues/events
Process involved in the registration of a new pesticide

Pesticides were first regulated in 1947 by the Federal Insecticide, Fungicide, and Rodenticide Act (FIFRA), which authorized the U.S. Department of Agriculture to oversee pesticide use. In 1954, the Federal Food, Drug and Cosmetic Act (FFDCA) authorized the Food and Drug Administration (FDA) to regulate pesticides in food products. Administration of pesticide regulation and registration was assigned to the EPA in the early 1970s. All pesticides must now go through an extensive process to be produced and made available for commercial use. Up to 142 tests must be performed before a pesticide receives an EPA product label registration. These tests include comprehensive health, safety, and environmental evaluations that can typically take 8 to 10 years and cost up to $50 million per pesticide.

Initially, an organic chemical proposed to be used as a pesticide is manufactured in small amounts to be tested in laboratory and possibly greenhouse studies that determine its ability to control pests, i.e., "biological activity." If successful, the organic chemical is evaluated for its toxicity to laboratory animals and potential for causing genetic damage. Those pesticides that will be used for crop protection and have met efficacy and safety standards are then tested in field studies. Field studies determine the appropriate rates and methods of application under valid conditions for crop production. Environmental impacts are also evaluated to examine the potential translocation, transformation, and persistence of the pesticide tested.

A decision to continue the development of a new pesticide comes after the initial few years of testing that determine if there is a scientific and economic basis for commercializing the pesticide. Millions of dollars and an extensive time commitment are required to continue testing the chemical properties, human and environmental concerns, prospective markets, efficacy, patent rights, manufacturing technology, and production costs of the potential pesticide. The EPA requires that considerable work be performed to verify the overall safety of the pesticide, which includes potential risks to the user, crop, the environment, and consumers. Product labels must contain information on proper application methods, dosage, and particular safety issues such as protective clothing, training, specific hazards, time requirements related to human exposure of treated fields, environmental warnings, and container disposal.

Once a potential organic chemical is considered acceptable for further testing, the manufacturer must conduct numerous long-term studies before the EPA will grant a product label registration. There are six general areas of testing. *Product chemistry* testing includes evaluating the active ingredients and composition of the finished formulated product. *Toxicology* testing in animals must determine the acute effects of single doses, chronic effects of long-term or lifetime exposures, effects on reproduction, mutagenic effects on genes and inherited traits, and potential carcinogenic effects from lifetime exposure. In addition, the possibility of *pesticide residues* in food and feed products must be evaluated by testing raw crops, processed food, and animal feed, meat, milk, poultry, and eggs. *Environmental concerns* are evaluated by researching pesticide transformations in soils, waters, and different plants, and potential translocation by runoff, leaching, and spray drift. *Ecological effects* to birds and fish must be assayed to ascertain potential acute and chronic toxicity impacts, and short- and long-term effects to nontarget plants must be characterized. Finally, determining *pesticide efficacy* under a wide variety of conditions is required.

Before registering a pesticide, the manufacturer must apply for an experimental use permit (EUP) from the EPA to conduct large-scale field testing. Previous results from chemical, toxicological, environmental, and ecological tests will be assessed by the EPA before issuing an EUP. Residue tolerance levels must be determined by the EPA if the pesticide-treated crop materials are to be used for animal or human consumption, and, if tolerance limits cannot be determined, all exposed material must be destroyed. Because small amounts of pesticide residue are common in treated crops, large-scale product testing must be able to quantify the tolerance level of residual materials. These tolérance levels determine the maximum pesticide residues in foods for human consumption as well as animal feed that will be allowable by law. Additional safety margins are established by the EPA so that the residual pesticide limits will be safe to consumers.

Residual tolerances are established by the EPA as authorized by the FFDCA. The manufacturer of a pesticide must present evidence to the EPA that establishes the residue levels are safe. Safety margins defined for a potential pesticide must be determined by scientific methods that can detect the original pesticide and its metabolic products. All methods used by the manufacturer to evaluate residual compounds must be further verified by government laboratories before tolerance levels are officially certified. Although pesticide tolerances in food and feed supplies are enforced by the FDA, it is the responsibility of the USDA to monitor meat, dairy, and poultry products for pesticide residues.

Risk assessment is also required for all new pesticides. Calculations that are used in the risk assessment process must be as accurate and precise as possible so that assumptions related to different scenarios are sufficiently sound. Therefore, it is important to know the types of hazards and the extent of exposure associated with a pesticide. Studies conducted on the toxicological effects can be used to determine hazard potentials and research involving pesticide characteristics and residue products facilitates the level of exposure related to the pesticide. Experiments such as feeding studies conducted on animals can be used to determine the level of "No Observable Effect Level" (NOEL) for consumable materials. As an additional safety measure, the NOEL is divided by a factor (e.g., 100 or 1000) to determine a reference dose (RfD) or acceptable daily intake (ADI) level. These values represent the maximum amount of the pesticide residue that should be ingested by an average person over a day or lifetime so that harmful effects do not occur. See Chapter 12 for more information on risk assessment.

After all the studies, calculations, and assessments are completed, and the pesticide product is deemed useful and safe, the EPA is given the multitude of information developed for registering the organic chemical as a pesticide. The registration packet material contains volumes of information that must be reviewed, verified, and approved by the scientific and administrative branches of the EPA Office of Pesticide Programs (OPP). After the pesticide label is approved by the EPA, commercial production and distribution for sale can proceed. The pesticide label is a legal document that provides information to the user as well as special precautions that protect consumers and the environment. Noncompliance with pesticide label directions is a federal violation that can result in civil and criminal penalties to violators.

8.3   Adverse effects

Ecosystem and human/animal exposure and processes involving organic chemicals occur via many pathways and reactions, as noted in Chapter 1 and Table 8.3. The potential for negative environmental impacts and human/animal health affects due to misuse or accidents involving organic chemicals is a concern because of the difficulty in many cleanup procedures and because of the direct and indirect consequences of misuse and accidents on surrounding ecosystems and various organisms. Care must be taken in the manufacturing, storage, transportation, and handling of organic chemicals. With adequate training and precautions, many incidences of contamination could be avoided.

Several noteworthy examples of adverse effects due to pesticide misuse have been detailed in the scientific literature. Many studies have cited the potential for residual levels of pesticides in soils and aquatic systems as the reason for pesticide toxicity to plants and fish. Bioaccumulation is also of concern since over time low levels of pesticides absorbed or ingested can accumulate to toxic levels in organisms, then in offspring, or in organisms higher on the food chain. Many of the pesticides are toxic to organisms other than those specifically targeted. At high enough levels, most pesticides can also be toxic to humans. One of the greatest concerns with the development and use of pesticides is their slow breakdown and ability to accumulate in organisms. However, it should be stated clearly that pesticides, if used properly, can increase food production and control insects (e.g., mosquitoes) that may have a devastating effect on the quality of animal and human life.

The mode of action of pesticides refers to the mechanism by which the pesticide kills or interacts with the target organisms. Based on their modes of biological action, pesticides are generally classified as either *contact* or *systemic*. Contact pesticides kill target organisms by weakening or disrupting cellular membranes which, in turn, results in the loss of cellular constituents. If the contact pesticide has an acute reaction with the target organism, death may be extremely rapid. Systemic pesticides must be absorbed or ingested by the target organism so that it may interfere with the organism's physiological (e.g., cell division, chlorophyll formation, tissue development) or metabolic (e.g., respiration, enzyme activity, photosynthesis) processes. Systemic pesticides are generally slow acting and may require days, weeks, or even longer periods of time before results become evident.

Table 8.3   Specific Organic Chemical Reactions Involving Different Ecosystems and Organisms

Reaction	Atmosphere	Hydrosphere	Soils	Vegetation	Humans and animals
Solubility	X	X	X	X	X
Hydrolysis	X	X	X	X	X
Sorption	X	X	X	X	X
Biodegradation	X	X	X	X	X
Photolysis	X	X	X	X	
Volatilization	X	X	X		
Accumulation			X		
Metabolism				X	X
Bioaccumulation				X	X
Respiration				X	X
Excretion					X

Source: Adapted from Ney (1995).

8.3.1 Human health

Many toxicological effects are related to organic chemicals. The simple alkane, alkene, and alkyne hydrocarbons are volatile gases that cause asphyxiate disorders, which are related to insufficient oxygen intake. Hydrocarbon liquids are known to cause dermatitis which is associated with the dissolution of fatty skin tissues and results in inflammation, drying, and scaly skin. Other problems associated with some of these substances are irritation, headaches, dizziness, and nervous system disorders that cause muscle weakness and sensory impairment of the hands and feet. Of the BTEX compounds, benzene is the most toxic because it is readily absorbed by fatty tissue. Benzene can damage bone marrow, cause skin irritation, headaches, fatigue, and more serious problems depending on the concentration and length of exposure. For example, inhalation exposure to 7 g/m^3 can cause acute poisoning in less than 1 h due to central nervous system failure; exposure to 60 g/m^3 can result in death within minutes. Chronic benzene poisoning from long-term exposure to low levels of benzene causes lowered blood counts of white cells and platelets, and bone marrow damage. Toluene is less toxic than benzene, but inhalation exposure results in headaches, nausea, dizziness, and at very high exposures leads to comas. Naphthalene, a fused double aromatic ring structure, can cause skin irritation, dermatitis, headaches, vomiting, and in extreme cases of poisoning death due to kidney failure. Some PAHs can be metabolized to products that are thought to be mutagens and are carcinogenic.

Methanol (CH_3OH) can be oxidized to products that affect the central nervous system, the optic nerve, and retinal cells. Individuals who mistakenly consume methanol as an alcoholic beverage can suffer permanent impairment or loss of vision, or in the case of acute exposures possibly death due to cardiac depression. Phenols are known to damage spleen, pancreas, and kidney organs and cause gastrointestinal disorders, kidney malfunctions, convulsions, and circulatory system failures. TCE and PCE are known to damage liver and kidney organs, the central nervous system, and are possible human carcinogens. Methyl isocyanate (CH_3–N=C=O), a highly toxic compound that damages the lungs, was responsible for 2,000 deaths and the injury of another 100,000 people in India when several tons were accidently released in 1984. One of the most toxic organic chemicals manufactured is nerve gas, an organophosphorus compound that attacks the central nervous system. Due to its lethal nature, and the fact that the liquid forms are readily adsorbed through skin, as little as 0.01 mg/kg, or a single drop, can result in death.

Pesticides are used to benefit humankind by controlling diseases such as malaria and typhus, weed infestation that reduces crop productivity, and insect outbreaks. However, high doses of some pesticides can be harmful to humans. Laboratory research has shown that high doses of certain pesticides given to animals can cause cancer, mutagenesis, neuropathy, and even death; small doses of some pesticides may result in minor impacts such as skin irritation and breathing problems. The pesticide concentrations given in many dose–response experiments are generally well above levels that are recommended by the manufacturers for the intended purpose of the pesticide. It should also be noted that many organic chemicals, including pesticides, are less harmful to humans than are some naturally occurring substances. Although the exposure of humans to pesticides is relatively low (except in cases of misuse or accidents), we must continue to be aware of the potential impacts of pesticides. Therefore, it is necessary for everyone working with pesticides to follow directions and take every precaution to protect themselves, other humans, and our environment from becoming contaminated.

Certain organic pesticides are also potentially hazardous to human health if improperly used. Paraquat, an organonitrogen herbicide, has been widely used for control of annual and broadleaf weed control in no-till systems and is a systemic toxicant that can enter the human body through inhalation, ingestion, or direct contact. Paraquat inhibits enzyme activity and impairs lung, kidney, liver, and heart functions and was cited as causing numerous human deaths. The well-known insecticide DDT (dichlorodiphenyltrichlorethane) was introduced in the 1940s to control malaria and typhus. Banned in the U.S. and most developed countries since 1973, DDT attacks the central nervous system, causing tremor, eye twitching, memory loss, and changes in personality. Other organohalide insecticides that have been banned in the U.S. include aldrin, dieldrin, heptachlor, and chlorodane, all of which can cause headaches, nausea, vomiting, and convulsions. Parathion, a phosphorothionate with a history of causing human health problems, was responsible for the deaths of 17 out of 79 Jamaicans who were exposed to contaminated flour in 1976. Symptoms of parathion poisoning include skin twitching, breathing problems, and ultimately respiratory failure from paralysis of the central nervous system; infant and adult deaths have resulted from doses of 2 and 120 mg, respectively. Two chlorophenoxy herbicides, 2,4-di- and 2,4,5-trichlorophenoxy acetic acids, also known as 2,4-D and 2,4,5-T, have been shown to cause human health problems. High concentrations of 2,4-D can cause nerve impairment, convulsions, and brain damage. Results of a study of Kansas farmers who worked with 2,4-D suggested they were six to eight times more likely to suffer from non-Hodgkin's lymphoma than nonexposed individuals. A mixture of 2,4-D and 2,4,5,T, also known as "Agent Orange," was widely used as a defoliant in the Vietnam War. It contained a by-product material tetrachlorodioxin (TCDD), which is generally called dioxin. Dioxin may be highly toxic and cause a skin disorder called chloracne, possible liver and tissue damage, and is a potential carcinogen.

8.3.2   Impacts on plants

Plants can be exposed to various organic pollutants through atmospheric deposition, contaminant spills, irrigation waters, and land-applied materials such as municipal biosolids and animal manures. Plant response is determined by the mode of action related to different organic chemicals, which is either through direct contact or by systemic uptake. Systemic pesticides are designed and manufactured to impact certain types of plants specifically, e.g., grasses vs. broadleaves, so that selective control can be maintained. As mentioned in the previous section, some herbicides (e.g., 2,4-D and 2,4,5-T) have been shown to cause human health problems when improperly and indiscriminately introduced into the environment.

   Plants that are sensitive to pesticides can show rapid signs of growth irregularity, loss in biomass, or death. Plants are capable of building up a resistance to certain pesticides if the pesticides are used frequently and at dosages that plants can tolerate. Higher concentrations of the pesticide would then be required to achieve a similar level of control. Pesticides other than herbicides, such as insecticides, can also affect plants that are not the specific target organism the pesticide was designed to control. Table 8.4 lists effects that certain herbicides and insecticides can have on nontargeted plants located within soil and water environments.

8.3.3   Groundwater contamination

Groundwater is used for several purposes. Potable water for greater than 50% of the U.S. population and for approximately 97% of rural residents comes from groundwater sources, and over 40% of agricultural irrigation water is also derived from groundwater sources. Groundwater quality is an issue that has generated much debate and concern, and along with surface water quality was the primary reason the Clean Water Act was implemented

Table 8.4 Effects of Some Herbicides and Insecticides on Nontargeted Plants in Soil and Water Environments

Pesticide type	Location	Effect
Herbicides		
Aromatic acid	Soil	Carryover in residue affects subsequent crop; soluble herbicides (e.g., picloram) can injure adjacent plants
	Water	Kills or inhibits some aquatic plants
Amides, anilines, nitriles, esters, carbamates	Soil	Some persistence resulting in residue carryover affecting subsequent crop
	Water	Surface erosion may transport herbicide to water bodies
Insecticides		
Organochlorine	Soil	Residue carryover can affect subsequent crops; transport to surface waters may affect aquatic plants
	Water	Contaminated waters can affect plants if waters used for irrigation
Organophosphates, carbonates, pyrethroid	Soil	Usually short lived, thus little effect to plants
	Water	Toxic to certain algae

Source: Madhun, Y. A. and Freed, V. H., in *Pesticides in the Soil Environment: Processes, Impacts, and Modeling*, H. H. Cheng, Ed., SSSA Book Series No. 2, Soil Science Society of America, Madison, WI, 429, 1990.

(see Chapter 1). Several definitions have been proposed for groundwater contamination, but the definition by Miller (1980), which states, "Groundwater contamination is the degradation of the natural quality of groundwater as a result of man's activities," places the blame completely and clearly on humans and their misuse of the environment. Contamination of groundwaters occurs through the downward leaching of organic chemicals, intentional discharge of wastes in subsurface wells, or contaminant spills of highly soluble, low-sorption organic pollutants. Several factors can influence groundwater contamination potential including sorption, biodegradation, hydrolysis, solubility, volatilization, and climatic parameters such as precipitation and evapotranspiration.

Contamination of groundwaters is receiving much attention; however, assessing the magnitude of the problem will be difficult unless the contaminant sources can be identified and prevented from further contaminating the environment. The Office of Technology Assessment identified 33 principal sources that have the potential for groundwater contamination, and grouped them into six categories. Table 8.5 lists the six categories of contaminant sources grouped according to the general nature of the source of the organic contaminant.

8.3.4   Impacts on microorganisms

Organic chemicals introduced into the environment can have a devastating effect on certain organisms, which may or may not be the intended purpose of the organic chemical. Soil and aquatic ecosystems contain a multitude of microorganisms, many of which are beneficial. Contamination of these ecosystems by organic chemicals, such as when high concentrations of pesticides are inadvertently used or when surface runoff containing pesticides is introduced into surface waters, can reduce microbial activity. However, in some situations enhancement of microbial activity may occur.

Table 8.5 Sources of Organic Chemicals with the Potential
for Contaminating Groundwater and Soil

Category	Sources	Description
I	Intentional discharge	Septic tanks and cesspools
		Injection wells (hazardous wastes)
		Land application (sludge, wastewater hazardous wastes)
II	Containment	Landfills and open dumps
		Surface impoundments
		Waste tailings and piles
		Animal burial sites
		Underground storage tanks
		Storage containers
III	Transportation	Pipelines
		Rail and truck
IV	Discharge from planned activities	Pesticide application
		Animal feed operations
		Urban runoff
		Atmospheric pollutants
V	Induced flow	Production wells (oil and gas)
		Construction excavation
VI	Natural	Groundwater–surface water interactions
		Wetlands
		Underground deposits (coal, oil, gas)

Source: Office of Technology Assessment (1984).

The theoretical response of soil microorganisms to different herbicides is shown in Figure 8.1, which could also be related to organic chemicals other than herbicides and to plant responses over time to various pesticides. "Initial application" in Figure 8.1 represents the first time the herbicide was used. "Subsequent applications" refers to the response of the ecological system to continued application of the herbicide. The initial herbicide effect can be summarized as follows: herbicide A = no inhibition of soil microorganisms, rapid degradation rate, increase in microbial populations; herbicide B = no inhibition of soil microorganisms, moderate degradation rate, increase in microbial populations; herbicide C = kills sensitive soil microorganisms, increase in herbicide-degrading microorganisms over time, slow degradation rate; herbicide D = initially herbicide kills sensitive microorganisms, herbicide metabolites build up and inhibit certain microorganisms, resistant microorganisms eventually degrade herbicide; herbicide E = kills sensitive microorganisms, resistant microorganisms increase but do not degrade herbicide. Upon further herbicide application, herbicides A, B, and C are rapidly degraded; herbicide D and its metabolites are degraded at a faster rate; and herbicide E concentrations build up due to lack of degrading microorganisms. These are just a few examples of how soil ecosystems and plants might respond to organic chemical inputs. In an extreme case of contamination, the number and activity of soil microorganisms may be reduced to essentially zero.

Xenobiotic organic chemicals are synthetic compounds that are foreign substances to biological communities such as plants and microorganisms. Because pesticides are primarily xenobiotic compounds, their degradation is the result of many factors that influence microbial activity and enzyme production in response to added organic chemical substrates. Enzymes are proteins that act as catalysts in biochemical reactions. Because xenobiotics are not natural products, enzymes responsible for the degradation of these materials generally recognize specific organic chemical groups or structures that are altered or

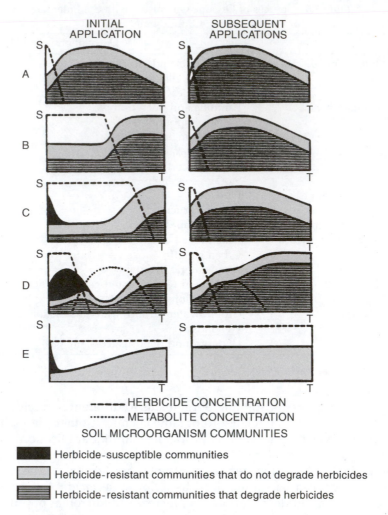

Figure 8.1 Theoretical response of soil microbial communities and degradation rates of herbicides (A through E) over time. Illustrations represent initial (left side) and subsequent (right side) applications. Time is represented on the abscissa and herbicide concentration on the ordinate axis. (Redrawn from Cullimore, 1971.)

degraded. Figure 8.1D illustrates an example of microbial degradation using specific enzymes to degrade herbicides, which are xenobiotic compounds, and metabolites produced from their partial degradation. Xenobiotics can be utilized as growth-supporting substances or cometabolized by microorganisms, primarily bacteria, with the last process resulting in the production of substances that can be innocuous or more toxic than the original compound.

8.4   Predicting the fate of organic chemicals in the environment

Reliable predictions of the movement and fate of organic chemicals in the environment (Figure 8.2) are important for determining the impact of organic chemicals on our environment. Proper use of pesticides (as well as all organic chemicals) requires a knowledge of factors that can influence their mobility and transformation, as well as human safety,

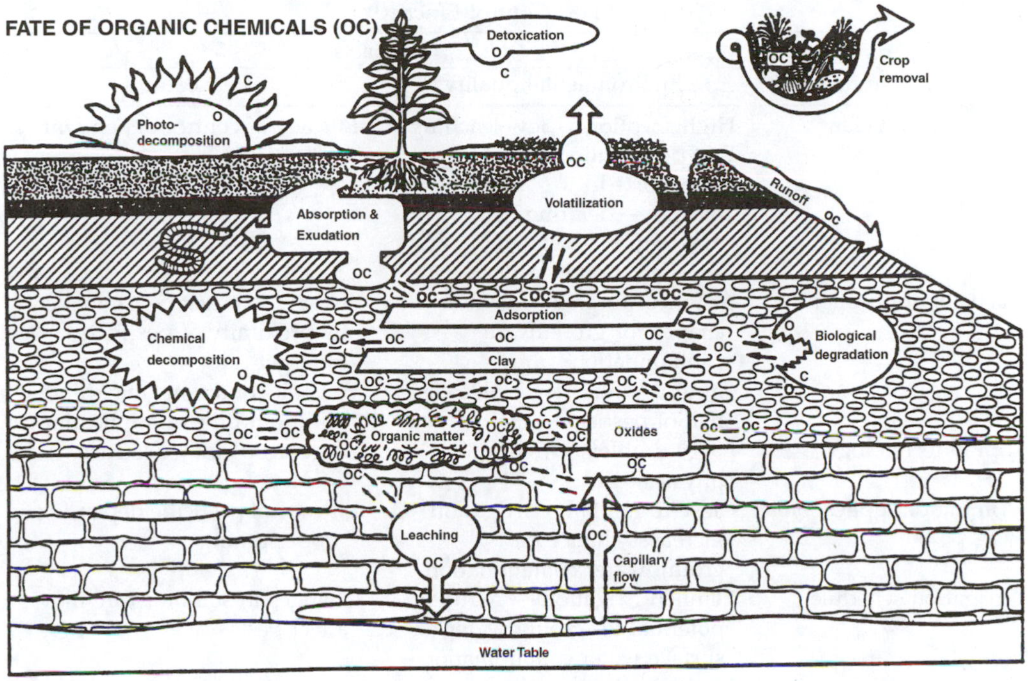

Figure 8.2 Processes and fate of organic chemicals (OC) in the environment. Both transfer and degradation processes are depicted in the figure; in transfer processes the OC is intact, whereas in degradation processes the OC is separated (O/C). (Adapted from Weber J. B. and Miller, C. T., in *Reactions and Movement of Organic Chemicals in Soils*, Soil Science Society of America, Madison, WI, 1989. With permission.)

environmental quality, and pesticide efficacy. Surrounding ecosystems can be impacted if conditions are conducive to pesticide drift, leaching, or surface runoff. Table 8.6 outlines some of the factors related to pesticide use that can impact environmental quality and pesticide effectiveness.

The mobility of organic chemicals and their ultimate degradation after being introduced into the soil environment are determined by chemical, physical, and biological processes. Each organic chemical has its own molecular structure that determines, at least partially, the degree to which it will react in soils. Some organic chemicals contain charged functional groups that enhance the chance of the chemical being adsorbed to soils; positively charged organic chemicals are adsorbed to the negatively charged sites on soil clays and organic matter (see Chapter 3 for more information on soil properties). Other organic chemicals may persist (e.g., recalcitrant organic compounds) in soils for long periods of time if chemical and microbial transformations are suppressed due to the toxicity of the organic chemical.

The transfer and degradation processes controlling the movement and fate of organic chemical are listed in Table 8.7. Physical drift is generally a concern during the application of pesticides, but could also occur at times of high winds or during rainstorms when pesticides are washed off vegetation. Small droplets have a greater potential for being transported by wind action than do large droplets. Other processes are generally related to what occurs to the organic chemicals in the soil environment. In the next sections, we will discuss many of these processes and other characteristics of organic chemicals that influence their fate and transport.

Table 8.6 Impacts of Pesticide Management Decisions on Both Environmental Quality and Pest Control Efficacy

Management decision	Impacts	
	Environmental quality	Pest control
Type of pesticide	High sorption — low leaching Fast degradation — not transported Toxicity — determines maximum allowable concentration in water	Efficacy of control dependent on pesticide
Type of application	Soil incorporated — greater chance for groundwater contamination Ground application — decreases risk of pesticide drift	Toxicity to target pest enhanced by proximity of pesticide to target Wet soil conditions may hinder optimum time of application
Spray technology	Large droplets — reduced physical drift	Large drops may decrease efficacy
Timing of application	Close to rainfall or irrigation — increases chance for groundwater contamination	Efficacy of pesticide depends on proper timing
Irrigation schedule	Planned irrigation — reduces potential for groundwater or surface water contamination	Changes in soil moisture may affect pests
Tillage practice	Minimum tillage effects — reduces runoff, requires increased levels of pesticide, greater risk of pesticide leaching and drift	Increases pest densities that require additional need for pesticides
Proper application procedures	Pesticide spills or misuse — increases water and soil contamination	No negative effect on pest control

Source: Shoemaker (1989).

8.4.1 Plant uptake

Organic chemicals, such as herbicides, are absorbed by plants either through their roots or aboveground foliage. Uptake is dependent on the organic chemical and the plant species. Seeds are also capable of adsorbing organic chemicals, which can occur either before or after seed germination. Adsorption of organic chemicals on the outer surface of seeds increases the chance of absorption, which can take place by nonmetabolic processes when seeds are imbibing water or through diffusion processes. Factors that influence seed uptake of organic chemicals are related to the properties of the chemical (concentration, structure, solubility, and diffusion rate), soil (temperature and pH), and seed (size, characteristics and permeability of the seed coat).

Plants are also capable of absorbing organic chemicals through their aboveground parts, which include stems, buds, and leaves. Succulent and perennial woody plants are noted for the uptake of herbicides through their stems. Buds are the primary targets of contact herbicides because entry of the herbicide into buds generally guarantees plant kill. Leaves absorb organic chemicals from both the upper and lower leaf surfaces; however, absorption is often faster through the lower surface because it has a thinner cuticle. Nonpolar organic chemicals enter into leaves more readily than do polar organic chemicals, or many inorganic constituents. The mechanism of entry for oily and aqueous solutions is believed to follow different pathways.

Table 8.7 Movement and Fate of Organic Chemicals in the Environment,
with Particular Reference to Pesticides

Process	Consequence	Factors
Transfer *(processes that relocate organic chemicals without altering their structure)*		
Physical drift	Movement of organic chemical due to wind action	Wind speed, size of droplet, distance to physical object
Volatilization	Loss of organic chemical due to evaporation from soil, plant, or aquatic ecosystems	Vapor pressure, wind speed, temperature
Adsorption	Removal of organic chemical by interacting with plants, soils, and sediments	Clay and organic matter content, clay type, moisture
Absorption	Uptake of organic chemical by plant roots or animal ingestion	Cell membrane transport, contact time, and susceptibility
Leaching	Translocation of organic chemical either laterally or downward through soils	Water content, macropores, soil texture, clay and organic matter content, rainfall intensity, irrigation
Erosion	Movement of organic chemical by water or wind action	Rainfall, wind speed, size of clay and organic matter particles with adsorbed organic chemicals on them
Degradation *(processes that alter the organic chemical structure)*		
Photochemical	Breakdown of organic chemicals due to the absorption of sunlight (i.e., UV light)	Structure of organic chemical, intensity and duration of sunlight, exposure
Microbial	Degradation of organic chemicals by microorganisms (see Chapter 3 for discussion on soil microorganisms)	Environmental factors (pH, moisture, temperature) nutrient status, organic matter content
Chemical	Alteration of organic chemical by chemical processes such as hydrolysis, and redox reactions	High and low pH, same factors as for microbial degradation
Metabolism	Chemical transformation of organic chemical after being absorbed by plants or animals	Ability to be absorbed, organism metabolism, interactions within the organism

Source: Marathon-Agricultural and Environmental Consulting Inc., Video Cassettes — Fate of Pesticides in the Environment, 1992.

Most organic chemicals can be readily taken up by roots and foliage and transported through plants by two systems — the symplast (living plant tissue) and apoplast (non-living plant tissue). Organic chemicals that enter the symplasts are exposed to enzymes capable of altering foreign substances. However, the symplast is enclosed by the apoplast, thus preventing direct contact between organic chemicals and the living tissue of the symplasts. The apoplast consists of cell walls and xylem that form an interconnected continuum throughout the plant. It is within the apoplast that toxic pesticides and other organic chemicals can be translocated over short and long distances in plants.

Once absorbed by plants, organic chemicals can be transformed by several reactions that alter the organic chemicals. For example, plants that are capable of absorbing herbicides, and other organic contaminants, may be able to transform or metabolize these organic chemicals to nonphytotoxic levels by one or more of the following reactions: *oxidation–reduction, hydrolysis, hydroxylation, dehalogenation, dealkylation, conjugation,* or *β-oxidation*. Each of these reactions alters the structure of the organic chemical and thus its chemical nature. Although plants are similar to microorganisms, insects, and mammals in their capability to

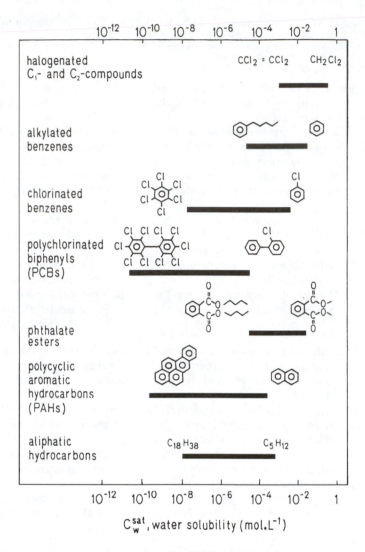

Figure 8.3 Variation in the range of water solubilities for specific classes of important organic chemicals. (Adapted from Schwarzenbach R. P. et al., *Environmental Organic Chemistry*, John Wiley & Sons, New York, 1993. With permission.)

metabolize organic chemicals, rates of plant transformations are generally slower, which in some cases can result in greater toxicity of the organic chemical to the plant.

8.4.2   Solubility

The solubility of an organic chemical is important to its fate and mobility, primarily because highly soluble chemicals tend to be rapidly distributed within soil and the hydrosphere. *Aqueous solubility* of organic chemicals is determined based on the total amount that dissolves in pure water at a specified temperature. When the aqueous solubility of an organic chemical is exceeded, a separate phase will exist in addition to the aqueous solution. All organic chemicals are soluble in water to some degree, although the amount dissolved is highly variable among the different organic compounds. Because water is a polar compound, and similar types of compounds tend to interact, it is not surprising that polar organic chemicals are generally more water soluble. Solubilities of common organic chemicals generally fall in the range of 1 to 100,000 mg/kg (weight of organic chemical per weight of pure water); however, many have higher solubilities. The difference between the least soluble and most soluble organic chemicals is approximately 1 billion-fold (see Figure 8.3 for the range in solubilities for specific groups of organic chemicals).

Solubility of organic chemicals in water is a function of temperature, pH, ionic strength (concentration of soluble salts), and the presence of other organic chemicals such as dissolved organic carbon (DOC). Most organic chemicals become more soluble with increasing temperature. The solubility of organic acids generally increases with increasing pH, whereas organic bases are expected to behave in an opposite manner. Soluble salts will commonly reduce organic solubility, which explains why organic chemicals tend to be less soluble in oceans than in fresh waters. Several studies have indicated there is usually an increase in the solubility of many poorly water-soluble organic chemicals with higher levels of DOC. Evidently there is an interaction between DOC and organic chemicals that enhances their solubility.

Various methods are used to estimate organic chemical solubility. Two frequently used methods are based on (1) chemical structure and (2) octanol–water partition coefficients. The former method was developed to estimate the aqueous solubility of particular groups of compounds, for example, with aliphatic and aromatic hydrocarbons. A significant amount of data are available on the relationship between aqueous solubility and octanol–water partition coefficients. *Octanol/water coefficients (K_{ow}) are determined by the following equation:*

$$K_{ow} = \frac{\text{amount of organic chemical in octanol (mg/L)}}{\text{amount of organic chemical in water (mg/L)}} \qquad (8.1)$$

The relationship between water solubility and K_{ow} is shown in Figure 8.4. As seen in this figure, there is an inverse relationship between solubility and K_{ow}.

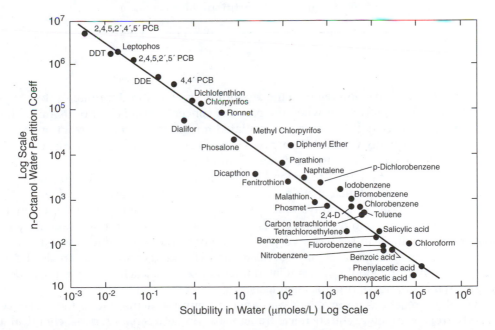

Figure 8.4   Relationship between octanol–water partition coefficients (K_{ow}) and solubility of several organic chemicals. Note the extensive range in solubilities of the organic chemicals. (From Chiou, C. T. et al., *Environ. Sci. Technol.*, 11, 475, 1977. With permission.)

Example problem 8.1

The solubility of an organic chemical is ten times greater in octanol than in water. Calculate the amount extracted from a 100 mL river water sample that contains 5 mg of the organic chemical using a 50 mL volume of octanol. First, consider the relationship defined in Equation 8.1.

$$K_{ow} = \frac{\text{amount in octanol (mg/L)}}{\text{amount in water (mg/L)}} = \frac{x \text{ mg/50 mL}}{(5-x) \text{ mg/100 mL}} \qquad (8.2)$$

using a K_{ow} of 10, the equation can be rearranged to

$$10 = \frac{x \text{ mg/50 mL}}{(5-x)\text{mg/100 mL}} \quad \text{and} \quad 10(5-x)\text{mg} = \frac{100 \text{ mL } x \text{ mg}}{50 \text{ mL}} = 2x \text{ mg} \qquad (8.3)$$

therefore,

$$(50 - 10x) \text{ mg} = 2x \text{ mg} \quad \text{or} \quad 50 \text{ mg} = 2x \text{ mg} + 10x \text{ mg} = 12x \text{ mg} \qquad (8.4)$$

so,

$$x = 50/12 \text{ mg} = 4.17 \text{ mg} \qquad (8.5)$$

and

$$(4.17/5)100 = 83\% \text{ partitioned into octanol} \qquad (8.6)$$

8.4.3 Half-life

The *half-life* ($t_{1/2}$) of a reaction refers to the amount of time required for half of the reactant to be converted into product or when the reactant concentration is half of its initial level. For organic chemicals, half-lives can be calculated for different types of reactions such as volatilization, photolysis (decomposition by sunlight), leaching potential (adsorption–desorption characteristics), and degradation (chemical and microbial). Half-life values are important for understanding the potential environmental impact of an organic chemical. For instance, if a highly toxic organic contaminant is accidently spilled into a lake and the rate of photolysis is rapid (suggesting a small $t_{1/2}$), the consequences may be minimal if the photolysis products are harmless. However, if a moderately toxic contaminant is spilled and it has a very slow rate of photolysis (indicating a large $t_{1/2}$), then the environmental impacts may be substantial. This kind of scenario can also be used to determine environmental impacts related to volatilization, leaching potential, and degradation characteristics of organic chemicals.

Hydrolysis reactions occur in soil and aqueous environments due to the presence of water. Hydrolysis involves the chemical transformation between an organic chemical and water that results in the breaking of one bond while forming a new C–O bond

$$\text{R–X} + \text{H}_2\text{O} \rightarrow \text{R–OH} + \text{H}^+ + \text{X}^- \qquad (8.7)$$

Table 8.8 Examples of Organic Functional Groups That Are Either Resistant
or Potentially Susceptible to Hydrolysis

Resistant	Susceptible
Alkanes, alkenes, alkynes	Alkyl halides
Benzene/biphenyls	Amides
PAHs	Amines
Halogenated aromatics/PCBs	Carbamates
Dieldrin/aldrin and related	Carboxylic acid esters
halogenated hydrocarbon pesticides	Epoxides
Aromatic amines	Nitriles
Alcohols	Phosphonic acid esters
Phenols	Phosphoric acid esters
Glycol	Sulfonic acid esters
Ethers	Sulfuric acid esters
Aldehydes	
Ketones	
Carboxylic acids	

Source: Harris, J. C., in *Handbook of Chemical Property Estimation Methods: Environmental Behavior of Organic Compounds*, W. J. Lyman et al., Eds., McGraw-Hill, New York, 1982a.

where R–X represents an organic chemical and X is a reactive functional group involved in the hydrolysis reaction. Hydrolysis is considered an important reaction that determines the fate of organic chemicals in soil and aquatic environments. Not all organic chemicals can undergo hydrolysis because many do not possess functional groups susceptible to hydrolysis (Table 8.8). Hydrolysis reactions are also generally pH dependent.

Half-lives of hydrolysis reactions that follow first-order kinetics at a constant pH can be described by the equation:

$$t_{1/2} = 0.693/k_h \qquad (8.8)$$

where k_h is the hydrolysis rate constant for the first-order kinetic reaction:

$$\partial C/\partial T = -k_h C \qquad (8.9)$$

in which C is concentration, T equals time, and $\partial C/\partial T$ represents the change in concentration with a change in time. The range in half-lives for the hydrolysis of several groups of organic chemicals is shown in Figure 8.5. Thus, for an organic chemical that hydrolyzes we can determine its concentration over time if we know its $t_{1/2}$.

Example problem 8.2

What are the concentration ratios and percentages of an organic chemical that hydrolyzes after one, two, and three half-life periods?

one $t_{1/2}$,	$C_f = 0.5C_i = 50\%C_i$	(8.10a)
two $t_{1/2}$,	$C_f = 0.5C_i/2 = 0.25C_i = 25\%C_i$	(8.10b)
three $t_{1/2}$,	$C_f = 0.25C_i/2 = 0.125C_i = 12.5\%C_i$	(8.10c)

or   number (x) of $t_{1/2}$ $C_f = (1/2)^x C_i = (1/2)^x 100(\%) C_i$ (8.10d)

where C_f and C_i are final and initial concentrations, respectively.

8.4.4   Volatilization

Volatilization of natural and synthetic organic chemicals in water is responsible for the transfer of organic chemicals from aquatic and soil environments into the atmosphere. In comparison with volatilization, the process of evaporation is related to organic substances that vaporize from an organic phase (e.g., gasoline spill) to the atmosphere. In some cases, volatilization is responsible for the gaseous movement of certain organic chemicals that have leached or been introduced into subsurface environments, i.e., the

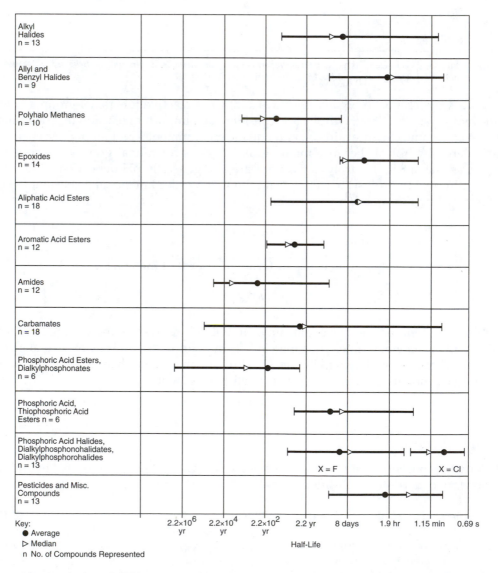

Figure 8.5   Hydrolysis half-lives ($t_{1/2}$) for several groups of organic chemicals. (Redrawn from Harris, 1982a; data from Mabey and Mills, 1978).

vadose zone. Information on the rate at which organic chemicals volatilize is important for understanding their persistence in soil and aqueous environments. For example, surface waters or soils contaminated by volatile organic chemicals may possibly be cleaned up by either evaporation or volatilization processes (see Chapter 11 for more information on remediation).

The volatilization of contaminants from waters depends on the chemical and physical properties of the organic chemical (solubility and vapor pressure), interactions with suspended materials and sediment, physical properties of the water body (depth, velocity, and turbulence), and properties of the water–atmosphere interface. The solubility of a gas in water is defined by its Henry's law constant, which is related to the partial pressure of the gas in a gas–liquid mixture, and is defined as:

$$K_H = P_{gas}/C_{aq} \tag{8.11}$$

in which K_H is Henry's law constant, P_{gas} is the partial pressure of the gas, and C_{aq} is the aqueous concentration of the gas. The K_H values for several classes of organic chemicals are shown in Figure 8.6. Large K_H values suggest the organic chemical has a strong tendency to escape from the water phase into the atmosphere. Rates of volatilization, expressed as half-lives, are also important and can vary from hours to years or more. Half-lives for the volatilization of TCE and the pesticide dieldrin from water have been estimated at 3 to 5 h for TCE and over a year for dieldrin. Table 8.9 lists some additional organic chemicals and their potential volatilization from water.

Factors that influence the volatility of organic chemicals from soils include intrinsic physiochemical properties of the chemical (vapor pressure, solubility, structure and nature of functional groups, and adsorption–desorption characteristics), concentration, soil properties (soil moisture content, porosity, density, organic matter and clay contents), and environmental factors (temperature, humidity, and wind speed). The initial step in the volatilization of some organic chemicals from soils is the ability to change from a solid or liquid in an aqueous solution to a vapor. After this, the vapor moves through the soil and disperses into the atmosphere by diffusion or turbulence. Laboratory and field measurements of volatilization half-lives for several pesticides range from 1.0 to 1.5 days for trifluralin to 42 to 45 days for DDT and atrazine.

8.4.5   Photolysis

Photochemical reactions involving sunlight are extremely important in determining the fate of contaminants in aquatic environments, and may also play a role in the degradation of organic chemicals at soil surfaces. In aquatic environments, photolysis can occur by direct or indirect processes. For direct photolysis, sunlight is absorbed directly by the organic chemical resulting in a chemical transformation. The rate of direct photolysis is dependent on sunlight intensity and overlapping spectral characteristics of solar radiation and the organic chemical. With indirect photolysis, other substances such as DOC, clay minerals, or inorganic elemental species absorb sunlight energy and either initiate a series of reactions that ultimately transform the organic chemical or transfer the excitation energy to the organic chemical.

Atmospheric ozone absorbs solar radiation below wavelengths of 290 nm. Therefore, direct photolysis by sunlight will not occur if the organic chemical in question does not absorb radiation at wavelengths above 290 nm. The intensity of sunlight that reaches the surface of the earth is determined by the thickness of the atmosphere and the angle of incident radiation, which is dependent on latitude, season, and time of day. Sunlight intensity is greatest in summer and least in winter.

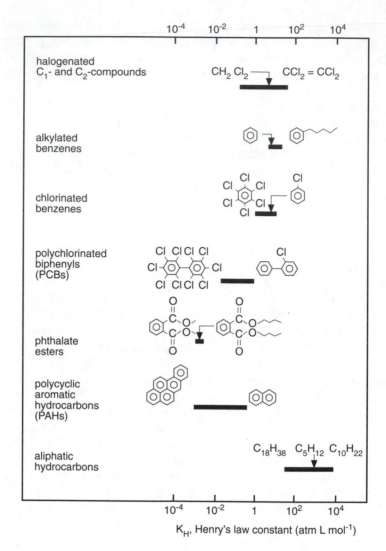

Figure 8.6 Variation in Henry's law constants for several classes of important organic chemicals. (From Schwarzenbach R. P. et al., *Environmental Organic Chemistry*, John Wiley & Sons, New York, 1993. With permission.)

Half-lives of organic chemicals that undergo photolysis in aqueous environments are generally in the range of hours to months. Photolysis half-life values for pesticides, PAHs, and some miscellaneous compounds are listed in Table 8.10. Unlike aqueous environments, photolysis reactions in soils are difficult to determine because of the heterogeneous nature of soils and the lack of sunlight penetration. Therefore, photolysis is primarily a surface phenomenon that can be prevented by incorporating the chemical into the soil.

8.4.6   Adsorption–desorption

Adsorption–desorption behavior between organic chemicals and soils and sediments is conceivably the most important process affecting organic contaminants. Understanding the interactions that occur between soils or aquatic sediments and organic contaminants is important in determining contaminant behavior, and therefore its movement and fate

Table 8.9   Examples of Organic Chemicals and Their Potential Volatility from Water

Volatility potential	Organic chemical	Half-life ($t_{1/2}$)
Low	Dieldrin	327 d
	3-Bromo-1-propanol	390 d
Medium	Penanthrene	31 h
	Pentachlorophenol	17 d
	DDT	45 h
	Aldrin	68 h
	Lindane	115 d
High	Benzene	2.7 h
	Toluene	2.9 h
	o-Xylene	3.2 h
	Carbon tetrachloride	3.7 h
	Biphenyl	4.3 h
	Trichlorethylene	3.4 h

Note: Half-lives given in days (d) or hours (h).

Source: Thomas, R. G., in *Handbook of Chemical Property Estimation Methods: Environmental Behavior of Organic Compounds*, W. J. Lyman et al., Eds., McGraw-Hill, New York, 1982.

Table 8.10   Half-Life Values for Several Organic Chemicals That Undergo Direct Photolysis

Class	Organic chemical	Half-life ($t_{1/2}$)
Pesticides	Trifluralin	~1 h
	Malathion	15 h
	Carbaryl	50 h
	Sevin	11 d
	Methoxychlor	29 d
	2,4-D, methyl ester	62 d
	Mirex	1 y
PAHs	Pyrene	0.7 h
	Benz[*a*]anthracene	3.3 h
	Phenanthrene	8.4 h
	Fluoranthene	21 h
	Naphthalene	70 h
Miscellaneous	Benz[f]quinoline	1 h
	Quinoline	5–21 d
	p-Cresol	35 d

Note: Half-lives given in days (d), hours (h), or years (y).

Source: Harris, J. C., in *Handbook of Chemical Property Estimation Methods: Environmental Behavior of Organic Compounds*, W. J. Lyman et al., Eds., McGraw-Hill, New York, 1982b.

in the environment. Predictive modeling of contaminant fate and transport requires reliable information on the sorption–desorption behavior of contaminants under variable conditions. Sorption of organic chemicals by clay and organic matter materials occurs by one or more of the following interactions: *van der Waals' forces, H-bonding, dipole–dipole interaction, ion-exchange, covalent bonding, protonation, ligand exchange, cation bridging, water bridging*, and/or *hydrophobic partitioning*.

Sorption of many contaminants by soils and sediments has been shown to be an effective means to reduce the mobility of a compound. Sorption can also affect the *bioactivity, persistence, biodegradability, leachability,* and *volatility* of organic chemicals. The type and nature of functional groups on an organic chemical largely determines its ability to be adsorbed. In soils, clay and metal oxide surfaces and organic matter are the dominant materials responsible for the sorption of organic contaminants. Clay surfaces can act as adsorption sites for one contaminant that in turn can influence the adsorption of other contaminants. An example of this is discussed below as a method for the containment of contaminants.

Modeling the sorption of organic chemicals by soils is frequently done by using adsorption isotherms. Sorption data are most commonly described by using either the Freundlich or the Langmuir equations, similar to that discussed for P in Chapter 5.

The Freundlich adsorption equation is

$$x/m = K_f C^{1/n} \tag{8.12}$$

where

x/m = mass of organic chemical adsorbed per unit weight of soil
K_f and n = empirical constants
C = equilibrium concentration of the organic chemical

The value of K_f is a measure of the extent of sorption. The linear form of the Freundlich equation is obtained by logarithmic transformation:

$$\log (x/m) = \log K_f + 1/n \log C \tag{8.13}$$

A plot of $\log (x/m)$ vs. $\log C$ should produce a straight line, with $1/n$ equal to the slope and $\log K_f$ the intercept.

The Langmuir adsorption equation is

$$x/m = \frac{K_l bC}{1 + K_l C} \tag{8.14}$$

where

x/m and C = the same as described above
K_l = adsorption constant related to binding strength
b = maximum amount of organic chemical that can be adsorbed by the soil

One of the linear forms of the Langmuir equation is:

$$\frac{C}{x/m} = \frac{1}{K_l b} + \frac{C}{b} \tag{8.15}$$

If a plot of $C/(x/m)$ vs. C is a straight line, then the adsorption data conform to the Langmuir equation, and b can be calculated from the slope and K_l from the intercept. Equations 8.12 and 8.14 are the general forms of the Freundlich and Langmuir equations, respectively. Equations 5.2 and 5.3 are the same except the variables have been written specifically for P adsorption.

For many organic chemicals, and especially nonpolar organic substances, sorption can be relatively constant, suggesting a linear trend, i.e., the amount sorbed is directly proportional to the solution concentration. The distribution constant, K_d, for these organic chemicals can be calculated using a modified Freundlich equation that considers $1/n$ to be approximately equal to 1:

$$K_d = \frac{x/m}{C} = \frac{\text{organic chemical adsorbed per unit weight of soil (mmol/kg)}}{\text{equilibrium concentration of the organic chemical (mmol/L)}} \quad (8.16)$$

Using the K_d and soil organic C (OC), another constant can be determined that is independent of soil type and is specific for the organic compound investigated.

$$K_{oc} = \frac{K_d}{f_{OC}} \quad (8.17)$$

where f_{OC} is the fraction of soil OC, e.g., %OC/100. The K_{oc} is essentially a coefficient that describes the distribution of the organic chemical between the aqueous and soil organic matter phases. Being relatively constant, K_{oc} values are often used to predict K_d values for soils with known organic C contents. The K_d value in turn defines the distribution of the contaminant between soil solid and solution phases.

Example problem 8.3

The sorption of a nonionizable hydrophobic organic chemical by soils with different organic C levels was found to be highly variable, as shown by the distribution coefficients (K_d) listed below. However, after normalizing to the organic C content the organic-C-corrected constants, K_{oc} were approximately the same.

Soil	%OC	K_d	log K_{oc}
1	3.7	35,000	5.98
2	2.8	25,000	5.95
3	8.3	85,000	6.01
4	1.2	10,000	5.92
5	11.8	125,000	6.03

There are two ways of determining an overall K_{oc} value for a particular organic chemical. The first method averages the individual K_{oc} values determined from separate soils, which in the example above would result in a log K_{oc} of 5.98. A more precise way of calculating K_{oc} values from a number of soils is to determine linear regression statistics for a plot of K_d vs. f_{OC} (e.g., %OC/100) that provide slope and correlation coefficients (r^2) that define the overall K_{oc} and fit of the data, respectively.

Recent studies have shown that certain organic chemicals, once adsorbed onto soil and clay materials, can provide an effective adsorption barrier to the leaching of other organic contaminants. The organic chemicals used for these studies are often organic cations from a group of organic chemicals (quaternary ammonium cations, also known as QUATS) that are generally inexpensive and widely used in such products as detergents, fabric softeners, antistatic sprays, and swimming pool additives. Modified soils and subsurface materials can be formed *in situ* by injecting QUATS into an area around contaminated sites, to prevent contaminant migration. In addition, organoclays can be prepared and injected as a slurry to form an effective barrier around sites as well. Since organoclays are capable of adsorbing petroleum constituents such as benzene, toluene, ethylbenzene, and *o-*, *m-*, and *p*-xylene (BTEX compounds), as well as a variety of other organic contaminants, they should be useful as a liner material around petroleum tank farms, underground storage tanks, and possibly for the treatment of petroleum-contaminated waters. Examples of organoclays are shown in Figure 8.7.

8.4.7   *Abiotic and biotic transformations*

Both abiotic (nonbiological) and biotic (biological) reactions, alone or in combination, are responsible for the transformations of organic chemicals in soil and aquatic environments. Under certain conditions abiotic reactions may dominate, whereas under other conditions biotic reactions may prevail. Degradation of organic chemicals is often assumed to occur by biotic processes; however, abiotic reactions may occur simultaneously. Many organic chemical transformations are mediated by microorganisms, but the actual reaction is an abiotic process.

The principal abiotic transformation reactions that occur in aquatic environments include *hydrolysis*, *oxidation–reduction* (redox), and *photolysis*; in sediments, hydrolysis and

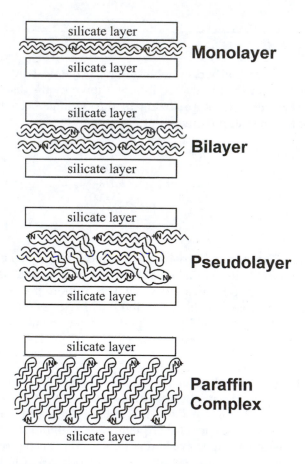

Figure 8.7   Examples of interlayer arrangements of organic cations in swelling 2:1 clays. The organoclays can be used to partition (remove from aqueous solution) organic contaminants in water. Organic clays are classified as monolayer, bilayer, pseudolayer, and paraffin complex. (Adapted from Jaynes and Vance, 1996.)

redox reactions are the dominant reactions. Oxidation reactions that take place in aquatic environments can be mediated by direct or indirect photolysis reactions, which depend on the organic chemical and substrates present. Nonphotolytic oxidation of organic chemicals can occur directly by reactions involving ozone, or via catalytic pathways with certain metals. Abiotic reduction reactions that influence organic chemical transformations may also be catalyzed by certain metal species, with iron (Fe) and manganese (Mn) being the most important. Redox reactions that occur in sediments follow a similar route as shown for soils (see Chapter 9 for further discussion of redox reactions).

In soils, abiotic transformations take place in the liquid phase (i.e., soil solution) and at the solid–liquid interface. In the soil solution, hydrolysis and redox reactions are the most common abiotic transformations, although a number of other reactions also occur. Clays, organic matter, and metal oxides are capable of catalyzing abiotic reactions that occur in soils. Some exchangeable cations, e.g., copper (Cu) and Mn, can also influence the transformation of organic chemicals. Hydrolysis and redox reactions again dominate the abiotic reactions that take place in soil ecosystems.

Microbial transformations of organic chemicals are classified as: (1) biodegradation (contaminant used as substrate for growth, i.e., metabolism); (2) cometabolism (contaminant is transformed by metabolic reactions without being used as an energy source); (3) accumulation (contaminant is incorporated into the microorganism); (4) polymerization or conjugation (contaminant is bound to another organic chemical); and (5) secondary effects of microbial activity (contaminant is transformed due to indirect microbial effects, i.e., pH, redox). Although these transformations are considered to be mediated by microorganisms, abiotic transformations are also involved, especially in the transformations related to categories 4 and 5.

Biodegradation is considered the primary mechanism by which organic chemicals are transformed to inorganic products such as CO_2, H_2O, and mineral salts. The metabolism of organic chemicals in soils by bacteria is typically greater than for other microorganisms. In natural ecosystems, biodegradation of organic chemicals is facilitated primarily by heterotrophic bacteria and actinomycetes, certain autotrophic bacteria, fungi including basidiomycetes and yeasts, and specific protozoa. Biodegradation can occur under aerobic and anaerobic conditions as shown for DDT in Figure 8.8.

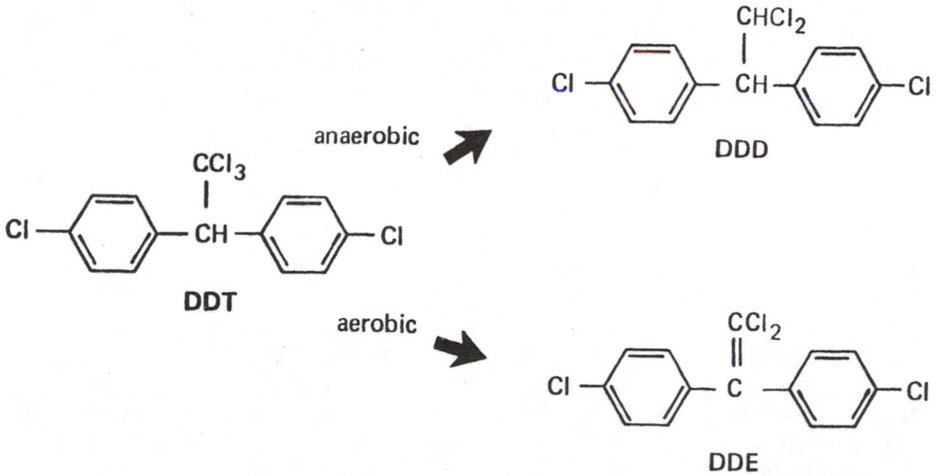

Figure 8.8   Example of anaerobic and aerobic products resulting from the biological transformation of the organic chemical DDT.

8.4.8   Response of organic chemicals in the environment

Once introduced into a soil or an aquatic ecosystem, organic chemicals can be translocated, transformed, or persist due to the factors described in the previous sections. Processes that result in the translocation of organic chemicals in the soil environment include plant uptake (e.g., absorption), runoff of dissolved or adsorbed species, volatilization and migration in the vapor phase, and leaching as a water-soluble chemical. Clearly, organic chemicals strongly sorbed to organic matter and other soil minerals will not migrate to a large extent unless the particle moves due to erosion or leaches downward as a colloid material. In surface waters, organic chemicals that are volatile may move rapidly into the atmosphere depending on climatic conditions and aquatic system properties. Transformation of organic chemicals by chemical, biological, and photolytic processes may alter their translocation and persistence in soil and surface water environments, which is extremely important for organic chemicals that become more toxic or impact future agricultural practices (e.g., next year's crop) as the modified or transformed substance.

To evaluate organic chemicals in the environment properly, one should have a relative appreciation of those factors that specifically influence the translocation, transformation, and persistence of these substances. Table 8.11 describes the general range of organic chemical properties that can potentially influence its environmental fate and transport. For example, a comparison of the pesticide DDT (0.002 mg/L solubility, K_{ow} of 960,000, K_{oc} of 238,000, and with a high bioconcentration factor) vs. the solvent TCE (1100 mg/L solubility, K_{ow} of 195, vapor pressure of 58 torr, and photochemically reacts) indicates DDT should be adsorbed, has a low runoff potential, and bioaccumulates, whereas TCE should have high leaching and runoff potential, is biodegradable, could volatilize, and can potential undergo photolytic transformation. Both chemicals can be

Table 8.11   Factors That Influence the Translocation, Transformation, and Persistence of Organic Chemicals and the Approximate Range or Rates That Describe These Environmental Processes

Property	Range in organic chemical properties		
Water solubility (ws)	<10 mg/L	10–1,000 mg/L	>1,000 mg/L
Octanol/water (K_{ow})	>1,000	500–1,000	<500
Sorption (K_{oc})	>10,000	1,000–10,000	<1,000
Hydrolysis ($t_{1/2}$)	>90 days	30–90 days	<30 days
Photolysis ($t_{1/2}$)	>90 days	30–90 days	<30 days
Vapor pressure (torr)	<0.000001	0.000001–0.01	>0.01
Environmental fate and transport of organic chemicals (with properties listed above)			
Soluble	Negligible	Variable	Yes
Hydrolyzes	Negligible	Variable	Yes
Photolyzes	Negligible	Variable	Yes
Volatilizes	Negligible	Variable	Yes
Adsorption potential	High	Intermediate	Low
Persistence potential	High	Intermediate	Low
Leaching potential	Low	Intermediate	High
Runoff potential	Low	Intermediate	High
Bioaccumulates	Yes	Variable	Negligible
Biodegradable	Slowly	Intermediate	Yes
Metabolized	Slowly	Intermediate	Yes

Source: Ney, R. E., *Fate and Transport of Organic Chemicals in the Environment: A Practical Guide*, 2nd ed., Government Institutes, Rockville, MD, 1995.

environmental contaminants that may enter the food chain. Further information on the fate and transport of organic chemicals in the environment can be found in the very useful practical guide by Ney (1995).

Environmental quality issues/events

Restoration of a contaminated wood-preservative site in Laramie, WY

Creosote is a complex mixture, derived from coal tar, that contains more than 250 compounds, of which about 85% are PAHs, 10% are phenolic compounds, and 5% are N-, S-, and O-heterocyclic compounds. Creosote and coal tar are dense (1.08 to 1.13 g/cm^3), nonaqueous-phase liquids (DNAPL) that can percolate beneath the groundwater table until an impermeable layer is reached. During the late 1970s, creosote solutions were used at approximately 188 of 631 wood-treating plants operating in the U.S. By the late 1980s there were 39 creosote- and coal tar–contaminated wood-preserving sites listed on the EPA Superfund National Priorities List (NPL), 12 of which were wood-preservative Superfund sites located in the western U.S. The Union Pacific Railroad (UPRR) tie-treating plant in Laramie, WY was one of these Superfund sites; however, unlike many Superfund cleanup projects, reclamation efforts at the Laramie Tie Plant have been totally funded by UPRR.

From 1886 to 1983 the UPRR managed the Laramie Tie Plant as a wood-preservative facility to treat railroad cross ties and snow fencing with creosote/oil and pentachlorophenol (PCP)/oil mixtures. While creosote was the primary wood preservative, smaller amounts of PCP were also used for a limited time after 1956. Retort drippage and spills during wood-treatment operations and wastewater discharges into low-lying impoundments at the site resulted in the accumulation of the DNAPL mixture and contamination of surface soils, subsurface media, and groundwaters. Unlike the many PAH compounds in creosote, PCP is more water soluble; PCP also has been detected in alluvium groundwaters at the Laramie tie site. Due to the hazardous nature of the wastes associated with the past Laramie Tie Plant activities, cleanup of the site was authorized under both the Resource Conservation and Recovery Act (RCRA) and the Comprehensive Environmental Response, Compensation, and Liability Act (CERCLA).

The Laramie Tie Plant is located on the Laramie River floodplain. Geologically, the site consists of a thin surficial layer of recent alluvial deposits that directly overlie bedrock units. The alluvial sediments consist of silts, sands, and gravels that typically range from 1.5 to 4 m in thickness. The alluvial sediments are underlain by the Morrison formation confining sequence, which consists of siltstones, shales, and fine sandstones, and the Sundance formation that comprises fine- to medium-grained sandstone. The Sundance formation is a secondary aquifer, although high salt contents make the water unsuitable for most purposes. Fortunately, most subsurface contamination at the site is limited to the alluvial sediments, although the Morrison and Sundance formations do contain some DNAPL contamination.

Remediation efforts

The first step in the Laramie Tie Plant closure and cleanup process was to build a dike along the Laramie River to protect the site from a 100-year flood. In 1985, the Laramie

River channel was realigned 50 m west of the site to permit the installation of a cutoff wall around a 140-acre area surrounding the contamination. Installation of a 3000-m-long soil/bentonite slurry cutoff wall was completed in 1987 to prevent lateral migration of contaminants from the site. Construction of the cutoff wall involved digging a trench 1 m wide and approximately 5 to 25 m deep depending on the depth to bedrock. The cutoff wall is part of the multicomponent contaminant isolation system (CIS) that was designed to minimize contaminant migration from the site. The CIS includes a water management system that maintains water levels within the CIS area at levels below those outside the site to ensure that any lateral water movement will be into the site, and an activated C water-treatment plant that treats the contaminated groundwater from within the site.

In 1987, feasibility studies were conducted to evaluate several remediation technologies including subsurface oil recovery and different treatment options such as soil washing and *in situ* bioremediation of subsurface and surface soils. Soil washing effectively cleaned the coarser sediments, but generated fluids containing stable oil emulsions, high contaminant concentrations, and high salinity that were extremely difficult to treat. Also, the fine-textured sediments and impermeable zones were left untreated. The initial tests indicated bioremediation of contaminated subsurface materials using native, aerobic microorganisms had the potential to be an effective process. The main requirement for the technical feasibility of subsurface bioremediation was meeting the oxygen demand of the process. At a flow rate of 15 million L of oxygenated water per day, full-scale subsurface bioremediation was projected to require 250 years or more. Surface bioremediation reduced contaminant concentrations; however, much of the high-molecular-weight PAHs and PCP persisted. Surface bioremediation was estimated to require 20 years or more to clean up a 1.2-m-thick layer of soil.

The water-flood oil recovery technique was found to be an effective means for recovering subsurface DNAPL oil from the alluvial deposits. In a 1989 pilot test, over 850,000 L of reusable preservative oil was recovered from a 1-ha subsurface area in 90 days. To date, water-flood oil recovery efforts have been successful in removing significant amounts of the DNAPL. The Laramie Tie Plant contamination included free DNAPLs and DNAPLs at immobile residual saturations, of which approximately 6.8 million L have been recovered by the water-flood oil recovery process. When oil recovery efforts are completed within the next few years, some of the DNAPL will remain in sediments beneath the site at immobile residual saturations. Equilibration of groundwaters with the remaining DNAPLs will maintain saturated aqueous concentrations of the contaminants. Groundwater pumped from within the CIS will therefore be treated and discharged for an indefinite period of time. Consequently, the contaminants at the site will be contained; however, the residual DNAPLs will continue to be a potential source of groundwater contamination.

Surface remediation

In June 1995, the EPA issued a final decision on the corrective actions recommended for the Laramie Tie Plant. A Corrective Action Management Unit (CAMU) was described that would consolidate the contaminated surface materials, in one area within the site. The proposed CAMU site would be the former location of the creosote/oil surface impoundments. Oil-contaminated concrete debris, 5000 m³ of contaminated soil from an area outside the CIS where PCP was used, and other small areas of contaminated soil will be excavated and transported to the CAMU site within the CIS. Eventually, a cap of impervious clay, synthetic material, or vegetation is expected to cover the surface impoundment area to prevent the infiltration of water. Much of the contamination will remain at the site after these reclamation efforts are completed; however, immobilization of the contaminants at the site will be assured by the CIS and water treatment process.

In April 1998, the City of Laramie and UPRR signed a memorandum of understanding (MOU) to develop a phytoremediation/greenbelt project cooperatively that includes cost-effective site restoration and beneficial use of part of the Laramie Tie Plant site. In the original RCRA negotiation, Wyoming Department of Environmental Quality (WDEQ) and the EPA underscored the importance of eventual site cleanup, which suggests additional strategies such as phytoremediation should be implemented at the site. Phytoremediation is expected to assist in the cleanup of the contaminated soils, subsurface materials, and groundwaters by physical, chemical, and biological processes. Because plants utilize large amounts of water, vegetative transpiration could reduce the volume of groundwater that requires treatment, which would reduce costs associated with operating the CIS water treatment facility. In addition to increased water use, plants used in the phytoremediation project will reduce wind and water erosion and provide aesthetic and botanical components to the Laramie Tie Plant ecosystem.

Plants selected for the phytoremediation/greenbelt project include trees, shrubs, grasses, and forbs that are adapted to Laramie's climate, can tolerate site conditions (e.g., contaminants, salts, soil properties), are perennial plant species, have phytoremediation capabilities, and include both shallow- and deep-rooted plants. Plants used in this project can enhance creosote and PCP degradation through increased microbial activity, particularly in the rhizosphere (area around the plant roots), improving aeration of the site, which will allow for greater chemical and biological oxidation, plant uptake, metabolism or degradation of organic contaminants, and cometabolism via biodegradation of root exudates and other organic plant and microbial substances. Specific vegetation selected for the phytoremediation project includes cottonwood, willow and hackberry trees; wild ryegrass and wheatgrass; and alfalfa, yellow sweet clover, and alsike clover. Other plant species will be planted primarily for improving aesthetics, botanical diversity, and stabilization of the site and will include sacaton and foxtail grasses; caragana, woods rose, buffaloberry, chokecherry, dogwood and serviceberry shrubs; and Colorado blue spruce and Rocky Mountain juniper evergreen trees. Some areas will be irrigated to maintain adequate water needs and for fertigation when necessary.

Conclusions

The great amount of effort and funds (over $50 million by the end of 2001) that have been and will continue to be (about $400,000 per year after 2001) expended by UPRR at the Laramie Tie Plant represents more of an ideal than a typical case. Conclusions regarding the practicality of reclamation technologies employed at the Laramie site will probably be followed at similar sites in the U.S. A large number of reclamation technologies have been field-tested at the UPRR Laramie tie treating plant. The field tests indicated that long-term containment (CIS including operation of water treatment facility) is a practical technology. Long-term containment will likely be the reclamation approach followed at similar sites in the U.S. Because the Laramie site is enclosed in a soil/bentonite cutoff wall, testing of new remediation techniques at the site will assist in developing future technologies for site remediation without the risk of off-site contamination. Use of phytoremediation practices is expected to enhance the long-term cleanup efforts of the Laramie Tie Plant, with plants providing a natural approach to water use, erosion control, and contaminant reduction.

8.5   Organic chemical analysis

To characterize the mobility and fate of organic chemicals, we must be able to monitor the compound of interest and the products derived from transformation processes that result in metabolites or degradative products. Chemical analysis is therefore an indispensable component of all environmental investigations dealing with the evaluation of risks associated with organic chemicals (see Chapter 12 for further discussion on risk assessment). Since the early 1960s, analytical techniques and instrumental advances have enhanced our capabilities to examine, evaluate, and predict the fate, mobility, and even toxicity of numerous organic chemicals that are used by agriculture, industry, and for improving our everyday lives.

Due to the extensive number and types of organic chemicals that are utilized on a daily basis, the many new organic substances that are manufactured yearly, and the different materials that can potentially contaminate various environments, the following discussion is very limited. Those interested in knowing more about these subjects should refer to the extensive literature related to organic analysis or consider courses in analytical, organic, and/or biochemistry. The basics involved in the sampling, processing, recognizing, and quantifying of organic chemicals will be presented in the following sections. In addition to determining the specific organic chemicals, one should always characterize the general properties and parameters that are pertinent to the ecosystems being studied. These include pH, electrical conductivity (EC) or total dissolved solids (TDS), organic matter (OM) or DOC contents, soil fertility (see Chapter 3), and other parameters specific to the ecosystem studied.

8.5.1   Sample collection and preparation

Collection and preparation of water, soil, and sediment samples were discussed in Chapters 2 and 3. Samples to be analyzed for organic chemicals must be properly collected, preserved, and stored to maintain their integrity and to ensure quality assurance/quality control throughout the analytical process. Certain samples may require additional steps to prepare them for analysis. If substances that interfere with the organic chemical analysis are present, they must be removed by purification methods, and if the organic chemical of interest is present at low levels, concentration methods must be used to increase the amount to above detection levels of the analytical technique employed. Soil and sediment samples generally require extraction of the organic chemicals, and because the various extraction methods used (e.g., acids, bases, solvents) tend to be nonselective, further cleanup and concentration are often required. Water samples can also be extracted using different solvents that are useful for concentration purposes (e.g., diethyl ether, *t*-butyl methyl ether, dichloromethane, and pentane). For soil and sediment samples, the use of water-miscible solvents (e.g., methanol, acetonitrile, dimethylformamide, and tetrahydrofuran) may be preferred because the organic chemical of interest can then be separated from the extracting solvent by using a water-immiscible solvent such as those used in extraction and concentration methods discussed above.

In addition to extraction and concentration procedures, cleanup and derivatization may also be required with some samples before analysis can be conducted. Some of the methods employed for the cleanup step include chromatographic techniques (i.e., thin-layer, gel permeation, and ion-exchange), dialysis membranes, reverse osmosis, and supercritical fluid extraction. Water samples are usually the easiest to extract because the organic chemicals are already in solution. Use of water-immiscible solvents, purging (volatile compounds), activated charcoal, hydrophobic resins (e.g., XAD adsorptive resins), and various liquid chromatography methods can be used to separate and purify the organic chemicals of interest. Organic chemicals in soils and sediments are more difficult to separate and purify because of the numerous inorganic and organic substances in these samples. For samples with nonvolatile organic chemicals, freeze-drying can be used, whereas with volatile organic chemicals, anhydrous sodium sulfate can be added; both of these samples can be extracted with water-immiscible solvents using a Soxhlet apparatus (e.g., reflux extractor of soluble constituents). A preextraction step may also be employed to remove the highly soluble, easily accessible organic chemicals, which can be combined with the Soxhlet-extracted materials before final analysis. Simple solvent extraction may be feasible for some samples with water-soluble organic chemicals of interest. For example, low-molecular-weight phenolic compounds in hydrosequences and developmental sequences of Michigan Spodosols were determined by extracting soils with sodium pyrophosphate ($Na_4P_2O_7$), acidifying the extracted solution to pH 2.5, followed by separating the phenolic organic chemicals using diethyl ether; Table 8.12 lists the organic compounds and the HPLC retention times determined in these studies. Identification of protocatechuic acid is especially noteworthy because of its ability to complex metal ions (e.g., Al^{3+} and Fe^{3+}) and enhance metal translocation in soils.

8.5.2　Identification and quantification of organic chemicals

Analytical methods commonly used for organic compound identification include gas chromatography (GC), high-pressure liquid chromatography (HPLC), mass spectrometry (MS) and GC-MS. Some additional analytical techniques that are used for specific studies on organic chemicals in environmental samples are ultraviolet (UV)-visible and infrared (IR) spectrophotometry, and nuclear magnetic resonance (NMR). Some organic chemicals, e.g., neutral compounds, can be analyzed directly using GC, HPLC, or GC-MS, whereas organic chemicals with functional groups such as hydroxyl, carboxylic acids, amino, and carbonyl groups may require derivatization procedures. With HPLC, derivatization may be needed only if UV detection is used and the organic chemicals of interest do not absorb in the UV range. For MS, UV-visible, IR, and NMR methods, specific functional groups, C compound characteristics, and bonding arrangements may be deduced either directly or indirectly.

Table 8.12　Phenolic Compounds Identified in a Study Investigating Low-Molecular-Weight Organic Chemicals in Spodosol Hydrosequences and Developmental Sequences

Compounds	Retention time (min)
Benzoic acids	
Protocatechuic acid (3,4 diOH)[a]	5.12
p-Hydroxybenzoic acid (4OH)	8.14
Vanillic acid (4OH, 3OCH₃)	11.07
Benzaldehydes	
p-Hydroxybenzaldehyde (4OH)	9.25
Vanillin (4OH, 3OCH₃)	14.31
Cinnamic acids	
trans *p*-Courmaric (4OH)	24.48
Ferulic (4OH, 3OCH₃)	34.29

Note: The retention times given for each compound were determined using HPLC with UV detection.

[a] Numbers and abbreviations in parentheses represent location and type of functional group arrangements on the organic chemicals.

Source: Adapted from Vance et al., 1985; 1986.

Organic chemicals can be characterized and quantified directly using UV-visible, IR, NMR, and MS methods; however, GC and HPLC separate organic substances, which are then analyzed and quantified using different types of detection sources. For GC a number of detectors are used, including flame ionization detection (FID) and electron capture detection (ECD), or the GC can be interfaced to an MS. The latter system can be extremely useful for the separation, characterization, and quantification of unknown organic chemicals if they are sufficiently volatile. However, due to the high cost of the instrumentation, GC-MS is not always available. FID and ECD detection are the two most common methods of detection with GC. FID detects a change in current flowing between two electrodes (polarizing and collector) placed in a flame when an organic compound is introduced. It is one of the most sensitive and widely used GC detectors, is sensitive to all organic compounds (10^{-12} g), and is not sensitive to inorganic compounds (e.g., H_2O, CO_2, CO, N_2, O_2). The EC detector is highly sensitive to organic halide, P, and N compounds, and detects organic compounds that capture electrons produced through the ionization of the carrier molecules by a beta source (β rays, see Chapter 7), which reduces the standing current. A major advantage of ECD is its selectivity and sensitivity (10^{-13} g) for organic halides and P and N compounds; compounds such as paraffins, simple hydrocarbons, amines, and alcohols are not electron-capturing species. For HPLC, several types of detectors are available, including UV-visible or fluorescence spectrophotometry and conductivity. An example of the use of HPLC with UV for detection of phenolic compounds extracted from soils is shown in Figure 8.9.

Once the organic compounds have been identified, the next step is to determine the concentration of the substance, i.e., quantify the amount present in the sample analyzed. A standard sample(s) should be characterized within the same batch of samples of unknowns to verify the specific compound is present and, with multiple standards of different concentrations, to ascertain the concentration in the sample analyzed. By using

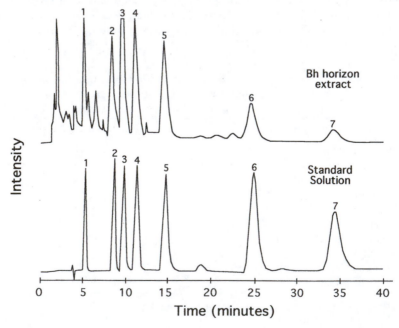

Figure 8.9   Typical HPLC chromatograph of sample and standard solutions. (Adapted from Vance et al., 1985.)

several standards, many instruments can be calibrated to measure the concentration of the unknown organic chemical directly. If an instrument cannot be calibrated, results such as peak area, peak height, absorption, fluorescence, conductivity, etc. for the instrument utilized can be evaluated using a computer spreadsheet to develop standard curves that calculate concentrations of the organic chemical of interest.

8.6   Alternative pest management strategies

Most people would agree that pest control is usually managed by using various types of organic pesticides. Integrated pest control programs, however, rely on a combination of *biological, chemical, mechanical,* and *cultural controls* to provide economic, ecological, and sociological benefits as we manage pest populations.

8.6.1   Integrated pest management

Integrated pest management, also known as IPM, is a program of pest control that relies on several practices to prevent pest outbreaks from occurring. Preferably, these practices should be compatible and each should augment the effectiveness of the other. Some of the components of an effective IPM program are soil preparation for the control of pests (i.e., weeds, microorganisms, and insects), chemical (pesticides) and biological pretreatment of soils for the control of weeds and insects, observations of insect activities both locally and regionally, better time management strategies, and understanding climatic conditions that are conducive to pest outbreaks.

In an IPM program, several practices must be followed to prevent pest problems. Practices commonly implemented in an IPM program include use of certified disease-, weed-, and insect-free seeds or plants; implementation of cultural practices such as crop rotation and sanitation measures; control of physical conditions (time of tillage, planting conditions, temperature and moisture of storage conditions for the prevention of diseases); utilization of chemical practices (pesticides, fumigants, seed and plant treatments, and use of disinfectants); and development of innovative biological control practices such as use of insect- and disease-resistant varieties (see the next section for additional discussion on biological control measures). Individually, these practices may not provide adequate protection for pest control; however, collectively these practices can minimize economic losses due to pest problems.

Integrated pest management programs are becoming more accepted within the agricultural production sector as individuals gain a greater understanding and appreciation of how nature works and as the concern for environmental quality grows. For example, a commonly used alternative agricultural practice is conservation tillage. Conservation tillage systems have become popular farming methods because these practices generally result in a reduction in soil loss due to wind and water erosion and an increase in the amount of water retention when crop residues are left on the soil surface. However, some conservation tillage practices actually enhance problems with weeds, diseases, insects, nematodes, rodents, and soil microorganisms. Consequently, biological control measures, use of resistant varieties, or altered cultural practices may be required along with pesticide applications to combat the problems developed by conservation tillage systems. In addition, as organisms become immune to pesticides, other forms of control, new and improved chemical treatments, and alternative plants such as transgenic Bt crops (plants cloned with *Bacillus thuringiensis* protein gene) must be evaluated for their sustainable contribution to agricultural production.

8.6.2   Biocontrol

Alternatives to the use of pesticides for controlling weeds and insects have gained popularity in recent times because they tend to require less input and are more sustainable. Biological control is an alternative to pesticide use that relies upon "natural enemies" (i.e., predators, parasites, pathogens, and herbivores) to suppress pest populations. In an ideal biological control program, pest populations are kept below levels where economic losses occur, without placing undue stress on the ecosystem. Hoy (1989) summarized the three most commonly employed biological control strategies as follows:

- *Classical:* Importation and establishment of exotic natural enemies to control exotic and, occasionally, native pests;
- *Conservation:* Actions to protect, maintain and/or increase the effectiveness of natural enemies; and
- *Augmentation:* Actions taken to increase populations or beneficial effects of natural enemies, which may not be self-sustaining.

Humans are primarily responsible for the dissemination of weeds and insects throughout the world. Many of the most troublesome weeds and insects in the U.S. (i.e., Russian thistle, St. John's wort (Klamath weed), johnsongrass, gypsy moth, screw worm, and Russian wheat aphid, to name a few) are a result of travel and commerce activities. Weeds and insects from exotic countries may proliferate if (1) they adapt to the new environment and (2) their natural enemies are not present. One approach for controlling weed and insect populations is to introduce biological controls from the pest's native region that can reduce or regulate the pest to a level that is economically, aesthetically, or environmentally acceptable. Natural enemies have been introduced to combat exotic weeds and insects that have displaced the native plant or insect species.

Biological control has been successfully used to control terrestrial and aquatic weeds. The biological control of the common prickly pear (*Opuntia inermis*) and spiny prickly pear (*O. stricta*) in Australia and St. John's wort or Klamath weed (*Hypericum perforatum*) in the U.S. are classic examples of terrestrial weed control by beneficial insects that feed on these weed species and not on agronomic plants. Herbivorous, e.g., white amur (*Ctenopharyngodon idella*), and nonherbivorous, e.g., carp (*Cyprinus carpio*), fish have been found to control aquatic weeds either by consumption or by uprooting them, respectively. Several examples of weeds common to the western U.S., and biological control agents showing promise in their control, are listed in Table 8.13.

Biological control of insects has also had some success. A significant example of biological insect control is the use of the vedalia beetle (*Rodolia cardinalis*) to control the cottonycushion scale (*Icerya purchasi*) that threatened the livelihood of the California citrus industry in the 1880s. Other pests that were at least partially controlled by beneficial insects include the European corn borer (*Ostrinia nubilalis*), European spruce sawfly (*Diprion hercyniae*), and Oriental fruit fly (*Dacus dorsalis*), to name a few. Successful biological control of pests can be accomplished on a small scale as well, such as in gardens or on small farms. Some of the more common natural enemies used for biological control of insects and mite pests are listed in Table 8.14.

Table 8.13   Some Biological Control Agents That Show Promise for Weed Control in the Western U.S.

Weed	Bioagent	Feeding habit and damage
Canada thistle	Ceutorhynchus litura	Weevil larvae mine leaves, stem, root crown, and root reducing overwintering survival
	Urophora cardui	Galls formed by fly larvae in stems act as metabolic sink
Diffuse and spotted knapweeds	Agapeta zoegana	Moth larvae mine roots killing small plants and reducing flowering in larger plants
	Cyphocleonus achates	Weevil larvae mine and gall vascular root tissue reducing plant vigor
	Larinus minutus	Weevil larvae feed in seedheads reducing seed production
	L. obtusus	
	Sphenoptera jugoslavica	Beetle larvae mine gall roots depleting carbohydrate reserves
	Urophora affinis	Fly larvae gall or feed in seedhead reducing seed production
	U. quadrifasciata	
Gorse	Apion ulicis	Weevil larvae eat seeds, adults eat foliage, mites pierce and extract cell contents from spines
	Tetranychus lintearius	and stems
Leafy spurge	Aphthona lacertosa	Beetle larvae feed on root hairs and young roots reducing storage of reserves and
	A. nigriscutis	water/nutrient uptake
	Spurgia esulae	Midge larvae feed and gall growing tips reducing flower and seed production
Musk thistle,	Rhinocyllus conicus	Weevil larvae consume developing seeds in head
Italian thistle,	Trichosirocalus horridus	Weevil larvae feed on growing tip of rosette killing weak plants or main stem of
plumeless thistle,		healthy plants
Scotch thistle		
Rush skeletonweed	Cystiphora schmidti	Midge larvae feed on stems and leaves reducing photosynthesis and flowering
	Eriophyes chondrillae	Galls form where mite feeds on buds resulting in plant death, stunting, or reduced seed
	Puccinia chondrillina	production
		Rust fungus infects all aboveground plant parts resulting in plant death, reduced
		photosynthesis, or stunting of stems
St. John's wort	Agrilus hyperici	Beetle larvae tunnel inside root killing most plants
	Chrysolina quadrigemina	Beetle larvae defoliate plants resulting in reduced foliage and root reserves
	C. hyperici	
Tansy ragwort	Longitarsus jacobaeae	Beetle larvae mine rosette roots and adults feed on leaves often killing the plant
	Pegohylemyia seneciella	Fly larvae consume seeds in developing seed heads
	Tyria jacobaeae	Moth larvae strip foliage and destroy flowers

Source: Rees, N. E. et al., Biological Control of Weeds in the West, Western Society of Weed Science, Bozeman, MT, 1996.

Table 8.14   Common Types of Natural Enemies Used in Biological Control
of Insect and Mite Pests

Natural enemy	Feeding habits/comments
Predators	
Lady beetles (Coccinellidae)	Adults and larvae feed on aphids, spider mites, insect eggs, and other soft-bodied insects; adults commonly colonize annual crops where suitable prey are present; several species commercially available
Ground beetles (Carabidae)	Adults and larvae generalist feeders on insects in and on soil; some feed on weed seeds; common in crop borders but cultivation limits abundance in crops
Rove beetles (Staphylinidae)	Adults and larvae feed on insects in and on soil including seedcorn maggot and onion maggot; common in crop borders
Green lacewings (Chrysopidae)	Larvae feed on aphids, small caterpillars, and beetles; adults feed primarily on nectar; adults commonly colonize annual crops; several species commercially available
Syrphid flies (Syrphidae)	Larvae feed on aphids and other soft-bodied insects; adults feed on flowers
Predatory bugs (Hemiptera)	"Beak" is used to pierce prey and suck out body fluids
Stink bugs (Pentatomidae)	Adults and nymphs of some species feed on insects including the Colorado potato beetle; some species are plant feeders
Damsel bugs (Nabidae)	Adults and nymphs feed on insects especially caterpillars and aphids; sometimes common in crops
Minute pirate bugs (Anthocoridae)	Adults and nymphs feed on thrips, spider mites, and insect eggs; sometimes common in crops; commercially available
Predatory mites (Phytoseiidae)	Feed primarily on spider mites; commercially available; fairly widespread, successful use as augmentative biocontrol agents in greenhouses and some outdoor crops
Spiders (Araneae)	Abundant generalist feeders on insects; hunt for prey on soil and plants or use webs to trap prey; often habitat specialists; very important as "natural" biocontrol agents
Parasitoids	
Tachinid flies (Tachinidae)	Females parasitize moths, beetles, sawflies, and other insects in diverse habitats; individual species often have highly restricted host ranges; often used in classical biocontrol
Parasitic wasps	
Braconid wasps (Braconidae) Ichneumonid wasps (Ichneumonidae) Chalcid wasps (Chalcidoidea)	Females parasitize a diverse array of arthropods including most of the important pest groups; also kill hosts through host feeding; diverse habitats; individual species often have highly restricted host ranges; extensively used in classical biocontrol; some commercially available for augmentation biocontrol
Insect diseases	
Viruses	Most common hosts are moths, butterflies, sawflies, and beetles; often highly host specific; some successfully used in classical and augmentation biocontrol; viral epidemics widespread in nature
Bacteria	Several bacteria (primarily *Bacillus* sp.) have been extensively researched for control of caterpillars, beetles, and mosquitoes; *B. thuringiensis* ("BT") is the most notable and is sold commercially for caterpillar and mosquito control; genes coding for a toxic protein in BT have been transferred to several crop species to provide constitutive defense against insect attack

Table 8.14 **(continued)**   Common Types of Natural Enemies Used in Biological Control
of Insect and Mite Pests

Natural enemy	Feeding habits/comments
Fungi	Infect a broad array of pest arthropods; "natural" outbreaks common; limited commercial availability to date for augmentative releases
Nematodes	Kill hosts by internal feeding or through mutualistic bacteria that cause septicemia in host; some species have broad host ranges and are used in augmentation biocontrol; others have more restricted host ranges and have been used in classical biocontrol
Protozoa	Includes species of *Nosema*, microsporidians that have been formulated in baits for control of locusts and grasshoppers

Source: Van Driesche, R. G. and Bellows, T. S., *Biological Control*, Chapman & Hall, New York, 1996.

Problems

8.1 List ten products that you use daily that contain organic chemicals. Based on what you have learned in this chapter, describe how the extraction, manufacturing, and/or processing of these materials may impact environmental quality.

8.2 What are the most probable causes of adverse effects of organic chemicals? What are some potential adverse affects due to natural causes that have impacted the atmosphere, surface waters, and groundwaters?

8.3 Human health issues involving organic chemicals have become more prevalent since the 1950s. Why are human health issues becoming more important as we develop new organic chemicals?

8.4 Soil microorganisms are important in the degradation of organic chemicals. How do certain types of soil microorganisms assist in the bioremediation of organic chemical–contaminated sites?

8.5 The K_{ow} can be used to predict the expected bioconcentration of an organic chemical that preferentially accumulates in organisms such as fish. Determine the level of DDT, in mg/kg, that would be bioconcentrated in the fatty tissue of fish if the surface water concentration for DDT was 0.000003 mg/L. Use Equation 8.1 and the K_{ow} for DDT of 960,000 to answer this question.

8.6 Determine the amount of material remaining after ten half-life periods for an organic chemical that hydrolyzes and has a half-life of 2 days and an initial concentration of 100 mg/kg.

8.7 Figure 8.6 illustrates the variation in Henry's law constants for several classes of organic chemicals. Using Equation 8.11 determine which of these groups of organic chemicals are preferentially distributed in the aqueous phase. Explain why.

8.8 The half-life photolysis rates of organic chemicals in surface waters can determine the potential bioaccumulation, volatilization, adsorption, and other reactions. Assuming only photolysis to be important, determine the concentration of malathion that would be expected after 60 days if 1 kg were accidentally spilled in a pristine pond.

8.9 Parathion was detected in a landfill site at a soil solution concentration of 2 mg/L. If the soil tested had an organic C content of 3% and the K_{oc} of parathion is 4800, what would be the expected level of parathion adsorbed?

8.10 In March 1989, the *Exxon Valdez* tanker ran aground in Prince William Sound, AK and released approximately 41 million L of crude oil that polluted 2200 km of Alaskan coastline. Of the methods used to clean up the contaminated beaches, fertilization with N and P was found to be the most successful. Explain why nutrient addition is sometimes needed in the remediation of organic chemicals and how N and P were useful for cleanup of the *Valdez* oil spill.

8.11 Why is it important to develop a sampling program before beginning the collecting of field samples? What are some reasons for including a QA/QC program in the evaluation of contaminated site investigations?

8.12 Inorganic and organic substances in river water samples can cause problems in the analysis of organic chemicals. What are some methods that can be used to remove these materials?

8.13 Pest management is an essential part of agricultural production. Although pesticides are widely used to control pests, other practices can reduce or eliminate the use of these materials. Describe three examples of alternatives to pesticide use.

8.14 Characterize the beneficial effects that can be obtained when using plants to remediate an organic chemical–contaminated site. What are some disadvantages to the use of plants in areas with groundwater that contains organic contaminants?

References

Bollag, J. M. and S. Y. Liu. 1990. Biological transformation processes of pesticides, in *Pesticides in the Soil Environment: Processes, Impacts, and Modeling*, H. H. Cheng, Ed., Soil Science Society of America, Madison, WI, 103–168.

Chiou, C. T., V. H. Freed, D. W. Schmedding, and R. L. Kohnert. 1977. Partition coefficient and bioaccumulation of selected organic chemicals, *Environ. Sci. Technol.*, 11, 475.

Cullimore, D. R. 1971. Interaction between herbicides and soil microorganisms, *Residue Rev.*, 35, 65–80.

Harris, J. C. 1982a. Rate of hydrolysis, in *Handbook of Chemical Property Estimation Methods: Environmental Behavior of Organic Compounds*, W. J. Lyman, W. F. Reehl, and D. H. Rosenblatt, Eds., McGraw-Hill, New York, chap. 7, 48 pp.

Harris, J. C. 1982b. Rate of aqueous photolysis, in *Handbook of Chemical Property Estimation Methods: Environmental Behavior of Organic Compounds*, W. J. Lyman, W. F. Reehl, and D. H. Rosenblatt, Eds., McGraw-Hill, New York, chap. 8, 43 pp.

Hoy, M. A. 1989. Intergrating biological control into agricultural IPM systems: reordering priorities, in *Proc. National Integrated Pest Management Symposium/Workshop*, Las Vegas, NV, Commun. Serv. NY, State Agricultural Experiment Station, Cornell University, Geneva, NY, 41–57.

Jaynes, W. F. and G. F. Vance. 1996. BETX sorption by organo-clays: co-sorption enhancement and equivalence of interlayer complexes, *Soil Sci. Soc. Am. J.*, 60, 1742–1749.

Madhun, Y. A. and V. H. Freed. 1990. Impact of pesticides on the environment, in *Pesticides in the Soil Environment: Processes, Impacts, and Modeling*, H. H. Cheng, Ed., SSSA Book Series No. 2, Soil Science Society of America, Madison, WI, 429–466.

Marathon-Agricultural and Environmental Consulting, Inc. 1992. Video cassettes — Fate of Pesticides in the Environment, Box 6969, Las Cruces, NM 88006.

Miller, D. W., Ed. 1980. *Waste Disposal Effects on Groundwater*, Premier Press, Berkeley, CA, 512.

Ney, R. E. 1995. *Fate and Transport of Organic Chemicals in the Environment: A Practical Guide*, 2nd ed., Government Institutes, Rockville, MD.

Office of Technology Assessment. 1984. Protecting the Nation's Groundwater from Contamination, U.S. Congress, Office of Technology Assessment, OTA-0-233, U.S. Government Printing Office, Washington, D.C.

Rees, N. E., P. C. Quimby, G. L. Piper, E. M. Coombs, C. E. Turner, N. R. Spencer, and L. V. Knutson, Eds. 1996. *Biological Control of Weeds in the West*, Western Society of Weed Science, Bozeman, MT.

Schwarzenbach, R. P., P. M. Gschwend, and D. M. Imboden. 1993. *Environmental Organic Chemistry*, John Wiley & Sons, New York.

Shoemaker, C. A. 1989. Integration of environmental concerns into IPM programs, in *Proceedings of the National IPM Symposium/Workshop*, Las Vegas, Commun. Serv. NY, State Agricultural Experiment Station, Cornell University, Geneva, NY, 121–128.

Thomas, R. G. 1982. Volatilization from water, in *Handbook of Chemical Property Estimation Methods: Environmental Behavior of Organic Compounds*, W. J. Lyman, W. F. Reehl, and D. H. Rosenblatt, Eds., McGraw-Hill, New York, chap. 15, 34.

Van Driesche, R. G. and T. S. Bellows. 1996. *Biological Control*, Chapman & Hall, New York.

Vance, G. F., S. A. Boyd, and D. L. Mokma. 1985. Extraction of phenolic compounds from a Spodosol profile: an evaluation of three extractants, *Soil Sci.*, 140, 412–420.

Vance, G. F., D. L. Mokma, and S. A. Boyd. 1986. Phenolic compounds in soils of hydrosequences and developmental sequences of Spodosols, *Soil Sci. Soc. Am. J.*, 50, 992–996.

Weber, J. B. and C. T. Miller. 1989. Organic chemical movement over and through soil, in *Reactions and Movement of Organic Chemicals in Soils*, B. L. Sawhney and K. Brown, Eds., Soil Science Society of America, Madison, WI, 305–334.

Supplementary reading

Baird, C. 1995. *Environmental Chemistry*, W. H. Freeman, New York.

Barcelona, M., A. Wehrmann, J. F. Keely, and W. A. Pettyjohn, Eds. 1990. *Contamination of Ground Water: Prevention, Assessment, Restoration*, Noyes Data Corp., Park Ridge, NJ, 213 pp.

Calvet, R. 1989. Adsorption of organic chemicals in soils, *Environ. Health Perspect.*, 83, 145–177.

Cheng, H. H., Ed. 1990. *Pesticides in the Soil Environment: Processes, Impacts, and Modeling*, Soil Science Society of America, Madison, WI.

Federici, B. A. 1998. Broadscale use of pest-killing plants to be true test, *Calif. Agric.*, 52, 14–20.

Larson, R. A. and E. J. Weber. 1994. *Reaction Mechanisms in Environmental Organic Chemistry*, Lewis Publishers, Boca Raton, FL.

Linn, D. M., T. H. Carski, M. L. Brusseau, and F. H. Chang, Eds. 1993. *Sorption and Degradation of Pesticides and Organic Chemicals in Soils*, SSSA Spec. Pub. No. 32, Soil Science Society of America, Madison, WI.

Lyman, W. J., W. F. Reehl, and D. H. Rosenblatt, Eds. 1982. *Handbook of Chemical Property Estimation Methods: Environmental Behavior of Organic Compounds*, McGraw-Hill, New York.

Manahan, S. E. 1993. *Fundamentals of Environmental Chemistry*, Lewis Publishers, Boca Raton, FL.

Neilson, A. H. 1994. *Organic Chemicals in the Aqueous Environment: Distribution, Persistence, and Toxicity*, Lewis Publishers, Boca Raton, FL.

Pepper, I. L., C. P. Gerba, and M. L. Brusseau. 1996. *Pollution Science*, Academic Press, New York.

Thomas, R. G. 1982. Volatilization from soil, in *Handbook of Chemical Properties Estimation Methods: Environmental Behavior of Organic Compounds*, W. J. Lyman, W. F. Reehl, and D. H. Rosenblatt, Eds., McGraw-Hill, New York, chap. 16, 50.

Wolfe, N. L., U. Mingelgrin, and G. C. Miller. 1990. Abiotic transformations in water, sediments, and soil, in *Pesticides in the Soil Environment: Processes, Impacts, and Modeling*, H. H. Cheng, Ed., Soil Science Society of America, Madison, WI, 103.

chapter nine

Biogeochemical cycles, soil quality, and soil management

Contents

9.1   Introduction: biogeochemical cycles and the environment

All of the nutrients, trace elements, organic chemicals, and airborne pollutants covered elsewhere in this book undergo physical, chemical, and biological transformations in soils. These transformations can increase, reduce, or even have no effect on the impact of these substances on other sectors of the environment. As we have learned more about these transformations, we have been able to develop conceptual models that express our understanding of the factors that control the cycling of elements and compounds in soils. Using these concepts we can then develop more quantitative models that predict which processes and pathways will predominate in a given system. From these concepts and models we develop the practical management programs needed to optimize the intended use of the soil, while protecting the quality of our environment.

Throughout this book we have discussed the cycling of elements and compounds in soils in great detail along with their interchange between the soil environment and other systems of interest, such as the atmosphere, hydrosphere, and food chain. In this chapter we focus on the broader interactions between biogeochemical cycling, ecosystem health, soil quality, and soil management.

9.1.1   *Biogeochemical cycles: definitions and uses*

A *biogeochemical cycle* can be defined as a conceptual description of the mechanisms by which an element or a compound is transformed within a defined system of interest, including the means by which the various forms are interchanged between the solid, liquid, and gaseous phases of that system. Biogeochemical cycling is, of course, not restricted to the soil environment, but includes geologic materials, biological organisms, air, and waters. These cycles are fundamentally global in nature, but we often view and attempt to manage them at smaller scales, such as within a watershed, or even within a city, a farm, or an individual field. Biogeochemical cycles summarize the major processes undergone by an element or compound at different scales and in different systems. They can also provide an overview of the environmental factors and management practices that control both transformations within a system and transport of an element or compound to another system. The global and soil nitrogen (N) cycles (see Figures 4.1 and 4.2) are good examples of the different scales of biogeochemical cycling of an element important in soil management. We understand the general processes operative in the global N cycle and their environmental impacts (e.g., groundwater contamination, ozone depletion, acid rain). From this understanding we seek to manipulate specific processes in the soil N cycle (e.g., mineralization, leaching, denitrification) to maximize soil productivity and minimize any negative effects of N on air, soil, or water quality.

From the perspective of soil science, we wish to use our knowledge of biogeochemical cycles to monitor and control the environmental fate and transport of many elements and compounds that occur naturally in soils as well as those that are intentionally (or unintentionally) added to soils. Earlier chapters have described these chemical and biological transformations (e.g., sorption–desorption, precipitation, mineralization–immobilization, oxidation–reduction) and the factors that affect the transport of an element or compound between different phases of the cycle via leaching, erosion, runoff, volatilization, or biological processes (e.g., plant uptake, microbial immobilization). Because we have this research-based knowledge, biogeochemical cycles often become the basis for more complex modeling efforts designed to identify the fate and transport of plant nutrients, nonessential elements, or organic molecules. From these concepts and models we identify the "best management practices" (BMPs) that optimize the intended use of the soil and at the same time minimize the potential for environmental degradation.

All too frequently, despite our concerns about the many possible impacts a potential pollutant may have on the environment, we lack the scientific knowledge, the technology, or the resources to control every possible transformation within a biogeochemical cycle, particularly as the scale of our effort increases. As a result, we often must prioritize our management efforts. Prioritization requires that we combine our understanding of these cycles with an assessment of the environmental "risk" associated with each transformation (see Chapter 12 for a complete discussion of the risk assessment process). We can then decide on an order of action that is most appropriate, and proceed.

Fortunately for us, for most elements and compounds, the types of transformations possible, and their rates, are limited by the range in properties of the system in which they are located. Thus, we often do not have to concern ourselves about many potential transformations and transport processes. There are many examples of this. Transformations that only occur under anaerobic conditions (e.g., denitrification) are common in wetlands, but highly unlikely in arid zone soils; leaching of phosphorus (P) is rare in most agricultural soils, but may become a problem in sandy soils used for intensive wastewater irrigation;

micronutrient deficiencies are less common in humid regions where acid soils predominate than in arid regions (calcareous soils) because metal solubility is greater in acidic soils; similarly, the environmental risks of metal-contaminated soils decrease when soils are limed; and biodegradation of organic pollutants may occur rapidly in surface soils, but slowly in subsoils where the nutrients and available carbon (C) required by soil microorganisms involved in the degradation process are found in very low concentrations. Clearly, understanding and quantifying the constraints on biogeochemical cycling of an element or a compound in different systems and how to prioritize our management efforts are important first steps in the development of environmentally sound soil management practices.

9.1.2 Biogeochemical cycles: bioregions, biomes, and ecosystems

Perhaps the broadest perspective that can be taken when considering the importance of biogeochemical cycles is that of the *bioregion*, a geographic area where land use is defined by the natural resources present and is thus limited by the soils, geology, topography, and climate characteristic of the region. A related concept is the *biome*, an ecological region with similar types of biological organisms and physical environments. The major biomes of the world are shown in Figure 9.1. Based on the properties of each biome (e.g., tundra, temperate forest, desert) we can anticipate characteristic patterns in biogeochemical cycling of elements and use research and management to identify the most appropriate, or inappropriate, land uses. On a smaller scale, the interrelationships and transfers between the distinct *ecosystems* within a biome must also be considered if we are to develop land management practices with minimal environmental impact (Figure 9.2). The term *ecosystem* means many things to many people; broad definitions are often similar to those

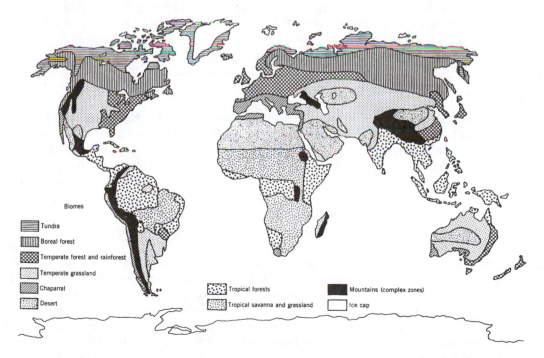

Figure 9.1 Geographic perspective on the major biomes of the world. (Adapted from Council on Environmental Quality, 1989.)

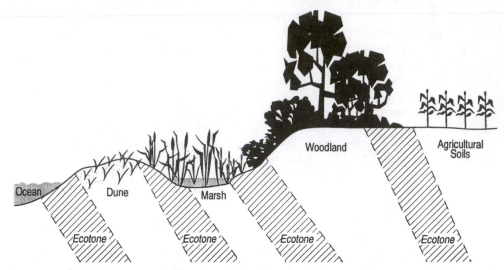

Figure 9.2 Interacting ecosystems in the landscape, showing the ecotones (transitional zones) that exist between individual ecosystems.

used for bioregions, but on a smaller and more manageable scale. Generally speaking, ecosystems refer to geographic areas with stable and reasonably similar biological communities, including the nonliving materials such as soil, air, and water. For our purposes an ecosystem refers to a region with sufficiently distinct biological and physical environments to be adaptable to similar approaches to land-use management. It is beyond the scope of this book to describe all biomes and ecosystems and the transfers of matter and energy that occur between them. We will focus primarily on the relationship between biogeochemical cycling in "managed soils," such as those used for agriculture, and their effects on nearby "natural ecosystems," such as wetlands.

9.1.3 Biogeochemical cycles: interacting ecosystems and ecotones

Many natural ecosystems such as upland forests, wetlands, and deserts are located in close proximity to soils used for agricultural production, cities and industry, mining, transportation systems, and waste disposal sites. Understanding how adjacent ecosystems interact with and impact one another requires a broader view of biogeochemical cycling. A familiar example of interacting ecosystems would be the agricultural production areas and natural wetlands located near urban areas. Agriculture provides the food and fiber for the municipality and is often relied upon to accept and manage many of the waste materials generated by the population and the industries in the urban ecosystem. The urban areas provide the market for agricultural products, the industrial and financial infrastructures that sustain agribusiness, and other "ecosystem outputs," such as education, scientific research, recreational activities, and goods and services. Wetlands perform many important functions for both of these ecosystems. They act as natural filtration zones for agricultural and urban runoff, they are important as a means to accept and retain water during storm events, thus preventing flooding, and they maintain diverse and unique habitats for wildlife and plants, thus sustaining biological diversity.

Increasingly, however, we find that the stabilities of natural ecosystems can be affected negatively by anthropogenic inputs, a ready example being our concerns about the effect of acidic deposition from the burning of fossil fuels on the ecology of forests and lakes. Similarly, in some cases we now use natural ecosystems to mitigate the environmental impact of other activities, such as when we rely on wetlands to purify polluted waters from agricultural or urban runoff. The sustainability of the natural wetland ecosystem when it interacts with these other ecosystems then becomes an environmental issue. Most recently, we have begun to use our knowledge of biogeochemical cycling to attempt to construct "natural" ecosystems. We do this for several reasons. In some cases we must replace a natural area that has been eliminated by other activities, such as when we remove a wetland or forest during road construction. We may also want to create a permanent buffer between human activity and a nearby natural area. A good example of this is the ongoing effort to construct a large wetland adjacent to the Everglades National Park in Florida that will filter P, and other pollutants, from agricultural runoff entering the Everglades. At a smaller scale we install riparian buffer zones between croplands and streams, rivers, and lakes or construct wetlands to filter pollutants from waters discharged by municipal wastewater plants. In each case the biogeochemical cycling of nutrients, non-essential trace elements, and organic compounds must be characterized in very different settings if we are to develop strategies that sustain both managed and natural areas.

We also need to be aware of the boundaries that exist between ecosystems, natural and human-made, that inhibit or facilitate transfers between the geographic areas we are capable of managing. The complex nature of most ecosystems means that these boundaries are often not distinct. We use the term *ecotone* to denote the transitional zones present between ecosystems. As noted above, wetlands are a good example of an ecotone as they represent an ecosystem that has some properties of the zones bordering it, the upland forest and the deep-water aquatic system (Figure 9.2). Another example of an ecotone might be *greenspace*, the parks, woodlots, and wetlands interspersed throughout a densely populated urban area that buffers many of the negative effects of an urban environment, such as noise, traffic, odors, heat, and storm water runoff.

9.2   Biogeochemical cycles and soil quality

One of the main goals of managing biogeochemical cycles is to sustain or improve the quality of our environment. The relationship between soil management and air and water quality is discussed elsewhere, particularly in Chapters 2, 10, and 11. As described in these chapters, we know that human activities can directly and often adversely affect air and water quality. Consequently, protection of water and air quality has long been a high priority of the general public and a key responsibility of the research, advisory, and regulatory communities in the U.S. Legislation defining and regulating "clean air" and "clean water" is well established throughout the world (see examples of U.S. laws in Table 1.3). The concept of "clean soils" and *soil quality*, however, is rather new, often highly controversial, and not as well defined from a scientific perspective.

9.2.1   Soil quality: definitions

Soil quality was defined by the Soil Science Society of America (SSSA, 1995) as "The capacity of a specific kind of soil to function within natural or managed ecosystem boundaries, to sustain plant and animal productivity, maintain or enhance water quality, and support human health and habitation." The SSSA conceptualized soil quality as "a three-legged stool, the function and balance of which requires an integration of three major components — sustained biological productivity, environmental quality, and plant and animal health" (Karlen et al., 1997). The National Research Council (1993) proposed a similar definition: "Soil quality is the capacity of the soil to promote the growth of plants, protect watersheds by regulating the infiltration and partitioning of precipitation, and prevent water and air pollution by buffering potential pollutants such as agricultural chemicals, organic wastes, and industrial chemicals."

9.2.2   Assessment of soil quality

By defining soil quality we begin to establish the parameters that are of importance when we manage soils. The obvious implication is that we wish to avoid land management practices that damage soil quality and thus negatively affect the capacity of soils to function as we believe they should. Recently, there has been a great deal of interest in developing specific "indicators" — measurable or observable soil properties and processes that together characterize the present quality of a soil. These indicators would be a means to assess quantitatively the extent of soil degradation from past activities (human and natural) and also serve as a means to track the success of our efforts to restore soil quality by improved management practices. Some indicators of agricultural soil quality, such as soil fertility testing, have been used for decades, as described in Chapter 3.

However, defining soil quality conceptually, let alone quantitatively, is a complex task. Some have argued that we should assess soil "health" in much the same manner as physicians evaluate human health, using the following six-step process: (1) identify symptoms; (2) identify and measure vital signs; (3) make a provisional diagnosis; (4) conduct tests to verify the diagnosis; (5) make a prognosis; and (6) prescribe a treatment. Inherent to such an approach is the presumed ability of those who manage soils to integrate whatever quantitative indicators of soil quality can be developed with more qualitative concepts such as diagnosis of symptoms and prescription of treatments.

If we are to follow such an approach, we will need a minimum data set (MDS) of measurable soil quality indicators that can be used to provide quantitative information on the capacity of a soil to function in a desired manner (Table 9.1 and Figure 9.3). To have value, soil quality indicators must be clearly shown to be well correlated with quantifiable soil functions, must respond in a measurable way to external change (natural or anthropogenic), must be adaptable for use by individuals with a range of backgrounds and skills, must be found in existing databases that are accessible and of value to soil quality assessment, and must be easily integrated into larger, ecosystem-scale, models, including socioeconomic models.

Some have suggested that we can develop a "soil quality index" (SQI) using "pedotransfer functions" (mathematical functions that relate parameters in the MDS to broader physical, chemical, and biological processes). We might, for example, develop a pedotransfer function that uses soil texture and organic matter content to estimate hydraulic conductivity or use data on soil bulk density, water-holding capacity, and pH to estimate crop rooting depth. By combining and integrating pedotransfer functions with societal goals we can separate soil quality into several distinct components (e.g., food and fiber production vs. water quality impacts), each with a different intrinsic value. It would then be possible to weight individual components of the overall SQI based on physical and social constraints such as climate, topography, hydrology, economics, and politics. If food production is of paramount performance, as opposed to air or water quality, efforts can be made to optimize soil quality for that end; in contrast, if protection of drinking water supplies and surface water quality are high social priorities, the SQI can be weighted to require intensive soil and water conservation practices, most likely at the expense of productivity (e.g., profits, crop yields).

This approach to soil quality seems a good fit for situations where the effects of soil use and management on environmental quality are sociopolitical issues. Consider the

Table 9.1 Example of a Minimum Data Set (MDS) of Soil Quality Indicators That Has Been Modified for Use with Land Application Programs Where Agricultural, Industrial, and Municipal By-Products Are Used

Soil quality indicators in the MDS	Rationale for inclusion of soil quality indicators in the MDS	Suggested additions to the MDS when by-products are used	Rationale for addition of indicators to the MDS
Physical Texture Topsoil depth Rooting zone depth Infiltration Bulk density Water-holding capacity	Indicators of retention and transport of water and chemicals, soil erosion, leaching, surface and subsurface runoff, and soil productivity; also related to water availability and useful in models that seek to integrate soil, landscape, and geographic variability into soil quality	Erodibility (RUSLE) Runoff potential Leachability index Compaction Heat capacity Porosity Soil color Drainage, MHW depth	Provide more direct, quantitative measures of the potential for transport of by-products or of solutes and soil particles from by-product-amended soils to water; aid in assessing the potential for loss from soils of volatile compounds which can affect air quality
Chemical Soil organic matter pH Electrical conductivity Extractable, N, P, and K	Define soil fertility, stability, and erosion extent, potential for N loss, biological and chemical activity, thresholds of microbial activity; useful as productivity and environmental quality indicators	CEC/AEC Sorption capacity Total, extractable, bioavailable, soluble, and desorbable nutrients and nonessential elements Environmental tests for N and P (DPS, PSNT, LCM, stalk nitrate)	Better assessment of potential of soil to retain or release elements and/or organic compounds to leaching or runoff waters; quantify buildup of elements and degree of saturation of soil sorption capacity, predict plant response to N and evaluate success of N management programs
Biological Microbial biomass C, N Mineralizable N Soil respiration	Microbial catalytic potential, repository for C and N, indications of effects on soil organic matter, measures of N supply and soil productivity, changes in biomass and total C pool	Microbial diversity Biodegradation potential in surface and subsoils Redox potential	Evaluate changes in microbial population diversity and size of various communities; assess capacity of entire soil profile to degrade organic pollutants in aerobic vs. anaerobic zones

Note: RUSLE = revised universal soil loss equation; MHW = mean high water table; CEC/AEC = cation/anion exchange capacity; DPS = degree of P saturation; PSNT = preside-dress soil nitrate test; LCM = leaf chlorophyll meter.

Source: Sims, J. T. and Pierzynski, G. M., in *Beneficial Uses of Agricultural, Industrial, and Municipal By-Products*, J. F. Power, Ed., Soil Science Society of America, Madison, WI, in press.

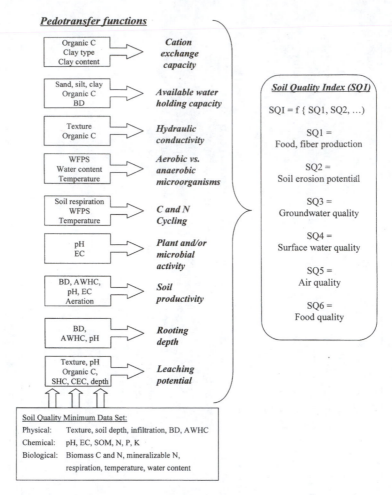

Figure 9.3   Use of soil quality indicators to assess soil processes and develop an SQI. (From Sims, J. T. and Pierzynski, G. M., in *Beneficial Uses of Agriculture, Industrial, and Municipal By-Products*, J. F. Power, Ed., Soil Science Society of America, Madison, WI, in press.)

use of by-products as soil amendments (discussed in more detail in Section 9.3). Local conditions (natural, economic, and political) often dictate the degree of soil quality that is acceptable when manures, biosolids, and the like are used as soil amendments, relative to other end uses for these materials. There are several reasons for this. First, if environmental protection costs are ignored, land application is almost always the least expensive approach to use (or dispose) of most by-products. Many areas do not have the economic wherewithal to implement alternatives to land application and, in some cases, such as with animal wastes, this is essentially the only option, regardless of the strength of the local or regional economy. This leaves local industries, municipalities, and even state governments with no realistic options for by-product use and compels local environmental agencies to develop management programs that minimize, but usually do not eliminate, the effect of by-products on soil and environmental quality. In contrast, in some areas concerns about air and water quality are paramount and the costs of alternative uses are more willingly borne by citizens and/or governments. This scenario would be most common in urbanizing, economically stable areas where the public relies upon more expensive technological or political solutions to deal with by-products. This may include more extensive treatment to reduce environmental risk or the development of public-supported infrastructures or policies that dispose of by-products by means other than land application. Examples include tertiary treatment at wastewater facilities, large-scale composting or pelletization facilities that are subsidized by local tax revenues to transport by-products from the area of generation to another region, and even strict zoning ordinances that prevent the growth of industries that generate by-products perceived to be offensive (e.g., the odors and wastes from poultry or swine operations) regardless of the cost to the local economy.

Clearly, the soil quality that is considered desirable or acceptable will vary between these two scenarios (lesser developed country, limited resources with food production, a high priority vs. an industrialized, wealthy country where environmental protection is a high priority). Historically, the acceptable balance between positive and negative effects of most land uses on the environment has been decided by local and state governments, within general rules and regulations established by federal agencies to protect air and water quality. It can be argued that throughout the decision-making process these governmental agencies should also include a systematic assessment of the short- and long-term impacts of land use on soil quality. In most cases decision making begins with an assessment of land-use goals for a defined geographic area. Once these goals are identified and agreed upon the next step would be to develop the minimum data sets and pedo-transfer functions needed to assess, quantitatively, the impacts of desired land uses on soil quality and the subsequent risks of these land uses to other sectors of the environment. From there we can derive the broader SQIs needed to evaluate more fully how the changes in soil quality associated with land use will affect food production and safety, air and water quality, and ecosystem health. This will help us to better weight the socioeconomic and political aspects of land use that are often as influential as the scientific and technological assessments we derive from such SQIs. There is also a need to develop an approach to monitor soil quality with time, integrating advances in research, improvements in technology, and changes in the politics and economics of land use. Finally, decisions taken to improve or sustain soil quality must occur in a broad social context as they often have adverse implications for some sector of the economy or environment. For instance, political and socioeconomic considerations in soil quality assessment will be of utmost importance for land application of by-products (animal manures, municipal biosolids) because of the environmental and economic tensions that arise when restrictions to land application are proposed (e.g., P-based management of animal manures; see Chapter 5) or alternatives are discussed (incineration, landfilling, ocean dumping). Given this, efforts to assess soil quality and use SQIs as management tools must have disciplinary breadth, going beyond the soil and agricultural sciences to include, for example, economists, political scientists, engineers, hydrologists, and ecologists.

9.3 Management of biogeochemical cycles: examples

The sheer number of individual biogeochemical cycles present in the soil alone is staggering. There are 103 elements in the periodic table, each with a distinct cycle. Some elements, C, hydrogen (H), oxygen (O_2), N, P, sulfur (S), cycle rapidly and ubiquitously because they are components of biological organisms (*biogenic* elements); others, such as the trace elements, are cycled at slower rates and in more localized geographic areas. Tens of thousands of synthetic organic compounds are produced in significant quantities by industries each year; many enter the soil environment either directly as manufactured products (e.g., pesticides), or as constituents of industrial and municipal wastes and wastewaters, or through accidental spills and leaks during transportation and storage. Each individual cycle is complex and often closely related to transformations undergone by other elements or compounds present. Nitrogen cycling, for example, can affect biodegradation of soils contaminated with organic chemicals because N is an essential nutrient for the microorganisms that degrade the chemical. Carbon cycling affects the availability of N, P, and S to plants and the mobility of trace elements in soils. Complicating the matter further is the fact that the bioregion or individual ecosystem in which the cycles are located determines which processes within a cycle predominate in the soil. Indeed, one of the more challenging aspects of soil management, for either agricultural or environmental purposes, is the highly site-specific nature of many management programs. Examples of this are plentiful: we cannot manage fertilizer N in exactly the same way in all soils because of the interactions between soil type and N losses; a reclamation strategy that successfully revegetates a mine spoil in the southeastern U.S. may fail completely in the mountains of West Virginia because of climatic differences; enhancing the biological degradation of gasoline from leaking underground storage tanks may be possible on-site in one state, while in another it may require excavation and incineration. If we have learned anything about soil management, it is the fact that managing biogeochemical cycling requires not only a good scientific understanding of the processes operative in the cycle, but ingenuity and innovation as well. Indeed, it could be reasonably argued that we do not manage these cycles at all; rather, we simply manipulate a limited number of factors to redirect one or more of the processes to our advantage. Perhaps this is the reason many individuals find environmental soil science an interesting scientific discipline — it provides an opportunity to adapt basic science (soil chemistry, physics, and microbiology) and practical knowledge to important and difficult problems faced by our society.

The following sections will describe some current approaches used to manage biogeochemical cycles under "real-world" conditions. We will consider comprehensive nutrient management planning for farms with and without animals, the use of municipal biosolids (sewage sludges) for agricultural crop production, and nutrient transformations in natural and/or constructed wetlands. The approaches used to manage these cycles are generalized and often vary considerably from one ecosystem to the next. Most represent the cooperative efforts of multidisciplinary teams, a characteristic feature of all good soil management programs.

9.3.1 *Comprehensive nutrient management planning*

Nutrient management planning begins with a broad understanding of the flux of nutrients into and out of agricultural systems. At the farm scale, nutrients enter the farm ecosystem as production inputs, such as feed, forage, bedding, livestock, fertilizer, municipal biosolids, N fixed by legumes, and by atmospheric deposition. Some of the nutrient inputs leave the farm in agricultural products, such as meat, milk, eggs, grain, hay, straw, silage, manure, and compost. And, because agricultural systems are not 100% efficient, some of the nutrients remain on the farm, accumulating in soils, animals, and perennial crops, or are lost from the farm by transport processes such as erosion, runoff, leaching, volatilization, and denitrification. On farms devoted primarily to livestock or poultry production, nutrients tend to concentrate, since more nutrients are brought onto these types of farms than leave in the plant and animal products that are sold (see Chapters 4 and 5 for discussions on nutrient mass balances for N and P). Many studies have shown that there are negative economic and environmental impacts associated with the accumulation and/or mismanagement of excess nutrients. Consequently, it is vital to develop and implement a *comprehensive nutrient management plan* (CNMP).

A CNMP is defined as a detailed, site-specific plan for a farm (or other enterprise) which recommends an approach for providing plants and animals with the nutrients needed to achieve economically optimum production goals while minimizing the negative impact of nutrients on the environment. A CNMP considers the amount, form, placement, and timing of application of animal wastes, fertilizers, municipal biosolids, and other nutrient sources and the storage or export of wastes and surplus nutrients. The foundation of a CNMP is the concept of a nutrient mass balance. Simply stated, in a well-designed CNMP the nutrient inputs to a farm are balanced, as closely as possible, by nutrient outputs. Nutrient accumulations in soils and losses to other sectors of the environment are minimized by the nutrient management practices detailed in the CNMP. A nutrient imbalance can result in economic losses, due to unnecessary expenditures on fertilizers and in environmental impacts, such as eutrophication of surface waters and contamination of drinking water supplies (see Chapters 3 through 5). As described in Table 9.2, and discussed in the next section, CNMPs must also address the specific steps needed for efficient management of nutrients and the development of monitoring strategies that can document the success, or failure, of each step in a CNMP.

9.3.2 Nutrient management for farms with and without animal production

Nutrient management on a farm begins with a quantitative assessment of the current balance between nutrient inputs to the farm and nutrient outputs from the farm. In essence, we seek to determine if farming practices are "overloading" some transformation in a biogeochemical cycle resulting in a negative effect on an adjacent ecosystem. Practically speaking, in most agricultural situations, only a limited number of biogeochemical cycles are actually "managed" by farmers or those involved in advising farmers. Agricultural research has shown that the most important and readily manipulated nutrient cycles from a crop production standpoint are N, P, and potassium (K). The C cycle must also be considered because of its role in controlling the mineralization or immobilization of N and P; farmers, however, rarely actively manage the cycling of C in soils. Proper management of soil pH through liming usually ensures that the calcium (Ca) and magnesium (Mg) cycles provide adequate amounts of these nutrients for plant growth without toxic effects from acidic cations — H, aluminum (Al), manganese (Mn); S and trace element cycles are normally of less concern for crop management except in certain well-understood, localized conditions. As an example, S deficiency and significant crop response to applications of S fertilizers are most common in humid regions on deep, sandy soils; trace element deficiencies are usually restricted to alkaline or overlimed soils — iron (Fe), Mn, zinc (Zn) — or high organic matter soils — copper (Cu).

A *nutrient budget* is a quantitative form of a biogeochemical cycle, one that estimates whether the current management practices on the farm are resulting in a nutrient deficit or excess. Nutrient budgets are normally constructed by first estimating the crop nutrient requirements at a realistic yield. Next, the contributions of soil nutrients are assessed by a comprehensive soil testing program, one that may even include subsoil testing for some nutrients that may be found below the "plow layer," but within the rooting zone of the crop. A good example of a nutrient where subsoil testing is important is S. Many studies have shown that the sulfate form of S (SO_4^{2-}) only leaches to moderate depths in soils and may be an important reservoir of plant-available S once roots penetrate to depths of 30 to 60 cm. Another example is subsoil testing for "residual" nitrate-N (NO_3-N) in areas with low rainfall. Late fall or spring deep soil tests (e.g., 1 to 1.5 m) for NO_3-N have been shown to identify accurately the amount of N that is available to this year's crop from a previous year's fertilizer or manure application. Results from soil tests provide not only an estimate of the amount of a nutrient present in the soil in a plant-available form, but also the likelihood of an economic crop response to fertilization with that nutrient, and the rate of the nutrient needed to obtain optimum growth for the specified crop. The difference between crop nutrient requirements and that available

Table 9.2   Overview of the Major Components of a Comprehensive
Nutrient Management Plan (CNMP)

Component	Specific information and action required
Farm identification and natural resource inventory	Operator's and owner's name, address and phone number, location and boundaries of all cropland owned and rented, individual field boundaries, field numbers and size, hectares of each crop type and "realistic yield goals," total amount of land where manure will be applied, animal types and number of animal equivalent units (AEUs) per hectare, soil types and slopes, areas where manure application should be limited, receiving waters and their designated uses, intakes to agricultural drainage systems, drinking water wells and springs, special protection groundwaters and surface waters, and areas of concentrated flow, such as ditches, grassed waterways, gullies, and swales
Soil testing	Routine agronomic soil tests usually include pH, lime requirement, organic matter, and plant-available P, K, Ca, and Mg; other special tests, such as the PSNT, S, trace elements (boron, Cu, Fe, Mn, Zn) soluble salts, texture, cation exchange capacity, P saturation, and P sorption capacity, may be useful in some cases; soil samples should represent each field or management unit and each soil type; in general, soil tests (except the PSNT which is done each year) should be repeated every 2 to 3 years, at the same time of year, and records should be kept to monitor changes in soil fertility due to variations in soil and crop management
Plant analysis	Plant tissue tests can help to diagnose nutrient deficiencies and are useful tools to determine if nutrient management practices need to be changed either during this growing season or for the next crop; two important new approaches to N testing for corn (and some other crops) that should be used each year are the LCM and the corn stalk nitrate test, which can provide both information on the need for N fertilization this year (LCM) and the success of an N fertilization program at the end of the year (corn stalk nitrate test)
Organic waste analysis	Animal wastes should be tested, at a minimum, for total N, ammonium-N, and total P and K; samples should be taken as close to the time of application as possible; if other organic wastes are land applied (e.g., municipal biosolids), they should (or must) be tested for nutrients and specific pollutants required by state and federal environmental agencies
Feed and crop management	In general, animal-to-land ratios that result in the efficient recycling of nutrients in manures on the farm and achieving nutrient mass balance; animal diets can be modified to reduce nutrient concentrations in manures (for example, diets including low-phytate corn and phytase enzymes can reduce P excretions by poultry and swine); the result is a manure with a N:P ratio that more closely matches that required by crops; crop rotation and cultural practices should also be implemented that maximize economic feed production and nutrient recycling
Nutrient application practices	Best management practices for nutrient applications are based upon "realistic crop yield" goals, cropping history, and results of soil, plant, and organic waste tests; once the most efficient nutrient rates are identified, select the most economic nutrient sources that meet crop needs and minimize losses to the environment; carefully assess nutrient application methods, timing, and placement, plant nutrient uptake patterns, and pathways of nutrient loss
Animal manure storage and barnyard management	All wastes and by-products (manures, composts, dead animals) should be handled and stored in a manner that does not adversely affect groundwater or surface water or create public health concerns; clean water should be diverted from contact with organic wastes, adequate storage capacity should be provided, runoff into surface waters and contaminant migration to groundwaters should be prevented, losses of nutrients to the atmosphere should be minimized, and treatment should be carried out to reduce odors, pathogens, and the attraction of vectors

Table 9.2 **(continued)**   Overview of the Major Components of a Comprehensive
Nutrient Management Plan (CNMP)

Component	Specific information and action required
Soil and water conservation practices	Conservation practices should be implemented that minimize the movement of soil, organic materials, nutrients, and pathogens from lands where fertilizers and organic wastes have been applied; measures that help to achieve these goals include forest riparian buffers, filter strips, vegetated field borders, grassed waterways, contour buffer strips, conservation tillage, rotational grazing and pasture management, and grassed waterways
Alternative uses for surplus nutrients	In areas where animal manure production and nutrient content exceeds crop nutrient needs, plans for utilizing surplus nutrients (manure excesses) are needed; some options include selling manure to farmers with a sufficient land base to use the manure nutrients, composting manures for use by home owners, or incinerating the manure for use in power generation; the chosen options should not be detrimental to the environment or public health, and must comply with existing laws that govern alternative uses; the plan should include information on the amount of manure which will be exported, the name and location of the importing operation, the nutrient content of the manure, and the planned use of the manure, including anticipated application rates
Monitoring and record-keeping	Accurate records and regular monitoring are needed to determine how well a CNMP is working; records should be kept of soil testing results, analyses of manure and other nutrient sources, and plant tissue testing, as well as records on crop yields, manure production, manure exports, nutrient application rates, and timing, location, and methods of application; these records will help show if nutrients are being managed in a manner which prevents losses to the environment while sustaining "realistic yields"; all nutrient management plans should be reviewed annually, and as part of the review process, significant changes in management or unforeseen circumstances that require a modification of the plan should be addressed (i.e., increases in the AEUs/ha, major decline in crop yields, animal disease outbreak, development of new BMPs, unforeseen severe weather conditions)

Source: Beegle et al. (1997) and Sims and Gartley (1996).

from the soil represents the amount that must be provided from external sources. At this point any other significant sources of nutrient inputs, such as those provided in irrigation waters, should also be taken into account when developing the nutrient budget. For farms without a significant animal production component, the final step in the nutrient budget process is to determine the most economical source of nutrients available and the most efficient application technique. However, if a farm has a large animal-producing facility, such as a poultry operation or a cattle feedlot, the next step is to account for the amount of nutrients available "on-farm" from animal wastes, since these materials are almost always applied to cropland. Methods to estimate the amount of N available from most animal manures are well established, and the approach used to calculate "plant-available N" in manures was described in Chapter 4. Less information is available on the availability of other elements in animal manures. However, most states provide reasonable estimates of the nutrient content of animal manures so that farmers can adjust fertilizer applications by proper crediting of manure nutrients, as shown in Table 9.3.

In many instances, the long-term use of animal wastes produces an excessive amount of certain nutrients in soils, particularly soils in close proximity to the site of waste generation (e.g., P, see Chapter 5). If this nutrient, or nonessential element, can have a significant environmental impact, the nutrient management plan must include strategies

Table 9.3 Average Concentrations (%, dry weight basis) of
Nitrogen, Phosphorus, and Potassium in Animal Manures

Animal type	N	P	K
Beef	3.25	0.96	2.08
Dairy	3.96	0.67	3.16
Poultry (layers)	4.90	2.08	2.08
Poultry (broilers)	4.00	1.69	1.90
Sheep	4.44	1.03	3.05
Swine	7.62	1.76	2.62
Turkey	5.96	1.65	1.94

Source: Sharpley, A. N. et al., in *Animal Waste Utilization: Effective Use of Manure as a Soil Resource*, J. Hatfield and B. A. Stewart, Eds., Ann Arbor Press, Chelsea, MI, 1998.

to minimize that impact. Conversely, if a nutrient deficit exists, the most efficient use of on-farm resources and fertilizers to optimize the supply of all nutrients to the crop must be identified. The resolution of problems of nutrient excess that are identified by a properly developed, farm-wide nutrient budget can be challenging because of the lack of economic alternatives to animal manure use other than land application reasonably close to the site of manure production. While the buildup of P (or other nutrients) to "excessive" levels in soils is most common with animal-based agriculture, it is by no means confined to that scenario. Other examples are vegetable farms with high soil test P levels from long-term overfertilization with commercial fertilizers and farms operated by municipalities or industries specifically to dispose of organic or inorganic wastes.

Implementation of a CNMP based on the nutrient budget requires efficient storage, handling, and application of the nutrients to the most appropriate sites on the farm (Table 9.2). The *site plan* shown in Figure 9.4 was developed for a dairy farm and illustrates many of the factors that must be considered to ensure that one ecosystem (the farm) does not adversely impact an adjacent ecosystem (a riparian zone and stream). It is an integrated approach to managing the major biogeochemical cycles of importance to this farm (nutrients) through development of a comprehensive plan to store, handle, and distribute nutrients from animal manures and fertilizers to a variety of crops. Environmental protection is considered by the use of storm-water runoff ponds, riparian corridors, and proper placement of animal and manure production facilities relative to drinking water wells.

9.3.3 Management of municipal biosolids

Urban areas produce large quantities of a wide variety of organic by-products that are suitable for land application. Examples include municipal biosolids (sewage sludges), composts of biosolids and wood by-products, municipal solid waste composts (without biosolids), yard waste (leaves, grass clippings) composts, and wastewaters from sewage or drinking water treatment plants. These materials are commonly applied to cropland, forests, turf, and ornamentals grown in landscapes and roadsides and in large-scale land reclamation projects. Land application of municipal by-products is almost always regulated by federal and state environmental agencies. Normally, permits must be obtained from these agencies following a review process that requires detailed, site-specific information on all aspects of the land application program including, but not limited to, soil type, crop rotation, groundwater and surface water properties, topography, odor control, and monitoring programs. Unlike agricultural situations where only a few biogeochemical cycles must be managed and monitored, environmental regulatory agencies require that management plans for municipal and industrial by-products carefully consider a large

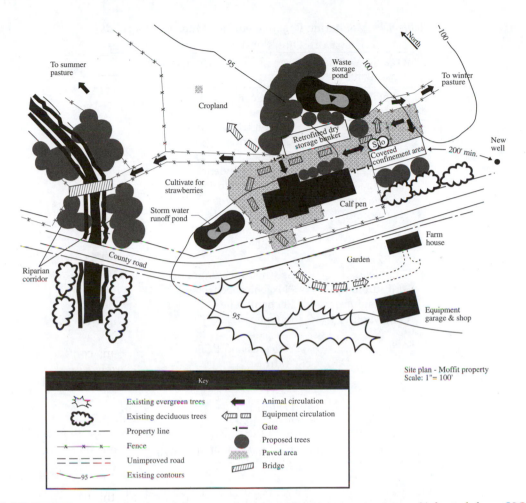

Figure 9.4 Site plan developed for an animal waste management system. (Adapted from U.S. Department of Agriculture, Soil Conservation Service, 1992.)

number of inorganic and organic constituents (Table 9.4). Management and monitoring programs must, therefore, often be based on our understanding of the biogeochemical cycles of dozens of elements or compounds. Most regulatory agencies use the *land-limiting constituent* (LLC) concept to determine both annual application rates and the total site-life, or length of time a by-product can be applied to a site. The LLC is the element or compound in a by-product that is perceived to present the greatest hazard and that should be used to determine the actual rate of by-product applied to a site this year and in total. Typical LLCs are nutrients, trace elements, and organic compounds. A simplified example of this approach to determine the LLC is illustrated in Table 9.5.

In the fall of 1992 the U.S. Environmental Protection Agency (EPA) released the "National Sewage Sludge Rule," developed under the national Clean Water Act (40 CFR, Part 503). Commonly known as the Part 503 rule, this document embodies a series of regulations that were developed from a comprehensive risk assessment of more than 15 years of research and management programs using municipal biosolids (sewage sludge) for farms, gardens, forests, and at dedicated sites (e.g., landfills). The Part 503 rule describes general and specific management practices for biosolids use in land application programs. For land application the general approach is to apply the municipal biosolids at an "agronomic rate" consistent with crop nutrient requirements. To ensure that this rate does

Table 9.4 Monitoring Requirements for Municipal Biosolids
in Delaware

Inorganic waste constituents[a]	Priority pollutants[b]
Total N	Volatile compounds
NH_4-N	Benzene
NO_3-N	Carbon tetrachloride
P	Chloroform
K	Toluene
Ca	Trichloroethylene
Mg	Vinyl chloride
Hg (mercury)	
Na (sodium)	Acid compounds
Cu	Pentachlorophenol
Ni (nickel)	Phenol
Zn	
Pb (lead)	Base/neutral compounds
Cd (cadmium)	Hexachlorobenzene
Cr (chromium)	Phenanthrene
CN (cyanide)	Pyrene
pH	
	Pesticides and PCBs
	Aldrin
	Chlordane
	2,4-D
	Dieldrin
	Heptachlor
	Toxaphene
	Polychlorinated biphenyls (PCBs)

[a] Total analysis required.

[b] Representative examples of each class given; in 1988 there were 126 priority pollutants identified by the EPA.

Source: Delaware Department of Natural Resources and Environmental Control (1988).

not apply excessive quantities of trace elements or other pollutants, monitoring of biosolids composition is required by permit. Four types of "pollutant limits" are included in the Part 503 rule (Table 9.6). *Ceiling concentration limits* are established to ensure that an excessively contaminated biosolid is not applied to soils, e.g., 85 mg/kg is the ceiling concentration for cadmium (Cd). *Pollutant concentration limits* are used to determine if a biosolid is of "exceptional quality (EQ)" (e.g., <39 mg/kg for Cd). Use of EQ biosolids still requires a permit but the requirements to track long-term loading rates or maintain records and monitoring are not as stringent. *Annual pollutant loading rates* (APLRs: maximum amount of a pollutant that can be applied to a site in any given year) and *cumulative pollutant loading rates* (CPLRs: total quantities of each pollutant that can be applied to a site) were also established in the Part 503 rule. CPLRs essentially define the total length of time a site can receive biosolids. For example, if a municipality desires to apply biosolids at rates appropriate for grain crops, and these rates provide 1.0 kg/ha/year of Cd, then the site can be used for 39 years.

The Part 503 rule also states that management practices to control runoff (e.g., buffer zones) be required and that landscapers or home owners be provided with detailed instructions on the proper means to use products derived from biosolids (e.g., composts)

Table 9.5   Example of the Approach To Determine the LLC for Land Application
of Municipal Biosolids

Parameter	Quantity generated (kg/year)	Site assimilative capacity (kg/ha/year)	Land area requirement (ha)
Total N	3000	400	8 [LLC]
P	2100	400	5
Ca	1.5	0.5	3
Cu	10	14	1
Ni	25	14	2
Pb	45	56	1
Zn	160	28	5

Assumptions:

1. Site-assimilative capacity for N based on crop uptake, ammonia volatilization, and requirement that nitrate loss in drainage waters will not exceed 10 mg NO_3-N/L.

2. Site-assimilative capacity for P recognizes the fact that, since biosolids are applied to meet crop N requirements, excess P (beyond crop requirements) will be applied. Conservation measures are thus required at the site to minimize P losses in runoff, erosion, and drainage.

3. Site-assimilative capacity for metals based on maximum cumulative metal-loading rate at the site, assuming a CEC of 10 cmol (+)/kg and a 20-year "site-life." Under current Delaware regulations these values are 10, 280, 280, 1120, and 560 kg/ha for Cd, Cu, Ni, Pb, and Zn. Using Cu as an example: site-assimilative capacity = 280 kg/ha ÷ 20 years = 14 kg/ha/year.

4. The LLC is defined as the constituent that requires the most land for safe utilization of the biosolids based on site-assimilative capacity. In this case the LLC will be N, which requires 8 ha.

Source: Delaware Department of Natural Resources and Environmental Control (1988).

for horticultural purposes. Careful recordkeeping, monitoring of biosolids for pollutant composition, and practices to ensure that threatened or endangered species are protected are also required under the Part 503 rule.

The implementation of this rule by the EPA is a good example of how our understanding of biogeochemical cycles can be used to develop management programs that identify and prioritize the risk of pollution. Beginning in 1984, the EPA reviewed data from studies involving from 200 to 400 pollutants that had been found in biosolids. A national research team recommended, based on scientific research on the transformations of these pollutants in the environment (i.e., their biogeochemical cycles), that 50 pollutants be reviewed more intensively. Further evaluation of research and other technical information resulted in the Part 503 rule, which now sets national limits for eight trace element pollutants for land application programs using biosolids (see Table 9.6). Chapter 12 provides specific details on how the risk assessment process was used in the development of the national regulations for biosolids.

Despite the decades of research that led to the development of the EPA Part 503 rule, concerns still exist that we are degrading soil quality by using municipal biosolids as soil amendments. A recent report from Cornell University entitled, "The Case for Caution: Recommendations for Land Application of Sewage Sludges and an Appraisal of the USEPA's Part 503 Sludge Rules," calls for reevaluation and revision of much of the 503 rule and for "further research on N release rates, the movement of metals and pathogens to groundwaters and surface water, the presence and impact of synthetic organic contaminants and of contaminants eliminated from U.S. EPA consideration due to inadequate data and ecological impacts (including soil organisms)" (Harrison et al., 1997). Of perhaps more immediate impact, the state of Maryland in 1998 passed the Water Quality Improvement Act, which requires that land application of animal manures *and* biosolids be based

Table 9.6 Pollutant Limits for Land Application for Municipal Biosolids — Criteria for Land Application Programs as Defined in the 1992 National Sewage Sludge 503 Rule of the EPA

Trace element	Concentration limit (mg/kg)		Loading rate limits	
	Ceiling	Pollutant	Cumulative (kg/site)	Annual (kg/ha/year)
Arsenic (As)	75	41	41	2.0
Cd	85	39	39	1.9
Cu	4300	1500	1500	75
Pb	840	300	300	15
Hg	57	17	17	0.85
Ni	420	420	420	21
Selenium (Se)	100	100	100	5.0
Zn	7500	2800	2800	140

Definitions of pollutant limits are as follows (*Source:* U.S. EPA, 1995):

Ceiling concentration limit: Maximum allowable concentration of a pollutant in biosolids that are to be land applied. Ceiling concentration limits are either the 99th percentile concentration of a pollutant as identified in the National Sewage Sludge Survey or the pollutant limits identified in the biosolids risk assessment process.

Pollutant concentration limit: Most stringent, risk-based pollutant limit in the 503 rule. Pollutant concentration limits define the "no-adverse-effect" concentration where biosolids can be safely land applied without the need for recordkeeping of cumulative pollutant loading rates. Biosolids that meet these concentration limits are sometimes referred to as "exceptional quality" biosolids and can be land applied as freely as other fertilizers and soil conditioners.

Cumulative pollutant loading rate (CPLR): CPLRs apply to biosolids with concentrations above the pollutant concentration limit. Permitting requirements mandate that accurate records be kept of the amount of pollutants applied to a site from biosolids subject to CPLRs. Once the CPLR is attained, biosolids can no longer be applied at the site. Even at the CPLR, risk assessment showed that the pollutant loading rate is protective of public health and the environment.

Annual pollutant loading rate (APLR): APLRs apply only to biosolids that are sold or given away in a bag or container and identify the maximum amount of a pollutant that can be applied to a site in 1 year. APLRs were derived by dividing CPLRs by 20, assuming that a 20-year site life where biosolids are land applied is reasonable. APLRs are used with bagged and containerized biosolids because of the difficulty in maintaining long-term records when biosolids are applied to situations such as lawns, home gardens, public contact sites (e.g., parkland).

on N and P by 2005. Many soils that have received long-term applications of biosolids now have soil test P concentrations above the values needed by crops. Consequently, this act may severely restrict the use of biosolids on cropland because of concerns about nonpoint-source pollution of surface waters by P. While major revision of the Part 503 rule is unlikely, some reevaluation of the limits and requirements of the rule may occur, particularly in individual states. A critical part of any such reevaluation, should it occur, will be the need for soil quality indicators to guide decision makers who must determine if biosolids are (1) degrading soil quality and thus irreparably damaging a natural resource; (2) creating soils that will degrade other sectors of the environment; (3) be unsafe to use for the production of food for humans and animals.

9.3.4 *Nutrient transformations in wetlands*

Wetlands were defined by the U.S. Fish and Wildlife Service in 1979 as "lands transitional between terrestrial and aquatic ecosystems where the water table is usually at or near the surface or the land is covered by shallow water." To be classified as a wetland, an area usually has to meet one or more of the following criteria: (1) predominantly support, at least periodically, hydrophytic (water-loving) vegetation; (2) have as its underlying substrate a predominantly undrained *hydric* soil; and (3) have as its underlying substrate nonsoil that is saturated or covered with shallow water at some time during the growing season of each year. The U.S. Natural Resources Conservation Service defines hydric soils as "a soil that in its undrained condition is saturated, flooded, or ponded long enough during the growing season to develop anaerobic conditions that favor the growth and regeneration of hydrophytic vegetation." The criteria for classification of wetlands are under intense scrutiny today because of the ecological value of wetlands in an undisturbed state and because of their potential value for other land uses once drained (e.g., agriculture, development, mining of peat). There are many types of wetlands: tidal salt- and freshwater marshes, inland freshwater marshes and swamps, peatlands, bogs, prairie "potholes," and riparian forests. Each is a unique ecosystem, but all share a number of vital functions that make wetland preservation a critical environmental issue. Among their more important functions, wetlands reduce erosion, provide control of floodwaters and storm-water run-off, maintain water quality by trapping sediments and pollutants, provide wildlife habitats and maintain biodiversity, produce food and timber, and act as an aesthetic buffer between urban and industrial areas. Humans alter wetlands by drainage, installation of navigational canals, dredging and filling, mining, and by point- or nonpoint-source pollution. Between the 1600s and the 1970s, ~30 to 50% of the wetlands in the U.S. were destroyed.

Biogeochemical cycles in wetlands are dominated by the hydrology of the ecosystem. Under saturated conditions O_2 becomes depleted or diffuses at such a slow rate that anaerobic transformations become the dominant processes in many biogeochemical cycles. From an environmental viewpoint, we often desire to use wetlands to reduce pollution of adjacent aquatic systems by upland land uses. Today, many cities, industries, and agricultural enterprises are beginning to investigate the use of constructed wetlands in upland areas to serve as natural wastewater treatment systems. In some cases these treatment systems can also produce harvestable products, such as water hyacinths, or be used for the production of fish (aquaculture). Hence, it is essential to understand the effects of anaerobiosis on chemical and biological reactions that involve pollutants. As many wetlands do not remain in an anaerobic state year-round, the effects of alternating wet–dry cycles on these reactions can be important as well. Similarly, most wetlands are not anaerobic throughout the soil profile; a shallow aerobic layer often exists in the upper few millimeters of the soil.

The factor controlling most important reactions in wetlands soils is the *redox potential* of the soil, a measure of the oxidation–reduction status. In aerobic soils the decomposition of organic matter (oxidation) produces electrons that are then accepted by O_2, forming water (reduction). When O_2 is absent or its rate of diffusion through the soil is very slow, as in wetland soils or lake sediments, other substances accept these electrons, resulting in the formation of end products other than water (Table 9.7). This was illustrated earlier with NO_3-N (see Chapter 4), where denitrification in wetlands converted a potential aquatic pollutant (NO_3-N) to gaseous N oxides.

The sequence of reduction in wetland soils is well understood; hence, a knowledge of the redox potential of a system can be used to predict the dominant electron acceptors present, as shown in Figure 9.5, which illustrates the transformations that occur when a soil is saturated with water. (Note that trends in this figure will vary with pH and temperature; Figure 9.5 assumes a pH of 7.0.) Available O_2 is depleted within 1 day; NO_3-N becomes the next substrate for electrons produced by anaerobic decomposition of organic matter, followed by Mn oxides and then Fe oxides. As the redox potential (E_h) declines, easily reducible solid forms of Mn disappear and exchangeable Mn^{+2} accumulates, followed by Fe^{+2} as Fe oxides are reduced. Only when all sources of NO_3-N, Mn^{+4}, and Fe^{+3} have been depleted, will the reduction of SO_4^{2-} to sulfide (S^{2-}) occur, followed by

Table 9.7 Redox Reactions of Primary Importance in Wetland Soils

Element	Elements or compounds involved in redox reaction		Redox potential for reaction (mv)[a]
	Oxidized species	Reduced species	
O	O	H_2O	700 to 400
	$[\frac{1}{2} O_2 + 2e^- + 2H^+ \rightleftharpoons H_2O]$		
N	NO_3^-	NH_4^+, N_2O, N_2	220
	$[NO_3^- + 2e^- + 2H^+ \rightleftharpoons NO_2^- + H_2O]$		
Mn	Mn^{4+} (manganic: MnO_2)	Mn^{2+} (manganous: MnS)	200
	$[MnO_2 + 2e^- + 4H^+ \rightleftharpoons Mn^{2+} + 2H_2O]$		
Fe	Fe^{3+} (ferric: $Fe(OH)_3$)	Fe^{2+} (ferrous: FeS, $Fe(OH)_2$)	120
	$[FeOOH + e^- + 3H^+ \rightleftharpoons Fe^{2+} + 2H_2O]$		
S	SO_4^{2-} (sulfate)	S^{2-} (sulfide: H_2S, FeS)	−75 to −150
	$[SO_4^{2-} + 8H^+ + 7e^- \rightleftharpoons \frac{1}{2} S_2^{2-} + 4H_2O]$		
C	CO_2 (carbon dioxide)	CH_4 (methane)	−250 to −350
	$[CO_2 + 8e^- + 8H^+ \rightleftharpoons CH_4 + 2H_2O]$		

[a] Redox potentials are approximate values and will vary with soil pH and temperature.

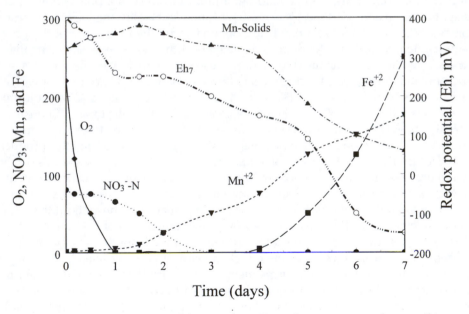

Figure 9.5 Schematic representation of the transformations than can occur when a soil is saturated with water and anoxic conditions develop. (Based on data from Turner and Patrick, 1968.)

the anaerobic degradation of organic C that results in the production of methane (CH_4), often referred to as "swamp or marsh gas." Should a soil such as this dry out and aerobic conditions be reestablished, the reactions will often reverse; NO_3-N will begin to accumulate as organic matter is mineralized and the soluble Mn and Fe will begin to form insoluble precipitates such as MnO_2 and $Fe(OH)_3$. For an element such as P, where sorption by Fe oxides is an important mechanism of retention in soils, alternating wet and dry cycles can affect both plant uptake and the potential for delivery of P to nearby aquatic systems that are sensitive to eutrophication. Research has shown that the development of anaerobic conditions in the lake sediments can reduce Fe oxides and Fe phosphates, resulting in a release of soluble P from these solid phases, increasing the likelihood of eutrophication.

Attempts to construct wetlands or riparian zones near agricultural fields to enhance denitrification should, therefore, consider the possibility that this could increase the release of soluble P into streams and rivers bordering these fields.

Other transformations that occur in wetlands, such as the fate of organic compounds (e.g., pesticides, hydrocarbons in storm-water runoff) or trace elements, have received less study; hence they are not as well understood as those involving plant nutrients. This will be of particular importance should the use of constructed wetlands as biological filtration zones for urban and industrial wastes increase in the future.

Problems

9.1 Contrast the biogeochemical cycles of N (Chapter 4), P (Chapter 5), Zn, and Pb (Chapter 7). Explain the major differences between these cycles with reference to (a) groundwater contamination; (b) surface water pollution; (c) potential impacts on human and ecosystem health.

9.2 Discuss the differences you would expect in a well-designed land management program for municipal biosolids in two distinctly different bioregions — for example, the arid, western vs. the humid, southeastern U.S..

9.3 Explain the relationships between the "minimum data set" of soil quality indicators given in Table 9.1 and the "pedotransfer functions" in Figure 9.3. Discuss the value of "pedotransfer functions" to land use management for (a) production agriculture and (b) land reclamation projects for strip-mined sites.

9.4 Contrast the difficulties in developing and implementing a comprehensive nutrient management plan (CNMP) for the following categories of nutrient users: (a) 500-ha cash grain farm growing corn, alfalfa, and wheat; (b) 100-ha poultry farm with 100,000 broiler capacity, growing soybeans (75 ha) and grain sorghum (25 ha); (c) municipality generating biosolids with an annual production of plant-available N of 8000 kg/year; (d) commercial lawn care company operating in the Chesapeake Bay Watershed (now required by law to have CNMPs).

9.5 Review the concept of an LLC and the example in Table 9.5. Explain how the LLC would change if: (a) to protect surface water quality, biosolids P applications were limited to a maximum annual application rate of 10 kg P/ha/year; (b) a major new metallurgical industry developed in the area which resulted in an increase in Zn generated by the plant to 560 kg/year; (c) the plant began to transport the biosolids it generated to a new area where CEC values were <5 cmol (+)/kg (assume the cumulative metal-loading rate permitted for soils in this lower CEC range is half that of the example in Table 9.5).

9.6 There is considerable interest in constructing wetlands to purify runoff or waste-waters from municipalities and industries. Discuss the following issues related to constructed wetlands:

a. In general, wetlands are expected to be more effective at removing N than P from agricultural runoff — why is this so?

b. What types of problems might be faced in sustaining a wetland constructed to treat storm-water discharge from a large shopping mall?

c. What would be the pros and cons of using wetlands designed to treat waste-waters from a sewage treatment plant for aquaculture (fish production)?

9.7 The changes in soil chemical properties following flooding are generalized in Figure 9.5. Explain the different trends in E_h that might occur if a flooded soil (a) had a very high nitrate concentration; (b) had a very low organic matter content (<1.5%); (c) had a high organic matter content and very low concentrations of reactive Fe and Mn; (d) was highly acidic (pH < 5.0).

9.8 One way to treat acid mine drainage (AMD; see Chapter 6 for detailed description) is to divert the drainage waters through large limestone drains, often constructed in channels or ditches. Most AMD is highly acidic, with high concentrations of soluble Fe. Occasionally, limestone drains fail because they become coated with Fe hydroxides and the lime no longer dissolves or dissolves too slowly to neutralize the acidity in the AMD (a process referred to as "armoring"). Explain why "armoring" occurs and why constructing a wetland in a limestone drain might correct this problem.

References

Beegle, D., L. E. Lanyon, and D. Lingenfelter. 1997. Nutrient Management Legislation in Pennsylvania: A Summary of the Final Regulations, Agron. Facts 40, Pennsylvania State University, University Park.

Council on Environmental Quality. 1989. Environmental Trends, U.S. Government Printing Office, Washington, D.C.

Delaware Department of Natural Resources and Environmental Control. 1988. Guidance and Regulations Governing the Land Treatment of Wastes, Division of Water Resources, DNREC, Dover.

Doran, J. W. and T. B. Parkin. 1996. Quantitative indicators of soil quality: a minimum data set, in *Methods for Assessing Soil Quality*, J. W. Doran and A. J. Jones, Eds., SSSA Spec. Pub. No. 49, Soil Science Society of America, Madison, WI, 25–38.

Harrison, E. Z., M. B. McBride, and D. R. Bouldin. 1997. The case for caution: recommendations for land application of sewage sludges and an appraisal of the USEPA's Part 503 sludge rules, Working paper, Cornell Waste Management Institute, Cornell University, Ithaca, NY.

Karlen, D. L., M. J. Mausbach, J. W. Doran, R. G. Cline, R. F. Harris, and G. E. Schuman. 1997. Soil quality: a concept, definition, and framework for evaluation, *Soil Sci. Soc. Am. J.*, 61, 4–10.

National Research Council. 1993. *Soil and Water Quality*, National Academy Press, Washington, D.C.

Sharpley, A. N., J. J. Meisinger, A. Breeuswma, J. T. Sims, T. C. Daniel, and J. S. Schepers. 1998. Impacts of animal manure management on ground and surface water quality, *Animal Waste Utilization: Effective Use of Manure as a Soil Resource*, in J. Hatfield and B. A. Stewart, Eds., Ann Arbor Press, Chelsea, MI, 173–242.

Sims, J. T. and K. L. Gartley. 1996. Nutrient Management Handbook for Delaware, Coop. Bull. No. 59, University of Delaware, Newark.

Sims, J. T. and G. M. Pierzynski. Assessing the impacts of agricultural, industrial, and municipal by-products on soil quality, in *Beneficial Uses of Agricultural, Industrial, and Municipal By-Products*, J. F. Power, Ed., Soil Science Society of America, Madison, WI (in press).

Soil Science Society of America (SSSA). 1995. SSSA statement on soil quality, *Agron. News*, June 7, Soil Science Society of America, Madison, WI.

Turner, F. T. and W. H. Patrick, Jr. 1968. Chemical changes in waterlogged soils as a result of oxygen depletion, in *Trans. IX Int. Cong. Soil Sci.* (Adelaide, Australia), 4, 53–56.

USDA Soil Conservation Service. 1992. Agricultural Waste Management Field Handbook, No. 651 of the National Engineering Handbook Series, U.S. Department of Agriculture, Landover, MD.

U.S. Environmental Protection Agency. 1995. A Guide to the Biosolids Risk Assessments for the EPA Part 503 Rule, EPA832-B-93-005, Office of Wastewater Management, Washington, D.C.

Supplementary reading

Bouma, J. 1997. Soil environmental quality: a European perspective, *J. Environ. Qual.*, 26, 26–31.

Constanza, R., B. G. Norton, and B. D. Haskell. 1992. *Ecosystem Health: New Goals for Environmental Management*, Island Press, Washington, D.C.

McBride, M. B. 1995. Toxic metal accumulation from agricultural use of sewage sludge: are USEPA regulations protective? *J. Environ. Qual.*, 24, 5–18.

Mitsch, W. J. and J. G. Gosselink. 1986. *Wetlands*, Van Nostrand Reinhold, New York.

Midwest Planning Service. 1985. Livestock Waste Facilities Handbook, Midwest Planning Serv. Rep. MWPS-18, 2nd ed., Iowa State University, Ames.

Schmidt, J. P. 1997. Understanding phytotoxicity thresholds for trace elements in land applied sewage sludge, *J. Environ. Qual.*, 26, 4–10.

Sims, J. T. 1995. Animal waste management, in *Encyclopedia of Agricultural Sciences*, Academic Press, New York, 185–201.

Sims, J. T., S. D. Cunningham, and M. E. Sumner. 1997. Assessing soil quality for environmental purposes: roles and challenges for soil scientists, *J. Environ. Qual.*, 26, 20–25.

U.S. Environmental Protection Agency. 1993. The Standards for the Use or Disposal of Sewage Sludge, Title 40 of the Code of Federal Regulations, Part 503. U.S. EPA, Washington, D.C.

U.S. Environmental Protection Agency. 1994. A Plain English Guide to the EPA Part 503 Biosolids Rule, USEPA/832/R-93/003, Office of Wastewater Management, Washington, D.C.

chapter ten

The atmosphere: global climate change and acid deposition

Contents

10.1 Introduction

"Air pollution is the presence of any substance in the atmosphere at a concentration high enough to produce an objectionable effect on humans, animals, vegetation or materials, or to alter the natural balance of any ecosystem significantly" (Wolff, 1999). Based on this definition, air pollution can be the result of solids, liquids, or gaseous materials. Air pollution has been a constant natural occurrence since the formation of the earth; however, only during postindustrial times have humans impacted the atmosphere on a global scale. Initially, localized air pollution problems resulted from human activities, but as human population growth increased, industrialization and natural resource utilization have resulted in impacts to our atmosphere as well as hydrosphere, biosphere, and terrestrial environments. Chapter 2 described the general properties of our atmosphere. This chapter will evaluate important phenomena related to anthropogenic changes in the atmosphere. Sources, impacts, and causes of two types of airborne pollution that have been major issues during the past 25 years will be examined. We will present details on atmospheric conditions that have resulted from human activity (e.g., global climate change and acidic deposition) and which have numerous effects and interrelationships with soil, plant, and aquatic ecosystems.

10.2. Global climate change

The prospect of *global climate change* comes from the phenomenon known as global warming. *Global warming* refers to the possibility that the global average air temperature may be increasing because the concentrations of various gases in the atmosphere are also increasing (see Chapter 2). These gases create the *greenhouse effect*, which allows the atmosphere to trap radiant energy that would otherwise radiate freely away from the earth, and therefore allows the atmosphere to retain heat and maintain a temperature suitable for life. In fact, the greenhouse effect allows life as we know it to exist. Without the greenhouse effect, the surface of the earth would be approximately 33°C colder than it is now. Global warming is often referred to as the greenhouse effect when in reality global warming represents an enhancement of the greenhouse effect. The *anthropogenic greenhouse effect* would be a more appropriate term for global warming.

10.2.1 The greenhouse effect

The wavelength of maximum emission for incoming solar radiation is 0.48 μm, representing a relatively high energy level. Conversely, the infrared emission from the surface of the earth has a wavelength of maximum emission of approximately 9.7 μm, representing a much lower energy level. Some incoming solar radiation has sufficient energy to pass through the atmosphere, but the reemitted infrared radiation, measured as heat, is absorbed by greenhouse gases, and the energy becomes trapped in the atmosphere. The gases that absorb the infrared radiation are called *radiatively active gases*. The higher the concentration of the greenhouse gases, the more heat is trapped. A simple example of the greenhouse effect can be found in an automobile parked outside on a sunny day with the windows closed. High-energy solar radiation passes through the glass and warms the interior surfaces. These surfaces then transfer that heat to the air inside the automobile and the warm air is trapped by the glass. The greenhouse gases act as the glass in our automobile example and trap some of the heat in the atmosphere. The mechanisms by which the air is heated are slightly different in the automobile from those in the atmosphere. In the automobile, most of the heat is transferred by the air directly contacting the warm surfaces, whereas in the atmosphere the gases absorb the emitted infrared radiation

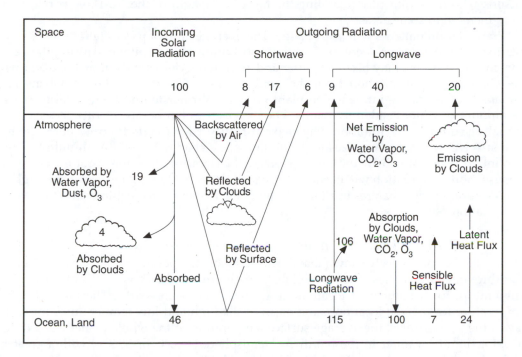

Figure 10.1 The global energy balance in units of percent of incoming solar radiation. (From MacCracken and Luther, 1985.)

to warm the air (although direct transfer of heat to the air also occurs in the atmosphere). Infrared radiation must travel an average of approximately 1 km in the atmosphere before it is absorbed by greenhouse gases.

The greenhouse effect can be shown from the global energy balance depicted in Figure 10.1. As shown in the figure, 100 units of incoming solar radiation enter the atmosphere and 31 units are reflected to space; 46 of the remaining units are absorbed by the surface and are partly used to directly heat the atmosphere (sensible heat flux) and to evaporate water (latent heat flux). The surface of the earth emits 115 units of infrared radiation, which is greater than the incoming solar radiation, and is evidence that the earth is still a cooling body. The fact that 100 units of infrared radiation are emitted downward from the atmosphere to the surface from absorption by clouds, water vapor, carbon dioxide (CO_2), and ozone (O_3), in addition to the 46 units absorbed by the surface directly from incoming solar radiation, allows the surface to warm more than it would by incoming solar radiation alone. The anthropogenic greenhouse effect causes an increase in the flux of infrared radiation downward to the earth's surface.

10.2.1.1 Temperature and precipitation changes

Unfortunately, climatic changes are likely to occur as the mean global temperature increases. These changes are as yet relatively unpredictable beyond the knowledge that an increased global temperature will influence the driving forces for our weather. Global warming obviously implies an increase in average temperature, but the changes are not as simple as adding a few degrees to the high and low temperature each day. For a given location, there will be more days each year that surpass some critical value (e.g., 38°C or ~100°F) and fewer days that would be considered extremely cold. The high extremes place the greatest burden on our resources as we run air conditioners and assist sick and elderly persons who have difficulty tolerating the heat. The length of the frost-free period will likely increase in some areas.

Precipitation patterns will also change. The average annual precipitation for a given location may increase or decrease, and the distribution of precipitation during the year may change. Winters may become wetter and summers drier as the atmosphere warms. The changes in precipitation have the potential to create the greatest problems for humans because of the influence on plant growth. Warmer temperatures will increase the amount of water needed by plants. The increases in temperature with decreases in annual or summer precipitation may make it difficult for plants to grow normally or even survive. Related to changes in precipitation are changes in relative humidity. The amount of water stored in the atmosphere will increase as the temperature goes up because more evaporation will occur and because air is able to hold more water at higher temperatures. These changes in humidity will be a driving force behind the changes in precipitation patterns.

10.2.1.2 Mean global temperature

It has yet to be conclusively shown that the mean global temperature is increasing because of anthropogenic activities. The mean global temperature is the mean of the mean annual temperature for a number of locations distributed across the earth. The mean annual temperature for a given location is the mean of the high and low temperatures for each day of the year. In 1998, the average surface temperature of the whole planet was 14.0°C (57.2°F), slightly warmer in the Northern Hemisphere (14.6°C or 58.3°F) with a greater landmass, and a little cooler (13.4°C or 56.1°F) in the Southern Hemisphere where there is more water. It is clear that the global mean temperature, as compared with the 30-year average from 1961 to 1990, has increased recently, as shown in Figure 10.2. Temperatures have increased approximately 0.6°C (1.1°F) in the last 100 years and the 4 warmest years on record have all occurred since 1990 (1998, 1997, 1995, and 1990 in descending order). Other evidence suggests that global climate change processes may increase nighttime temperatures more than daytime temperatures. Still, the observations on temperature are within the natural variability of the climate and there is further disagreement on whether the trends of the past two decades will persist. There have been times when the climate has been both warmer and cooler than present conditions without any influence from humans. By the guidelines of the scientific method, discussed in Chapter 1, one could not conclude that anthropogenic global warming is real yet. This fact has been used by some to justify a lack of action aimed at reducing the atmospheric loadings of greenhouse gases. It is also difficult to appreciate the potential change in our climate due to a change of only a few degrees Celsius. To put this into perspective, consider that the climate has warmed only 5°C (9°F) since the last ice age.

10.2.2 Greenhouse gases

It is known that the atmospheric concentrations of the primary greenhouse gases have increased since preindustrial times. The Industrial Revolution began in England about 1760 and then spread to other countries. The primary greenhouse gases include CO_2, methane (CH_4), nitrous oxide (N_2O), O_3, chlorofluorocarbons (CFCs), hydrochlorofluoro-carbons (HCFCs), hydrofluorocarbons (HFCs), perfluorocarbons (PFCs), and sulfur hexafluoride (SF_6) (Table 10.1). Water can also be considered a greenhouse gas since it too absorbs and releases radiation. The first four gases are naturally occurring, although humans have had a profound impact on their atmospheric concentrations. The remainder are products solely of human activities.

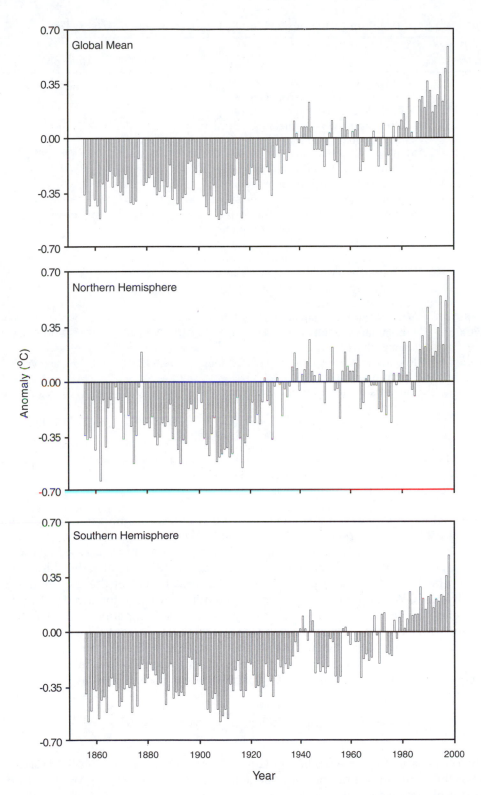

Figure 10.2 Mean annual temperature anomalies for the whole earth and for the Northern and Southern Hemispheres for the period 1856 to 1998. The units on the *y*-axis are the departure from the average for the period 1961 to 1990. (From Carbon Dioxide Information Analysis Center, 1999.)

Table 10.1 Characteristics of the Primary Greenhouse Gases

Greenhouse gas(es)	Comments	GWP[a]	Atmospheric lifetime
CO_2	Natural and anthropogenic sources; fossil fuel use primary anthropogenic source	1	50–200 years
CH_4	Natural and anthropogenic sources; rice culture and enteric fermentation primary anthropogenic sources	21	12 years
N_2O	Natural and anthropogenic sources; N fertilizer use, biomass burning, and fossil fuel combustion primary anthropogenic sources	310	120–150 years
O_3	Stratospheric O_3 blocks ultraviolet radiation; acts as a greenhouse gas in the troposphere; produced by reaction with NO_2, CH_4, and hydrocarbons in the atmosphere	—[b]	Hours to days
CFCs	Used as refrigerants and blowing agents; being phased out because of the stratospheric O_3-depleting potential	3,700–5,400	Days to 400 years
HCFCs	Preliminary substitute for CFCs; also being phased out	90–1,650	1–40 years
HFCs	Replacement for CFCs and HCFCs; long atmospheric lifetime	140–12,000	1–265 years
PFCs	Released during Al production; long atmospheric lifetime	6,500–9,200	3,200–55,000 years
SF_6	Used as a dielectric in electrical transmission and distribution systems and in Mg casting	23,900	3,200 years

[a] Global warming potential over a 100-year time period.

[b] Complex dependence as a function of altitude.

The influence of a given greenhouse gas on the overall anthropogenic greenhouse effect is a product of the potency of the gas and its concentration in the atmosphere. The potency refers to how efficient a gas is at absorbing heat and is expressed as the global warming potential (GWP). The GWP is a relative scale that compares the ability of a gas to absorb heat in relationship to CO_2, one of the least potent of the greenhouse gases. Generally, the GWP is the ratio of warming potential (radiative forcing) for 1 unit mass of a greenhouse gas to the same mass of CO_2 over a specified period of time. A 100-year timeframe is used most often as the standard. To compare the contributions of the greenhouse gases to the anthropogenic greenhouse effect, the emissions can be expressed on the basis of the equivalent C input to the atmosphere, generally as a million metric tons of C equivalent (MMTCE). The MMTCE for a given gas is calculated as

$$MMTCE = (gas\ emissions\ in\ Tg/year) \times (GWP) \times \frac{12}{44} \qquad (10.1)$$

where 12/44 represents the fraction of C in CO_2. With this calculation, 1 MMTCE from CO_2 has the same impact on the anthropogenic greenhouse effect as 1 MMTCE from any other greenhouse gas.

Although CO_2 is the least potent greenhouse gas, it contributes the most to the greenhouse effect because of its relatively high atmospheric concentration, which is a result of large emissions. Atmospheric emissions of CO_2 in the U.S. in 1997 were approximately 1500 MMTCE compared to 180 MMTCE for CH_4, 110 MMTCE for N_2O, and 40 MMTCE for the combination of HFCs, PFCs, and SF_6. Emissions of CFCs and HCFCs are decreasing rapidly. These compounds have been found to contribute to the decline of stratospheric O_3 and are being phased out as part of the Montreal protocol, an international treaty that addresses the release of O_3-depleting substances worldwide (see Chapter 2).

Water vapor is also a very important radiatively active gas, although the concentrations of water vapor in the atmosphere generally do not change as a direct result of human activities. A warmer climate, however, will induce greater evaporation worldwide and the additional water vapor will likely contribute to greater warming.

Another important characteristic of a greenhouse gas is its atmospheric lifetime. The atmospheric lifetime is 50 to 200 years for CO_2, approximately 12 years for CH_4, 120 to 150 years for N_2O, and varying times for the remaining greenhouse gases. The atmospheric lifetime is factored into the GWP, but it also suggests how successful efforts to reduce greenhouse gas emissions might be in the short run. If CO_2 emissions declined dramatically, we would still realize the effects of CO_2 already in the atmosphere for some time, whereas reductions in CH_4 emissions might improve the situation within a few decades. The PFCs and SF_6 are problematic because their atmospheric lifetimes are all in excess of 3200 years.

Global circulation models (GCMs) are used to predict the effects of increasing concentrations of greenhouse gases on components of the climate. The models are extremely complex and require considerable computing power to run. The most recent advance in GCMs allows a coupling of the atmosphere with the effects of the oceans. Typically, the models will compare the climate with the equivalent of a doubling of atmospheric CO_2 concentration, a condition that will likely occur before the middle of this century. Table 10.2 presents predicted changes from two GCMs for temperature and precipitation with a doubling of CO_2 in the atmosphere for four regions in the U.S. over winter and summer. The models are consistent in that all predict at least a 2.5°C increase in mean temperature. The models are not as consistent in predicting changes in annual precipitation, with as much as an 18 cm difference between the models for the Northwest during the winter season. As average temperatures increase, decreases in precipitation

Table 10.2 Predicted Seasonal Changes in Mean Temperature and Precipitation in Response to an Approximate Doubling of CO_2 Equivalent in the Atmosphere by Two Coupled Ocean-Atmosphere Global Circulation Models for Four Regions in the U.S.

	Temperature (°C)		Precipitation (mm/day)	
	HADCM2[a]	CGCMI[a]	HADCM2	CGCMI
Winter				
Central Plains	3.5	6.0	0.5	−0.5
Great Lakes	3.5	4.5	0.5	0.0
Southeast	2.5	3.0	1.0	−0.75
Northwest	2.5	3.5	3.0	5.0
Summer				
Central Plains	2.5	3.0	0.0	−0.5
Great Lakes	2.5	3.0	0.25	−0.5
Southeast	2.5	3.0	0.5	−0.75
Northwest	2.5	3.0	0.0	0.0

[a] HADCM2 is the Hadley Centre's Second-Generation Coupled Ocean-Atmosphere Global Circulation Model; CGCM1 is the Canadian Global Coupled Model, Version 1.

Source: Doherty and Mearns (1999).

become very important. The information also illustrates that the effects of any anthropogenic greenhouse gas will vary by region. It should be noted, however, that there is still a great deal of uncertainty in the area of climate modeling.

Soils can be a source or a sink for CO_2, CH_4, and N_2O and these gases will be discussed in more detail in the following sections. Soils play no role in the cycling of O_3, CFCs, HCFCs, PFCs, or SF_6 and comments on these gases will be very brief. Efforts are already being made to reduce emissions of CFCs and HCFCs. Unfortunately, their replacements are also potent greenhouse gases. Since O_3 is a pollutant in the lower atmosphere, efforts are already under way to reduce O_3 inputs to the atmosphere, and this will have beneficial effects toward reducing global warming.

10.2.2.1 Carbon dioxide

Carbon dioxide is present in the atmosphere at the highest concentration, by a factor of at least 200, compared with the other greenhouse gases, and it therefore has the highest relative contribution (approximately 60%) to the anthropogenic greenhouse effect. It is the least potent of the greenhouse gases with its contribution to global warming primarily a result of its high atmospheric concentration. The dramatic increase in the atmospheric level of CO_2 since the Industrial Revolution is shown in Figure 10.3. These data are a combination of CO_2 concentrations in samples of air collected at the Mauna Loa Observatory since 1959 and of air bubbles trapped in ice cores. A more thorough appreciation of the reasons CO_2 concentrations are increasing in the atmosphere can be obtained by examining the global C cycle (Figure 10.4). For the time period immediately prior to large-scale human population increases, atmospheric CO_2 concentrations were essentially constant. The primary CO_2 inputs to the atmosphere were CO_2 released from the oceans and from the decomposition of dead plant material. The primary mechanisms for removal of CO_2 from the atmosphere were plant uptake and absorption by the oceans. The inputs

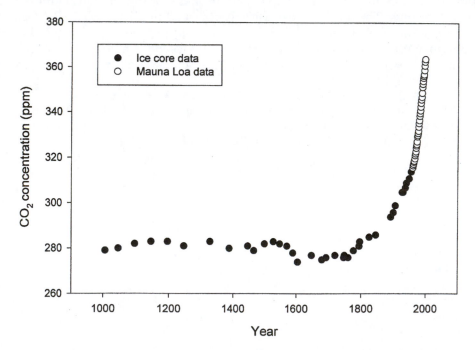

Figure 10.3 Atmospheric CO_2 concentrations from 1006 to 1997 as determined from air trapped in ice and from actual air samples. (From Carbon Dioxide Information Analysis Center, 1999.)

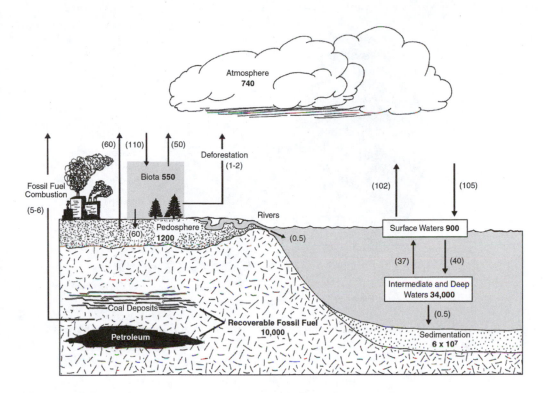

Figure 10.4 The global C cycle (in Gt/year). (From Moore, 1988.)

were approximately balanced with the outputs and all other C pools were relatively stable. Human activities have resulted in the use of C stored as recoverable fossil fuels, biomass, and soil organic matter that has been emitted into the atmosphere, after it has been converted to CO_2, at a faster rate than it can be removed by plant growth or by adsorption into the oceans. Atmospheric CO_2 concentrations have increased from approximately 280 parts per million by volume (ppmv) in preindustrial times to 365 ppmv today.

A summation of the C transfer rates in Figure 10.4 shows that the net transfer of C into the atmosphere is 3 to 5 Gt/year (Gt = 10^9 Mg). The 6 to 8 Gt/year placed into the atmosphere from combustion of fossil fuels and deforestation is slightly offset by a net transfer of C into the oceans at 3 Gt/year. Humans have some control over several components of the C cycle including deforestation, combustion of fossil fuels, C lost to the atmosphere because of soil erosion, and, to a lesser extent, the amount of C taken up by growing plants. Carbon used for plant growth is stored both aboveground and in the soil while the plant is alive. Some of the C remains in the soil as organic matter after the plant dies and decomposes. The U.S. contributes more CO_2 to the atmosphere than any other single nation (Figure 10.5). Nearly all of this CO_2 is from energy usage with some 84% of our energy needs met with energy sources that release C (Figure 10.6).

The role of soils in the global cycling of C is quite diverse. First, soil and detritus contain 1200 Gt of C, which is one of the larger pools of C in the global C cycle. Plants fix atmospheric CO_2 during photosynthesis. Soil microganisms decompose dead plants, returning a portion of the C to the atmosphere as CO_2, and retaining a portion of the C as soil organic matter. Clearly, the significance of soils in C storage should not be overlooked. In fact, when soils are taken out of natural vegetation and used for agricultural production, they will typically lose 25 to 50% of their organic matter, unless properly managed. When

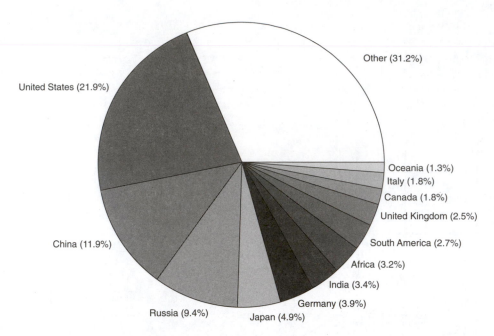

Figure 10.5 Global CO_2 emissions by country. (From Carbon Dioxide Information Analysis Center, 1999.)

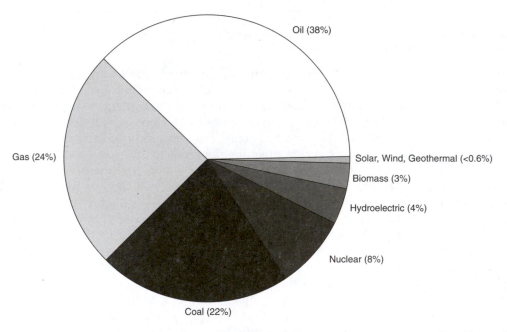

Figure 10.6 Energy use by primary energy source in the U.S. (From U.S. Department of Energy, 1999.)

soil erosion occurs, there is also a loss of C in the eroded material. Worldwide, an estimated 25 billion tons of soil are lost to erosion each year and, assuming 4% organic C content, that represents a potential C input of 1 Gt/year or 20 to 30% of the net increase in atmospheric C each year. Fortunately, not all of that C reaches the atmosphere.

Strategies for dealing with increasing CO_2 levels in the atmosphere fall into two general categories: reducing CO_2 inputs to the atmosphere or increasing the amount of

C stored in various C pools. Reducing CO_2 inputs to the atmosphere can be accomplished in a variety of ways. Increasing fuel efficiency will reduce the amount of fossil fuel required to produce the same amount of energy or perform the same amount of work. Generating electricity with nuclear fuel, solar energy, or wind or water power would greatly reduce the amount of coal combustion that is currently required. The use of biomass fuels also falls into the category of reducing CO_2 inputs to the atmosphere even though they produce CO_2 when burned. The difference between biomass fuels and fossil fuels is that biomass fuels recycle CO_2 that was already in the atmosphere, whereas fossil fuels release CO_2 that was removed from the atmosphere a long time ago. In essence, biomass fuels, which are considered renewable resources, reduce the introduction of new CO_2 into the atmosphere by producing energy without utilizing fossil fuels. Slowing deforestation will also reduce CO_2 to the atmosphere as will reducing soil erosion, as discussed above.

Carbon sequestration is a relatively new concept that encompasses a wide variety of technologies ranging from promoting the storage of C in the terrestrial ecosystem (mostly in soil) to capturing CO_2 emitted from fossil fuel combustion. In all cases the amount of C stored in pools other than the atmosphere must be increased for the strategy to be effective. The soil is one such pool, but the oceans, geologic formations, and vegetation are other possibilities. Current estimates are that C sequestration could remove 1 Gt C/year from the atmosphere before 2050 if a concerted effort were made to increase vegetation; to reduce C emissions from power plants, automobiles, and other petroleum-consuming processes; and to prevent soil erosion.

For the terrestrial ecosystem, revegetation (forests and grasslands), increasing soil C levels, and the promotion of beneficial reuse of organic waste materials (animal manures, municipal biosolids, etc.) have the potential to increase the amount of stored C. Net sequestration by the terrestrial ecosystem is currently at approximately 2 Gt/year, but some scientists believe this could be doubled through reforestation, conservation tillage, and crop residue management. The reader is reminded of the magnitude of the amounts of C that need to be considered. Significant changes in the size of any C pool will require some degree of global cooperation, and the most likely outcome will be a decrease in the rate of increase of CO_2 levels in the atmosphere.

The remaining C sequestration approaches are more experimental at this time. Improved technologies for capturing CO_2 need to be developed and the concept of injecting captured CO_2 into the oceans or geologic formations should be evaluated. Carbon dioxide injection is already practiced in oil-extraction operations, but issues related to the total assimilation capacity and the length of time the CO_2 would remain trapped require further attention.

Overall, the easiest actions to implement are the "no regret" type of actions. These reflect changes that should be made regardless of the global warming problem. There are a number of reasons, for example, that soil erosion should be prevented or that fuel efficiencies should be improved. If global warming turns out to be less serious than is now believed, the resources spent on those actions would not be wasted.

The difficulties in predicting the effects of global warming on the climate have been discussed briefly. Adding to the complexity of the situation are the potential responses of plants themselves to the increased CO_2 concentrations, without regard to the effects of changing climate on plant productivity. Most plants fall into one of two general categories based on differences in their biochemical pathways for fixing atmospheric CO_2. The C_3 plants (e.g., wheat, soybeans, cotton) take CO_2 from the air into mesophyll cells where the Calvin cycle converts it to sucrose. The C_4 plants (e.g., corn, sorghum, millet) also take CO_2 from the air into mesophyll cells but the fixed CO_2 (as a C_4 molecule) is transported to the bundle sheath cells where the Calvin cycle takes place. This "pumping" of CO_2

from the mesophyll cells into the bundle sheath cells maintains a greater CO_2 concentration gradient between the atmosphere and the mesophyll cells, which increases the ability of the C_4 plants to assimilate CO_2. Thus, the C_4 plants are more efficient users of CO_2.

The C_3 plants, being less efficient users of CO_2, will actually increase their photosynthetic rate in response to elevated CO_2 concentrations, provided that other growth factors are not limiting. Thus, one could predict increased growth from C_3 plants as atmospheric CO_2 levels increase. The C_4 plants will also respond positively to increased CO_2 concentrations, although not to the extent that C_3 plants would. Both C_3 and C_4 plants tend to use water more efficiently when CO_2 levels are higher. This is because the stomates (openings in leaves that provide the means of gas and water exchange between plants and the atmosphere) do not open as much in the CO_2-enriched environment, and less water is lost during photosynthesis. This effect is more pronounced with the C_4 plants. In the case of C_3 plants, there is also an increase in plant growth without an increase in water use. Both C_3 and C_4 plants can be more efficient in their use of nutrients, especially N, which is a positive response.

Unfortunately, water and nutrients are already the primary limiting factors for plant growth across much of the earth, and this will likely remain the case as the climate changes. Clearly, if large-scale increases in plant productivity were possible due to increasing atmospheric CO_2 levels, then atmospheric CO_2 concentrations would not be as high as they are today. Increased biomass production has, however, likely reduced the magnitude of the increase in atmospheric CO_2 concentrations. Water may even become more of a limiting factor due to increasing temperatures.

10.2.2.2 Methane

Methane is produced by methanogenic bacteria anytime organic matter decomposes anaerobically. These conditions are typically found in flooded soils, in the digestive systems of ruminant animals, and when organic wastes are stored or handled as liquids, as is often done with animal manures and municipal biosolids. Humans have increased CH_4 concentrations in the atmosphere by increasing the amount of C that is metabolized anaerobically. Similar to CO_2, atmospheric CH_4 concentrations were relatively constant prior to large-scale human population of the earth. Wetlands and native ruminant animals were the primary sources of CH_4 at that time. Today, rice culture, domestic ruminant animals, landfills, and our own use of CH_4 as a fuel source provide additional CH_4 to the atmosphere (Figure 10.7). Atmospheric CH_4 concentrations have increased from approximately 790 parts per billion by volume (ppbv) in preindustrial times to 1700 ppbv (or 1.7 ppmv) today. Figure 10.8 provides a breakdown of global annual emissions of CH_4 with time and by source. Of all the greenhouse gases, CH_4 emissions are the most closely tied to the production of food on a global basis. It is estimated that one third of global CH_4 emissions is from rice culture and from the digestive processes and waste handling associated with domesticated livestock alone.

Methane is a more potent greenhouse gas than CO_2. With an atmospheric concentration of approximately 1700 ppbv, CH_4 contributes 15% to the anthropogenic greenhouse effect. The unique thing about CH_4 is the short atmospheric lifetime of 12 years compared with CO_2, N_2O, or CFCs (Table 10.1). The fate of atmospheric CH_4 is not completely understood. It is known that soil bacteria responsible for nitrification can oxidize CH_4 as well as NH_4^+. Therefore, aerated low-N soils can be a CH_4 sink. Most CH_4 is dissipated via reaction with tropospheric hydroxy (OH), however. The short atmospheric lifetime of CH_4 also suggests that efforts to reduce CH_4 emissions could be effective in slowing the increase in the rate of global warming.

Efforts to directly reduce CH_4 emissions resulting from food production will not be easily implemented because of increasing population and demand for food in areas of

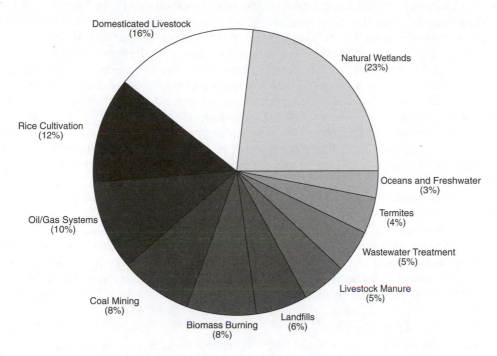

Figure 10.7 Sources of atmospheric methane emissions. (From U.S. Environmental Protection Agency, 1999.)

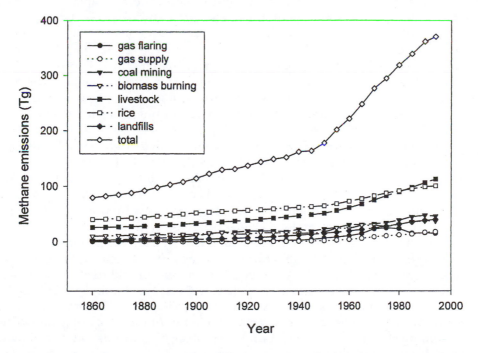

Figure 10.8 Global methane emissions from 1860 to 1994. (From Carbon Dioxide Information Analysis Center, 1999.)

the world where rice is a major staple. In addition, large sectors of the world population have low-quality diets and, historically, diet improvement has come about by increasing the proportion of meat in the diet. This would suggest increasing the numbers of domestic animals in some areas of the world. Research may help reduce the amount of CH_4 produced per unit of rice produced. This could be accomplished by breeding rice with a higher grain-to-straw ratio, which results in less straw available for anaerobic decomposition, or by increasing the proportion of upland rice varieties grown. Other water-use strategies may reduce CH_4 emissions from rice production. Research may also reduce CH_4 emissions from livestock, particularly ruminants. The CH_4 emitted from these animals reflects a loss of feed energy and is an inefficiency. Efforts to increase feed efficiency or to increase the rate of weight gain by livestock will decrease the amount of CH_4 emitted.

Direct efforts at reducing CH_4 emissions will need to focus on natural gas distribution systems, fossil fuel production methods, and landfills and other waste-handling methods. It has been estimated that some natural gas distribution systems lose as much as 3% of the gas through leakage. Oil exploration and coal mining also release large amounts of CH_4 as do landfills and some organic waste–handling systems. In each of these examples there are preventable losses of CH_4, and methods developed for capturing or retaining CH_4 will benefit the global climate problem.

10.2.2.3 Nitrous oxide

Nitrous oxide contributes approximately 5% to the greenhouse effect. Concentrations have increased in the past 200 years from approximately 280 to 310 ppbv. Nitrous oxide is produced in soils through both biotic and abiotic means. Various microbial pathways produce N_2O, including nitrification and denitrification. Nitrous oxide is released from all soils, regardless of cultivation or fertilization, although N fertilization will increase N_2O emissions because of the greater amount of substrate available for nitrification and denitrification. Emissions of N_2O from soils can be stimulated by land-use conversion, particularly the conversion of land from any native vegetation to agricultural production. As discussed previously, soil organic matter levels tend to decrease when soils are first cultivated. Some of the N in the organic matter will be lost as N_2O. Nitrous oxide is also produced when reduced-N compounds in fossil fuels or other biomass materials are oxidized during combustion. As a result of increases in the activities described above, the concentration of N_2O in the atmosphere has increased compared with preindustrial times.

Nitrous oxide is a more potent greenhouse gas than CO_2 or CH_4, although it contributes only 5% to the anthropogenic greenhouse effect. Previously described means for reducing CO_2 emissions, such as reducing fossil fuel use and conserving soil organic matter, will also reduce N_2O emissions because N is an integral component of fossil fuels, biomass, and soil organic C. Direct efforts toward reducing N_2O emissions will have to address fertilizer N use. Fertilizer-derived N_2O emission is a complicated process that can be influenced by factors such as fertilizer type, N rate, application method, tillage, soil moisture, temperature, soil organic C content, and microbial activity. Overall, a reduction in fertilizer-derived N_2O emissions will require increases in N use efficiency, which will occur to some extent with elevated atmospheric CO_2 levels. Methods for increasing N use efficiency were discussed in Chapter 4. At present, fertilizer-derived N_2O emissions are estimated to account for 2.5% of global N_2O emissions, assuming that approximately 1.0% of fertilizer N is lost as N_2O. Nitrous oxide emissions due to fertilizer use would be expected to increase worldwide since N fertilizer use is increasing, particularly in countries with high population densities and increasing food demands.

10.2.3 Uncertainties and complexities

Within the general topic of climate change, the unknown far outweighs the known. Beginning with the climate itself, it is generally accepted that the climate will change, although the nature of the changes are largely unpredictable. One of the reasons for the uncertainty is the large number of feedback mechanisms. That is, a change in one thing will induce additional changes in other parameters such that the net effect of the first change becomes difficult to predict. A simple example was given earlier when it was noted that we know that the greenhouse gases are radiatively active and warm the atmosphere. However, it cannot be said with certainty that continued increases in the concentrations of the gases will cause a continued proportional increase in warming. It is useful to discuss these feedbacks because they influence some of the proposed actions for responding to global climate change. Evaluating uncertainties also assists in appreciating the complexity of the global warming problem. Bear in mind that feedbacks can be both positive and negative. Several examples will be offered to illustrate this complexity.

Undoubtedly water vapor will play a role in potential climate changes. As the climate warms, there will be greater evaporation because more sensible heat is available and warmer air can hold more water than cooler air. The higher concentration of water vapor in the atmosphere may produce more cloud cover. If that cloud cover occurs frequently as a uniform layer (stratus type) at low altitudes over large areas, the clouds may have a net cooling effect because they reflect shortwave radiation back into space. If the cloud cover occurs frequently as a broken layer (cumulus type), the additional water vapor will likely absorb heat and contribute to warming. The additional water vapor will also increase the OH concentration in the troposphere. Hydroxyls can react with both CH_4 and O_3, reducing their concentrations, and acting as a positive feedback for atmospheric warming.

The increasing temperature will have a multitude of effects. Warmer soils will induce oxidation of C and the total amount of C stored in soils may decrease, resulting in a further increase in atmospheric CO_2 concentrations. Warmer air temperatures may also allow the northward expansion of forested regions. Provided that the reforestation is not accompanied by an equivalent amount of deforestation elsewhere, this could increase CO_2 sequestration. Warmer oceans may support higher phytoplankton populations, which absorb CO_2, but warmer water holds less CO_2 itself. Adsorption of heat by the oceans makes less heat available for warming the atmosphere, but could have adverse effects on marine ecosystems.

One aspect of global climate change that is often forgotten is the potential adaptability of the population to the changing climate. The changes are sometimes considered in a catastrophic context — productive lands suddenly transformed to desertlike areas. Climate change will occur gradually, and the population and food production practices will have to adapt. The growing of grain crops may need to take place at higher latitudes in the Northern Hemisphere. Canada and northern areas of the former Soviet Union and China could become more important grain-producing areas than they are today. The shift in production would occur gradually over a number of years or even decades. In addition, crop breeding and biotechnology may produce crops that are more heat and drought resistant. These points are not intended to downplay the significant economic and social issues associated with a shifting agriculture, but rather to illustrate that adaptations are possible.

As the net effects of the increase in anthropogenic greenhouse gas concentrations in the atmosphere are impossible to predict at this time, so are the net effects of any actions taken to reduce the emissions of any of the gases into the atmosphere. Most experts would agree that significant climatic changes have already been induced, although we do not yet know what they are. This was illustrated in Table 10.2. It is fairly certain that the climate has warmed or will warm as a whole and that the resulting net effects will not likely be favorable. Therefore, any realistic efforts made to reduce emissions of greenhouse gases would seem worthwhile regardless of the uncertainties.

In December 1997, many countries in the world participated in the Kyoto Conference, held in Kyoto, Japan, in an attempt to reach global consensus on a strategy for reducing greenhouse gas emissions. The resulting Kyoto Protocol states that global emissions of greenhouse gases will be reduced to an average of 5.2% below 1990 emission levels by the period 2008 to 2012. The U.S. agreed to 7% below 1990 levels, the European Union to 8%, and Japan to 6%, while other countries have less stringent requirements. The gases covered include CO_2, CH_4, N_2O, HFCs, PFCs, and SF_6. The agreement also allows credit for C sinks, although the mechanism for doing so has not been worked out. This global effort is encouraging, but as of this writing the U.S. has not formally ratified the agreement.

10.3 Acidic deposition

Although acidic deposition has been recognized as a problem for over a century, recent studies suggest there is still much to be learned. As one of the most important, highly publicized, and controversial aspects of atmospheric pollution, acidic deposition has been a major research focus for the last 25 years. A major reason for the increased interest has been the involvement of human emotions combined with media sensationalism that enhances public concern about environmental issues. For example, to many the term *acid rain* evokes frightening images of industrial pollution literally raining down upon the earth and causing serious and perhaps irreversible environmental damage. Technically, the term *acidic deposition* — which includes not only rainfall, but acidic fogs, mists, snowmelt, gases, and dry particulate matter — is a more precise description of the problem. The primary origin of acidic deposition is the emission of sulfur dioxide (SO_2; Chapter 8) and N oxides (NO_x; Chapter 4) when fossil fuels are burned for energy production.

Adverse human and animal health effects are related to high exposures of acidic aerosols. Air quality can be reduced due to increased S-containing particulate matter that can be deposited as dry particles or dissolved to produce acidic substances. Acidic deposition can also impact structures such as buildings, sculptures, and monuments that are constructed of the weatherable materials limestone, marble, bronze, and galvanized steel. Although acid soil conditions are known to influence the growth of plants, agricultural impacts related to acidic deposition are less of a concern because of the buffering capacity of most of these ecosystems. When acidic substances are deposited in lakes, streams, forests, and other natural ecosystems, either by wet or dry deposition, a number of adverse environmental effects are believed to occur. Among the most serious are direct damages to vegetation, particularly forests, and changes in soil and surface water chemistry that can adversely affect plant, animal, and aquatic life. The relationship among acidic deposition–related emissions and the products, pathways, and human and environmental effects are shown in Figure 10.9.

In this section we will examine political policies and treaties, address the sources and geographic distribution of acidic deposition, describe the more common types of human, material, and ecosystem problems caused by acidic deposition, and give an overview of some remediation efforts currently used to reverse the effects of acidification.

10.3.1 Legislative acts and programs

The issue of acidic deposition crosses state and national boundaries because, although acidic compounds can be deposited short distances from the source of pollution, they may also be transported hundreds of kilometers before being returned to the earth in rainfall or other forms. It has been estimated, for example, that 30 to 40% of the S deposited in the northeastern U.S. originated in the industrial Midwest. On an international scale, the U.S. and

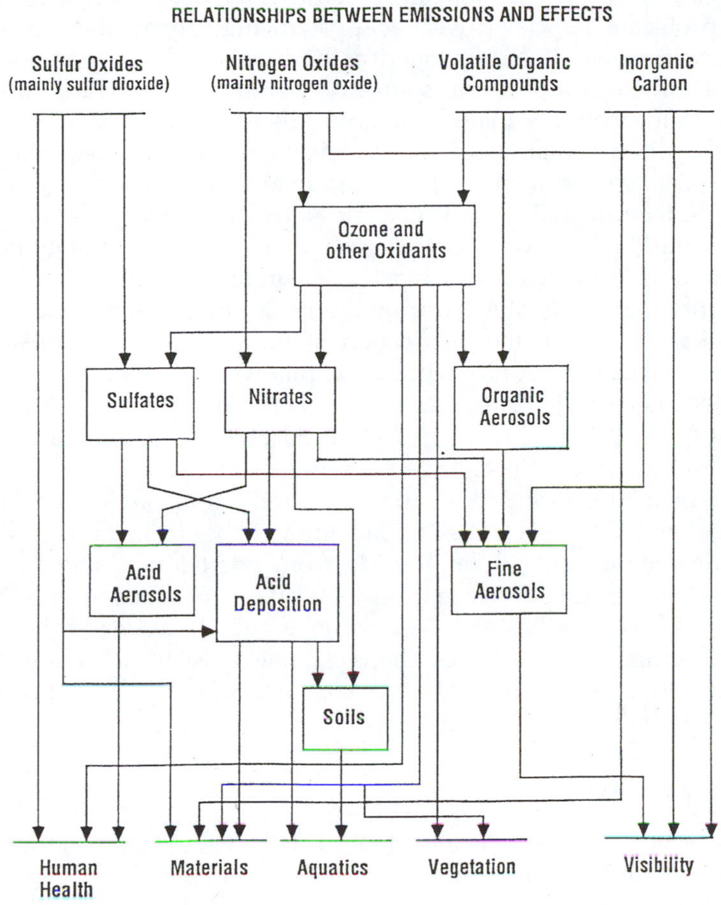

Figure 10.9 Relationships among acidic deposition-related emissions, products, pathways, and human and environmental effects. (From NAPAP Report to Congress, 1992.)

Canada have debated the sources of acidic deposition and the economics of reversing acidification in eastern North America for years and have only recently developed a strategy to address the problem. Similarly, in Europe, the small size of many countries means that emissions in one industrialized area can readily affect forests, lakes, and cities in another nation. A study estimated that 17% of the acidic deposition falling on Norway originated in Britain and that 20% of the acidic deposition in Sweden came from Eastern Europe. Recent political changes and ongoing economic instability in Eastern European countries that burn high-S coal for energy generation may mean even more delays in implementing international agreements to reduce emissions of acid-producing compounds.

Most industrialized countries have concluded that the issue is serious enough to proceed without final resolution of all scientific issues, a process that could take decades. For example, by 1988, 21 European nations had signed a "Long-Range Transboundary Air Pollution" agreement that targeted reductions in SO_2 emissions. Progress has been slow because of the lack of alternative energy sources to those that produce SO_2 and NO_x and the billions of dollars needed to further reduce emissions of these gases from current sources. Unfortunately, as with many other environmental issues, solutions to the acidic deposition problem are complex, expensive, and long term in nature. In addition, a major cause of a significant level of uncertainty associated with acidic deposition is the fact that other changes in atmospheric chemistry and global climate change have paralleled the increase in acidic deposition.

In the 1960s and 1970s, both European and North American scientists identified large-scale, regional problems associated with acid-producing sources that resulted in severe impacts to the environment. In North America, the metal-smelting operations in Sudbury, Canada were identified as a principal source of SO_2 in acid deposition that killed local vegetation and influenced air quality for great distances downwind of the polluting sources (see the environmental quality issues/events box on the restoration of Sudbury, Canada, p. 372). Canada and the U.S. developed federal programs to study the effects of acid deposition in the mid-1970s; the Canadian Network for Sampling Precipitation was initiated in 1976 and the U.S. National Atmospheric Deposition Program began in 1978.

The 10 years of coordinated research activities under the EPA National Acid Precipitation Assessment Program (NAPAP) during the 1980s, involving hundreds of prominent international scientists, led to the publication of numerous reports related to "Acidic Deposition: State of the Science and Technology" that were mandated by the Acid Precipitation Act of 1980. Later in 1992 an NAPAP Report to Congress was developed in accordance with the 1990 Amendment to the 1970 Clean Air Act, which summarized the findings of the NAPAP and presented the expected benefits of the Acid Deposition Control Program. Proposed mandates included an annual 9.1 million Mg reduction in point-source SO_2 emissions below 1980 levels and a national limit of about 13.6 million Mg (8.5 million Mg from electric utilities and 5.1 million Mg from point-source industrial emissions), which is an approximate 40% decrease from 1980 levels. A reduction in NO_x of about 1.8 million Mg from 1980 levels has also been set as a goal; however, while NO_x has been generally on the decline since 1980, projections estimate a rise in NO_x emissions, primarily from gasoline motors, after the year 2000. In 1980, the U.S. levels of SO_2 and NO_x emissions were 23.6 and 22.6 million Mg, respectively.

10.3.2 Sources and distribution

Typical sources of acidic deposition include coal- and oil-burning electric power plants, automobiles and other vehicles, and large industrial operations (e.g., smelters) (Figure 10.10). Once the S and N gases enter the earth's atmosphere they react very rapidly with moisture in the air to form sulfuric (H_2SO_4) and nitric (HNO_3) acids, respectively. The pH of natural rainfall in equilibrium with atmospheric CO_2 is about 5.6; however, as shown in Figure 10.11c for the U.S., the average pH of rainfall is as low as 4.2 in some areas. The nature of acidic deposition, that is, the relative percentages of H_2SO_4 and HNO_3 present, is controlled largely by the geographic distribution of the sources of SO_2 and NO_x. In the U.S., H_2SO_4 is the main source of acidity in precipitation in the Midwest and Northeast because coal-burning electric utilities in these areas and Canada also emit large quantities of SO_2. In the western U.S., utilities and industry burn coal with a lower S content; HNO_3 may be of greater concern, particularly in densely populated areas such as California where cars and other vehicles that burn gasoline are major sources of NO_x. The geographic nature of SO_2 and NO_x emissions and the acidity of precipitation in the U.S. are clearly shown in Figure 10.11.

Unquestionably, emissions of SO_2 and NO_x have increased in the 20th century (Table 10.3), due to accelerated industrialization in developed countries and antiquated processing practices used early this century that still function in some undeveloped countries. However, after a decade of research during the 1980s, the EPA NAPAP concluded there were some definite impacts due to acidic deposition that warranted remediation. Results from NAPAP research will be examined in the following sections that describe effects of acidic deposition to humans, structures, and ecosystems.

In simplest terms, acids are substances that tend to donate protons (hydrogen ions, H^+) to another substance in a chemical reaction. Acids are often classified as strong or weak, with strong acids tending to dissociate (lose H^+) completely in water and weak acids undergoing

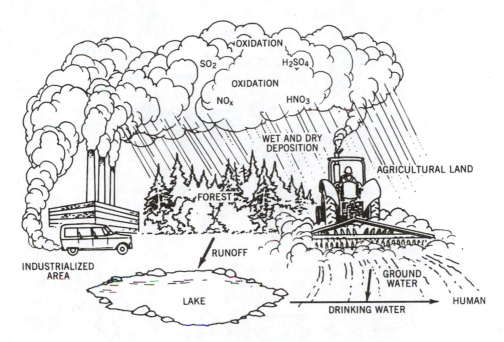

Figure 10.10 Generalized process of the production and distribution of acidic deposition from source to different ecosystems. (Adapted from U.S. Environmental Protection Agency, 1980.)

only partial dissociation. In addition to H_2SO_4 and HNO_3, other common strong acids are hydrochloric (HCl) and phosphoric (H_3PO_4); weak acids include carbonic (H_2CO_3), acetic (CH_3COOH), and boric (H_3BO_3). (Note that only H_2SO_4 and HNO_3 are important components of acidic deposition.) The term *pH* is used to indicate the relative acidity of a solution (or soil) and is defined as the negative logarithm of the activity of the H^+ in solution [pH = $-\log(H^+)$]; hence, the greater the H^+ concentration in a solution, the lower the pH. Pure water has a pH of 7.0, natural rainfall about 5.6, and severely acidic deposition less than 4.0. The pH of most soils range from 3.0 to 8.0. For most crop production systems, recommended pH values range from 6.0 to 7.0; however, the recommended pH range for organic soils is 5.0 to 6.0.

Example problem 10.1

Uncontaminated rainwater should have a pH around 5.6. According to the pH scale, water with equal amounts of H^+ and OH^- has a pH level of approximately 7. The reason rainwater pH is lower than 7 is due to CO_2 chemistry and the formation of carbonic acid (H_2CO_3). The acidity of uncontaminated rainwater (pH 5.6) is

$$pH = 5.6 = -\log (2.5 \times 10^{-6} \, M \, H^+) \qquad (10.2)$$

An acid rain with a pH of 4 would contain

$$pH = 4.0 = -\log (1 \times 10^{-4} \, M \, H^+) \qquad (10.3)$$

or 0.0001 M H^+.

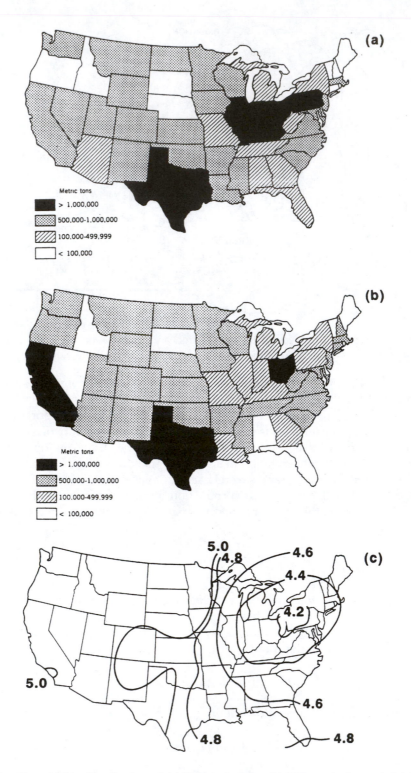

Figure 10.11   Geographic distribution of (a) SO_2 emissions; (b) NO_x emissions; and (c) acidity of precipitation in the U.S. (Adapted from Council on Environmental Quality, 1989; U.S. Environmental Protection Agency, 1988; World Resources Institute, 1988.)

Table 10.3 Historical U.S. Trends in the Emissions of Nitrogen and Sulfur Oxides (million Mg/year)

Year	Nitrogen oxides	Sulfur dioxides
1900	2.37	9.06
1910	3.72	15.7
1920	4.68	19.2
1930	7.27	19.1
1940	6.69	18.1
1950	9.16	20.3
1960	12.8	20.2
1970	19.6	28.3
1975	21.0	25.4
1980	22.6	23.5
1985	21.3	21.1
1990	21.6	21.0
1991	21.6	20.4
1992	21.9	20.2
1993	22.2	19.8
1994	22.6	19.3
1995	21.7	16.8
1996	21.2	17.3

Source: U.S. Environmental Protection Agency, National Air Quality and Emissions Trends Report (1996).

When acids are added to soils or waters, the decrease in pH that occurs depends greatly on the *buffering capacity* of the system. Buffering refers to the ability of a system to maintain its present pH by neutralizing added acidity. Clays, organic matter, oxides of aluminum (Al) and iron (Fe), and calcium (Ca) and magnesium (Mg) carbonates (limestones) are the components responsible for pH buffering in most soils. Acidic deposition, therefore, will have a greater impact on sandy, low-organic-matter soils than those higher in clay, organic matter, and carbonates. In fresh waters, the primary buffering mechanism is the reaction of dissolved bicarbonate ions with H^+, i.e.,

$$H^+ + HCO_3^- \rightleftharpoons H_2O + CO_2 \tag{10.4}$$

10.3.3 Human health effects

Both direct and indirect effects on human health have been attributed to acidic deposition. Although few direct human health problems have been reported, the result of long-term exposure to acidic deposition copollutants such as O_3 and NO_2, which are respiratory irritants, can cause pulmonary edema that increases fluids in the lungs and can lead to death in severe cases. Individuals with respiratory ailments such asthma, emphysema, or bronchitis are more sensitive to low-level O_3 exposures that can be manifested as headaches, dry cough, or irritation of the mucous membrane, as well as eye and nose irritation. Sulfur dioxide at high levels is also a known respiratory irritant, but is generally absorbed high in the respiratory tract so that it does not impact the lungs or sensitive alveoli. Particulate matter in the atmosphere can accentuate respiratory problems as was noted in the United Kingdom when coal use was the dominant fuel source for domestic heating. This phenomenon, termed *London Fog*, was suspected to be the cause of an increase in the average death rate during a severe week of pollution in 1952.

Indirect human health effects due to acidic deposition are more important than are direct effects. Some of the concerns center around contaminated drinking water supplies and consumption of fish that contain potential toxic metal levels. With increasing acidity (e.g., lower pH levels), metals such as mercury (Hg), Al, cadmium (Cd), lead (Pb), zinc (Zn), and copper (Cu) become more bioavailable. However, acidic deposition impacts to drinking water supplies have been limited and are associated primarily with cisterns and some surface water sources. The greatest human health impact may be the consumption of fish exposed to elevated Hg levels that bioaccumulate in the food chain. Freshwater pike and trout have been identified as the fish species that contain the highest average concentrations of Hg. In an NAPAP survey of 25,000 individuals that represented the U.S. population, 47 exceeded the maximum Hg daily intake of 30 µg. Only 23 of these individuals consumed high amounts of freshwater fish. Therefore, the most susceptible individuals to impacts from acidic deposition and its precursor substances would be those who live in an industrial area, have respiratory problems, drink water from a cistern, and consume a significant amount of freshwater fish in their diet.

As mentioned above, a long-term urban concern is the possible impact of acidic deposition on municipal water systems that rely on surface waters to provide drinking water, such as in the Catskills of New York. Many municipalities make extensive use of Pb and Cu piping in water-distribution systems. For instance, questions have recently been raised about human health effects related to the slow dissolution of some metals (Pb, Cu, Zn) from older plumbing materials when exposed to more acidic waters. However, problems associated with metal toxicities from drinking water supplies impacted by acidic deposition are rare.

10.3.4 Material and cultural resource impacts

Different types of materials and cultural resources can be impacted by air pollutants that are either the precursor constituents or actual products of acidic deposition (Table 10.4). While the actual corrosion rates for most metals have decreased since the 1930s, data from three U.S. sites indicate wet and dry acidic deposition may account for 31 to 78% of the dissolution of galvanized steel and Cu. In urban or industrial settings, increases in atmospheric acidity can dissolve carbonates (e.g., limestone, marble) in buildings and other structures, resulting in considerable aesthetic and economic damage. Deterioration of stone products by acidic deposition is caused by (1) erosion and dissolution (e.g., dissolution and loss of material or surface details); (2) alterations (blackening of stone surfaces); and (3) spalling (cracking and spalling of stone surfaces as a result of accumulations of alternation crusts). Painted surfaces can be discolored or etched by both wet and dry acidic deposition, and there may also be degradation of organic binders used to strengthen paints. A variety of other substrates are impacted by the consequences of acidic deposition (Table 10.4).

Several examples are available that describe the deterioration of different materials from acidic deposition. A marble column at the Field Museum in Chicago has eroded 10 µm in 70 years, which is three times the NAPAP erosion rate determined in NAPAP studies. Statues erected at Gettysburg National Monument Park in the 1880s dissolved 10 times faster (approximately 8 to 9 µm/m of rain) where there were complex shapes as compared with flat, vertical objects. In the northeastern U.S., corrosion rates for Cu and galvanized steel were 0.4 and 0.6 µm/year in rural areas and 0.9 and 1.5 µm/year in urban areas, respectively, which was due to increased exposure from SO_2 and acidity. Painted surfaces of residential structures are particularly susceptible to damage from acidic deposition, with the states in the Atlantic Coast region of the U.S. suffering the highest economic loss ($1.2 billion in 1984).

Table 10.4 Material Damage Due to Different Atmospheric Pollutants

Materials	Impacts	Principal pollutants	Other environmental factors	Mitigation measures
Metals	Corrosion, tarnishing	SO_2, other acid-forming gases	Moisture, air, salt, particulate matter	Surface plate or coat, replace with corrosion-resistant material, remove to controlled environment
Building stone	Surface erosion, soiling, black crust formation	SO_2, other acid-forming gases	Mechanical erosion, particulate matter, moisture, salts, temperature changes, vibrations, CO_2, microbes	Clean, impregnate with resins, remove to controlled environment
Ceramics, glass	Surface erosion, surface crust formation	Acid-forming gases, especially those containing fluoride	Moisture	Protective coatings, replace with more resistant materials, remove to controlled atmosphere
Paints, organic coatings	Surface erosion, discoloration, soiling	SO_2, hydrogen sulfide, O_3	Moisture, sunlight, particulate matter, mechanical erosion, microbes	Repainting, replacement with more resistant materials
Paper	Embrittlement, discoloration	SO_2	Moisture, physical wear, acidic materials introduced in manufacturing	Use synthetic coatings, store in controlled environment, deacidify, encapsulate, impregnate with organic polymers
Photographic materials	Microblemishes	SO_2	Particulate matter, moisture	Remove to controlled atmosphere
Textiles	Reduced tensile strength, soiling	SO_2 and NO_x	Particulate matter, moisture, light, physical wear, washing	Replace, use substitute materials, impregnate with polymers
Textile dyes	Fading, color change	NO_x, ozone	Light, temperature	Replace, use substitute materials, remove to controlled environment
Leather	Weakening, powdered surface	SO_2	Physical wear, residual acids introduced in manufacturing	Remove to controlled environment, consolidated with polymers, or replace
Rubber	Cracking	O_3	Sunlight, physical wear	Add antioxidants to formulation, replace with resistant materials

Source: Linthurst (1984).

10.3.5 Ecosystem effects of acid deposition

Acidic deposition can have many environmental impacts on different ecosystems. Prior to discussing the mechanisms involved and some approaches to reverse the effects of acidification, it is important to examine the nature of acidity in soil, vegetation, and aquatic environments. In many natural systems, the damage from acidification is often not directly due to the presence of excessive H^+, but is instead caused by changes in other elements when the system becomes more acidic (e.g., pH decreases). Examples include increased solubilization of metal ions that can be toxic to plants and animals such as Al^{3+} and some trace elements (e.g., manganese (Mn^{2+}), Pb^{2+}), more rapid losses of basic cations (e.g., Ca^{2+}, Mg^{2+}), and the creation of unfavorable soil environments for soil microorganisms important in many nutrient cycles. Information related to ecosystem responses with increased acidic deposition will be discussed according to the different environments impacted.

10.3.5.1 Soils

Soils become acidic by a number of different processes (see Chapter 3 for additional details on the soil environment). In humid regions, one of the natural processes associated with weathering that results in the acidification of soils is caused by precipitation exceeding evapotranspiration. As mentioned earlier, "natural" rainfall is acidic (pH of ~5.6) and continuously adds a weak acid (H_2CO_3) to soils. This acidification results in a gradual leaching of basic cations (Ca^{2+} and Mg^{2+}) from the uppermost soil horizons, leaving Al^{3+} as the dominant exchangeable cation. Exchangeable Al^{3+} is in equilibrium with soluble Al^{3+} in the soil solution that can react with water to produce H^+ and thus acidify the soil, as discussed below and shown in Figure 10.12:

$$Al^{3+} + H_2O \rightleftharpoons Al(OH)^{2+} + H^+ \qquad (10.5)$$

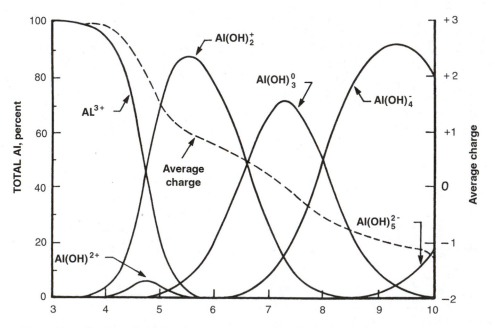

Figure 10.12   Distribution of soluble forms of Al as a function of soil pH. (From Marion, G. M. et al., *Soil Sci.*, 121, 76, 1976. With permission.)

$$Al(OH)^{2+} + H_2O \rightleftharpoons Al(OH)_2^+ + H^+ \tag{10.6}$$

$$Al(OH)_2^+ + H_2O \rightleftharpoons Al(OH)_3^0 + H^+ \tag{10.7}$$

Example problem 10.2

A soil has an exchangeable Al^{3+} acidity of 20 mmol/kg at pH 4.5. We want to add enough liming material to the site to reduce the exchangeable Al^{3+} acidity to 5 mmol/kg. How much $CaCO_3$ would be needed to neutralize this amount of acidity? First, we must calculate the concentration of Al acidity that must be neutralized, i.e.,

$$20 \text{ mmol/kg} - 5 \text{ mmol/kg} = 15 \text{ mmol/kg} \tag{10.8}$$

A mole of $CaCO_3$ can neutralize 2 mol of H^+ that originates from the Al acidity due to hydrolysis reactions. Because 1 mol of $CaCO_3$ weighs 100 g, we will need 50 g to neutralize each mole of acidity (or 50 mg/mmol acidity). If we assume the amount of soil required to be amended at the site is approximately 2,240,000 kg/ha, we can determine the amount of $CaCO_3$ needed by first calculating the amount of acidity to be neutralized as

$$2{,}240{,}000 \text{kg/ha} \times 15 \text{ mmol/kg} = 33{,}600{,}000 \text{ mmol/ha} \tag{10.9}$$

and then determining the kg of $CaCO_3$ needed by:

$$33{,}600{,}000 \text{ mmol/ha} \times 50 \text{ mg } CaCO_3/\text{mmol} \times 1 \text{ kg}/1{,}000{,}000 \text{ mg} = 1{,}680 \text{ kg} \tag{10.10}$$

Other natural processes that contribute to soil acidification include plant and microbial respiration that produces CO_2 and thus H_2CO_3 (see Chapter 3), mineralization and nitrification of organic N (see Chapter 4), and the oxidation of FeS_2 in soils disturbed by mining or drainage (see Chapter 6). However, most of the acidity in soils between pH 4.0 and 7.5 is due to the hydrolysis of Al^{3+}. In extremely acidic soils (pH < 4.0), strong acids such as H_2SO_4 are a major component of soil acidity.

Acidic deposition results in the addition of strong acids to soils. Under certain conditions this can result in the leaching of basic cations (Ca^{2+} and Mg^{2+}) and an increased solubility of Al^{3+}. However, the inorganic and organic chemistry of this system is highly complex (Figure 10.13); despite an extensive research effort, much remains to be learned about the specific mechanisms involved and the situations where acidic deposition has a significant environmental impact. In a practical sense, acidic deposition will have a greater effect on forest soils than agricultural or urban soils because in the latter situations we routinely and inexpensively counteract the effects of all acidifying processes by liming. Although it is possible to lime forest soils, which is done frequently in some European countries, the logistics and cost often preclude this as a routine management practice except in areas severely impacted by acidic deposition.

The rate and extent of increased acidification beyond that caused by naturally acidic rainfall depend both on the buffering capacity of the soil and the use of the soil. The issue is complicated by the fact that many of the areas subjected to the greatest amount of acidic deposition are also areas where considerable natural acidification occurs. For example, forested soils in the northeastern U.S. are developed on highly acidic, sandy

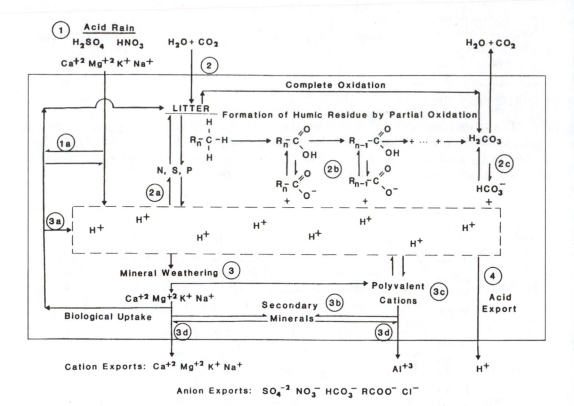

Figure 10.13 Effects of acidic deposition on soils. (1, 1a) Inputs of acid rain add H^+, acidic anions, and basic cations; (2) biological processes such as mineralization (2a) of C, N, S, P produce H^+ but also form weak organic acids that can consume H^+ (2b) or ultimately result in loss of H_2O and CO_2 from soil following complete oxidation (2c); (3) weathering of primary minerals produces cations for (3a) plant uptake, (3b) secondary minerals, (3c) acidic, and (3d) basic cations; (4) export of positive charge from soils as leachable cations (H^+, Al^{3+}, Ca^{2+}, Mg^{2+}, K^+, Na^+) must be balanced electrically by anion export (SO_4^{2-}, NO_3^-, Cl^-, HCO_3^-, organic acids). (Adapted from Krug and Frink, 1983.)

parent material that has undergone tremendous changes in land use in the past 200 years that contributed to natural acidification. However, clear-cutting and burning by the first Europeans to settle in the area has been almost completely reversed and many areas are now totally reforested. Soil organic matter that accumulates with the reforestation represents a natural source of acidity and pH buffering that must be balanced with anthropogenic sources of acidity (e.g., acidic deposition). Similarly, greater leaching or depletion of basic cations by plant uptake in increasingly reforested areas must consider the significant inputs of these same cations in precipitation. Many studies have examined the changes in chemical properties of soils following long-term acidic deposition and changing land-use patterns. While it is beyond the scope of this chapter to discuss this research in detail, several excellent reviews are cited in the references listed at the end of this chapter.

Excessively acidic soils are undesirable for several reasons. Direct phytotoxicity from soluble Al^{3+} or Mn^{2+} can seriously injure plant roots, reduce plant growth, and increase plant susceptibility to pathogens. The relationship between Al^{3+} toxicity and soil pH is complicated by the fact that in certain situations organic matter can form complexes with Al^{3+} that reduce its harmful effects on plants. Acid soils are usually less fertile because they lack important basic cations such as potassium (K^+), Ca^{2+}, and Mg^{2+}. Leguminous plants may fix less N_2 under very acidic conditions because of reduced *rhizobial* activity and greater soil adsorption of molybdenum (Mo) (as the MoO_4^- anion — a key component of nitrogenase that is an important enzyme in the N fixation process) by clays and Al and Fe oxides. Mineralization of N, phosphorus (P), and S can also be reduced because of the lower metabolic activity of bacteria at low pH levels. It should be noted that many plants and microorganisms adapt to or even prefer very acidic conditions (e.g., pH < 5.0). Examples include ornamentals such as azaleas and rhododendrons and food crops such as cassava, tea, blueberries, and potatoes. In fact, considerable efforts in plant breeding and biotechnology are directed toward developing Al- and Mn-tolerant plants that can survive in bioregions such as the Tropics where highly acidic soils are commonplace.

10.3.5.2 Agricultural crops

A 1979 report by the U.S. and Canada Research Consultation Group stated that "there is every indication that acid rainfall is deleterious to crops" and that there is "the potential for widespread economic damage to a number of field crops." Impacts to agricultural ecosystems, however, have been minor, and, in fact, acid deposition at times has been shown to be beneficial. Direct impacts would be associated with aboveground plant communities, whereas indirect effects could be manifested through degradation of soil quality. Ozone pollution, which often accompanies atmospheric acid substances, is thought to be primarily responsible for crop damage, particularly in the northeastern U.S.. Ozone crop damage is especially problematic at 0.10 ppmv O_3 levels (when compared to 0.025 ppmv O_3 levels), with crop yield losses ranging from < 20% (kidney beans, corn, wheat, and cotton under drought conditions), 20 to 50% (soybeans, spinach, and cotton under irrigated conditions), and >50% (peanuts, lettuce, and turnips).

Acidic deposition contains N and S, which are important plant nutrients. Therefore, foliar applications of acidic deposition at critical growth stages can be beneficial to plant development and reproduction. Generally, controlled experiments require the simulated acid rain to be pH 3.5 or less in order to produce plant injury. Remember that unpolluted rainfall has a pH of about 5.6, which suggests that the amount of acidity needed to damage some plants is 100 times greater than natural rainfall. Crops that responded negatively (e.g., decreased growth and yields) in simulated acid rain studies have included garden beets, broccoli, carrots, mustard greens, radishs, and pinto beans; however, there are cultivar differences among the different crops. Positive responses to acid rain have been identified with alfalfa, tomato, green pepper, strawberry, corn, lettuce, and some pasture grass crops; again there are cultivars of each that show no response to the simulated acid rain tests.

As mentioned earlier, most agricultural lands are maintained at pH levels that are optimal for crop production. In most cases the ideal pH is around pH 6.0 to 7.0; however, pH levels of organic soils are usually maintained at closer to pH 5.0. Because agricultural soils are generally well buffered, the amount of acidity derived from atmospheric inputs is not sufficient to significantly alter the overall soil pH. Nitrogen and S soil inputs from acidic deposition are beneficial, and with the reduction in S atmospheric levels mandated by 1990 Amendments to the Clean Air Act, the S fertilizer market has grown, according to the Sulfur Institute. The amount of N added to agricultural ecosystems as acidic deposition is rather insignificant in relation to the 100 to 300 kg N/ha/year required by agricultural crops. It is generally assumed that acidic deposition is unlikely to impact properly managed agricultural ecosystems and that the N and S additions are beneficial.

10.3.5.3 Forest ecosystems

Perhaps the most publicized and visible issue related to acidic deposition has been its effects on forest vegetation. Widespread forest decline has been reported in areas where significant acidic deposition has occurred. In Europe, it has been estimated that as much as 35% of all forests have been affected by acidic deposition. Similarly, in the U.S. many important forest ranges such as the Adirondacks in New York, the Green Mountains in Vermont, and the Great Smoky Mountains in North Carolina have experienced sustained decreases in tree growth for several decades and show serious visible damage. As noted earlier, conclusive evidence that forest decline or dieback is caused solely by acidic deposition is lacking and complicated by the many known interactions between acidification and other environmental or biotic factors that influence tree growth. However, it is known that O_3 can cause crop and tree damage, and NAPAP research has confirmed that acidic deposition has contributed to a decline in high-elevation red spruce in the northeastern U.S.. In addition, *N saturation* of forest ecosystems from atmospheric N deposition is believed to result in increased plant growth, which in turn increases water and nutrient use followed by deficiencies that can cause chlorosis and premature needle drop as well as increased leaching of base cations from the soil environment. The following will provide a brief summary of some means by which acidic deposition can directly affect trees and other plants (Figure 10.14); indirect effects on plant growth due to soil acidification, as summarized in Section 10.3.5.1, can also be significant.

Wet and dry acidic deposition on leaves and needles may enter directly through plant stomates. If the deposition is sufficiently acidic (pH ~3.0), damage can also occur to the waxy cuticle on the surface of leaves and needles, increasing the potential for direct injury of exposed leaf mesophyll cells. Foliar lesions are one of the most common symptoms of plants subjected to simulated acidic precipitation. Gaseous compounds such as SO_2 and SO_3 present in acidic mists or fogs can also enter leaves through the stomates, form H_2SO_4 upon reaction with H_2O in the cytoplasm of leaf cells, and disrupt many pH-dependent metabolic processes. Several studies have shown increased necrosis of leaves and needles when exposed to high levels of SO_2 gas. The exact mechanisms by which acidification of leaf cells causes injury are not known and certainly vary among species. Physiological changes associated with exposure to simulated acidic deposition include collapsed epidermal cells, eroded cuticles, loss of chloroplast integrity and decreased chlorophyll content, loosening of fibers in cell walls and reduced cell membrane integrity, and changes in osmotic potential that cause a decrease in cell turgor.

Considerable research has been directed at the secondary effects on vegetation exposed to acidic deposition, specifically the likelihood of increased disease or insect damage. The general hypothesis is that any injury to a leaf surface will promote the survivability of pathogenic organisms and their entry into the plant. Studies have shown that leaves exposed to acidic precipitation are more "wettable," a key condition for the establishment of a pathogenic population. Lesions on acidified leaf surfaces are openings that permit ready entry of plant pathogens, in much the same manner as natural wounds. Other effects of acidic deposition on leaves may also predispose them to disease or insect damage. Leaching of nutrients and organic compounds and disruption of photosynthesis and nutrient metabolism weaken the resistance of the plant to any type of stress.

There may also be an increased likelihood of root diseases. In addition to the damages caused by exposure to H_2SO_4 and HNO_3, roots can be directly injured or have their growth rate impaired by increased concentrations of soluble Al^{3+} and Mn^{2+} in the rhizosphere. Root exudates (organic compounds secreted from root cells) have been shown to increase when aboveground plant metabolism is affected by acidic deposition on leaves and stems. Changes in the amount and composition of these exudates can then alter the activity and population diversity of soilborne pathogens. The general tendency associated with

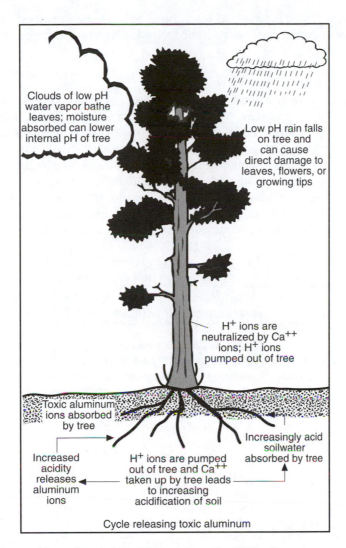

Clouds of low pH water vapor bathe leaves; moisture absorbed can lower internal pH of tree

Low pH rain falls on tree and can cause direct damage to leaves, flowers, or growing tips

H^+ ions are neutralized by Ca^{++} ions; H^+ ions pumped out of tree

Toxic aluminum ions absorbed by tree

Increasingly acid soilwater absorbed by tree

Increased acidity releases aluminum ions

H^+ ions are pumped out of tree and Ca^{++} taken up by tree leads to increasing acidification of soil

Cycle releasing toxic aluminum

Figure 10.14 Effects of low-pH rainfall and trees in soil acidification processes. In soils with inorganic Al-to-Ca ratios of greater than 1.0, competition between these two species favors Al. Because Ca is a secondary nutrient required for tree growth, low-pH soils and increased Al:Ca levels can result in decreased forest health. (From Bush, M. B., *Ecology of a Changing Planet*, Prentice-Hall, Upper Saddle River, NJ, 1997. With permission.)

increased root exudation is an enhancement in microbial populations due to an additional supply of carbon (energy). Chronic acidification can also alter nutrient availability and uptake patterns, and thus aboveground plant growth and yield.

Long-term studies conducted at the Hubbard Brook Experimental Forest in New Hampshire suggest acidic deposition has resulted in significant leaching of basic cations that are important for acid neutralization and plant nutrition. Consequently, forest growth and biomass production of primarily red spruce trees in the White Mountains has dropped to approximately zero since 1987. With the decrease in acidic deposition and reduction in about 80% of the airborne basic cations, mainly Ca^{2+} but also Mg^{2+}, from 1950 levels, researchers at Hubbard Brook suggest forest growth has slowed because soils are not capable of weathering at a rate that can replenish these essential plant nutrients. In Germany, acidic deposition was implicated in the loss of soil Mg^{2+} as an accompanying cation associated with the downward leaching of SO_4^{2-}, which resulted in forest decline. Several European countries have used helicopters to fertilize and lime forests in an effort to improve forest health.

10.3.5.4 Aquatic ecosystems

Ecological damage to fresh waters from acidic deposition is a serious and increasingly well-documented global problem. As with forests, the weight of scientific evidence suggests that a number of interrelated factors associated with acidic deposition are responsible for many of the undesirable changes occurring in aquatic ecosystems. Acidification of fresh waters is, however, not a new phenomenon. Studies of lake sediments suggest that increased acidification began in the mid-1800s, although the process has clearly accelerated since the 1940s. Current studies indicate there is significant S mineralization in forest soils impacted by acidic deposition and that the SO_4^{2-} levels in adjacent streams remain high, despite a decrease in the amount of atmospheric S deposition. Intensive scientific research

Table 10.5 Chemical Properties of Precipitation, Throughfall, Leachate, and Stream Waters in Two European Watersheds

Watershed description	pH	H+	Ca2+	Total SO$_4^{2-}$	Al (µg/L) Total	Al (µg/L) Soluble
Loch Chon (Scotland)						
Subject to high acidic deposition; rainwater partially neutralized by vegetation and in soil profile; leachate strongly influenced by calcium-rich geology; streams support fish						
(meq/m²/year)						
Input						
Above canopy	—	48	—	132	—	—
Below canopy	—	24	—	138	—	—
(µeq/L)						
Concentration						
Precipitation	4.5	29	11	53	—	—
Throughfall	4.8	15	50	92	—	—
Leachate from O horizon	4.2	68	—	122	—	—
Leachate from BC horizon	4.4	37	—	89	—	—
Stream water	5.1	8	72	93	129	47
Kelty (Scotland)						
Large inputs of sulfate and acidic deposition; spruce canopy contributes to acidification; high sulfate uptake by trees, but little buffering of acidity by soils and geologic material; many streams cannot support fish						
(meq/m²/year)						
Input						
Above canopy	—	72	—	140	—	—
Below canopy	—	80	—	230	—	—
(µeq/L)						
Concentration						
Precipitation	4.5	29	11	53	—	—
Throughfall	4.2	57	45	164	—	—
Leachate from O horizon	4.0	100	—	115	—	—
Leachate from BC horizon	4.1	80	—	73	—	—
Stream water	4.5	34	48	93	140	62

Source: Mason, B.J., *Acid Rain: Its Causes and Effects on Inland Waters*, Oxford University Press, New York, 1992.

on aquatic ecosystems has been conducted since the early 1970s, resulting in hundreds of technical papers and dozens of comprehensive books on the subject (see the references section of this chapter). The subject is complex, and the magnitude of the effects of acidic deposition on fresh waters has been shown to vary widely among geographic regions. This section reviews the major factors that control the extent of freshwater acidification, the most common ecological effects, and some approaches to remediate the problem.

Geology, soil properties, and land use are the main determinants of the effect of acidic deposition on aquatic chemistry and biota, as shown in Table 10.5 for two European watersheds. Lakes and streams located in areas with calcareous geology (e.g., limestone) resist acidification more than those where granite and gneiss are the predominant geologic materials. Soils developed from calcareous parent materials also tend to be deeper and more buffered against acidification than the thin, acidic soils common to granitic areas. Land management also affects freshwater acidity. Forested watersheds tend to contribute more acidity than those dominated by meadows, pastures, and agronomic ecosystems. Trees and other vegetation in forests have been shown to "scavenge" (retain on leaves and stems) acidic compounds in fogs, mists, and atmospheric particulate matter. These acidic compounds are later delivered to forest soils when rainfall leaches them from the surfaces of the vegetation. Rainfall below forest canopies (e.g., *throughfall*) is usually more acidic than ambient precipitation. Silvicultural operations that disturb soils in forests as part of planting, fertilizing, draining, and harvesting trees can increase acidity by stimulating the oxidization of organic N and S, and reduced-S compounds such as FeS_2. Conversely, runoff and leachate from watersheds dominated by well-limed agricultural soils can act to neutralize acidity in lakes and streams. However, if agricultural soils are fertilized with ammoniacal N fertilizers and not limed properly, they can also contribute to freshwater acidification. Other factors that can influence freshwater acidification include rainfall intensity and duration, topography, and hydrogeology, all of which act to determine the direction and rate of water flow (and thus acidic compounds) through soils and parent material to freshwaters.

Example problem 10.3

The amount of liming material needed to neutralize the acidity from different sources varies based on the H^+ associated with the acid-producing material. Examples include the requirement of approximately 360 kg/ha of lime (as $CaCO_3$) needed to neutralize the acidity from 100 kg N/ha as urea, $(NH_2)_2CO$. The reaction of urea added to soils and its subsequent nitrification is as follows:

$$(NH_2)_2CO + 2H_2O + H^+ \rightleftharpoons (NH_4)_2CO_3 \rightleftharpoons 2NH_4^+ + HCO_3^- \qquad (10.11)$$

$$HCO_3^- + H^+ \rightleftharpoons H_2O + CO_2 \qquad (10.12)$$

$$2NH_4^+ + 4O_2 \rightleftharpoons 4H^+ + 2NO_3^- + 2H_2O \qquad (10.13)$$

which results in the net formation of 2 mol H^+ (e.g., 2 g H^+) for each mole of urea (60 g urea/mol urea or 28 g urea-N/mol urea) and therefore,

$$100 \text{ kg urea-N/ha} \times 2 \text{ kg } H^+/28 \text{ kg urea-N} = 7.1 \text{ kg } H^+/\text{ha} \qquad (10.14)$$

$$7.1 \text{ kg } H^+/\text{ha} \times 50 \text{ kg } CaCO_3/\text{kg } H^+ = 357 \text{ kg } CaCO_3/\text{ha} \qquad (10.15)$$

However, only 50 kg/ha of lime (as $CaCO_3$) would be needed to neutralize the acidity of 100 cm of rain with a pH of 4.0.

$$10{,}000 \ m^3/ha \times 1{,}000 \ L/m^3 = 10{,}000{,}000 \ L/ha \qquad (10.16)$$

$$10{,}000{,}000 \ L/ha \times 0.0001 \ mol \ H^+/L = 1{,}000 \ mol \ H^+/ha \qquad (10.17)$$

$$1{,}000 \ mol \ H^+/ha \times 50 \ g \ CaCO_3/mol \ H^+ = 50{,}000 \ g \ (or \ 50 \ kg) \ CaCO_3/ha \quad (10.18)$$

A number of ecological problems arise when freshwaters are acidified below pH 5.5, and particularly below pH 4.0. Decreases in biodiversity (number of different species present) and primary productivity (actual numbers and biomass) of phytoplankton, zooplankton, and benthic invertebrates commonly occur (Figure 10.15). Decreased rates of biological decomposition of organic matter have occasionally been reported, which can then lead to a reduced supply of nutrients as mineralization slows. Microbial communities may also change, with fungi predominating over bacteria. Proposed mechanisms to explain these ecological changes center around physiological stresses caused by exposure of biota to higher concentrations of Al^{3+}, Mn^{2+}, and H^+ and lower amounts of available Ca^{2+}. One specific mechanism suggested involves the disruption of ion uptake and the ability of aquatic plants to regulate sodium (Na^+), K^+, and Ca^{2+} export and import from cells.

Acidic deposition can also affect fish, amphibians, and mammals. Widespread evidence exists for declining fish populations in acidified lakes and, under conditions of extreme acidity, of fish kills (Figure 10.15). In general, if the water pH remains above 5.0, few problems are observed; from pH 4.0 to 5.0 many fish are affected, and below pH 3.5 few fish can survive. The major cause of fish kill is the direct toxic effect of Al^{3+}, which interferes with the role Ca^{2+} plays in maintaining gill permeability and thus respiration. Aluminum toxicity can be a serious problem and can result in major fish kills if any climatic condition, such as heavy rains or rapid snowmelt, substantially accelerates the flow of Al^{3+}-laden water to streams or lakes within the watershed. Calcium has been shown to mitigate the effects of Al^{3+}, but in many acidic lakes the Ca^{2+} levels are inadequate to overcome Al^{3+} toxicity. Low pH values also disrupt the Na^+ status of blood plasma in fish. Under very acidic conditions, H^+ influx into gill membrane cells both stimulates excessive efflux of Na^+ and reduces influx of Na^+ into the cells from the external waters. Excessive loss of Na^+ can cause mortality in fish. Other indirect effects of acidification on fish survival include reduced rates of reproduction, high rates of mortality early in life or in reproductive phases of adults, and migration of adults away from acidic areas. Amphibians are affected in much the same manner as fish, although they are somewhat less sensitive to Al^{3+} toxicity. Birds and small mammals often have lower populations and lower reproductive rates in areas adjacent to acidified fresh waters. This may be due to a shortage of food due to smaller fish and insect populations or to physiological stresses caused by consuming organisms with high Al^{3+} concentrations.

10.3.6 Reversing the effects of acidic deposition

The environmental damage caused by acidic deposition will be difficult and extremely expensive to correct. In the long term, reversal of these effects can only be accomplished by reducing S and N emissions. Some approaches to achieve this include burning less fossil fuel; using cleaner energy sources (e.g., low-S coal or wind and solar systems); and

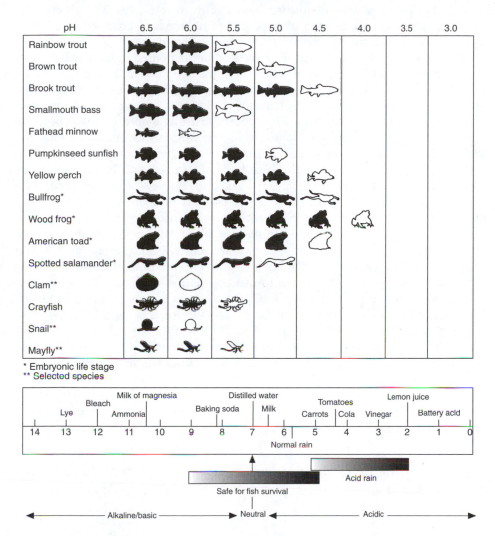

Figure 10.15   Examples of the sensitivity of aquatic species to different pH levels. Note that the solid image represents a pH level that the aquatic species survives and the species outline indicates conditions unfavorable for survival. Examples of various substances and their pH levels are also shown for comparison. (From Bush, M. B., *Ecology of a Changing Planet*, Prentice-Hall, Upper Saddle River, NJ, 1997. With permission.)

designing more efficient "scrubbers" to reduce the amount of these gases emitted by utilities, industries, and vehicles. There is a general consensus among the scientific community that reducing emissions will result in slow, but eventual improvement in acidified ecosystems, particularly fresh waters. For the present, despite the firm conviction of most nations to reduce acidic deposition, it appears that the staggering costs of such actions will delay implementation of this approach for many years. The U.S. Congress estimated that to reduce SO_2 emissions alone by 9 million Mg (10 million tons) would cost as much as $3.6 billion over a 5-year period; the EPA estimated that 20-year costs could be as much as $33 billion. Annual costs associated with damages attributed to acidic deposition impacts to structural, ecological, and environmental systems in the northeastern U.S. have been approximated to be $5 billion due to corrosion of buildings, bridges, and other structures ($2 billion), forest ecosystem damage ($1.75 billion), agricultural ecosystems impacts ($1 billion), and loss of tourism and fishing revenue ($0.25 billion). The 1990 Amendments to the Clean Air Act are expected to reduce acid-producing air pollutants, e.g., SO_2 and NO_x, that are common contaminants released in energy-related operations such as electric power plants. One cleanup method implemented in the 1990 Amendments was that of emission allowances that set the amount of S that can be emitted by a utility. The allowances are based on the historical fuel use and SO_2 emissions of a utility, with each allowance representing 1 ton of SO_2 that can be bought, sold, or banked for future use.

Short-term remedial actions for acidic deposition are available and have been successful in some ecosystems. Liming of lakes and some forests (also fertilization with trace elements and Mg^{2+}) has been practiced in European counties for over 50 years and is now done in many other areas of the world. Hundreds of Swedish and Norwegian lakes have been successfully limed in the past 25 years. The effectiveness of liming depends mainly on the residence time of water in the lake. Lakes with short mean residence times (see Chapter 2) for water retention may need annual or biannual liming; others may need to be limed every 5 to 10 years. As an example, ten small Adirondack lakes were limed in 1983 to 1984, and the percent survival of brook trout increased from about 10% at pH 4.5 to 60 to 80% at pH 5.5 to 6.5. By 1986, however, three of the limed lakes were sufficiently reacidified to cause significant declines in fish populations. The logistics of liming lakes and forests are formidable. Aerial application of lime via planes and helicopters is often required; hence application costs often far exceed the expense of the liming materials. Liming may have some negative effects as well. Much of the vegetation in forested areas is well adapted to acidic soils; liming (or overliming) may alter the distribution of species in these ecosystems in an unpredictable and perhaps undesirable manner.

Environmental quality issues/events

Restoration efforts of an industrial region impacted by sulfur dioxide, heavy metals, and acidic deposition: Sudbury, Canada

---

The Sudbury region in Canada, located about 200 km east of the upper peninsula of Michigan, has the largest known concentration of mineable nickel (Ni) ore in the world, estimated at 20 million Mg. Sudbury ore also contains significant economic levels of Cu, platinum (Pt), palladium (Pd), osmium (Os), iridium (Ir), rhodium (Rh), ruthenium (Ru), gold (Au), silver (Ag), selenium (Se), and tellurium (Te), as well as trace amounts of Zn, Pb, arsenic (As), and cobalt (Co). Minerals in the ore are primarily sulfides with high-grade ores yielding 7 to 10% Ni and Cu combined.

Sudbury mineral resources were discovered during the construction of the Canadian transcontinental railway in 1883. In 1888 the first roast yard and smelting operation was established in the area and used to obtain Cu; at this time Ni was considered an impurity. The roasting process involved piling crushed ore on logs and burning the material for 2 months. The intensity of the fire resulted in ore combustion that caused the oxidation of sulfides and release of gaseous SO_2. There were 11 roast yards used during 1888 to 1929, with most of the yards located near the nine smelters that operated in the Sudbury region. Practically all the woody materials in the general vicinity of the roast yards were used in the roasting process. Because an estimated 10 million Mg of SO_2 were released at ground level from the roasting yards, it has been suggested that the extensive destruction of vegetation and toxic soils was due to these early activities.

Smelting operations, however, resulted in much more widespread damage of the surrounding ecosystem because the fumes contain Ni and Cu particles and significant amounts of SO_2. It was realized in the early 1940s that smelter emission had caused extensive damage to the forests in the Sudbury region. Smoke from the smelters could be seen 120 km away and the smell of S was noticed at least 60 km from the source emissions. Zones of damage around the three major smelters were described in the earlier 1970s as either barren areas that amounted to approximately 17,000 ha or large, semibarren areas surrounding the barren areas, which were estimated to be another 72,000 ha. It has also been suggested that more than 7000 lakes within the Sudbury region were also affected by pollutants derived from smelting activities.

The environmental damage due to mining and processing the Sudbury region ore is probably the most highly recognized destruction of a North American ecosystem; Noril'sk in central Siberia is considered the world's most polluted area (10,000 km^2) and the largest point source of SO_2. Sudbury damage is the result of impacts related to activities that have occurred over more than 100 years, and involved SO_2 and trace elements (Cu, Ni, Al) that combined create low-pH environments with highly bioavailable toxic metals (see Chapters 6 and 7 for further details). The 1960s brought about an intense pressure to reduce S emissions from smelting operations and in the 1970s there was a rapid decline in the amount of S that impacted local air quality. Measures that have resulted in improved air quality include construction of the world's tallest smokestack (381 m) to increase the dispersal of smelter emissions, closing some of the older smeltering and processing plants, reducing S emissions, and halting smelter operation during weather conditions that did not allow dispersal of emission plumes.

During the 1970s and 1980s, several studies were conducted to evaluate terrestrial and aquatic recovery due to the reduction in S and trace element pollution. While the recovery was slower than expected, primarily because of SO_2 fumigation (e.g., high SO_2 concentrations in localized areas) caused by climatic conditions and long-term consequences of high acidity and toxic metals, changes were apparent particularly in areas of minor pollution. Fertilization, liming, and introduction of new plant species with acid or metal tolerances have accelerated the recovery process. Liming was found to be important for increasing the soil pH, forming Ni and Cu precipitates, reducing metal uptake due to increased Ca^{2+} and Mg^{2+} competition, and preventing Al^{3+} toxicity. One of the goals for the revegetation program was to use minimal amounts of amendments and low seeding rates so that increased plant colonization and diversity of native species would occur. Aspen trees appear to be rapidly invading the test sites with birch trees following liming treatments. Other plant species indigenous to the Sudbury region that are revegetating treated lands include oaks, willows, and various wildflowers and grasses. Over time, there has been evidence of plant succession and there is hope that biotic communities are adapting to the changing soil and environmental conditions and will ultimately produce a diverse ecosystem.

Significant lake impacts due to smelter operations included losses of sport fish species such as lake trout, brook trout, and smallmouth bass, the aquatic trophic level species zooplankton, phytoplankton, and benthic invertebrates, as well as waterfowl and amphibians. Liming of Sudbury lakes resulted in aquatic habitat improvements via increased pH and reduced toxic element concentrations. Lakes that have been limed and stocked with bass and lake trout have shown promise for the recovery of recreational sportfishing activities. In addition, with increases in fish populations and invertebrate prey, the breeding habitats for waterfowl has improved, and as the ecosystem improves, aquatic species, plants, and animals are expected to increase with the reduction in S and trace element emissions.

While over $1 billion has been spent on reducing S emissions and environmental restoration activities, there is still much to be done in the Sudbury region. Both industry and municipalities have worked together to revegetate more than 4000 ha of barren lands and have planted over 3 million trees. Additional improvements will undoubtedly be made to the impacted aquatic ecosystems by more liming programs that bring the pH levels of affected lakes to near 6.0. The enormous efforts that have been expended to improve the environmental quality of the region over the past 25 years have resulted in Sudbury being named one of Canada's ten best cities according to a prominent Canadian magazine. More noteworthy was the Governments Honor Award received in 1992 at the United Nation's Earth Summit in Rio de Janeiro, Brazil, which recognized the outstanding achievements attained with the Sudbury municipal land reclamation program.

—Source: Gunn (1995)

Problems

10.1 For each of the major greenhouse gases, describe measures that might be taken to reduce atmospheric concentrations. How would these measures impact society?

10.2 For a fictitious greenhouse gas with a global warming potential (GWP) of 50 and an annual emission of 5 Tg, calculate the million metric ton carbon equivalent (MMTCE) value and compare it with those for CO_2, CH_4, and N_2O.

10.3 Describe the difference between the following terms: acid rain, acid precipitation, acid deposition, acid-forming substances, acidity.

10.4 What are the primary sources that contribute to acidic deposition? Where are these sources located? What forms of atmospheric substances are released from these sources?

10.5 Calculate the difference in rainwater acidity between that of the Rocky Mountains (pH = 6.0) and rainfall in Florida (4.7), Tennessee (4.5), and Ohio (4.2).

10.6 There are several reasons acid soils are harmful to plants, animals, and possibly humans. What are some of the problems associated with acid soils, how are they manifested, and what can be done to correct these problems?

10.7 Why have forest ecosystems been affected more by acidic deposition than agricultural ecosystems? What is N saturation and why might it be a greater concern in the future?

10.8 If a lake containing 3 million m^3 of water has a pH of 4.50, what is the amount of acidity (H^+) that would need to be neutralized if the lake pH were to be limed to a pH of 5.60? How much pure $CaCO_3$ would be required to neutralize the acidity?

10.9 Describe various methods that can be implemented to restore damaged soils, forests, and lakes. List some potential problems that could occur in these ecosystems if the restoration practices do not consider the impacts that may result from altering the whole ecosystem.

10.10 Use the data in Table 10.3 to hypothesize why NO_x and SO_2 have increased in the U.S. since 1900. What may have caused the increases and decreases in concentrations during the 20th century?

References

Bush, M. B. 1997. *Ecology of a Changing Planet*, Prentice-Hall, Upper Saddle River, NJ.
Carbon Dioxide Information Analysis Center. 1999. Trends Online: A compendium of data on global climate change. Oak Ridge National Laboratory, U.S. Department of Energy, Oak Ridge, TN.

Charles, D. F., Ed. 1991. *Acidic Deposition and Aquatic Ecosystems*, Springer-Verlag, New York.

Council on Environmental Quality. 1989. Environmental Quality, 18th and 19th Annual Reports, U.S. Government Printing Office, Washington, D.C.

Doherty, R. and L. O. Mearns. 1999. A comparison of simulations of current climate from two coupled atmosphere–ocean GCMs against observations and their evaluations of future climates, Report to the NIGEC National Office, National Center for Atmospheric Research, Boulder, CO, 47 pp.

Gunn, J. M., Ed. 1995. *Restoration and Recovery of an Industrial Region: Progress in Restoring the Smelter-Damaged Landscape near Sudbury, Canada*, Springer-Verlag, New York.

Krug, E. C. and C. R. Frink. 1983. Effects of acid rain on soil and water, Bull. 811, Connecticut Agricultural Experimental Station, New Haven.

Linthurst, R. E., Ed. 1984. The Acidic Deposition Phenomenon and Its Effects: Critical Assessment Review Papers, Vol. 2, Effects Sciences, EPA-600/8-83-016B, U.S. Environmental Protection Agency, Washington, D.C.

Marion, G. M., D. M. Hendricks, G. R. Dutt, and W. H. Fuller. 1976. Aluminum and silica solubility in soils, *Soil Sci.*, 121, 76–82.

Mason, B. J. 1992. *Acid Rain: Its Causes and Effects on Inland Waters*, Oxford University Press, New York.

MacCracken, M. C. and F. M. Luther, Eds. 1985. Projecting the Climatic Effects of Increasing Carbon Dioxide (1985), DOE/ER-0237, U.S. Department of Energy, Washington, D.C.

Moore, B. 1988. Presentation to the Global Warming Round Table, sponsored by the U.S. Department of Energy, Washington, D.C.

National Acid Precipitation Assessment Program. 1992. Report to Congress, U.S. Government Printing Office, Pittsburg, PA, 130 pp.

Smith, W. H. 1999. Acid rain, in *The Wiley Encyclopedia of Environmental Pollution and Cleanup*, R. A. Meyers and D. K. Dittrick, Eds., John Wiley & Sons, New York, 9–15.

U.S. Department of Energy. 1999. Carbon Sequestration: State of the Science, U.S. Department of Energy, Office of Science, Washington, D.C.

U.S. Environmental Protection Agency. 1980. Acid Rain, EPA-600-79-036, U.S. Government Printing Office, Washington, D.C.

U.S. Environmental Protection Agency. 1988. Environmental Progress and Challenges: EPA's Update, EPA-230-07-88-033, U.S. Government Printing Office, Washington, D.C.

U.S. Environmental Protection Agency. 1996. National Air Quality and Emissions Trends Report, U.S. Environmental Protection Agency, Washington, D.C.

U.S. Environmental Protection Agency. 1999. The Draft 1999 Inventory of U.S. Greenhouse Gas Emissions and Sinks (1990–1997), U.S. Environmental Protection Agency, Washington, D.C.

Wolff, G. T. 1999. Air pollution, in *The Wiley Encyclopedia of Environmental Pollution and Cleanup*, R. A. Meyers and D. K. Dittrick, Eds., John Wiley & Sons, New York, 48–65.

World Resources Institute. 1988. *World Resources, 1988–1989*, Basic Books, New York.

Supplementary reading

Alawell, C., M. J. Mitchell, G. E. Likens, and H. R. Krouse. 1999. Sources of stream sulfate at the Hubbard Brook Experimental Forest: long-term analyses using stable isotopes, *Biogeochemistry*, 44, 281–299.

Conservation Foundation. 1987. *State of the Environment: A View Toward the Nineties*, Conservation Foundation, Washington, D.C.

Forster, B. A. 1993. *The Acid Rain Debate: Science and Special Interests in Policy Formation*, Iowa State University Press, Ames.

Heij, G. J. and J. W. Erisman, Eds. 1995. Acid Rain Research: Do We Have Enough Answers? *Proceedings of a Speciality Conference*, Studies in Environmental Science 64, Elsevier Science, New York.

Kamari, J. 1990. *Impact Models to Assess Regional Acidification*, Kluwer Academic, London.

Kennedy, I. R. 1992. *Acid Soil and Acid Rain*, John Wiley & Sons, New York.

Likens, G. E., C. T. Driscoll, and D. C. Buso. 1996. Long-term effects of acid rain: response and recovery of a forest ecosystem, *Science*, 272, 244–246.

Linthurst, R. A. 1984. *Direct and Indirect Effects of Acidic Deposition on Vegetation*, Butterworth, Stone-
     ham, MA.

Reuss, J. O. and D. W. Johnson. 1986. *Acid Deposition and the Acidification of Soils and Waters*, Springer-
     Verlag, New York.

Schlesinger, W. H. 1997. *Biogeochemistry: An Analysis of Global Change*, 2nd ed., Academic Press, San
     Diego, CA.

Watson, R. T., M. C. Zinyowera, and R. H. Moss, Eds. 1996. *Climate Change 1995 — Impacts, Adaptations
     and Mitigation of Climate Change: Scientific-Technical Analyses*, published for the Intergovernmen-
     tal Panel on Climate Change, Cambridge University Press, New York.

chapter eleven

Remediation of soil and groundwater

Contents

11.1 Introduction

Remediation refers to processes or methods for treating contaminants in soil or water such that they are contained, removed, degraded, or rendered less harmful. For water it is groundwater that is most often remediated, although one may consider removal of substances from surface water as remediation. There are several subcategories of remediation. *Site remediation* in often used for processes that contain a contaminant but do not necessarily directly affect the contaminant, whereas *soil* or *water remediation* generally refers to processes that directly treat the medium and affect the contaminant in some way. *In situ remediation* refers to treatment of soil or water in place while *ex situ remediation* involves physical removal and treatment of either soil or water.

There are numerous reasons a remedial action may be necessary. The most obvious is that the soil or water has been contaminated and the natural inclination is to return that resource to its original, uncontaminated state. The implication is that the contamination may cause or is causing harm (we would call it pollution if it were actually causing harm) to some organism. Of course, not all contaminated sites are remediated just because they are contaminated. Three common reasons for remedial action are a risk assessment that suggests an unacceptable level of risk (see Chapter 12), direct evidence of human or ecological harm, and the violation of regulatory limits for contaminant concentrations in soil, food, or water. A human health risk assessment will use some guidelines for exposure that may suggest remediation is warranted. For noncarcinogenic substances the anticipated dose of the contaminant may exceed the reference dose for that substance or the cancer risk assessment may exceed the 10^{-4} to 10^{-6} risk guidelines for a potential carcinogen. An ecological risk assessment may suggest that soil or water contaminant concentrations exceed a toxicological threshold for one or more species. Direct evidence of human or ecological harm might include blood lead (Pb) concentrations in children that exceed a critical value or the lack of vegetation on a contaminated site. Regulatory limits could include soil that fails the *Toxicity Characteristic Leaching Procedure* (TCLP) test or contaminant concentrations in groundwater that exceed the *maximum contaminant levels* (MCL) for drinking water purposes.

The *Superfund* process provides a mechanism for the possible cleanup of contaminated sites. An abbreviated version is presented in Figure 11.1. Site discovery can come about in a variety of ways. Routine testing of drinking water, for example, may turn up unacceptable concentrations of contaminants and the source of the contamination must be located. Once the source is found, a preliminary assessment is performed where the nature and extent of the contamination is determined. *Principal responsible parties* (PRPs) are also identified. If there is a strong likelihood that the site will require remediation, it will be placed on the *National Priorities List* (NPL). Often there is opposition to a site being "listed" because a stigma is associated with a Superfund site. The remedial investigation/feasibility study is a much more detailed examination of the site, which will include risk assessments and suggestions of remedial alternatives. The PRPs are actively involved at this point. The

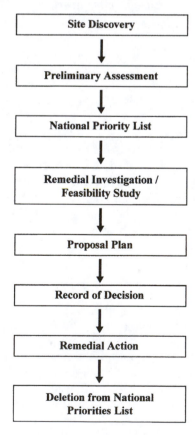

Figure 11.1 Flow diagram for the Superfund process.

proposed plan is a listing of all possible options for remediation (including no action) with cost estimates, how each addresses the human and ecological risks, and the preferred alternative. The *record of decision* (ROD) follows and represents the actual plan that will be used for the remediation. The remedial action then takes place and includes a post-treatment assessment of the effectiveness of the remediation, generally after 5 years. If the remedial action successfully addresses the risk, the site will be removed from the NPL. The bulk of the work is performed by contractors who are paid by the EPA, states, and the PRPs. Each step generates documentation that is subject to a public comment period before it is accepted. The general public has many opportunities for input and the EPA may provide funds for technical advice for citizen groups.

One might question the overall goal of a remedial action. Again, the obvious goal is to reduce actual or potential human or ecological risk from the contaminants. On a more practical scale, the goal might be merely to treat soil to pass the TCLP test or containment or isolation of the contaminant to prevent further dispersal, if there is no immediate threat to humans or the environment. When soil or water is treated to remove contaminants, the question often raised is how clean it needs to be. Specifically, how low do the contaminant concentrations have to be before the cleanup is considered successful. The intuitive answer is that the contaminant concentrations should be reduced to background levels or those expected in uncontaminated soil or water. In reality, such concentrations are difficult to obtain. Further, as a remedial process reduces the contaminant concentration in soil or water it becomes more difficult and expensive as the concentration decreases. Eventually, the cost of further efforts must be balanced against the benefits and the contaminant concentrations will not likely be at background levels before the costs become unreasonable. This had led to greater emphasis on risk and the reduction of risk to an acceptable level rather than a focus on contaminant concentrations alone.

The cost of remediation is an important issue. The Superfund program provided funds for cleanup of NPL sites when no legally responsible and viable parties can be identified. In this case, the taxpayers pay the bill for the cleanup. Often, the cost for cleanup of NPL sites is shared between the state and federal government and the PRPs. For NPL sites on military bases or Department of Energy facilities, those agencies pay for the remediation, although it is the taxpayer who ultimately pays the price. Private industry pays for its own sites. Regardless of who pays the bill, excessive costs will be reflected in increased taxes or the poor financial status of private companies. Again, the cost vs. benefits approach must be used. Cost must also be weighed as various remediation options are considered for a given site. On a more philosophical note, one could consider the effectiveness of money spent on remediation to produce small improvements in risk vs. spending the money addressing other significant social needs with greater potential benefits.

For organizational purposes, it is useful to separate remediation processes according to the general classes of contaminants. The most general separation would be inorganic contaminants vs. organic contaminants, which is what we will use in this chapter. Inorganic contaminants such as metals cannot be degraded into simple compounds, do not tend to form vapors, and do not tend to leach to any great extent. Organic contaminants, on the other hand, may be susceptible to degradation by a number of processes, may volatilize, and may or may not adsorb strongly to soils or aquifer materials. As will be shown, remediation processes used for one class may not be appropriate for the other. Many sites are contaminated by more than one substance so more than one approach may be necessary for remediation. Since organic contaminants have such widely varying properties, the category is often split into several smaller categories based on volatility, polarity, solubility, or presence of aromatics or halogens. Such a division is beyond the scope of this chapter.

It is difficult to estimate the number of sites that need remediation. The Superfund program maintains the NPL in the U.S., but many sites are not on the NPL for a number of reasons. As of April 1999 over 1200 sites were on the NPL. It is also estimated that 300,000 out of 750,000 known underground storage tanks in the U.S. have leaked. Many of those tanks contained petroleum hydrocarbons of some type. Worldwide, the magnitude of the problem is much larger.

Another important factor in remediation technologies is that innumerable techniques and approaches are in use or have been tested to some degree. It is simply beyond the scope of this book to present a complete list of options for each category of contaminants. An attempt is made to present technologies that have been used or that show promise. The reader is referred to the supplementary reading list for a more detailed discussion of the subject.

11.2 Inorganic contaminants

11.2.1 Soil

If one considers the influence of excessive soil concentrations of inorganic contaminants on soil quality, then saline, sodic, and acidified soils would be in need of remediation as well as those soils contaminated by inorganic substances with known harmful effects on various organisms. Generally, the former categories are not considered under soil remediation and only those elements with known harmful effects on various organisms will be considered here. In the context of this book that would include those elements discussed under trace elements, including radionuclides.

Figure 11.2 presents a decision tree for possible remedial action for a site contaminated with inorganic subtances. The first decision is whether or not to take action. A decision to take no action can be a conscious one and is often used for sites posing no immediate threat to the environment and for which there is little chance of movement for the contaminants. If remedial action will be taken, the next decision is whether the approach will be *in situ* or *ex situ*. Sites that are small and highly contaminated lend themselves to *ex situ* remediation, particularly those involving radionuclides where the half-lives of the isotopes are long. Conversely, large sites having low levels of contamination lend themselves to *in situ* remediation. As compared with sites contaminated with organic chemicals, large sites contaminated with inorganic substances are much more common because of large-scale contamination associated with the mining and smelting of trace elements.

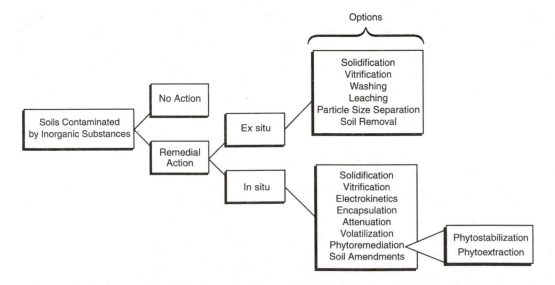

Figure 11.2 Decision tree for soils contaminated with inorganic substances.

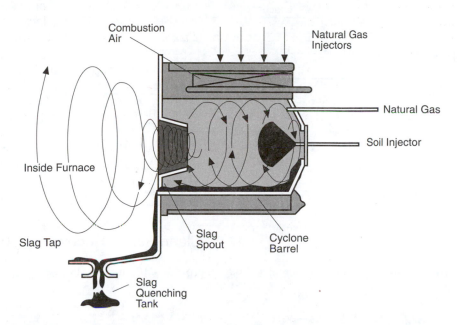

Figure 11.3 Schematic of an experimental soil vitrification process. (From EPA, 1992.)

Most of the six options under *ex situ* remediation have a strong engineering component. Solidification involves the mixing of solidifying agents with the contaminated soil to produce a concrete-like product with low permeability and with durability under multiple freeze and thaw cycles. Permeabilities can be as low as 10^{-9} cm/s. Solidification reduces TCLP concentrations to acceptable levels and is suitable for stabilizing radioactive waste prior to storage in a nuclear waste repository. A significant disadvantage is that solidification increases waste volume by 10 to 30%. A related process is vitrification, which also creates a product that has low permeability, high durability, and with a reduced TCLP concentration. Vitrification involves heating the waste to as much as 2000°C to produce a melt that hardens into a glasslike material, as shown in Figure 11.3. One significant advantage to vitrification is that many organic contaminants will also be oxidized during the process, although there may be a need for emission control for waste gases. Volume may be reduced by 25 to 35%. The process is also energy intensive. Overall, both processes prevent further interaction between the contaminants and the environment and have been particularly suited for treating wastes containing radionuclides.

Washing, leaching, and particle size separation are all related processes that attempt to remove contaminants from the soil. Washing and leaching employ surfactants, acids, bases, chelates, or ion-exchange resins to remove inorganic contaminants from the soil matrix. Particle size separation removes the small particles (approximately the clay-sized, or <2 μm, fraction) from the soil matrix and exploits the fact that the concentration of the inorganic substances is generally higher in the finer-sized fractions. Some processes may use both washing and particle size separation, as shown in Figure 11.4.

Soil removal is a commonly employed technique when contaminated soils are found in residential areas (Figure 11.5). The concept is straightforward in that the contamination is removed, although in practice this is a difficult process because of the need to store the excavated material and the need to find clean replacement soil. Typically, 15 cm of soil is removed and the contaminant levels are checked in the remaining soil. If contaminant levels still exceed the target values, another 15 cm is removed. This process is repeated until as much as 60 cm of soil has been removed, depending on the requirements for the

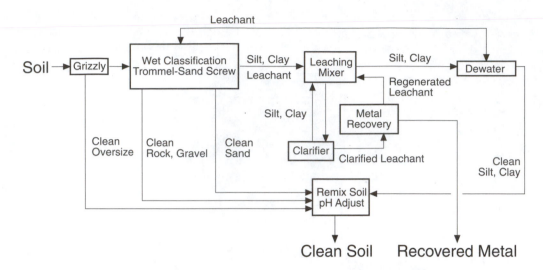

Figure 11.4 Schematic of an experimental soil washing process. (From EPA, 1992.)

Figure 11.5 Soil removal from around a home in Joplin, MO. (Photograph credit: Jim Mueller.)

site. If contaminant levels still exceed target levels after the maximum allowable depth has been reached, a nondegradable mesh material may be put into place prior to refilling with clean soil. The mesh allows roots and water to pass and acts as a warning to anyone who might dig at the site after the remediation is complete.

Ex situ processes have several disadvantages that bear mentioning. First, conservation of mass still applies and any inorganic substances removed from contaminated soils must still be disposed of or recovered, and this can add substantial cost to the process. Washing, for example, generates large volumes of wastewater that must be treated. A second major disadvantage is that sites contaminated with inorganic substances are frequently large in size. One can estimate that each hectare of land excavated to 60 cm generates approximately 8100 Mg of soil that needs to be treated. Some trace element–contaminated areas exceed 1000 ha in size, so the mass of material quickly becomes cumbersome.

In situ processes have been the focus of increased interest because of the problems outlined above. There are *in situ* versions of solidification and vitrification that produce a similar product as described above, but the solidified or vitrified monolith is formed in place. This approach is still somewhat experimental, but offers the advantage of not having to remove the end products for storage at another location. Electrokinetics is another experimental approach that involves the use of a cathode and an anode placed in the soil. A current is passed between the two electrodes, causing the migration of ions toward the electrodes. This is facilitated by the application of fluids that enhance the solubilities of the inorganic contaminants. The end result of the process is the contaminants are contained in a smaller volume of soil, which reduces overall treatment costs.

Encapsulation is occasionally used for small sites and involves covering the site with a layer of material that has a very low permeability, such as compacted clay or concrete, to minimize percolation of water through the site and to prevent windblown dust from the site. This process is analogous to that used for closing landfills. An example from Butte, MT is shown in Figure 11.6. Attenuation simply means dilution of the contaminated soil with uncontaminated material such as soil, municipal biosolids, or coal fly ash. For contamination that is confined to the soil surface, attenuation can be achieved by mixing with uncontaminated soil that lies beneath the contaminated layer. Volatilization can potentially be used for elements such as arsenic (As), mercury (Hg), or selenium (Se), which can exist in methylated, volatile forms. This experimental technique has shown the greatest promise with Se. Microbes can enhance the formation of volatile dimethylselenide and conditions can be controlled to promote this microbial activity.

Phytoremediation, and the two subcategories of phytostabilization and phytoextraction, have received considerable attention in recent years. As the name implies, phytoremediation involves the use of plants for remediating a contaminated site. Phytostabilization refers to

Figure 11.6 Encapsulation technology in use in Butte, MT. (Photograph credit: Gary M. Pierzynski.)

the use of plants to vegetate a site and prevent loss of contaminated material through wind or water erosion. The plants may be grasses, forbs, or trees. Leaching losses may also be reduced because the plants reduce net percolation of water through the site because of their transpirational needs. Phytoextraction is an experimental approach that attempts to remove contaminants from soil by plant uptake. Biomass enriched in the contaminants is removed from the site and burned. The ash contains the contaminants in a much smaller volume, and valuable metals can actually be recovered.

Of the two categories of phytoremediation, phytostabilization has gained the greatest acceptance and has been used extensively. The classic example would be a site contaminated by trace elements that has little or no vegetation (see Figure 7.2). Evidence of severe erosion can generally be found at such sites and indications of trace element problems can often be found in floodplains many miles from the site due to movement and deposition of contaminated sediments by water. There are many possible reasons for the lack of vegetative growth. Trace element phytotoxicities are a problem for elements such as zinc (Zn), copper (Cu), or nickel (Ni). The soils or mine spoil materials may also be poor growth media and are nutrient deficient, have abnormally high or low pH, or have poor soil physical properties. These problems must be corrected before vegetation can be established and maintained over long periods of time.

Various soil amendments have been used to improve the prospects of establishing and maintaining vegetation on contaminated sites. Fertilizers can be used to correct nutrient deficiencies and liming materials can be used to increase soil pH to acceptable levels. Organic waste materials such as municipal biosolids or animal wastes can add plant nutrients, improve soil physical properties, dilute contaminant concentrations, and can reduce trace element bioavailability to plants, thus decreasing problems with phytotoxicity. Figure 11.7 illustrates the effectiveness of various soil amendments in reducing Zn bioavailability in a metal-contaminated alluvial soil. In this situation the indicator of bioavailable Zn was that extracted from the soil with a 0.5 M KNO_3 solution. The treatments included the addition of lime, phosphorus (P), cattle manure, biosolids, poultry litter, or various combinations of limestone and cattle manure. The treatments produced KNO_3-extractable Zn concentrations ranging from 3.7 to 63 mg/kg with a corresponding range of soybean tissue Zn concentrations of 318 to 1153 mg/kg. The manipulation of the bioavailable Zn levels was done without changing the total Zn concentration of the soil.

The concept of phytoextraction was developed from observations by botanists that certain plant species, called *hyperaccumlators*, had very high concentrations of certain elements. Hyperaccumulators have been found for cobalt (Co), Cu, Ni, Pb, Se, uranium (U), and Zn. A plant tissue concentration of 1000 mg/kg would be indicative of hyperaccumulation for all but Zn, where concentrations as high as 10,000 mg/kg are possible.

The ability of a given plant species to remove an element from the soil would be a product of the biomass produced and the concentration of that element in the biomass, as shown below:

$$\frac{1000 \text{ mg Ni}}{\text{kg biomass}} \times \frac{2000 \text{ kg biomass}}{\text{ha year}} \times \frac{1 \text{ kg}}{10^6 \text{ mg}} = \frac{2.0 \text{ kg Ni removed}}{\text{ha year}} \qquad (11.1)$$

One can calculate the approximate mass of an element present in 1 ha of soil using a depth, bulk density, and the soil contaminant concentration. You can then estimate the number of harvests required to remove a given amount of an element from that hectare. For example, assuming a bulk density of 1.35 g/cm³, we have

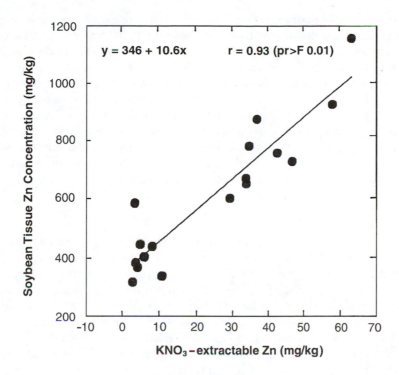

Figure 11.7 Relationship between relative yield and soybean tissue Zn concentrations in a metal-contaminated alluvial soil. The variation in tissue Zn concentrations was a result of changes in soil bioavailable Zn levels induced by various soil amendments. (Data from Pierzynski and Schwab, 1993.)

$$\frac{135,000 \text{ kg soil}}{\text{ha cm}} \times 30 \text{ cm} \times \frac{100 \text{ mg Ni}}{\text{kg soil}} \times \frac{1 \text{ kg}}{10^6 \text{ mg}} = \frac{405 \text{ kg Ni}}{\text{ha}} \tag{11.2}$$

which would require 203 years to remove at 2.0 kg Ni/ha year. This example illustrates one of the problems with this approach, i.e., low removal rates. This is a result of low contaminant concentrations in the biomass or low biomass production. Overall, the hyper-accumulators do not produce large quantities of biomass. One long-range solution to this problem is to use plant breeding or biotechnology to develop plants that hyperaccumulate and produce more biomass. Another experimental solution is to use plants that produce high quantities of biomass but do not hyperaccumulate. Liquids containing chelating agents that solubilize metals in soils are added to the soil while the plants are actively growing, causing increased uptake by the plants. The high metal concentrations in the plant induce a phytotoxic response, but that is not a concern since the crop is harvested shortly afterward anyway. To date, there have been no large-scale remediation efforts performed with phytoextraction, although it remains an area of active research.

Soil amendments are often used as part of phytostabilization efforts, as described above, but they can also be the primary method of remediation themselves. One example referred to in Chapters 7 and 12 is the use of P amendment for Pb-contaminated soil. The proposed mechanism is the formation of insoluble Pb phosphates in the soil, as shown below:

$$5Pb^{2+} + 3H_2PO_4^- + H_2O \rightleftharpoons Pb_5(PO_4)_3OH + 7H^+ \tag{11.3}$$

Since ingestion of Pb-contaminated soil is one of the primary routes of exposure for humans, these Pb phosphates must be so insoluble that they do not dissolve in the digestive system but pass through without being absorbed into the body (i.e., have a low bioavailability). The use of P may also reduce the bioavailability of other trace elements as well, although the mechanisms are not as well defined. There are many other soil amendments that have been studied and that may reduce the bioavailability of inorganic contaminants to plants, animals, or humans. These include liming materials, iron (Fe) and aluminium (Al) (oxy)hydroxides, zeolites, clays, ferrous sulfate, and alkaline biosolids, the obvious advantage for remediation being that soil amendments are relatively easy to add to soil *in situ*, even over large areas. It is likely that the use of soil amendments will become increasingly popular as part of remedial activities.

11.2.2 Water

For many of the inorganic contaminants, particularly the cationic metals and some oxyanions, movement through soil is restricted and large-scale contamination of groundwater is not common. There are instances where shallow aquifers actually intersect existing or previous mining activities and significant contamination of groundwater has occurred. Similarly, in watersheds where vegetative cover is limited, there may be significant movement of inorganic contaminants to surface waters. Often, aquatic life criteria are exceeded in such surface waters but remediation generally involves processes described above for soils to limit movement of inorganic contaminants and no direct remediation of water is performed. If either surface water or groundwater is used for drinking water purposes, some remedial action may be necessary.

The remedial alternative used most often for water contaminated with inorganic substances is abandonment of the source. If private wells are supplying homes, then the installation of a public water system can solve the problem. Alternatively, bottled water may be used for cooking and direct consumption. If a public water system has difficulty meeting drinking water standards, new wells may have to be drilled outside of the contaminated areas or new surface water sources may need to be located. Dilution of contaminated water with clean water will also allow the finished water to meet standards.

Large-scale purification systems are generally not practical for public water systems. Water-softening processes may reduce the concentrations of inorganic contaminants somewhat in the finished water. Purification systems for individual homes are quite effective and have been used in many remedial actions. Water softeners for individual homes, reverse osmosis systems, and distillation may reduce contaminant concentrations to acceptable levels. The difficulty with this approach for a remedial action is that the systems require regular maintenance to remain effective. It is difficult to ensure that all of the required maintenance is performed, even if that maintenance is provided at no cost.

11.3 Organic contaminants

Many of the remediation options that were described for inorganic contaminants would also be appropriate for organic chemicals. However, organic contaminants have several characteristics that make additional remediation options available. Two such characteristics are biodegradability and volatility. Biodegradability presents opportunities for degradation of organic contaminants through biological activity, usually by microorganisms. Volatilization exploits the high vapor pressure of some compounds and allows the removal of organic contaminants from soil or water as vapors. The term *volatile organic chemicals* (VOCs) is often used to describe this group of organic contaminants. There are other subcategories of organic chemicals that are often used. These are grouped together because of similar chemical properties or because the collection of organic chemicals often occurs together as contaminants. Examples include PAHs (polycyclic aromatic hydrocarbons), BTEX (benzene, toluene, ethylbenzene, and xylene), and PCBs (polychlorinated biphenyls).

Several key properties of organic chemicals are important in their fate and transport in the environment as well as in the approach used for remediation, regardless of whether the contamination is with soil or water. Biodegradability, polarity, solubility, volatility, and the tendency to adsorb to soils or aquifer materials all need to be considered for remediation. Some compounds lend themselves to biodegradation while others, particularly halogenated or aromatic substances, can be quite resistant to biodegradation. Polarity and solubility are closely related in that polar substances tend to have high solubilities, and vice versa. Highly soluble substances can be more mobile in soils and can also be more accessible to microorganisms for biodegradation. The tendency to adsorb to soils or aquifer materials also greatly influences mobility and accessibility for microorganisms. This characteristic is also related to polarity. Nonpolar substances generally have a high affinity for the naturally occurring organic C in soils and are strongly adsorbed.

Soils and water contaminated with organic substances often involve more than one compound. Therefore, more than one remediation method may be necessary to address all of the contamination.

11.3.1 Soil

Figure 11.8 presents the decision tree for soils contaminated with organic substances. As with the inorganic contaminants, no action can be a conscious decision. If remedial action is to be undertaken, there are *ex situ* and *in situ* options. Solidification, vitrification, washing, leaching, soil removal, and encapsulation for both *ex situ* and *in situ* approaches are similar to that described for the inorganic contaminants.

The remaining *ex situ* options include incineration, aeration, and bioremediation. Incineration is similar to vitrification in that the soil is heated, but incineration is performed with the specific objective of oxidizing the organic chemicals, preferably all the way to CO_2 and

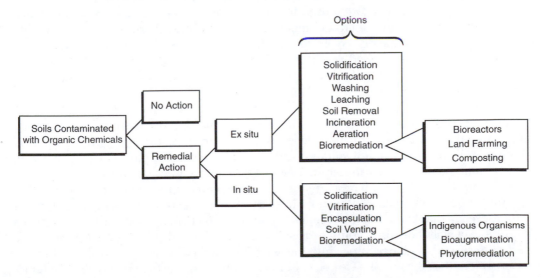

Figure 11.8 Decision tree for soils contaminated by organic chemicals.

H₂O. This approach has been used for soils contaminated with highly toxic organic chemicals such as dioxins. Aeration is used for VOCs and can be performed many different ways. Aeration can be passive as is often used with soils contaminated with gasoline around a leaking underground storage tank. In this situation the tank is removed and the soil is simply exposed to the atmosphere until most of the VOCs have evaporated. Aeration can also be accomplished by forcing air through a volume of soil to facilitate the removal of VOCs.

Bioremediation takes advantage of the ability of organisms to degrade organic chemicals into simpler, and one hopes less toxic, substances. Generally, this is accomplished with microorganisms, although plants have some ability to metabolize organic contaminants. Microorganisms may use the organic compounds as a source of food, since they contain carbon (C), or they may cometabolize the organic contaminants. Cometabolism means that the organisms convert the organic compounds to other substances as a by-product of their normal biochemical processes and are not necessarily using the contaminants as a source of energy.

There are many types of microorganisms. For bioremediation under aerobic conditions, the bacteria and, to a lesser extent, fungi play the largest role in degradation of organic substances. Under anaerobic conditions the role is almost entirely filled by bacteria. Of particular importance is that an organism or consortium of organisms exist that can degrade the organic substance or substances in question. If a soil is contaminated by an organic, the process of natural selection may result in a species or group of species with an enhanced ability to degrade that organic. Experimentally, genetically engineered microorganisms (GEMs) have been developed for the specific purpose of bioremediation. Ideally, one could inoculate the soil with these organisms to facilitate degradation. Since bioremediation requires active growth of microorganisms, it follows that the basic needs of the organisms must be met in order for degradation of organic contaminants to take place as rapidly as possible. This includes appropriate temperature, moisture, pH, nutrient, and redox conditions, as well as a usable source of energy.

The use of bioreactors is one way to ensure that optimal growing conditions are maintained during bioremediation. *Bioreactors* are enclosed vessels into which contaminated soil is placed and microbial growth is promoted for the purpose of bioremediation. Often the soil is kept as a slurry. Nutrient solutions can easily be added and temperature and aeration are easily controlled. Once the desired reduction in contaminant concentrations has been attained, a new batch of soil is introduced into the vessel.

Land farming is a method often used for organic substances such as pesticides, which are normally released into the environment in small amounts. If soil has become highly contaminated with a pesticide, say, from a spill, the soil is spread over land such that the amount of the pesticide applied is within the normal application rates for that pesticide (Figure 11.9). The fate and transport would then be the same as an application of that pesticide according to the label.

Composting can also be used for soils contaminated by organics. The soil alone will not compost, but it will when mixed with compostable organic wastes such as grass, leaves, animal manures, or municipal biosolids. The composting process involves a period of high microbial activity that generates heat and stabilizes the organic C in the material. This high microbial activity can help to degrade certain organic substances.

Soil venting and bioremediation are the remaining choices for *in situ* remediation that have not yet been described. Soil venting (also called vapor extraction) is similar to *ex situ* aeration except that the soil is not removed and a means of forcing air into, or a means of applying a vacuum to, a volume of soil must be devised. Generally, this involves inserting screened or slotted pipes into the soil. This method works for VOCs and sometimes requires that the waste gases be captured and treated, particularly when the remediation is taking place in a residential area.

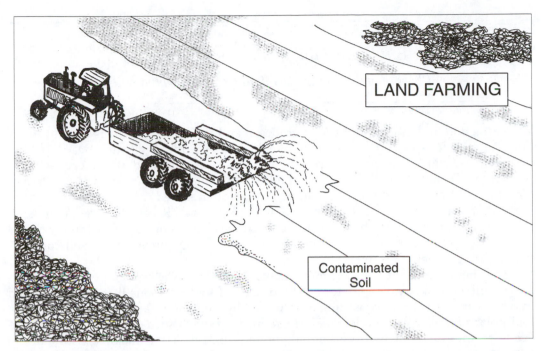

Figure 11.9 Land farming as a bioremediation option for soils contaminated by organic chemicals.

In situ bioremediation can be approached three different ways: use of indigenous soil microorganisms, *bioaugmentation*, and *phytoremediation*. Degradation of some organic contaminants will proceed by the action of indigenous soil microorganisms. Sometimes this process can be hastened by supplementation with nutrients or water to ensure active microbial growth over extended periods of time. Bioaugmentation refers to the inoculation of soil with microorganisms with an enhanced ability to degrade the organic substance in question. As described above, these organisms can be isolated from native populations or be genetically engineered. Once an organism with the desired capabilities has been isolated, the population can be increased so that an inoculum can be prepared for use. Unfortunately, applications of bioaugmentation have had limited success. Generally, when the organisms are placed in their new environment, they tend to die off within a few weeks. The reasons for this are not clear, but may be related to adaptability to the site conditions or to some poorly understood microbial ecology issues.

In situ phytoremediation for organic contaminants is quite different from phytoremediation for inorganic substances. Both involve the use of plants for remediation, but the mechanisms are not the same. Phytoremediation for organic chemicals involves promoting plant growth on contaminated sites with the goal of reducing the contaminant concentration in the soil. The plants accomplish this in several ways. The most important mechanism is likely related to the influence of plant roots on soil microbial activity. There is a small volume of soil surrounding each root that has enhanced microbial activity compared with the bulk soil. This zone is called the *rhizosphere*, or the volume of soil directly influenced by growing plants. The roots exude C-containing materials that the microbes use as sources of energy. In addition, the roots draw water and some nutrients and contaminants toward them. Thus, the rhizosphere is a more favorable environment for microbial growth than the bulk soil and this enhanced microbial activity is thought to hasten the degradation of organic contaminants. The plants themselves can also metabolize some organic contaminants and some VOCs can be moved from the soil to the atmosphere via transpiration.

Environmental quality issues/events
Phytoremediation of petroleum-contaminated soils

The extraction, processing, and use of crude oil and its refined products present many opportunities for soil contamination. Crude oil spills occur at well sites; near pipelines and storage facilities; during transport by rail, truck, or ship; and at refinery facilities. Similar opportunities exist for accidental release of refined products such as gasoline or diesel or jet fuels. In addition, these materials are often stored in underground storage tanks that may leak and contaminate the underlying soil. Recent EPA regulations on underground storage tanks have greatly reduced the chances of leaks from tanks located on commercial properties, but many existing storage sites already have soil contaminated with petroleum products.

The chemical identity of the contaminants depends on the type of petroleum product. Crude oil is a complex mixture of thousands of hydrocarbons that can be divided into three general classes consisting of saturated hydrocarbons, aromatic hydrocarbons, and polar organic compounds. Similarly, refined products such as gasoline have hundreds to thousands of hydrocarbon compounds. Some of the major constituents in gasoline are n-butane, n-pentane, isopentane, benzene, toluene, and xylene. The term *total petroleum hydrocarbons* (TPH) is often used to describe a wide range of petroleum hydrocarbon constituents measurable by a particular type of analysis. When dealing with petroleum-contaminated soils, one can measure the concentration of individual petroleum constituents or the TPH concentration. Many of the petroleum constituents are toxic or carcinogenic for humans and other organisms. The TPH concentration has little to do with risk but can be used as an indicator of the degree of contamination and the success of a remediation process.

Phytoremediation, as a type of bioremediation, is a promising approach for remediation of soil contaminated with petroleum hydrocarbons. Many of the less complex constituents of petroleum products readily undergo bioremediation by soil microorganisms in unvegetated soil. However, there is a significant recalcitrant fraction that is very slow to degrade. The TPH concentrations may decline rapidly for a short time after a petroleum spill onto soil, but eventually will reach a level from which further decreases are unlikely or will be very slow. At this point, the use of plants may help reduce TPH concentrations further. Some of the advantages of phytoremediation are that it may be performed without disturbing the soil, it can be less expensive than the alternatives, it is powered by solar energy, and it has the potential to attain lower contaminant concentrations than other remediation options.

The mechanism by which plants facilitate the degradation of petroleum hydrocarbons in soils is not clearly understood. As was discussed in the text, there are several possible mechanisms. The enhanced growth of microorganisms in the rhizosphere is one such mechanism. Two other mechanisms are the cometabolism of the contaminants by the plants or the loss of volatile compounds via transpiration. Microbial densities in the rhizosphere can be ten times higher than in unvegetated soil. In the case of TPH, cometabolism and volatile losses are not likely, and microbial degradation in the rhizosphere is the most likely mechanism by which petroleum compounds are degraded.

Phytoremediation of soils contaminated with petroleum hydrocarbons has been attempted at several locations. Data from one such site are presented in Table 11.1. The soil from this site was contaminated with diesel fuel and had been placed in a bi600mediation cell. The soil came from a refueling facility for ships located on Craney Island near Norfolk, VA, one of the largest such facilities operated by the U.S. Navy. The bioremediation cell was 0.2 ha in size and was equipped with a leachate collection system. A 45-cm layer of contaminated soil was placed on top of 90 cm of sand. The plots were established in the soil layer and consisted of four treatments with six replications. The treatments were the type of vegetation and consisted of an unvegetated control and plots vegetated with tall fescue (*Festuca arundinaceae*), white clover (*Trifolium repens*), or bermudagrass (*Cynodon dactylon*).

There was some degradation of TPH in the unvegetated treatment over time. After 6 months there was no significant difference in TPH concentration between treatments, but this was not unexpected since the plants were relatively small and the root systems were not well established. After 12 months the fescue treatment had significantly higher degradation compared to the other treatments and by 24 months all vegetated treatments had significantly higher degradation compared with the unvegetated treatment. With further development it is likely that phytoremediation of petroleum-contaminated soils can be a viable and less costly option, both *in situ* and *ex situ*, compared with currently accepted practices. Current estimates are that phytoremediation is two to three times less expensive than bioremediation without plants and ten times less expensive than placing the soil in a landfill.

Table 11.1 Phytoremediation of Petroleum Contaminated Soils from the Craney Island Fuel Facility, Norfolk, VA

Treatment	6 months	12 months	18 months	24 months
Unvegetated	12 a[a]	21 b	17 b	31 c
Clover	11 a	29 ab	34 a	50 a
Fescue	9 a	33 a	34 a	45 ab
Bermuda	13 a	27 ab	30 ab	40 b

(% Petroleum degradation)

[a] Means followed by the same letter in a column are not significantly different using least significant difference (LSD) at $P < 0.05$.

Note: Special thanks to A. Paul Schwab, Purdue University, for supplying the information for this box. The research was funded by the U.S. Department of Defense's Advanced Applied Technology Demonstration Facility (AATDF) housed at Rice University and was conducted in conjunction with industrial cooperators.

11.3.2 Water

Groundwater contaminated by organic chemicals is a serious problem for a number of reasons. The number of sites with known or potential groundwater contamination is large considering the number of NPL sites and the estimated number of leaking underground storage tanks. Once contamination has occurred, it is very difficult to remediate because the contaminated materials may lie a considerable distance below the surface, contaminated groundwater cannot easily be isolated, the contaminant also interacts with the aquifer material, contaminant may be present in the unsaturated zone that has not yet reached the groundwater, and the groundwater is moving, albeit slowly. If one considers the volume of water that can be contaminated by a relatively small volume of an organic chemical, it seems necessary to clean up contaminated sites since a very important resource is at risk. One 208-L drum (55 gal) of trichloroethylene (TCE) could raise the TCE concentration of 60×10^9 L (16×10^9 gal) of water from 0 to 5 µg/L, the MCL for TCE in drinking water. This is calculated as follows. The density of TCE is 1.46 kg/L so 208 L has a mass of

$$208 \text{ L} \times \frac{1.46 \text{ kg}}{\text{L}} = 303.7 \text{ kg} \tag{11.4}$$

$$\text{Contaminated volume} = 303.7 \text{ kg TCE} \times \frac{10^9 \text{ µg}}{\text{kg}} \times \frac{\text{L}}{5 \text{ µg}} = 60.7 \times 10^9 \text{ L} \tag{11.5}$$

If a site is found to have groundwater contaminated with organic chemicals, several important pieces of information are needed to determine the urgency and overall approach to the remediation. As with soil remediation, the identity and chemical properties of the organic chemicals are important. One additional property that is needed is the density of the chemical. For those that are immiscible in water and less dense than water, one would expect the contaminant plume to remain at the top of the aquifer. Conversely, those chemicals that are immiscible in water and more dense than water will attempt to sink to the bottom of the aquifer with the end result being that the substance will likely be dispersed throughout the aquifer. The direction and approximate velocity of groundwater flow are also needed. Obviously, the contaminant plume will follow the direction of flow and the velocity may provide an estimate of how far the plume has traveled. This is particularly important if there are wells using water from the aquifer nearby. Fortunately, groundwater does not move quickly unless there is karst geology. The average pore-water velocity (also known as average linear velocity or seepage velocity) for water in an aquifer can be estimated with

$$v = \frac{K}{\phi} \times \frac{\Delta h}{\Delta l} \tag{11.6}$$

where v = velocity (cm/s), K = saturated hydraulic conductivity (cm/s), ϕ = volumetric porosity (unitless), and $\Delta h/\Delta l$ = hydraulic gradient (unitless). The saturated hydraulic conductivity is a property of the aquifer and characterizes the ease with which water moves through the material. Gravel would have a much higher K value than clay. The hydraulic gradient is the change in elevation at the top of the water table (Δh), relative to some reference value, across a given distance (Δl). As the hydraulic gradient increases, v also increases. Water runs down a steep hill faster than it does down a hill that is not as steep. The potential interaction between the contaminant and the aquifer material may also be considered. If the contaminant adsorbs to the aquifer material, the velocity will be less than that of the water. Often one is interested in how long it will take a contaminant to move from a given point to the nearest well using water from the aquifer. These calculations are generally performed as a worst-case scenario assuming no interaction between the contaminant and the aquifer material and, if ranges of values are given for K, the maximum value is used. For example, calculate the length of time required for a contaminant to move 2 km in groundwater when K is estimated to range from 10^{-4} to 10^{-6} cm/s, ϕ = 0.30, and the hydraulic gradient is 0.14:

$$\text{Time} = 2 \text{ km} \times \frac{10^5 \text{ cm}}{\text{km}} \times \frac{0.30}{0.14 \times 10^{-4} \text{ cm/s}} \times \frac{1 \text{ year}}{31.536 \times 10^6 \text{ s}} = 136 \text{ year} \tag{11.7}$$

or one can calculate the distance moved in 1 year:

$$\text{Distance} = 0.14 \times \frac{10^{-4} \text{ cm/s}}{0.30} \times \frac{31.536 \times 10^{6} \text{ s}}{\text{year}} = \frac{1472 \text{ cm}}{\text{year}} \qquad (11.8)$$

or about 48.3 ft in 1 year or 10^{-6} miles per hour. These examples illustrate that under many conditions groundwater does not move very fast and there is little need to rush into remedial action. There is adequate time for careful planning.

Figure 11.10 represents the decision tree for groundwater contaminated by organic chemicals. As before, no action can be a deliberate decision. If a contaminant plume is small and there are no wells nearby, then a decision to wait may be allowed. Considering the speed at which remediation technologies are being developed or improved, one could certainly make a case for waiting for a more effective or less expensive alternative to be adopted.

Containment is used for small contaminant plumes and can be either hydraulic or physical barriers. A hydraulic barrier uses injection wells downgradient from the contaminant plume to reduce the hydraulic gradient across the site to very low values. The net effect is to reduce the groundwater velocity to near zero, thus containing the contaminants. This can be used as a temporary measure until other alternatives are put into place. The goal of physical containment measures is the same except that the barriers are made of solid materials that must be installed. Slurry walls consist of trenches filled with clay mixtures that have a very low permeability (or saturated hydraulic conductivity). Sheet piling is generally made of steel sheets that are driven into the ground, and grout curtains are installed like slurry walls except the grouting material hardens. Physical barriers are prone to leaking, are difficult to install and maintain, and are poor long-term solutions to groundwater contamination.

Pump-and-treat methods are among the most popular groundwater remediation methods used. The basic process is that a well or wells are placed into or near the contaminant plume. Water is pumped from these wells and treated with a variety of processes to reduce the contaminant concentration. The treated water is often injected back into the aquifer upgradient from the plume to help force the contaminant to the

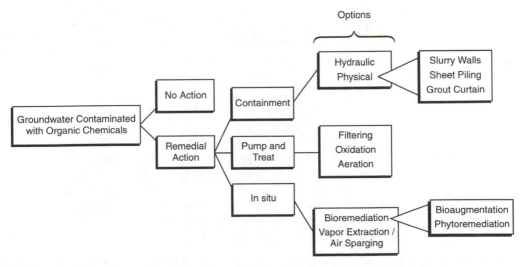

Figure 11.10   Decision tree for groundwater contaminated by organic chemicals.

extraction wells or downgradient from the plume to help form a hydraulic barrier to help contain the plume. The treatment processes can include filtering, oxidation, or aeration. Filtering can be through charcoal or biofilters. Oxidation of organic chemicals can be promoted by the addition of oxygen gas or hydrogen peroxide, combustion, or ultraviolet light. Aeration works well for VOCs and can simply involve allowing the water to fall through an aeration tower. Figure 11.11 shows a schematic of a pump-and-treat process where treated water is injected into the aquifer both up- and down-gradient from the contaminant plume. Both injection wells help force the contaminants to the extraction well.

In situ remediation of groundwater would be the most desirable way to address contaminated groundwater, but it is one of the most technically challenging approaches. Bioremediation can be either as bioaugmentation or phytoremediation. Bioaugmentation involves injecting microorganisms into the aquifer in an attempt to induce microbial breakdown of the contaminants. Unfortunately, aquifers are generally cold and have limited nutrients, O_2, and C compounds to act as energy sources. Even when the missing ingredients are added, the microorganisms have difficulty surviving and performing the desired function. Phytoremediation has been used successfully when groundwater is close to the surface and the contaminant concentrations are not too high. This generally involves using deep-rooted trees that thrive in wet environments, such as poplars or cottonwoods. The environment is often adjacent to streams, rivers, or lakes with the groundwater flowing toward the body of water. The tree roots intercept some of the water and use it for transpiration. In the process they can degrade or volatilize some of the contaminants.

Vapor extraction/air sparging involves forcing air into an aquifer or using vapor extraction wells under a vacuum, or a combination of both, to try and get VOCs out of the water, aquifer material, and vadose zone. The captured vapors can be treated with biofiltration, adsorption onto charcoal, or combustion.

A significant limitation of the pump-and-treat methods and the *in situ* approach is the difficulty in getting contaminant concentrations down to low levels. Typically, as a remediation process begins, the contaminant concentrations in the water decrease rapidly but eventually level off and will not go lower without significant additional effort. If this plateau in concentration is not as low as needed, the remediation may not be considered successful.

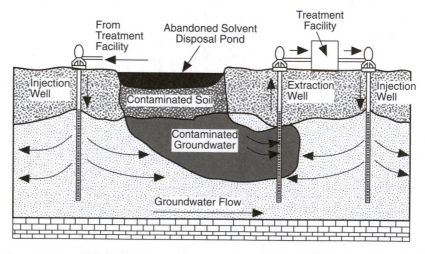

Figure 11.11   Schematic of well placement for a pump-and-treat process for groundwater contaminated by leachate from an abandoned solvent disposal pond. (Drawing by Sarah Blair.)

Problems

11.1 A Zn hyperaccumulator plant has a Zn concentration of 6000 mg/kg and produces 4000 kg biomass/year. Calculate the annual Zn removal rate for the plant species.

11.2 A contaminated site has 500 mg Cu/kg in the upper 30 cm of soil. Calculate the required annual removal rate for Cu if you wanted to use phytoextraction on this site and complete the remediation in 20 years. Assume a soil bulk density of 1.35 g/cm³. If a Cu hyperaccumulator had 1000 mg/kg Cu in the tissue, how much biomass would need to be produced each year to complete the remediation in 20 years?

11.3 Discuss advantages and disadvantages of *in situ* vs. *ex situ* remediation.

11.4 Describe the various manifestations of phytoremediation. How do they differ for inorganic vs. organic contaminants?

11.5 Describe the various manifestations of bioremediation. What are some of the limitations and challenges of this approach?

11.6 The density of vinyl chloride (C_2H_3Cl) is 0.91 kg/L and the MCL for drinking water is 2 μg/L. Calculate the volume of water that could be contaminated by 1 L of C_2H_3Cl if the concentration of C_2H_3Cl in the water was raised from 0 to 2 μg/L. Repeat the calculation for ethylbenzene (C_8H_{10}) with a density of 0.87 kg/L and a MCL of 70 μg/L.

11.7 The density of carbon tetrachloride (CCl_4) is 1.59 kg/L and the MCL for drinking water is 5 μg/L. Calculate the volume of CCl_4 required to raise the CCl_4 concentration of 1 L of water by 5 μg/L.

11.8 An aquifer has a hydraulic conductivity of 10^{-5} cm/s, a porosity of 0.35, and a hydraulic gradient of 0.05. Calculate the length of time (in years) required for groundwater to move 1 km.

11.9 An aquifer has a hydraulic conductivity of $10^{-3.5}$ cm/s, a porosity of 0.40, and a hydraulic gradient of 0.019. Calculate the average pore-water velocity in centimeters per second and miles per hour. Determine how far the groundwater travels in 1 year (in cm/year).

11.10 Calculate the mass of soil in the bioremediation cell described in the environmental quality issues/events box assuming a bulk density of 1.20 g/cm³. Calculate the cost of remediating the soil with phytoremediation at $15/Mg (metric ton or 1000 kg), with bioremediation at $45/Mg, and with landfilling at $150/Mg.

References

Pierzynski, G. M. and A. P. Schwab. 1993. Bioavailability of zinc, cadmium, and lead in a metal-contaminated alluvial soil, *J. Environ. Qual.*, 22, 247–254.

U.S. Environmental Protection Agency. 1992. The Superfund Innovative Technology Evaluation Program: Technology Profiles, 5th ed., EPA/S40/R-92/077, U.S. Government Printing Office, Washington, D.C.

Supplementary reading

Deuel, L. E., Jr. and G. H. Holliday. 1997. *Soil Remediation for the Petroleum Extraction Industry*, PennWell, Tulsa, OK, 242 pp.

Eweis, J. B., S. J. Ergas, D. P. Y. Chang, and E. D. Schroeder. 1998. *Bioremediation Principles*, McGraw-Hill, New York, 296 pp.

Iskandar, I. K. and D. C. Adriano, Eds. 1997. *Remediation of Soils Contaminated with Metals*, Science Reviews, Middlesex, U.K., 255 pp.

Kuo, J. 1999. *Practical Design Calculations for Groundwater and Soil Remediation*, CRC Press, Boca Raton, FL, 263 pp.

Pepper, I. L., C. P. Gerba, and M. L. Brusseau, Eds. 1996. *Pollution Science*, Academic Press, San Diego, CA, 397 pp.

Riser-Roberts, E. 1998. *Remediation of Petroleum Contaminated Soils*, CRC Press, Boca Raton, FL, 542 pp.

chapter twelve

Risk assessment

Contents

12.1 Introduction

Risk is defined as the probability or chance of injury, loss, or damage. This is the most general definition of risk that includes situations ranging from financial losses, to storm damage to buildings, to human health effects from exposure to pollutants. This chapter summarizes the basic concepts used in risk assessment and illustrates how the risk assessment process is used in writing environmental regulations with an emphasis on contaminated soils.

Our definition of risk includes effects on any organism, although the tendency is to focus the discussion on human health effects. Morbidity and mortality, both immediate and delayed, are the two general effects of concern. Risk can be quantified as the number of negative outcomes (injury, loss, or damage) divided by the number of organisms exposed to the risk. Some risks are very easy to quantify. For example, the risk from dying in an automobile is relatively easy to determine. The negative outcome is generally immediate and the cause is not in question. The exposed population is also relatively easy to identify. It would be reasonable to assume all people are exposed to the risk since nearly everyone drives or rides in automobiles. One could further refine our definition of the exposed population by normalizing for the number of miles driven. Other risks are extremely difficult to quantify because neither the negative outcome nor the exposed population is easy to identify and quantify. Exposure to potentially carcinogenic substances would be a good example. The negative outcome may not occur for many years after exposure and all cases of a given type of cancer cannot be related to a single causative factor.

Table 12.1 provides risk information for a number of causes of death for all age groups in the U.S. For the ten leading causes of death, the annual individual risk was calculated by dividing the number of deaths for each cause in 1995 by the U.S. population at that time (approximately 263,000,000). This assumes the entire population is exposed to each risk which, of course, is not always true. In other words, the *exposed population* is a subset of the *total population*. Not everyone drives or rides in automobiles and even fewer travel by commercial aviation, and these individuals are not exposed to those risks. The assumption is more valid for causes such as unintentional injuries or diseases such as diabetes or asthma where everyone is exposed to the risk to some extent. Nevertheless, such simple calculations allow people to appreciate the relative magnitude of the risk of death by various causes. People can look at the numbers in Table 12.1 and make lifestyle changes that reduce their risk to those causes of death. Of course, the numbers would look different if we considered individual age groups. Unintentional injuries are the leading cause of death for people from 1 to 34 years old, as one might expect for this relatively healthy age group, but behavior early in life can enhance your chances of death from a number of causes later in life.

The general public can make a number of responses to risks. The risk can be avoided or eliminated, as has been the case for chlorofluorocarbons (CFCs) used as propellants in aerosol cans or with the pesticide DDT for which substitutes are available. The cause of the risk can be regulated or modified to reduce the frequency or magnitude of the negative outcome. For example, we could build flood-control structures or limit the amount of a substance, e.g., nitrogen (N), phosphorus (P), or trace elements, that could be applied to soils. The vulnerability of the exposed population can be reduced. In this case, the cause

Table 12.1 Individual Risk from Various Causes of Death in the U.S.

Cause of death	Annual individual risk	Ratio[a]
Ten leading causes of death[b]		
Heart disease	2.8×10^{-3}	357
Cancer	2.1×10^{-3}	476
Cerebrovascular	6.0×10^{-4}	1,667
Bronchitis, emphysema, asthma	3.9×10^{-4}	2,564
Accidents and injuries	3.5×10^{-4}	2,857
Pneumonia and influenza	3.2×10^{-4}	3,125
Diabetes	2.3×10^{-4}	4,348
HIV	1.6×10^{-4}	6,250
Suicide	1.2×10^{-4}	8,333
Liver disease	9.6×10^{-5}	10,417
Other causes of death		
Motor vehicle accidents	1.6×10^{-4}	6,250
General aviation[c]	2.9×10^{-6}	344,800
Commercial aviation[c]	5×10^{-7}	2,000,000
Lightning	3.7×10^{-7}	2,702,000

[a] One death per number of people in population.

[b] Based on 1995 data.

[c] For the 10-year period ending in 1997.

Source: Data from the Centers for Disease Control and National Transportation Safety Board, 1999.

of the risk is not changed but the people potentially affected by it may receive advance warning and be able to reduce their losses. Several postevent strategies are often used. Better ambulance service can be implemented in response to traffic accidents, thus increasing the chance for surviving an accident; however, the chance of being in an accident does not change. Insurance also provides financial reimbursement for financial losses.

Risk assessment is a process by which we attempt to determine the probability and magnitude of injury, loss, or damage that may result from a potential health hazard. *Risk management* is a process by which economic, political, legal, and ethical ramifications of the results of risk assessment are considered. Regulatory decision making is based on both risk assessment and risk management. Environmental issues can be very emotional because they potentially involve human health and economic impacts. Overall, risk assessment provides a scientific basis for the creation of environmental regulations, although there is no guarantee that risk assessment results will actually be used in the regulatory decision-making process.

There are a number of philosophical issues that need to be presented in relation to quantifying risk and the risk assessment process. The first is the concept of *negligible risk* vs. *absolute* or *zero risk*. Negligible risk implies that there is some very small but nonzero chance that the negative outcome will occur, whereas absolute or zero risk implies that there is no chance the negative outcome will occur. Some use the phrase *acceptable level of risk* for something that has negligible risk. The regulation of cancer-causing substances is an excellent example. Exposure to some potentially carcinogenic substances is regulated such that the increased risk of cancer due to the exposure is on the order of 10^{-5} to 10^{-6}, meaning that 1 out of 100,000 or 1 out of 1,000,000 people exposed might develop cancer. Compared with the risk values in Table 12.1, these risk levels are quite small and would be difficult to detect statistically compared with background cancer rates (about 0.03 over a lifetime). This level of risk is often called *de minimis* risk, or a risk too small to be of concern to society. Alternatively, some might say that in a city of 1 million there would be an increase of 1 to 10 preventable cancer cases. Now if the exposure to the potentially carcinogenic substance resulted in some benefits for society, or would cost a great deal of money to prevent, some might consider 10^{-5} to 10^{-6} an acceptable level of risk and the resulting regulations would be considered a negligible risk standard. If regulations were such that no exposure was the only acceptable scenario, the regulations would be considered an absolute or zero risk standard.

Generally, regulations use the negligible risk approach. One notable exception was the Delaney clause, which used an absolute risk standard for pesticide residues in processed food. Legislation such as the Delaney clause was drafted in times when detection limits for determining the concentrations of potential carcinogens or other contaminants were quite high, on the order of milligrams (10^{-3} g), with concentrations below that considered to be zero. Today, concentrations are routinely measured at the nanogram (10^{-9} g) level, so what used to be considered zero might now easily be quantified. As detection limits get lower, it becomes more difficult to rationalize the use of absolute risk standards since they cannot be obtained. Thus, the Delaney clause was recently replaced with a negligible risk standard in the Food Quality Protection Act (1996), which has tolerance levels for pesticide residues.

A *cost–benefit approach* is often used for environmental issues. A chemical company cannot obtain a label for a new pesticide unless it proves that the pesticide has some benefits compared with pesticides currently on the market and that the environmental costs and risks are acceptable (see Chapter 8, environmental quality issues/events: Process involved in the registration of a new pesticide, p. 277). Alternatively, one can consider the cost of cleaning up a contaminated site vs. the benefits of preventing human health problems or ecological damage. Eventually, a point is reached where additional money spent on cleanup yields little additional benefits, even though the site may not be as clean as some would like it to be.

One must also be cautioned against becoming too comfortable with the risk numbers and concluding that any risk at the 10^{-5} to 10^{-6} level or less is acceptable because of the comparison with higher risks associated with other activities. In general, we must strive to keep risk levels low and avoid unnecessary exposure to potentially harmful substances as a safety margin. For example, scientists are just beginning to appreciate the toxicological effects of exposure to chemical mixtures. In some cases exposure to two substances at low levels can have a greater effect than exposure to either substance individually.

There are two major reasons we would be concerned with risk assessment for soil contaminants. First, if the concentration of a substance in soil is deliberately increased, will any organism experience an unacceptable increase in risk? Second, what increased risks are realized by organisms because of soil contamination that has already occurred? Recall from Figure 1.1 that there are both direct and indirect ways for organisms to be exposed to substances in soils that can cause harm. Therefore, an increase in risk can be realized from the ingestion of soil itself (e.g., by children or grazing livestock) or indirectly from the consumption of groundwater or surface water contaminated by a substance present in soil or from the consumption of crops grown in, or livestock exposed to, the contaminated soil.

12.2 Risk perception

A discussion of risk assessment would not be complete without consideration of the general public's perception of risk as compared with that of the scientific community. As you might suspect, the perception of risk by these two groups can be quite different and their response may also be contradictory. Surveys have revealed that people generally feel that they face more risks today than people faced in the past. This is despite the fact that average life expectancy has increased in recent times. Let us first consider risk perception by the general public.

It is well established that laypeople overestimate the frequency of rare causes of death while underestimating the frequency of more common causes of death, as illustrated in Table 12.2. Far fewer people die as a result of tornadoes, floods, pregnancy, or botulism than are estimated to die by laypeople. The actual number of deaths per year in the U.S. from any of these causes is less than 1000. Likewise, far more people die from asthma, stomach cancer, diabetes, and strokes than were estimated. More than 100,000 people die from strokes per year, for example. This information indicates that familiarity with the cause of death induces a bias in perception of the risk. A death as a result of botulism is quite unusual and would likely be reported by the news media. Since people are not familiar with botulism as a cause of death, the noteworthy status of the death would raise their consciousness of the cause and could make them perceive that botulism is more prevalent than the number of cases reported. A death as a result of a stroke would not be

Table 12.2   Over- and Underestimated
Frequency of Death as Judged by Laypeople

Overestimated	Underestimated
Tornadoes	Asthma
Floods	Stomach cancer
Pregnancy	Diabetes
Botulism	Stroke

Source: Slovik, P. et al., *Environment*, 21, 14, 1979.

Table 12.3 Factors Associated with Risk Perception by the General Public

Risk perception factor	Subfactors for high or low risk perception	
	High	Low
Dread	Uncontrollable	Controllable
	Globally catastrophic	Not globally catastrophic
	Fatal consequences	Consequences not fatal
	Not equitable	Equitable
	Involuntary	Voluntary
Knowledge	Effects not observable	Observable effects
	Unknown to those exposed	Known to those exposed
	Effect delayed	Effects immediate
	New risk	Old risk

Source: Slovik, P., *Science*, 236, 280, 1987.

a newsworthy item and, over time, the awareness of the magnitude of this cause of death would diminish.

Other factors that influence risk perception by the general public are outlined in Table 12.3. The two primary factors involved in risk perception by the general public are *dread* and *knowledge* (whether the risk is known or unknown). Dread is an intense fear that something bad might happen. Within each major factor are subfactors that can enhance or reduce our perception of risk. An accident or activity that invokes dread and is relatively unknown will be perceived as the most risky, whereas an accident or activity that does not invoke dread, and is familiar, will be perceived as the least risky. This has led to the observation that people tend to have the greatest fear of the least likely things. Obviously, many accidents or activities fall somewhere in the middle in that they may invoke dread but are familiar to us, or vice versa. Whether the risk is undertaken voluntarily or not is also an important consideration. A person smoking a cigarette is at much greater risk from tobacco smoke than nonsmokers subjected to passive smoke, yet the nonsmokers will have a greater objection to the risk than does the smoker.

Control is also a major subfactor. When we feel we have control of an activity our perception of the risk is much lower. Driving an automobile is a good example. If you needed to travel 2000 km you could roughly estimate from Table 12.1 that you would be 320 times safer flying to your destination than driving. Many people would feel safer driving, however, because they are in control of the automobile (and probably think they are good drivers). When you fly, nearly all aspects of the trip are out of your control. In addition, the thought of an airplane crash invokes dread as often there are no survivors. In the period from 1989 to 1998, an average of approximately 130 people per year were killed in commercial aviation accidents in the U.S. while over 40,000 were killed each year in automobile accidents. Despite these figures, many people are still not convinced that they should wear their seatbelts while driving. The classification scheme shown in Table 12.3 identifies many of the important considerations for how risk is perceived by laypeople and allows one to predict qualitatively how the public will perceive a risk.

An additional factor that helps explain variations in risk perception by the public over time is that reports of accidents or events serve as signals. This simply means that the public's awareness of a risk, however small that risk may be, is greatly increased by a major event or accident. The Three Mile Island and Chernobyl nuclear accidents made the public's perception of risks from nuclear energy increase dramatically and have played a major role in slowing or stopping the use of nuclear energy in the U.S. In the future, we may have to rely more on nuclear energy as a means of addressing global climate change from greenhouse gases. Concerns over contamination of Times Beach, MO by dioxin heightened the public's awareness of the dangers of this organic chemical. The crash of a commercial airliner is certainly going to get media coverage and will at least cause a temporary reduction in the number of people willing to fly; yet the total number of people killed in automobile accidents goes relatively unnoticed in society. Technological advances present a dichotomy of sorts for risk perception. Such advances can increase the perceived risk because they are not understood by the general public while at the same time the public often relies upon technological advances as a salvation from risk.

The public's perception of risk places regulatory agencies in somewhat of a quandary. While they are public agencies designed to respond to public needs and concerns, they cannot justify the allocation of scarce resources to problems that are perceived to be much worse than reality.

12.3 Carcinogenicity

The risk assessment procedures that are followed are influenced by whether or not the substance has been shown to be carcinogenic. The EPA has a classification scheme for carcinogenicity based on human and animal evidence. This scheme is outlined in Table 12.4. Substances in Group A are known human carcinogens (e.g., radon, vinyl chloride), Group B refers to probable human carcinogens, Group C refers to a possible human carcinogen, Group D refers to something that is unclassified because of inadequate data, and Group E refers to a substance with evidence of noncarcinogenicity. For regulatory purposes Groups A and B are collectively called Category I, Group C is called Category II, and Groups D and E are referred to as Category III.

As an example of how the classification may influence the regulatory process one can use the maximum contaminant level goals (MCLG) for drinking water. The MCLG is the desired maximum concentration for a substance in drinking water, which considers all its potential harmful effects. The MCLG for any Category I substance is zero while those for Category II and III substances are nonzero values calculated in various ways. A concentration of zero is not obtainable and the actual drinking water standard is called the maximum contaminant level (MCL). The regulations require that the MCL be as close to the MCLG as possible with the best available technology for removing that substance from the water. Obviously, this process tends to force the MCL to be lower for Category I substances than for Category II or III substances.

Cancer death rates in the U.S. have received some attention in recent years because the overall rate has been increasing for some time. According to the American Cancer Society, the cancer rate (number of cancer deaths per 100,000 people) increased by 7% for men and 8% for women when comparing rates for 1971 to 1973 to those in 1991 to 1993

Table 12.4 Carcinogenic Categorization of Substances Based on Animal and Human Data

Human evidence	Animal evidence				
	Sufficient	Limited	Inadequate	No data	No evidence
Sufficient	A	A	A	A	A
Limited	B1	B1	B1	B1	B1
Inadequate	B2	C	D	D	D
No Data	B2	C	D	D	E
No Evidence	B2	C	D	D	E

Note: Group A substances are known human carcinogens; Group B1 are probable human carcinogens based on limited evidence of carcinogenicity from human epidemiological studies; Group B2 are probable human carcinogens based on sufficient evidence of carcinogenicity from animal studies; Group C are possible human carcinogens; Group D are not classified due to inadequate data; and Group E have evidence of noncarcinogenicity.

Table 12.5 Change in Cancer Death Rates (per 100,000 population) When
Comparing 1971–1973 to 1991–1993

Cancer site	Sex	Rates 1971–1973	Rates 1991–1993	Percent change
All sites	Male	204.5	219.0	7
	Female	132.0	142.0	8
Cervix	Female	5.6	2.9	−48
Hodgkin's disease	Male	1.9	0.9	−63
	Female	1.1	0.4	−64
Kidney	Male	4.3	5.1	19
	Female	1.9	2.3	21
Larynx	Male	2.8	2.5	−11
	Female	0.3	0.5	67
Liver	Male	3.4	4.6	35
	Female	1.8	2.1	17
Lung	Male	61.3	73.5	20
	Female	12.7	32.9	159
Melanoma	Male	2.0	3.2	60
	Female	1.3	1.5	15
Non-Hodgkin's lymphoma	Male	5.8	8.1	40
	Female	3.9	5.3	36
Prostate	Male	21.4	26.8	25
Stomach	Male	10.4	6.6	−37
	Female	5.0	3.0	−40
Testicular	Male	0.7	0.2	−71

Source: American Cancer Society, 1999.

(Table 12.5). Some believe that increasing cancer death rates are a result of exposure to anthropogenic chemicals in our environment and are a signal that more restrictive environmental regulations are needed. While one cannot say for certain that our environment is responsible for the increasing cancer rates, evidence suggests there are other factors that at least partly explain the changes. One is that the average age of the population is increasing. Cancer death rates increase with age so the overall increase in cancer rates is partially a reflection of an older population. There are also some types of cancer associated with smoking that have increased because of increasing numbers of people who have smoked for long periods of time and because the number of women who smoke has dramatically increased. The rates for lung and larynx cancers for women increased 159 and 67%, respectively, which impacts the overall rate. And, of course, cancer treatments have made tremendous advances in the last 20 years. Still, the death rates for some types of cancer have increased without explanation, and therefore environmental factors cannot be ruled out. Not all the news is bad, as more recent data suggest that cancer death rates began to decline after 1992.

12.4 Risk assessment

Risk assessments are complicated endeavors that span many areas of expertise and invariably involve the use of assumptions and seemingly arbitrary safety factors that are sources of criticism. Risk assessments can be for human health, ecological risk, or a combination of the two. Several recent examples on a national scale include risk assessments for emissions from incinerators, use of cement kiln dust as an agricultural liming agent, and land application of municipal biosolids. The latter was mentioned previously and will be discussed in more detail in Section 12.4.1. Risk assessments are also often performed for various contaminated sites and are more local in scope. The risk assessment for emissions from incinerators, for example, had to consider inorganic and organic contaminants and was required to take into account dispersion of contaminants into the atmosphere; potential human health and ecological risks from direct inhalation of contaminants; wet and dry deposition of contaminants onto soil, water, and plants; uptake of contaminants by plants from soil or by direct deposition; potential human health and ecological risks from contaminants in plants; the influence of contaminants on aquatic organisms; and potential human health and ecological risks from contaminants in the aquatic organisms.

The overall goal of a risk assessment is to consider existing or possible contamination of soil, air, water, or sediments; trace all possible routes of exposure for organisms of concern; determine the dose received by organisms of concern; and determine the potential negative impacts of that dose on the organisms of concern. The general process of risk assessment consists of one or more of the following steps: *hazard identification, exposure assessment, dose–response assessment*, and *risk characterization*.

Hazard identification is a qualitative assessment of a substance which determines that exposure to a specific substance causes harm. One way for a substance to be found harmful would be through simple epidemiological studies in which the frequency of disease was found to be higher in an exposed group compared with an unexposed group. Under the Toxic Substances Control Act (TSCA), any new substance is considered to be a hazard until proven otherwise. The combination of exposure assessment, dose–response assessment, and risk characterization are sometimes referred to as quantitative risk assessment.

12.4.1 Exposure assessment

Exposure assessment is the process by which the identity of the organisms exposed to a contaminant and the dose received are determined. All possible means of exposure to a contaminant and the relative contribution of each route of exposure to the dose of the recipient are investigated. The *dose* is the amount of the contaminant ingested or inhaled. One needs to consider how the organism can receive a dose as a starting point. Humans and animals, for example, can be exposed via inhalation, dermal exposure, or by ingestion (Figure 1.1). Therefore, air quality, chance of dermal contact, and the possibility of the pollutant entering the digestive tract must be considered in exposure assessment. Plants have analogous pathways for exposure. Plants extract nutrients, contaminants, and water from soils, and are sensitive to changes in soil composition. Plants respire and photosynthesize and therefore respond to changes in air quality. Substances can also be absorbed on the waxy surfaces of leaves, similar to dermal exposure in humans, and dust fall and quality of soil and water contacting the leaves must be considered.

An excellent example of exposure assessment relative to potential soil contamination is that done for the biosolids disposal regulations approved in December 1992. These regulations have far-ranging implications. Of interest to this discussion are the limits for the total amount of trace elements that can be applied to soils via land application of municipal biosolids (see Table 9.6). Table 12.6 lists the exposure pathways considered in the risk assessment process for these regulations. While these pathways were used specifically to consider land application of municipal biosolids, in a general sense they could be appropriate for anything contained in or applied to soils (e.g., pesticides, animal manures, fertilizers). The number or types of pathways would vary, of course, depending on the fate and transport mechanisms of the pollutant. Each pathway has an organism as an endpoint and the potential negative effects on the organism are the concern.

Table 12.6 Pathway Models for Land Application of Municipal Biosolids

Pathway	Description of HEI
1: Biosolids → soil → plant → human	Individuals with 2.5% of all food produced on amended soils
2: Biosolids → soil → plant → human	Home gardeners with 1000 Mg/ha, 60% garden foods for lifetime
3: Biosolids → soil → human child	Ingested biosolids product, 200 mg/day
4: Biosolids → soil → plant → animal → human	Farms, 45% of home-produced meat
5: Biosolids → soil → animal → human	Farms, 45% of home-produced meat
6: Biosolids → soil → plant → animal	Livestock feeds, 100% on amended land
7: Biosolids → soil → animal	Grazing livestock, 1.5% biosolids in diet
8: Biosolids → soil → plant	Phytotoxicity, strong acidic amended soil but with limestone added to prevent natural aluminum and manganese toxicity
9: Biosolids → soil → soil biota	Earthworms, microbes, in amended soil
10: Biosolids → soil → soil biota → predator	Shrews (*Sorex araneus* L.), 33% earthworms in diet, living on site
11: Biosolids → soil → airborne dust → human	Tractor operator
12: Biosolids → soil → surface water → human	Subsistence fishers
13: Biosolids → soil → air → human	Farm households
14: Biosolids → soil → groundwater → human	Well water on farms, 100% of supply

Source: Chaney, R. L. et al., Soil root interface: ecosystem health and human-food-chain protection, in P. H. Huang et al., Eds., *Soil Chemistry and Ecosystem Health*, SSSA Spec. Pub. No. 52, Soil Science Society of America, Madison, WI. 1998.

One of the steps in exposure assessment for the biosolids regulations was to identify the most sensitive pathway for each trace element. The most sensitive pathway refers to the pathway by which an adverse effect could occur at the lowest soil contaminant concentration. Examples include pathway 8 for boron (B), zinc (Zn), copper (Cu), and nickel (Ni) where phytotoxicities limit plant growth before negative effects on other organisms are thought to occur; pathway 3 for lead (Pb) and fluorine (F) where direct consumption of soil causes the greatest problems; pathway 6 for molybdenum (Mo) and selenium (Se) where ruminant animals suffer molybdenosis or selenosis, respectively, because of plant uptake of these elements that occurs without the plants themselves experiencing phytotoxicities or before other organisms are at risk; and pathway 1 for cadmium (Cd) where food chain transfer to humans is the greatest concern. The fate and transport mechanisms for these trace elements in soils clearly play a role in determining which pathway ends up being the most sensitive. For example, Pb is strongly sorbed by soils to the extent that plant uptake and, consequently, food chain transfer to humans or animals is of little importance, whereas direct consumption of soil is of great concern. Cadmium, on the other hand, is readily taken up by plants and is easily moved from the soil into the food chain.

In conjunction with the determination of the most sensitive pathway is the definition of highly exposed individuals (HEIs), those persons with exposure greater than that of 95% of the population and with the greatest likelihood of suffering the greatest harm at the lowest dose of the pollutant. The HEI represents a subdivision of a group of organisms. The basis for the division might be gender, age, cigarette smoking, dietary habits, sources of food, or the area where a person resides. The HEIs for each pathway used for the biosolids disposal regulations are also given in Table 12.6. Other examples of HEIs include infants, for NO_3^- in drinking water, or predatory birds or people whose diet is composed mainly of fish, for DDT. The premise behind the HEI concept is that if regulations protect the most exposed and sensitive individuals in the most sensitive pathway, then all other individuals will be protected as well. In practice, an HEI is not used for all risk assessment calculations. Where the HEI is used, however, the selection of an appropriate HEI is important. It is easy to construct a scenario for an HEI that is so restrictive that there are no individuals within the population who fit the full description of the HEI. When this occurs, the calculations are for a risk that cannot exist and the results are overly protective.

To determine dietary exposure for humans, an accurate determination of what is consumed by people must be made. This is accomplished with a total diet study. The total diet study incorporates a survey of consumers that determines how much of various food groups are consumed by various gender and age categories of the population. If the composition of one or more of the food groups changes, then the effect of that change on the population can be estimated. Data are also available on average water consumption by adults and children. A value of 2 L/day is often assumed for adults, although this is likely an overestimate for most individuals. Overestimating factors such as water or food consumption is another way to build safety into the risk assessment process. Soil ingestion by children has also been extensively studied. Most children between the ages of 1 and 6 will consume less than 0.03 g of soil/day with the contribution from indoor dust vs. outside soil ranging from 0 to 100%. Some children with pica (habitual ingestion of nonfood items), on the other hand, have been shown to ingest in excess of 8 g of soil/day averaged over 10 days of measurement.

Ecological risk assessments present unique challenges for exposure assessment primarily because knowledge is lacking on the behavior and feeding habits of many species of wildlife. In addition, there are many possible species to consider and a complete risk assessment for each is not possible. Representative or indicator species may be selected to simplify the process. For example, exposure assessment for birds may be accomplished by using a representative species from those that eat seeds, insects, fish, or mammals. In risk assessments for contaminated sites, animals may be trapped and their exposure assessed by tissue analysis. Food chain modeling may also be employed. An example of a commonly modeled pathway would be seeds, mouse, and predatory birds where the concentration of the contaminant is followed up the food chain. For lipophilic substances such as DDT, polychlorinated biphenyls (PCBs), or methyl mercury ($Hg(CH_3)_2$) tissue concentrations are biomagnified between trophic levels with predators at the greatest risk.

12.4.2 Dose–response assessment

In a general sense, dose–response assessment establishes the relationship between the amount of a substance that an organism receives (the dose) and the effects on that organism (the response). Responses can be favorable or unfavorable. For the risk assessment process, dose–response assessment may take information from the exposure assessment and determine the effects on the exposed organisms.

The dose–response curve forms the conceptual basis for dose–response assessment. These curves can have several general shapes, as shown in Figure 12.1. Figure 12.1a indicates no response to the dose, Figure 12.1b is a linear response, Figure 12.1c is a threshold response, and Figure 12.1d is an asymptotic response. Note that the curves do not have to pass through the origin, indicating that the response may occur in the population without exposure to the substance in question. In other words, other factors may induce the response in addition to the substance being studied.

The type of response that is measured can vary considerably. Often, the type of response seems quite obvious for a risk assessment, such as an increase in cancer incidence as the dose increases or mortality rates for insects exposed to increasing doses of a contaminant. The responses may not always be as obvious, however, as sometimes the response is indirect. Changes in blood chemistry or tissue composition may be the

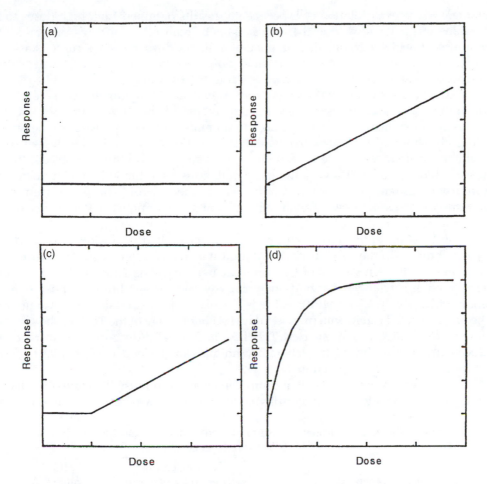

Figure 12.1 Four general shapes for dose–response curves. Curve (a) reflects no response, curve (b) is a linear response, curve (c) is a threshold response, and curve (d) is an asymptotic response.

indicator used for a particular contaminant for animals or humans rather than an actual disease or set of clinical symptoms. In ecological risk assessments the response may be changes in species diversity, density, or productivity.

A value that is sometimes calculated from dose–response experiments is the *no observable adverse effect level* (NOAEL). The NOAEL is the highest dose that can be administered without a statistically significant increase in adverse response. A linear dose–response curve, for example, predicts an increase in response with each incremental increase in dose, no matter how small the increment. The increase in response when the dose is greater than zero, however, must become large enough to be statistically significant. The NOAEL is a measure of how much the dose must be increased before a response occurs or before a threshold is surpassed. The NOAEL value is used in some risk assessment calculations.

As precise as the definition for the NOAEL is, few dose–response experiments actually use doses near the NOAEL value such that the NOAEL can be determined with a great deal of confidence. In fact, the doses used in animal experiments are typically much higher than humans are expected to receive. The primary reason for this is that the increase in response seen at the low doses expected in humans is so small that it would take an enormous number of laboratory animals exposed to the low doses over extended periods of time to obtain statistically significant results. Thus, a major problem in dose–response assessment is the extrapolation of experimental results obtained from high doses to laboratory animals to the low doses that will actually be realized by humans. Several models may be used for this extrapolation. In general, a linear dose–response curve is assumed for carcinogenic substances and a threshold dose–response curve is assumed for noncarcinogenic substances, unless evidence exists to the contrary.

The extrapolation of results obtained from animal studies to humans is a second major problem in dose–response assessment. It is an accepted belief that substances known to cause cancer in humans will cause cancer in animals, but the opposite has never been conclusively shown. There are obvious ethical questions involved with dose–response experiments, primarily because we cannot perform studies with human subjects. In addition, epidemiological studies with human populations have limited value in identifying quantitative dose–response relationships. Thus, the use of laboratory animals for dose–response assessment, with all of the apparent shortcomings, is a necessary component of risk assessment.

One interesting approach to extrapolating from high dose to low dose and from animals to humans is the Ames test. The Ames test is named after its creator, Bruce Ames, a biochemist at the University of California at Berkeley. The Ames test is a means of ranking possible carcinogenic hazards in humans by calculating a human exposure/rodent potency (HERP) value. The HERP value is calculated by expressing the average daily intake of a suspected carcinogen (mg/kg/day) as a percentage of the TD_{50} for the substance in the most sensitive rodent species. The TD_{50} is the daily dose of the substance that produced tumors in 50% of the laboratory animals at the end of a standard lifetime. Example HERP values are given in Table 12.7.

The Ames test has been somewhat controversial and is not used in regulatory decision making. It does provide some interesting comparisons, however. The relative risk from

Table 12.7   Human Exposure/Rodent Potency Index Values
for Selected Substances

Daily human exposure	Carcinogen dose (per 70 kg person)	HERP value (%)
Chlorinated tap water, 1 L	Chloroform, 83 µg	0.001
Contaminated well water, 1 L	Trichloroethylene, 2800 µg	0.004
Chlorinated pool water, 1 h (child)	Chloroform, 250 µg	0.008
Conventional home air (14 h/day)	Formaldehyde, 598 µg	0.6
PCBs, daily dietary intake	PCBs, 0.2 µg	0.0002
DDE, DDT, daily dietary intake	DDE, 2.2 µg	0.0003
EDB, daily dietary intake	EDB, 0.42 µg	0.0004
Bacon, cooked (100 g)	Dimethylnitrosamine, 0.3 µg	0.003
Peanut butter, one sandwich (32 g)	Aflatoxin, 64 ng	0.03
Mushroom, one raw (15 g)	Mixture of hydrazines	0.1
Beer, 345 ml (12 oz)	Ethyl alcohol, 18 mL	2.8
Wine, 250 ml (8.5 oz)	Ethyl alcohol, 30 mL	4.7
Phenobarbital, one sleeping pill	Phenobarbital, 60 mg	16
Clofibrate (average daily dose, for reducing cholesterol)	Clofibrate, 2000 mg	17
Formaldehyde, worker's average daily intake	Formaldehyde, 6.1 mg	5.8
EDB, worker's average daily intake (high exposure)	EDB, 150 mg	140

DDE = dichlorodiphenyldichloroethylene.

Source: Data from Ames et al. (1987).

some environmental pollutants, e.g., trichloroethylene (TCE), PCBs, or pesticide residues, e.g., DDT, ethylene dibromide (EDB), is less than that from certain pharmaceuticals or from natural carcinogens that are in our food and drink. We also see that the HERP values for two prescription drugs that are commonly taken over extended periods of time are relatively high as are the values for a worker's exposure to formaldehyde or EDB. From the discussion of risk perception, it is apparent that the perceived risk from dietary exposure to PCBs, DDT, or EDB would be much greater than that for the consumption of peanut butter, chlorinated tap water, or alcoholic beverages, while the Ames test would indicate otherwise for the real risks based on carcinogenicity.

12.4.3 Risk characterization

Risk characterization combines the results from exposure assessment and dose–response assessment for determining the management practices or cleanup procedures that produce acceptable exposures to the receptor organisms. It is often a back-calculation, starting with an acceptable exposure to an organism and working back to the management practices or media concentrations (e.g., levels in soil or water) that produce the maximum acceptable exposures. Such calculations may form the basis for regulatory action.

A value that is often used in risk characterization is the reference dose (RfD). The RfD is the daily intake of a chemical which, if taken during an entire lifetime, will be without appreciable risk. The units of the RfD are typically mg/kg body weight/day. The RfD can be calculated as the NOAEL divided by a safety factor on the order of 10 to 1000, as is the case when animal data are being extrapolated to humans. The safety factor adjusts for comparisons between species and for studies using chronic, subchronic, or acute exposures. Several examples might best illustrate applications of the risk characterization process.

Atrazine is a herbicide commonly used to control weeds in corn and sorghum. One question that has been debated is whether atrazine should be classified as a Group B2 or Group C carcinogen. One study with one strain of rat found an increased incidence of mammary tumors after exposure to atrazine, but the EPA did not feel this was sufficient evidence to place atrazine in Group B2. Several studies determined NOAEL values for atrazine ranging from 0.48 to 5.0 mg/kg/day. The responses monitored in these studies included discrete myocardial degeneration in dogs to second-generation lower pup weights in rats. The EPA will typically use the lowest NOAEL for the calculation of the RfD. In this case, the RfD was calculated as 0.0048 mg/kg/day (the NOAEL divided by a safety factor of 100). From the RfD, the drinking water equivalent level (DWEL) was calculated as follows:

$$\text{DWEL} = \frac{\text{RfD} \times \text{BW}}{I_w} \tag{12.1}$$

where BW is the body weight of an adult (70 kg is typically used) and I_w is the drinking water consumed by the person in a day (2 L/day). The DWEL value is the maximum allowed drinking water atrazine concentration assuming the person obtains all of his or her atrazine from the water. In this case the DWEL is 0.168 mg/L.

The next step is to calculate the MCLG, which is calculated one of two ways. The preferred way is to adjust the DWEL based on the amount of the chemical obtained from other sources, such as food and air. If this information is not available (the usual case), then the MCLG is calculated assuming that 20% of the chemical is obtained from drinking water. An additional safety factor of 10 is used for atrazine since it is a Group C carcinogen. This results in an MCLG of:

$$\frac{0.168 \text{ mg/L} \times 0.2}{10} = 0.003 \text{ mg/L} \tag{12.2}$$

The best available technologies allow the MCL to be the same as the MCLG for atrazine and the drinking water standard became 0.003 mg/L (3 ppb). Any public drinking water supplier whose average atrazine concentration exceeds 3 ppb must take steps to reduce it. Atrazine can be removed from water with charcoal filtering, although the process is expensive.

Considerable debate has taken place over atrazine. Some parties feel that sufficient evidence exists to ban atrazine altogether. Atrazine users feel that it is an important tool (effective and inexpensive) in crop production and that atrazine is safe when used properly. The Federal Insecticide, Fungicide and Rodenticide Act (FIFRA) is the legislation that covers the labeling of atrazine. It stipulates that atrazine should perform its intended function without unreasonable adverse effects on the environment. Unreasonable adverse effects are described as any unreasonable risk to humans or the environment, taking into account the economic, social, and environmental costs and benefits of the use of atrazine. Clearly, the law provides a need to balance risks or costs against benefits. The weight of evidence is on the side of benefits for atrazine for the moment. The cost issue is important in this case because of the high cost of charcoal filtering. Public water supplies that use a high proportion of surface water as a source may have problems meeting the 3 ppb atrazine standard if atrazine is used extensively in the drainage area of the water source. In one unique case, a pesticide management area was established in Kansas around a reservoir that places more stringent rules on the use of atrazine than required by the product label. Compliance is voluntary at this point and atrazine levels have declined. If the concentrations had not declined, the rules may have become mandatory.

A second example of risk characterization pertains to the allowable cumulative application rate of a contaminant to soil via land application of municipal biosolids, as was described previously in Chapter 9 and Table 9.6. The reference application rate (RP, kg/ha) is calculated as follows:

$$RP = \frac{\left(\dfrac{\text{RfD} \times \text{BW}}{\text{RE}} - \text{TBI}\right) \times 10^3}{\sum \text{UC}_i \times \text{DC}_i \times \text{FC}_i} \tag{12.3}$$

where RP, RfD, and BW are as defined before, TBI is the total background intake rate of the contaminant (mg/day), RE is the relative effectiveness of ingestion exposure (unitless, generally assumed to be 1 unless evidence exists to the contrary), 10^3 is a conversion factor (μg/mg), UC_i is the uptake response slope for the food group i (μg/g dry weight per kg of contaminant applied per ha), DC_i is the daily dietary consumption of the food group i (g dry weight/day), and FC_i is the fraction of food group i assumed to originate from biosolids-amended soil (unitless, assumed to be 2.5% for the population as a whole). The value of DC_i is determined from the total diet study. While the actual calculations are cumbersome and will not be repeated here, the reader should appreciate several aspects of Equation 12.3. The equation is written with a human as the receptor organism, although it is appropriate for any organism, provided the appropriate data are available. The numerator determines the allowable effective dose of the contaminant to the receptor organism above the background level. The denominator determines the increase in the contaminant concentration in each food group as the contaminant is added to the soil and factors in how much of that food group is consumed and the proportion of that food group that originates from soils impacted by the contaminant. In the case of direct soil ingestion, the denominator is changed to reflect the change in soil contaminant concentration as the contaminant is added to the soil and the soil ingestion rate. Additional safety factors may also be added.

Recall from Chapter 7 that trace element contamination of soils is essentially permanent. Therefore, it must be recognized that the municipal biosolids disposal regulations allow a certain amount of irreversible soil contamination to occur. This is in contrast to the Western European philosophy that argues that soils have a finite capacity to absorb pollutants without negative consequences and that humans should not use up any of that capacity (i.e., soil concentrations should not be allowed to change). The U.S. regulations were written with a risk vs. benefit philosophy. The risk assessment process indicates the increase in risk can be regulated to insignificant levels and there are considerable benefits for land-applying biosolids. Land application of biosolids can save money compared with other disposal options. For example, placing the material in a landfill has become prohibitively expensive in recent years and has associated risks such as leachate generation and the potential for contamination of groundwaters and surface waters. Incineration or ocean dumping are no longer considered environmentally acceptable. The biosolids also contain plant nutrients and organic matter that can benefit soil physical properties and soil fertility. Finally, national regulations might actually reduce risk in areas where biosolids disposal was not previously regulated. More recent research in Europe suggested that metals in biosolids can impact soil microbial activity at soil concentrations below that allowed by U.S. regulations, although such effects may occur under only very unique situations. Such impacts on the soil environment were not considered in the risk assessment for biosolids because such data were not available at the time. Whether the risk assessment should be expanded to include these effects is a value judgment that must be made by society.

Results from human health risk assessments for noncarcinogenic substances are often expressed in terms of a *hazard index*. The hazard index is simply the dose received divided by the reference dose for that substance. If the hazard index exceeds 1, then the risk is unacceptable because the reference dose has been exceeded. For substances that are regulated as carcinogens, a *slope factor* is used. The slope factor is determined from a dose–response curve and represents the increase in the incidence of cancer per unit increase in dose. When the dose is determined by the previous steps, one can estimate the resulting cancer incidence rate or risk and decide whether that level is acceptable or not. Typically, cancer risks above 10^{-6} to 10^{-4} are considered too high. Figure 12.2 shows a hypothetical dose–response curve for a possible carcinogenic substance in drinking water. Linear and threshold response curves are shown for comparison as are the actual range of doses used in animal studies and the anticipated range of human exposure. Often the linear response is assumed, as it is a more conservative estimate of risk, despite toxicological evidence that a threshold response may be more realistic. This figure also illustrates the two problems with dose–response assessment that were discussed earlier.

Once an algorithm is developed for risk for a given exposure pathway, parameters can be easily manipulated to evaluate the impact on the assessment of risk. The algorithm includes all of the assumptions that were utilized in the risk assessment. Assumptions are an integral part of risk assessment and are used to simplify the process, to account for missing or inadequate information, and to incorporate safety factors into the process. For example, if one is concerned about the impact of contaminant uptake by plants on the consumers of those plants a value will have to be used for plant contaminant concentration. Assuming data are available, one could select an arithmetic mean, a geometric mean, or the 90th, 95th, or 99th percentile values. Each choice will impact the conservatism of the

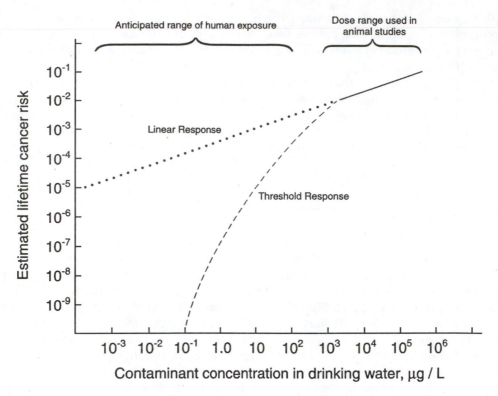

Figure 12.2   Hypothetical dose–response curve for a carcinogenic substance showing the actual doses used in animal studies and the anticipated doses that humans will receive. The assumed dose–response relationships at low doses illustrate the potential impact of the assumptions on risk.

risk estimate, with a 99th percentile value being more conservative than a 90th percentile value. Risk calculated with a stated set of assumptions and a defined algorithm are called *deterministic risk estimations. Probabilistic risk estimations* are sometimes used to augment deterministic estimations. A probabilistic risk estimation often uses a Monte Carlo simulation where the algorithm is run with ranges of values (based on a statistical distribution) for input variables. Each input variable is systematically varied across a reasonable range along with all other combinations of ranges for the remaining variables such that risk calculations are made for all possible combinations. This could alert the risk assessor to input variable combinations that produce unrealistic risk estimates. If no such unrealistic estimates are produced, it lends credibility to the deterministic approach that was used. This also helps improve confidence in risk estimations made with limited data.

12.5 *Uncertainty*

The risk assessment process is placed in perspective in Figure 12.3. Risk assessments are conducted in response to public pressures, existing environmental policies, or both. Once the process is complete, risk management takes over and eventually some regulatory action is implemented. This may be a cleanup level for a contaminated site or a new or revised environmental regulation. An important feature is the feedback loop. Once we have attempted to clean up a site or to enforce regulations, we may approach the risk assessment process differently in the future. Our perceptions of risk may change, we may become more accepting of present conditions, or we may have a better appreciation of the success or failure of regulations.

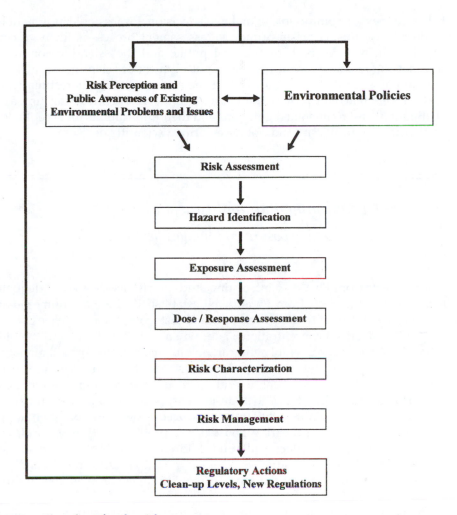

Figure 12.3   Flowchart for the risk assessment process.

   Both the general public and the scientific community can be critical of the risk assessment process and the resulting regulations. The general public often wants unrealistic levels of safety and, at times, seems to be unaware of the fact that a risk-free society cannot exist. The desired levels of risk are much lower than the risks realized from most everyday activities. Such reasoning led to the passage of the Delaney clause, which prohibited the use of cancer-causing agents as food or feed additives. The overall concept of the Delaney clause was sound except that it provides an absolute risk standard that does not take into account probability, costs, or benefits. The general public is also uncomfortable with arbitrary safety factors that it may not perceive as safe enough. In addition, increases in risk are often stated in terms of a few additional negative outcomes per 100,000 or 1,000,000 people exposed. This leaves the impression that preventable negative outcomes are deliberately allowed to occur.

   The scientific community is critical of regulations that are written with limited or suspect data. Indeed, it is unusual to have the perfect data set, including dose–response information at the levels of actual exposure. The previously mentioned problems of extrapolation from high dose to low dose and from animals to humans have caused considerable debate. Both extrapolations require making predictions beyond the range of the data, which is generally not acceptable within the scientific method. Scientists always want more data before they are comfortable in making a decision. Professional judgment is used often and, because of this, criticism of the risk assessment process is likely to continue indefinitely. It is helpful for both the general public and the scientific community to understand that the risk assessment process is not in quest of scientific truth. The goals are to estimate risk such that it is unlikely that the actual risk is underestimated and to develop regulations that will protect the environment without incurring unreasonable expense. Ultimately, environmental regulations and their implications for acceptable levels of risk are a social–political issue and not necessarily a scientific decision.

Environmental quality issues/events

Lead exposure and human health

Human health effects from Pb are an interesting study in risk assessment. Adequate data are available; the health effects from Pb are well established; there are numerous routes of exposure; exposure is easy to detect with a blood test; computer models exist that can be used to predict blood Pb concentrations; and, when soil Pb is involved, the bioavailability of that Pb can vary considerably and can possibly be reduced by soil management.

Hazard identification is not an issue with Pb as it is clear that exposure to Pb has negative consequences on human health. Young and unborn children are the most susceptible to Pb poisoning. Difficulties include premature birth, decreased birth weight, decreased mental abilities, learning difficulties, reduced growth in young children, and damage to the nervous system (hearing and balance problems), kidneys, and immune system. Acute Pb poisoning can affect adults and children alike; symptoms include abdominal cramping, elevated blood pressure, weakness in the extremities, anemia, and possible hearing loss. By far the problem of the greatest magnitude is chronic exposure to Pb in the environment by children.

Exposure assessment has identified numerous routes of exposure for children. In a home environment Pb can be ingested or inhaled from the diet, air, water, dust, soil, or leaded paint residues. If any of these media are enriched in Pb, the overall exposure can exceed critical values and the possibility of negative health consequences increases. Dietary exposure is generally not an issue unless the child consumes locally grown food that was produced in a high-Pb environment. Home gardening in a community impacted by Pb smelting activities would be one example. There is little uptake of Pb by the plants themselves, but the consumption of unwashed or unpeeled vegetables (especially leafy vegetables and root vegetables such as potatoes, carrots, beets, or radishes) can lead to considerable Pb ingestion by the consumer. Inhalation of Pb from the air is also generally not an issue unless it is an urban area where leaded gasoline is still in use or there are Pb-enriched particulates present in the air. Ingestion of Pb from water can be a significant source of Pb. Even without obviously contaminated drinking water sources, water in contact with old plumbing systems can have elevated background Pb exposure. Young children are in close contact with dust because they are frequently on the floor and exhibit hand-to-mouth activities. House dust mainly comprises carpet and fabric fibers, skin cells, animal dander, paint residues, and soil particles. If any of these components are enriched in Pb, the house dust will be as well. Adults, children, and pets constantly bring Pb-enriched particles into the home on their shoes or fur. Similarly, if the soil around a home has an elevated Pb concentration, there is a high likelihood that children will ingest some in their outdoor play activities. Leaded paint on the exterior or interior of the home will increase soil and dust Pb concentrations, but children can also directly ingest paint chips and increase their Pb intake.

The dose–response relationships for Pb intake are not clearly defined because of the many means of exposure and the numerous potential health effects. Blood Pb concentrations serve as a good indicator of Pb exposure and are often used to infer health effects. Some studies suggest, for example, that IQ test scores in children will decrease by 2 points for each 10 μg/dL increase in blood Pb concentrations. Overall the current health guidelines suggest that blood Pb concentrations in children not exceed 10 μg/dL and that corrective actions be taken if the concentration exceeds 15 μg/dL. The average blood Pb concentration of children in the U.S. is less than the critical value. Estimates are that more than 50% of children in some inner-city areas exceed this guideline and surveys have shown over 20% exceedance in some Pb-contaminated areas. Cessation of Pb in gasoline and solders has dramatically lowered blood Pb concentrations in the general public since about 1980.

Risk characterization is handled with the EPA Integrated Exposure Uptake Biokinetic (IEUBK) model, which takes into account all of the exposure routes for Pb in children, the internal adsorption/uptake mechanisms, and the various means of elimination of Pb from the body, and then calculates the resulting blood Pb concentrations. The components of the model are illustrated in Figure 12.4. One can use the IEUBK model with site-specific inputs to determine which exposure route results in the greatest impact on blood Pb concentrations in children. That information can then be used to determine which corrective actions would have the greatest benefit in reducing risk. Of particular interest in this book are the soil and dust input parameters. Both the Pb concentration and the bioavailability in these media can be specified in the model. As was shown in Table 7.7, the bioavailability of Pb in soils and other contaminated media varies greatly. Numerous risk

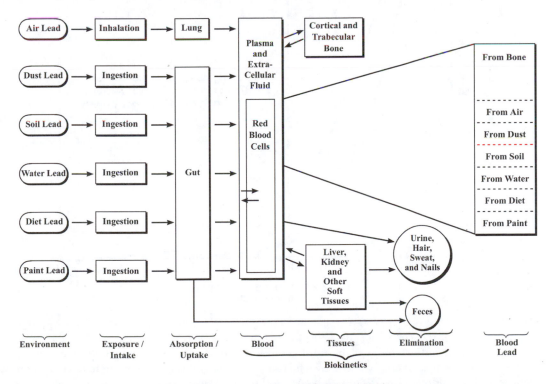

Figure 12.4   Schematic of the IEUBK model. (From U.S. EPA, 1994.)

assessments have shown soil and dust Pb to be major factors for Pb exposure by children. Often the remediation strategy used for Pb-contaminated soils in residential areas (see Chapter 11) is excavation and replacement with clean soil. Research is also under way in an attempt to find ways to reduce Pb bioavailability with soil amendments *in situ* so that excavation is not needed. The IEUBK model can be used to assess the effects of the reductions in Pb bioavailability.

For contaminated sites the goal of remediation is often to have less than 5% of the children with greater than 10 µg/dL blood Pb concentration. Figure 12.5 shows the blood Pb concentration ranges for a number of situations. Situation 1 has 800 mg Pb/kg soil with 30% absolute bioavailability, which estimates 17.5% of the children 6 to 72 months old would have >10 µg/dL blood Pb concentration. Default values were used for all remaining input variables. Situations 2 and 3 produce similar results in that approximately 5% of the children exceed 10 µg/dL blood Pb concentration. Situation 2 used the model to determine the soil Pb concentration that would give the desired blood Pb distribution. In this case a soil Pb concentration of 375 mg/kg reduced the percentage of children exceeding 10 µg/dL from 17.5 to 5%. If this were a community with Pb-contaminated soil, the results suggest that soil with a Pb concentration exceeding 375 mg/kg should be removed. Situation 3 uses 800 mg/kg soil Pb again, but reduces the absolute bioavailability to 15% (a 50% reduction). Here, similar results are obtained as compared with situation 2, but the implications are that an *in situ* remediation would be able to accomplish the

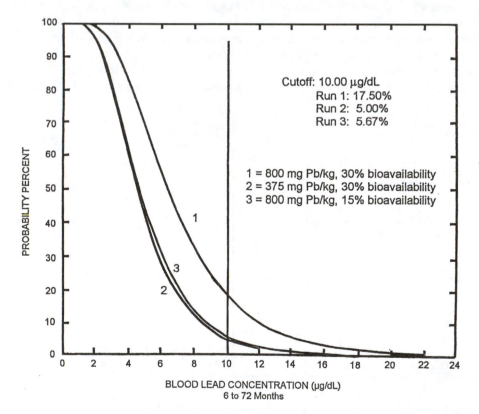

Figure 12.5   IEUBK model output showing the influence of soil Pb bioavailability on the proportion of children 6 to 72 months of age who have >10 µg/dL blood Pb concentration. Curve 1 assumes 800 mg Pb/kg soil and 30% bioavailability while curve 3 uses 800 mg Pb/kg and 15% bioavailability. Curve 2 uses 30% bioavailability and shows that soil Pb cannot exceed 375 mg/kg to have no more than 5% of the children with >10 µg/dL blood Pb concentration.

same goal. An *in situ* remediation would have tremendous cost advantages and would be less disruptive than soil excavation, but people would have to accept that the Pb was still present, but not in a harmful form.

Problems

12.1 There is a clear difference in actual risks vs. how those risks are perceived by the general public and politicians. What are some of the advantages and disadvantages of this discrepancy? Should regulatory agencies actively work at educating people about this difference so that the public's perception of risk corresponds to actual risks?

12.2 Consider a nuclear reactor meltdown, an automobile accident, the crash of a commercial airliner, and the release of genetically engineered organisms into the environment. Rank the risk of death from each according to the perceived risk by the general public. Estimate the actual risk of death from each and compare the ranking to that for risk perception.

12.3 Using Equations 12.1 and 12.2, recalculate the MCLG for atrazine in drinking water assuming: (a) the lowest NOAEL value found in animal studies was 1.0 mg/kg/day; (b) a safety factor of 10 was used to calculate the RfD instead of a factor of 100; and (c) that 80% of the atrazine a person receives is from drinking water.

12.4 Describe how the determination of the MCL for atrazine in drinking water would be impacted if atrazine were a Category I substance instead of a Category II substance.

12.5 Perform unit analysis on Equation 12.3 and demonstrate that the units for RP are kg/ha.

12.6 Describe the differences between an absolute risk standard and a negligible risk standard. Discuss the advantages and disadvantages of each.

12.7 Using Equation 12.3, calculate RP when RfD = 0.001 mg/kg day, BW = 70 kg, RE = 1.0, TBI = 0.0161 mg/day, and $\Sigma UC_i \times DC_i \times FC_i$ = 2.5 µg ha/kg day. (Data from Chaney et al., 1998.)

References

Ames, B. N., R. Magaw, and L. S. Gold. 1987. Ranking possible carcinogenic hazards, *Science*, 236, 271–280.

Chaney, R. L., S. L. Brown, and J. S. Angle. 1998. Soil root interface: ecosystem health and human food-chain protection, in P. M. Huang, D. C. Adriano, T. J. Logan, and R. T. Checkai, Eds., *Soil Chemistry and Ecosystem Health*, SSSA Spec. Pub. No. 52, Soil Science Society of America, Madison, WI, 279–311.

Slovik, P. 1987. Perception of risk, *Science*, 236, 280–285.

Slovik, P., B. Fischhoff, and S. Lichtenstein. 1979. Rating the risks, *Environment*, 21, 14–20.

U.S. Environmental Protection Agency. 1989. Development of Risk Assessment Methodology for Land Application and Distribution and Marketing of Municipal Sludge, EPA/600/6-89/001.

U.S. Environmental Protection Agency. 1994. Guidance Manual for the Integrated Exposure Uptake Biokinetic Model for Lead in Children, U.S. EPA, Washington, D.C.

Supplementary reading

Cothern, C. R. 1996. *Handbook for Environmental Risk Decision Making: Values, Perceptions, and Ethics,* CRC Press, Boca Raton, FL.

Glickman, T. S. and M. Gough, Eds. 1990. *Readings on Risk,* Resources for the Future, Washington, D.C.

Hallenback, W. H. and K. M. Cunningham. 1986. *Quantitative Risk Assessment for Environmental and Occupational Health,* Lewis Publishers, Chelsea, MI.

McColl, R.S., Ed. 1987. *Environmental Health Risks: Assessment and Management,* University of Waterloo Press, Waterloo, Ontario, Canada.

Newman, M. C. and C. L. Strojan, Eds. 1998. *Risk Assessment: Logic and Measurement,* Ann Arbor Press, Ann Arbor, MI.

U.S. Environmental Protection Agency. 1987. The Risk Assessment Guidelines of 1986, EPA/600/8-87/045.

Appendix

Table A.1 The Elements and Their Symbols, Atomic Numbers, and Atomic Weights Based on the Assigned Relative Atomic Mass of $^{12}C = 12$

Name	Symbol	Atomic number	Atomic weight
Actinium	Ac	89	227.028
Aluminum	Al	13	26.98154
Americium	Am	95	(243)[a]
Antimony	Sb	51	121.75
Argon	Ar	18	39.948
Arsenic	As	33	74.9216
Astatine	At	85	(210)
Barium	Ba	56	137.33
Berkelium	Bk	97	(247)
Beryllium	Be	4	9.01218
Bismuth	Bi	83	208.9804
Boron	B	5	10.81
Bromine	Br	35	79.904
Cadmium	Cd	48	112.41
Calcium	Ca	20	40.08
Californium	Cf	98	(251)
Carbon	C	6	12.011
Cerium	Ce	58	140.12
Cesium	Cs	55	132.9054
Chlorine	Cl	17	35.453
Chromium	Cr	24	51.996
Cobalt	Co	27	58.9332
Copper	Cu	29	63.546
Curium	Cm	96	(247)
Dysprosium	Dy	66	162.50
Einsteinium	Es	99	(254)
Erbium	Er	68	167.26
Europium	Eu	63	151.96
Fermium	Fm	100	(257)
Fluorine	F	9	18.998403
Francium	Fr	87	(223)
Gadolinium	Gd	64	157.25
Gallium	Ga	31	69.72
Germanium	Ge	32	72.59
Gold	Au	79	196.9665
Hafnium	Hf	72	178.49
Helium	He	2	4.00260
Holmium	Ho	67	164.9304
Hydrogen	H	1	1.0079
Indium	In	49	114.82
Iodine	I	53	126.9045
Iridium	Ir	77	192.22
Iron	Fe	26	55.847

continued

Table A.1 (continued) The Elements and Their Symbols, Atomic Numbers, and Atomic Weights Based on the Assigned Relative Atomic Mass of $^{12}C = 12$

Name	Symbol	Atomic number	Atomic weight
Krypton	Kr	36	83.80
Lanthanum	La	57	138.9055
Lawrencium	Lr	103	(260)
Lead	Pb	82	207.2
Lithium	Li	3	6.941
Lutetium	Lu	71	174.967 ± 0.003
Magnesium	Mg	12	24.305
Manganese	Mn	25	54.9380
Mendelevium	Md	101	(257)
Mercury	Hg	80	200.59
Molybdenum	Mo	42	95.94
Neodymium	Nd	60	144.24
Neon	Ne	10	20.179
Neptunium	Np	93	237.0482
Nickel	Ni	28	58.70
Niobium	Nb	41	92.9064
Nitrogen	N	7	14.0067
Nobelium	No	102	(259)
Osmium	Os	76	190.2
Oxygen	O	8	15.9994
Palladium	Pd	46	106.4
Phosphorus	P	15	30.97376
Platinum	Pt	78	195.09
Plutonium	Pu	94	(244)
Polonium	Po	84	(209)
Potassium	K	19	39.0983
Praseodymium	Pr	59	140.9077
Promethium	Pm	61	(145)
Protactinium	Pa	91	231.0359
Radium	Ra	88	226.0254
Radon	Rn	86	(222)
Rhenium	Re	75	186.2
Rhodium	Rh	45	102.9055
Rubidium	Rb	37	85.4678
Ruthenium	Ru	44	101.07
Samarium	Sm	62	150.4
Scandium	Sc	21	44.9559
Selenium	Se	34	78.96
Silicon	Si	14	28.0855
Silver	Ag	47	107.868
Sodium	Na	11	22.98977
Strontium	Sr	38	87.62
Sulfur	S	16	32.06
Tantalum	Ta	73	180.9479
Technetium	Tc	43	(97)
Tellurium	Te	52	127.60
Terbium	Tb	65	158.9254
Thallium	Tl	81	204.37
Thorium	Th	90	232.0381
Thulium	Tm	69	168.9342

Table A.1 **(continued)**   The Elements and Their Symbols, Atomic Numbers, and Atomic Weights Based on the Assigned Relative Atomic Mass of $^{12}C = 12$

Name	Symbol	Atomic number	Atomic weight
Tin	Sn	50	118.69
Titanium	Ti	22	47.90
Tungsten	W	74	183.85
Uranium	U	92	238.029
Vanadium	V	23	50.9415
Xenon	Xe	54	131.30
Ytterbium	Yb	70	173.04
Yttrium	Y	39	88.9059
Zinc	Zn	30	65.38
Zirconium	Zr	40	91.22

[a] Anthropogenic element or one that occurs only in minute quantities in nature.

Table A.2   Conversion Factors for SI and Non-SI Units

To convert Column 1 into Column 2, multiply by	Column 1 SI Unit	Column 2 non-SI Unit	To convert Column 2 into Column 1 multiply by
Length			
0.621	kilometer, km (10^3 m)	mile, mi	1.609
1.094	meter, m	yard, yd	0.914
3.28	meter, m	foot, ft	0.304
1.0	micrometer, μm (10^{-6} m)	micron, μ	1.0
3.94×10^{-2}	millimeter, mm (10^{-3} m)	inch, in	25.4
10	nanometer, nm (10^{-9} m)	Angstrom, Å	0.1
Area			
2.47	hectare, ha	acre	0.405
247	square kilometer, km² (10^3 m)²	acre	4.05×10^{-3}
0.386	square kilometer, km² (10^3 m)²	square mile, mi²	2.590
2.47×10^{-4}	square meter, m²	acere	4.05×10^3
10.76	square meter, m²	square foot, ft²	9.29×10^{-2}
1.55×10^{-3}	square millimeter, mm² (10^{-3} m)²	square inch, in²	645
Volume			
9.73×10^{-3}	cubic meter, m³	acre-inch	102.8
35.3	cubic meter, m³	cubic foot, ft³	2.83×10^{-2}
6.10×10^4	cubic meter, m³	cubic inch, in³	1.64×10^{-5}
2.84×10^{-2}	liter, L (10^{-3} m³)	bushel, bu	35.24
1.057	liter, L (10^{-3} m³)	quart (liquid), qt	0.946
3.53×10^{-2}	liter, L (10^{-3} m³)	cubic foot, ft³	28.3
0.265	liter, L (10^{-3} m³)	gallon	3.78
33.78	liter, L (10^{-3} m³)	ounce (fluid), oz	2.96×10^{-2}
2.11	liter, L (10^{-3} m³)	pint (fluid), pt	0.473

To convert Col. 2 into Col. 1, multiply by	Column 1 SI Unit	Column 2 non-SI Unit	To convert Col. 1 into Col. 2, multiply by
Mass			
2.20×10^{-3}	gram, g (10^{-3} kg)	pound, lb	454
3.52×10^{-2}	gram, g (10^{-3} kg)	ounce (avdp), oz	28.4
2.205	kilogram, kg	pound, lb	0.454
0.01	kilogram, kg	quintal (metric), q	100
1.10×10^{-3}	kilogram, kg	ton (2000 lb), ton	907
1.102	megagram, Mg (tonne)	ton (U.S.), ton	0.907
1.102	tonne, t	ton (U.S.), ton	0.907
Yield and rate			
0.893	kilogram per hectare, kg/ha	pound per acre, lb/acre	1.12
7.77×10^{-2}	kilogram per cubic meter, kg/m^3	pound per bushel, lb/bu	12.87
1.49×10^{-2}	kilogram per hectare, kg/ha	bushel per acre, 60 lb	67.19
1.59×10^{-2}	kilogram per hectare, kg/ha	bushel per acre, 56 lb	62.71
1.86×10^{-2}	kilogram per hectare, kg/ha	bushel per acre, 48 lb	53.75
0.107	liter per hectare, L/ha	gallon per acre	9.35
893	tonnes per hectare, t/ha	pound per acre, lb/acre	1.12×10^{-3}
893	megagram per hectare, Mg/ha	pound per acre, lb/acre	1.12×10^{-3}
0.446	megagram per hectare, Mg/ha	ton (2000 lb) per acre, ton/acre	2.24
2.24	meter per second, m/s	mile per hour, mi/h	0.447
Specific surface			
10	square meter per kilogram, m^2/kg	square centimeter per gram, cm^2/g	0.1
1.000	square meter per kilogram, m^2/kg	square millimeter per gram, mm^2/g	0.001
Pressure			
9.90	megapascal, MPa (10^6 Pa)	atmosphere	0.101
10	megapascal, MPa (10^6 Pa)	bar	0.1
1.00	megagram per cubic meter, Mg/m^3	gram per cubic centimeter, g/cm^3	1.00
2.09×10^{-2}	pascal, Pa	pound per square foot, lb/ft^2	47.9
1.45×10^{-4}	pascal, Pa	pound per square inch, lb/in^2	6.90×10^3

continued

Table A.2 (continued)   Conversion Factors for SI and Non-SI Units

To convert Column 1 into Column 2, multiply by	Column 1 SI Unit	Column 2 non-SI Unit	To convert Column 2 into Column 1 multiply by
Temperature			
$1.00\ (K - 273)$	Kelvin, K	Celsius, °C	$1.00\ (°C + 273)$
$(9/5\ °C) + 32$	Celsius, °C	Fahrenheit, °F	$5/9\ (°F - 32)$
Energy, work, quantity of heat			
9.52×10^{-4}	joule, J	British thermal unit, Btu	1.05×10^{3}
0.239	joule, J	calorie, cal	4.19
10^{7}	joule, J	erg	10^{-7}
0.735	joule, J	foot-pound	1.36
2.387×10^{-5}	joule per square meter, J/m^2	calorie per square centimeter (langley)	4.19×10^{4}
10^{5}	newton, N	dyne	10^{-5}
1.43×10^{-3}	watt per square meter, W/m^2	calorie per square centimeter minute (irradiance), $cal/cm^2\ min$	698
Transpiration and photosynthesis			
3.60×10^{-2}	milligram per square meter second, $mg/m^2\ s$	gram per square decimeter hour, $g/dm^2\ h$	27.8
5.56×10^{-3}	milligram (H_2O) per square meter second, $mg/m^2\ s$	micromole (H_2O) per square centimeter second, $\mu mol/cm^2\ s$	180
10^{-4}	milligram per square meter second, $mg/m^2\ s$	milligram per square centimeter second, $mg/cm^2\ s$	10^{4}
35.97	milligram per square meter second, $mg/m^2\ s$	milligram per square decimeter hour, $mg/dm^2\ h$	2.78×10^{-2}
Plane angle			
57.3	radian, rad	degrees (angle), °	1.75×10^{-2}
Electrical conductivity, electricity, and magnetism			
10	siemen per meter, S/m	millimho per centimeter, mmho/cm	0.1
10^{4}	tesla, T	gauss, G	10^{-4}

Water measurement

9.73×10^{-3}	cubic meter, m³	acre-inches, acre-in.	102.8
9.81×10^{-3}	cubic meter per hour, m³/h	cubic feet per second, ft³/s	101.9
4.40	cubic meter per hour, m³/h	U.S. gallons per minute, gal/min	0.227
8.11	hectare-meters, ha-m	acre-feet, acre-ft	0.123
97.28	hectare-meters, ha-m	acre-inches, acre-in.	1.03×10^{-2}
8.1×10^{-2}	hectare-centimeters, ha-cm	acre-feet, acre-ft	12.33

Concentrations

1	centimole per kilogram, cmol/kg (ion-exchange capacity)	milliequivalents per 100 grams, meq/100 g	1
0.1	gram per kilogram, g/kg	percent, %	10
1	milligram per kilogram, mg/kg	parts per million, ppm	1

Radioactivity

2.7×10^{-1}	becquerel, Bq	curie, Ci	3.7×10^{10}
2.7×10^{-2}	becquerel per kilogram, Bq/kg	picocurie per gram, pCi/g	37
100	gray, Gy (absorbed dose)	rad, rd	0.01
100	sievert, Sv (equivalent dose)	rem (roentgen equivalent man)	0.01

Plant nutrient conversion

	Elemental	Oxide	
2.29	P	P_2O_5	0.437
1.20	K	K_2O	0.830
1.39	Ca	CaO	0.715
1.66	Mg	MgO	0.602

Source: American Society of Agronomy, Madison, WI. Reprinted with permission.

Table A.3   Journals and Periodicals Related to the Environmental Sciences

Advances in Soil Science
Agricultural Chemistry
Agrochimica
Agronomy Journal
Ambio
American Society of Agronomy Abstracts
Analyst
Analytical Chimica Acta
Analytical Letters
Analytical Chemistry
Atmospheric Environment
Canadian Journal of Fisheries and Aquatic Science
Canadian Journal of Chemistry
Canadian Journal of Soil Science
Communications in Soil Science and Plant Analysis
CRC Critical Review of Environmental Contamination
Crop Science
Environmental Geology and Water Science
Environmental Letters
Environmental Research
Environmental Science and Technology
Environmental Toxicology and Chemistry
Estuarine, Marine and Coastal Shelf Sciences
Geochimica et Cosmochimica Acta
Geoderma
Geological Society of America Bulletin
Journal of Chemical Ecology
Journal of Chemical Education
Journal of Environmental Quality
Journal of Environmental Health
Journal of Atmospheric Chemistry
Journal of Environmental Science and Health
Journal of Environmental Engineering (Division of ASCE)
Journal of the Geological Society (London)
Journal of Soil Science
Journal of the Air Pollution Control Association
Journal of the Association of Official Analytical Chemists
Journal of Environmental Science Technology
Journal of Great Lakes Research
Journal of the Water Pollution Control Federation
Journal of Environmental Toxicology and Contamination
Journal of Marine Research
Journal of the American Water Works Association
Journal of the Indian Chemical Society
Journal of Soil & Water Conservation
Limnology and Oceanography
Marine Chemistry
Marine Environmental Chemistry
Nature
Science
Soil Science
Soil Science Society of America Proceedings

Table A.3 (continued)   Journals and Periodicals Related to the Environmental

Talanta
Water, Air, and Soil Pollution
Water Research
Water, Science and Technology
Weed Science Journal

Table A.4 Internet Web Sites Related to the Information Contained in This Book

Site	Internet address
U.S. Government agencies	
U.S. Department of Agriculture (USDA)	http://www.usda.gov
USDA Agricultural Research Service	http://www.ars.usda.gov
USDA Natural Resources Conservation Service	http://www.nrcs.usda.gov
USDA National Agricultural Statistics Service	http://www.nass.usda.gov/census
USDA National Agricultural Library	http://www.nalusda.gov
USDA Cooperative State Research, Education, and Extension Service	http://www.reeusda.gov
USDA Economic Research Service	http://www.econ.ag.gov
USDA Foreign Agricultural Service	http://www.fas.usda.gov
U.S. Geological Survey	http://www.usgs.gov
U.S. Environmental Protection Agency	http://www.epa.gov
Trends Online: A Compendium of Data on Global Climate Change	http://www.cdiac.esd.ornl.gov/trnds/htm
Agency for Toxic Substances and Disease Registry	http://www.atsdr.cdc.gov/atsdrhome.html
Professional organizations	
American Society of Agronomy	http://www.agronomy.org
Crop Science Society of America	http://www.crops.org/
Soil Science Society of America	http://www.soils.org/
American Chemical Society	http://www.acs.org/
Environmental advocacy groups	
Audubon Society	http://www.audobon.org
Chesapeake Bay Trust	http://www2.ari.net/home/cbt/
Sierra Club	http://www.sierraclub.org
National Resources Defense Council	http://www.nrdc.org
EnviroWEB: link to many environmental groups	http://www.enviroweb.org
Home pages for authors of *Soils and Environmental Quality*	
G. M. Pierzynski	http://www.ksu.edu/agronomy/FACULTY/PIERZYNSKI
J. T. Sims	http://bluehen.ags.udel.edu/envirosoil/
G.F. Vance	http://nature.uwyo.edu

Table A.5   Prefixes Used in the SI System of Units

Prefix	Symbol	Order of magnitude
yocto	y	10^{-24}
zepto	z	10^{-21}
atto	a	10^{-18}
femto	f	10^{-15}
pico	p	10^{-12}
nano	n	10^{-9}
micro	m	10^{-6}
milli	m	10^{-3}
centi	c	10^{-2}
deci	d	10^{-1}
deka	da	10^{1}
hecto	h	10^{2}
kilo	k	10^{3}
mega	M	10^{6}
giga	G	10^{9}
tera	T	10^{12}
peta	P	10^{15}
Exa	E	10^{18}
zetta	Z	10^{21}
yotta	Y	10^{24}

Index

A

U